Newnes Engineering Science Pocket Book

Newnes Engineering Science Pocket Book

Third edition

John Bird

BSc(Hons), CEng, CMath, FIMA, MIEE, FCollP, FIEIE

CRC Press
Taylor & Francis Group
Boca Raton London New York

CRC Press is an imprint of the
Taylor & Francis Group, an **informa** business

A SPON PRESS BOOK

First published as the *Newnes Engineering & Physical Science Pocket Book* 1993
Second edition 1996
Third edition as the *Newnes Engineering Science Pocket Book* 2001

First published 2001 by Spon press
This edition published 2011 by Spon Press

Published 2019 by CRC Press
Taylor & Francis Group
6000 Broken Sound Parkway NW, Suite 300
Boca Raton, FL 33487-2742

© 2001 by John Bird
CRC Press is an imprint of Taylor & Francis Group, an Informa business

First issued in paperback 2019

No claim to original U.S. Government works

ISBN 13: 978-0-367-44722-9 (pbk)
ISBN 13: 978-0-7506-4991-9 (hbk)

**Visit the Taylor & Francis Web site at
http://www.taylorandfrancis.com**

**and the CRC Press Web site at
http://www.crcpress.com**

British Library Cataloguing in Publication Data
A catalogue record for this book is available from the British Library

Typeset by Laser Words, Madras, India

Contents

Preface

Newnes Engineering Science Pocket Book is intended to provide students, technicians, scientists and engineers with a readily available reference to the essential engineering science formulae, definitions and general information needed during their studies and/or work situation — a handy book to have on the bookshelf to delve into as the need arises.

The text is divided, for convenience of reference, into three main sections embracing general engineering science, mechanical engineering and physical science and electrical engineering science.

The text assumes little previous knowledge and is suitable for a wide range of courses of study. It will be particularly useful for students studying for NVQ's and GNVQ's, technician certificates and diplomas, for GCSE and A levels, and for Engineering degrees.

John Bird
University of Portsmouth

Part One General Engineering Science

1 SI Units

Units

The system of units used in engineering and science is the Système Internationale d'Unités (International system of units), usually abbreviated to SI units, and is based on the metric system. This was introduced in 1960 and is now adopted by the majority of countries as the official system of measurement.

The basic units in the S.I. system are listed below with their symbols:

Quantity	Unit
length	metre, m
mass	kilogram, kg
time	second, s
electric current	ampere, A
thermodynamic temperature	kelvin, K
luminous intensity	candela, cd
amount of substance	mole, mol

Prefixes

S.I. units may be made larger or smaller by using prefixes that denote multiplication or division by a particular amount. The six most common multiples, with their meaning, are listed below:

Prefix	Name	Meaning	
T	tera	multiply by 1 000 000 000 000	(i.e. $\times 10^{12}$)
G	giga	multiply by 1 000 000 000	(i.e. $\times 10^{9}$)
M	mega	multiply by 1 000 000	(i.e. $\times 10^{6}$)
k	kilo	multiply by 1 000	(i.e. $\times 10^{3}$)
m	milli	divide by 1 000	(i.e. $\times 10^{-3}$)
μ	micro	divide by 1 000 000	(i.e. $\times 10^{-6}$)
n	nano	divide by 1 000 000 000	(i.e. $\times 10^{-9}$)
p	pico	divide by 1 000 000 000 000	(i.e. $\times 10^{-12}$)

Length, area, volume and mass

Length is the distance between two points. The standard unit of length is the **metre**, although the **centimetre, cm, millimetre, mm and kilometre, km**, are often used.

$$1 \text{ cm} = 10 \text{ mm}, 1 \text{ m} = 100 \text{ cm} = 1\,000 \text{ mm and } 1 \text{ km} = 1\,000 \text{ m}$$

Area is a measure of the size or extent of a plane surface and is measured by multiplying a length by a length. If the lengths are in metres then the unit of area is the **square metre, m^2**

$$1 \text{ m}^2 = 1 \text{ m} \times 1 \text{ m} = 100 \text{ cm} \times 100 \text{ cm} = 1\,0000 \text{ cm}^2 \text{ or } 10^4 \text{ cm}^2$$

$$= 1\,000 \text{ mm} \times 1000 \text{ mm} = 1\,000\,000 \text{ mm}^2 \text{ or } 10^6 \text{ mm}^2$$

Conversely, $1 \text{ cm}^2 = 10^{-4} \text{ m}^2$ and $1 \text{ mm}^2 = 10^{-6} \text{ m}^2$

Volume is a measure of the space occupied by a solid and is measured by multiplying a length by a length by a length. If the lengths are in metres then the unit of volume is in **cubic metres, m^3**

$$1 \text{ m}^3 = 1 \text{ m} \times 1 \text{ m} \times 1 \text{ m}$$

$$= 100 \text{ cm} \times 100 \text{ cm} \times 100 \text{ cm} = 10^6 \text{ cm}^3$$

$$= 1000 \text{ mm} \times 1000 \text{ mm} \times 1000 \text{ mm} = 10^9 \text{ mm}^3$$

Conversely, $1 \text{ cm}^3 = 10^{-6} \text{ m}^3$ and $1 \text{ mm}^3 = 10^{-9} \text{ m}^3$

Another unit used to measure volume, particularly with liquids, is the **litre, l**, where $1 \text{ l} = 1000 \text{ cm}^3$

Mass is the amount of matter in a body and is measured in **kilograms, kg.**

$$1 \text{ kg} = 1000 \text{ g (or conversely, } 1 \text{ g} = 10^{-3} \text{ kg)}$$

and $1 \text{ tonne (t)} = 1000 \text{ kg}$

Derived SI Units

Derived SI units use combinations of basic units and there are many of them. Two examples are:

$$\text{Velocity} \qquad - \text{ metres per second (m/s)}$$

$$\text{Acceleration} - \text{ metres per second squared (m/s}^2)$$

Charge

The **unit of charge** is the coulomb (C) where one coulomb is one ampere second. (1 coulomb $= 6.24 \times 10^{18}$ electrons). The coulomb is defined as the quantity of electricity which flows past a given point in an electric circuit when a current of one ampere is maintained for one second. Thus,

charge, in coulombs $\boxed{Q = It}$

where I is the current in amperes and t is the time in seconds.

Force

The **unit of force** is the newton (N) where one newton is one kilogram metre per second squared. The newton is defined as the force which, when applied to a mass of one kilogram, gives it an acceleration of one metre per second squared. Thus,

force, in newtons $\boxed{F = ma}$

where m is the mass in kilograms and a is the acceleration in metres per second squared. Gravitational force, or weight, is mg, where g $= 9.81$ m/s^2

Work

The **unit of work or energy** is the **joule (J)** where one joule is one newton metre. The joule is defined as the work done or energy transferred when a force of one newton is exerted through a distance of one metre in the direction of the force. Thus

work done on a body, in joules, $\boxed{W = Fs}$

where F is the force in newtons and s is the distance in metres moved by the body in the direction of the force. Energy is the capacity for doing work.

Power

The **unit of power** is the watt (W) where one watt is one joule per second. Power is defined as the rate of doing work or transferring energy. Thus,

power, in watts, $\boxed{P = \dfrac{W}{t}}$

where W is the work done or energy transferred, in joules, and t is the time, in seconds. Thus,

energy, in joules, $\boxed{W = Pt}$

Electrical potential and e.m.f.

The **unit of electric potential** is the volt (V), where one volt is one joule per coulomb. One volt is defined as the difference in potential between two points in a conductor which, when carrying a current of one ampere, dissipates a power of one watt, i.e.

$$\text{volts} = \frac{\text{watts}}{\text{amperes}} = \frac{\text{joules/second}}{\text{amperes}} = \frac{\text{joules}}{\text{amperes seconds}} = \frac{\text{joules}}{\text{coulombs}}$$

A change in electric potential between two points in an electric circuit is called a **potential difference**. The **electromotive force (e.m.f.)** provided by a source of energy such as a battery or a generator is measured in volts.

2 Density

Density is the mass per unit volume of a substance. The symbol used for density is ρ (Greek letter rho) and its units are kg/m^3

$$\text{Density} = \frac{\text{mass}}{\text{volume}} \quad \text{i.e.} \quad \boxed{\rho = \frac{m}{V}} \quad \text{or} \quad \boxed{m = \rho V} \quad \text{or} \quad \boxed{V = \frac{m}{\rho}}$$

where m is the mass in kg, V is the volume in m^3 and ρ is the density in kg/m^3

Some **typical values of densities** include:

Aluminium	2700 kg/m^3	Steel	7800 kg/m^3
Cast iron	7000 kg/m^3	Petrol	700 kg/m^3
Cork	250 kg/m^3	Lead	11 400 kg/m^3
Copper	8900 kg/m^3	Water	1000 kg/m^3

For example, the density of 50 cm^3 of copper if its mass is 445 g is given by:

$$\textbf{density} = \frac{\text{mass}}{\text{volume}} = \frac{445 \times 10^{-3}\text{kg}}{50 \times 10^{-6}\text{m}^3} = \frac{445}{50} \times 10^3$$

$$= \mathbf{8.9 \times 10^3 kg/m^3} \text{ or } \mathbf{8900\ kg/m^3}$$

Similarly, the volume, in litres, of 20 kg of paraffin oil of density 800 kg/m^3 is given by:

$$\text{volume} = \frac{m}{\rho} = \frac{20\text{ kg}}{800\text{ kg/m}^3} = \frac{1}{40}\text{m}^3 = \frac{1}{40} \times 10^6\text{ cm}^3 = 25\,000\text{ cm}^3$$

$$1 \text{ litre} = 1000 \text{ cm}^3 \text{ hence } 25\,000 \text{ cm}^3 = \frac{25\,000}{1000} = \textbf{25 litres}$$

The **relative density** of a substance is the ratio of the density of the substance to the density of water, i.e.

$$\boxed{\text{relative density} = \frac{\text{density of substance}}{\text{density of water}}}$$

Relative density has no units, since it is the ratio of two similar quantities. Typical values of relative densities can be determined from above (since water has a density of 1000 kg/m^3), and include:

Aluminium	2.7	Steel	7.8
Cast iron	7.0	Petrol	0.7
Cork	0.25	Lead	11.4
Copper	8.9		

The relative density of a liquid may be measured using a **hydrometer.**

For example, the relative density of a piece of steel of density 7850 kg/m^3, given that the density of water is 1000 kg/m^3, is given by:

$$\textbf{relative density} = \frac{\text{density of steel}}{\text{density of water}} = \frac{7850}{1000} = \textbf{7.85}$$

3 Scalar and Vector Quantities

Scalars and Vectors

Quantities used in engineering and science can be divided into two groups:

(a) **Scalar quantities** have a size (or magnitude) only and need no other information to specify them. Thus, 10 centimetres, 50 seconds, 7 litres, 3 kilograms, 25°C, £250, 10 cm^3 volume and 10 joules of energy, are all examples of scalar quantities.

(b) **Vector quantities** have both a size or magnitude **and** a direction, called the line of action of the quantity. Thus, a velocity of 50 kilometers per hour due east, an acceleration of 9.81 meters per second squared vertically downwards, a force of 15 newtons at an angle of 30 degrees, and a north-westerly wind of 15 knots are all examples of vector quantities.

The **speed** of a body can be stated without reference to the direction of movement of that body. Thus, speed is a scalar quantity. If, however, we specify the direction of motion as well as the speed of the body, the quantity is then termed the **velocity** of the body. Velocity is thus a vector quantity.
A **weight** of, say, 20 newtons, might initially appear to be a scalar quantity; however, weight also has a direction, i.e. downwards (towards the center of the earth). Thus, weight is a vector quantity.
When we say a man has walked 7 km we give no indication of direction. Thus, **distance** is a scalar quantity. If, however, the man walks 4 km westwards, then 3 km northwards as shown in Figure 3.1, his final position at C is 5 km away from his initial position at *A* (by Pythagoras' theorem). This change in position is called **displacement**. Thus 7 km is the distance walked, and 5 km in a direction N37°W is a vector quantity.

Summarising, a quantity that has magnitude and direction is a vector quantity, whereas a quantity that has magnitude only is a scalar quantity.

Vector Representation

A vector may be represented by a straight line, the length of line being directly proportional to the magnitude of the quantity and the direction of the line being

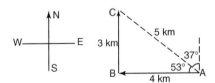

Figure 3.1

in the same direction as the line of action of the quantity. An arrow is used to denote the sense of the vector, that is, for a horizontal vector, say, whether it acts from left to right or vice-versa. The arrow is positioned at the end of the vector and this position is called the 'nose' of the vector. Figure 3.2 shows a velocity of 20 m/s at an angle of 45° to the horizontal and may be depicted by **oa** = 20 m/s at 45° to the horizontal.

To distinguish between vector and scalar quantities, various ways are used. These include:

(i) **bold print,**

(ii) two capital letters with an arrow above them to denote the sense of direction, e.g. $\overrightarrow{AB}$, where A is the starting point and B the end point of the vector,

(iii) a line over the top of letters, e.g. $\overline{AB}$ or $\bar{a}$

(iv) letters with an arrow above, e.g. $\vec{a}$, $\vec{A}$

(v) underlined letters, e.g. $\underline{a}$

(vi) $xi + jy$, where i and j are axes at right-angles to each other; for example, $3i + 4j$ means 3 units in the i direction and 4 units in the j direction, as shown in Figure 3.3

(vii) a column matrix $\begin{pmatrix} a \\ b \end{pmatrix}$; for example, the vector **OA** shown in Figure 3.3 could be represented by $\begin{pmatrix} 3 \\ 4 \end{pmatrix}$

Thus, in Figure 3.3, $\mathbf{OA} \equiv \overrightarrow{OA} \equiv \overline{OA} \equiv 3i + 4j \equiv \begin{pmatrix} 3 \\ 4 \end{pmatrix}$

Thus, **OA** represents a vector quantity, but OA is the magnitude of the vector **OA**. Also, positive angles are measured in an anticlockwise direction from a horizontal, right facing line, and negative angles in a clockwise direction from this line — as with graphical work. Thus 90° is a line vertically upwards and −90° is a line vertically downwards.

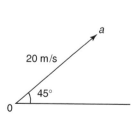

Figure 3.2

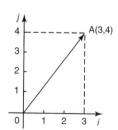

Figure 3.3

4 Atomic Structure of Matter

Elements

There are a very large number of different substances in existence, each substance containing one or more of a number of basic materials called elements.

'An **element** is a substance which cannot be separated into anything simpler by chemical means'. There are 92 naturally occurring elements and 13 others, which have been artificially produced.

Some examples of common elements with their symbols are: Hydrogen H, Helium He, Carbon C, Nitrogen N, Oxygen O, Sodium Na, Magnesium Mg, Aluminium Al, Silicon Si, Phosphorus P, Sulphur S, Potassium K, Calcium Ca, Iron Fe, Nickel Ni, Copper Cu, Zinc Zn, Silver Ag, Tin Sn, Gold Au, Mercury Hg, Lead Pb and Uranium U.

Atoms

Elements are made up of very small parts called atoms. 'An **atom** is the smallest part of an element which can take part in a chemical change and which retains the properties of the element'.

Each of the elements has a unique type of atom.

In atomic theory, a model of an atom can be regarded as a miniature solar system. It consists of a central nucleus around which negatively charged particles called **electrons** orbit in certain fixed bands called **shells**. The nucleus contains positively charged particles called **protons** and particles having no electrical charge called **neutrons**.

An electron has a very small mass compared with protons and neutrons. An atom is electrically neutral, containing the same number of protons as electrons. The number of protons in an atom is called the **atomic number** of the element of which the atom is part. The arrangement of the elements in order of their atomic number is known as the **periodic table**.

The simplest atom is hydrogen, which has 1 electron orbiting the nucleus and 1 proton in the nucleus. The atomic number of hydrogen is thus 1. The hydrogen atom is shown diagrammatically in Figure 4.1(a). Helium has 2 electrons orbiting the nucleus, both of then occupying the same shell at the same distance from the nucleus, as shown in Figure 4.1(b).

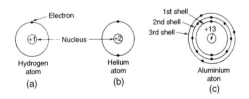

Figure 4.1

The first shell of an atom can have up to 2 electrons only, the second shell can have up to 8 electrons only and the third shell up to 18 electrons only. Thus an aluminium atom which has 13 electrons orbiting the nucleus is arranged as shown in Figure 1(c).

Molecules

When elements combine together, the atoms join to form a basic unit of new substance. This independent group of atoms bonded together is called a molecule. 'A **molecule** is the smallest part of a substance which can have a separate stable existence'.

All molecules of the same substance are identical. Atoms and molecules are the **basic building blocks** from which matter is constructed.

Compounds

When elements combine chemically their atoms interlink to form molecules of a new substance called a compound. 'A **compound** is a new substance containing two or more elements chemically combined so that their properties are changed'.

For example, the elements hydrogen and oxygen are quite unlike water, which is the compound they produce when chemically combined.

The components of a compound are in fixed proportion and are difficult to separate. Examples include:

(i) water H_2O, where 1 molecule is formed by 2 hydrogen atoms combining with 1 oxygen atom,

(ii) carbon dioxide CO_2, where 1 molecule is formed by 1 carbon atom combining with 2 oxygen atoms,

(iii) sodium chloride NaCl (common salt), where 1 molecule is formed by 1 sodium atom combining with 1 chlorine atom, and

(iv) copper sulphate $CuSO_4$, where 1 molecule is formed by 1 copper atom, 1 sulphur atom and 4 oxygen atoms combining.

Mixtures

'A **mixture** is a combination of substances which are not chemically joined together'. Mixtures have the same properties as their components. Also, the components of a mixture have no fixed proportion and are easy to separate. Examples include:

(i) oil and water

(ii) sugar and salt

(iii) air, which is a mixture of oxygen, nitrogen, carbon dioxide and other gases

(iv) iron and sulphur

(v) sand and water

Mortar is an example of a mixture — consisting of lime, sand and water.

Compounds can be distinguished from mixtures in the following ways:

(i) The properties of a compound are different to its constituent components whereas a mixture has the same properties as it constituent components.
(ii) The components of a compound are in fixed proportion whereas the components of a mixture have no fixed proportion.
(iii) The atoms of a compound are joined, whereas the atoms of a mixture are free.
(iv) When a compound is formed, heat energy is produced or absorbed whereas when a mixture is formed little or no heat is produced or absorbed.

Solutions

'A **solution** is a mixture in which other substances are dissolved'.

A solution is a mixture from which the two constituents may not be separated by leaving it to stand, or by filtration. For example, sugar dissolves in tea, salt dissolves in water and copper sulphate crystals dissolve in water leaving it a clear blue colour. The substance that is dissolved, which may be solid, liquid or gas, is called the **solute**, and the liquid in which it dissolves is called the **solvent**. Hence **solvent + solute = solution**.

A solution has a clear appearance and remains unchanged with time.

Suspensions

'A **suspension** is a mixture of a liquid and particles of a solid which do not dissolve in the liquid'.

The solid may be separated from the liquid by leaving the suspension to stand, or by filtration. Examples include:

(i) sand in water
(ii) chalk in water
(iii) petrol and water

Solubility

If a material dissolves in a liquid the material is said to be **soluble**. For example, sugar and salt are both soluble in water.

If, at a particular temperature, sugar is continually added to water and the mixture stirred there comes a point when no more sugar can dissolve. Such a solution is called **saturated**. 'A solution is saturated if no more solute can be made to dissolve, with the temperature remaining constant'.

13

'**Solubility** is a measure of the maximum amount of a solute which can be dissolved in 0.1 kg of a solvent, at a given temperature'. For example, the solubility of potassium chloride at 20°C is 34 g per 0.1 kg of water, or, its percentage solubility is 34%

(i) Solubility is dependent on temperature. When solids dissolve in liquids, as the temperature is increased, in most cases the amount of solid that will go into solution also increases. (More sugar is dissolved in a cup of hot tea than in the same amount of cold water.) There are exceptions to this, for the solubility of common salt in water remains almost constant and the solubility of calcium hydroxide decreases as the temperature increases.

(ii) Solubility is obtained more quickly when small particles of a substance are added to a liquid than when the same amount is added in large particles. For example, sugar lumps take longer to dissolve in tea than does granulated sugar.

(iii) A solid dissolves in a liquid more quickly if the mixture is stirred or shaken, i.e. solubility depends on the speed of agitation.

Crystals

A crystal is a regular, orderly arrangement of atoms or molecules forming a distinct pattern, i.e. an orderly packing of basic building blocks of matter. Most solids are crystalline in form and these include crystals such as common salt and sugar as well as the metals. Substances that are non-crystalline, are called amorphous, examples including glass and wood. **Crystallisation** is the process of isolating solids from solution in a crystalline form. This may be carried out by adding a solute to a solvent until saturation is reached, raising the temperature, adding more solute and repeating the process until a fairly strong solution is obtained, and then allowing the solution to cool, when crystals will separate. There are several examples of crystalline form that occur naturally, examples including graphite, quartz, diamond and common salt.

Crystals can vary in size but always have a regular geometric shape with flat faces, straight edges and having specific angles between the sides. Two common shapes of crystals are shown in Figure 4.2. The angles between the faces of the common salt crystal (Figure 4.2(a)) are always 90° and those of a quartz crystal (Figure 2(b)) are always 60°. A particular material always produces exactly the same shape of crystal.

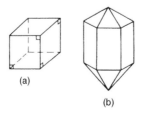

(a)

(b)

Figure 4.2

14

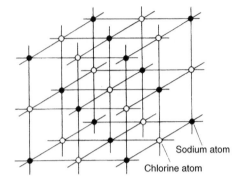

Sodium atom

Chlorine atom

Figure 4.3

Figure 4.3 shows a crystal lattice of sodium chloride. This is always a cubic shaped crystal being made up of 4 sodium atoms and 4 chlorine atoms. The sodium chloride crystals then join together as shown.

Metals

Metals are **polycrystalline** substances. This means that they are made up of a large number of crystals joined at the boundaries, the greater the number of boundaries the stronger the material.

Every metal, in the solid state, has its own crystal structure. To form an **alloy**, different metals are mixed when molten, since in the molten state they do not have a crystal lattice. The molten solution is then left to cool and solidify. The solid formed is a mixture of different crystals and an alloy is thus referred to as a **solid solution**. Examples include:

  (i) brass, which is a combination of copper and zinc,
 (ii) steel, which is mainly a combination of iron and carbon,
(iii) bronze, which is a combination of copper and tin.

Alloys are produced to enhance the properties of the metal, such as greater strength. For example, when a small proportion of nickel (say, $2\% - 4\%$) is added to iron the strength of the material is greatly increased. By controlling the percentage of nickel added, materials having different specifications may be produced.

A metal may be hardened by heating it to a high temperature then cooling it very quickly. This produces a large number of crystals and therefore many boundaries. The greater the number of crystal boundaries, the stronger is the metal.

A metal is annealed by heating it to a high temperature and then allowing it to cool very slowly. This causes larger crystals, thus less boundaries and hence a softer metal.

5 Chemical Reactions

Introduction

A **chemical reaction** is an interaction between substances in which atoms are rearranged. A new substance is always produced in a chemical reaction.
Air is a mixture, and its composition by volume is approximately: nitrogen 78%, oxygen 21%, other gases (including carbon dioxide) 1%.

Oxygen

Oxygen is an odourless, colourless and tasteless element. It is slightly soluble in water (which is essential for fish), has a boiling point of $-183°C$ (i.e. 90 K), a freezing point of $-219°C$ (i.e. 54 K) and has approximately the same density as air. Oxygen is a strongly active chemical element and combines with many substances when they are heated.

Uses of oxygen include: chemical processing, metal cutting and welding processes to give a very hot flame when burnt with other gases, and for divers, mountaineers, fire-fighters using breathing apparatus and for medical use in hospitals.

If a substance, such as powdered copper, of known mass, is heated in air, allowed to cool, and its mass remeasured, it is found that the substance has gained in mass. This is because the copper has absorbed oxygen from the air and changed into copper oxide. In addition, the proportion of oxygen in the air passed over the copper will decrease by the same amount as the gain in mass by the copper.

All substances require the presence of oxygen for burning to take place. Any substance burning in air will combine with the oxygen. This process is called **combustion**, and is an example of a chemical reaction between the burning substance and the oxygen in the air, the reaction producing heat. The chemical reaction is called **oxidation**.

An element reacting with oxygen produces a compound that contains only atoms of the original element and atoms of oxygen. Such compounds are called **oxides**. Examples of oxides include: copper oxide CuO, hydrogen oxide H_2O (i.e. water) and carbon dioxide CO_2

Rusting

Rusting of iron (and iron-based materials) is due to the formation on its surface of hydrated oxide of iron produced by a chemical reaction. Rusting of iron always requires the presence of oxygen and water.

Any iron or steel structure exposed to moisture is susceptible to rusting. This process, which cannot be reversed, can be dangerous since structures may be weakened by it. Examples of damage caused by rusting may be found in steel parts of a motor vehicle, the hull of ships, iron guttering, bridges and similar structures. Rusting may be prevented by:

(i) painting with water-resistant paint
(ii) galvanising the iron
(iii) plating the iron (see chapter 42, page 218)
(iv) an oil or grease film on the surface

Chemical Equations

To represent a reaction a chemical shorthand is used. A symbol represents an element (such as H for hydrogen, O for oxygen, Cu for copper, Zn for zinc, and so on) and a formula represents a compound and gives the type and number of elements in the compound. For example, one molecule of sulphuric acid, H_2SO_4, contains 2 atoms of hydrogen, 1 atom of sulphur and 4 atoms of oxygen. Similarly, a molecule of methane gas, CH_4, contains 1 atom of carbon and 4 atoms of hydrogen.

The rearrangement of atoms in a chemical reaction is shown by **chemical equations** using formulae and symbols.

For example:

(a) $S + O_2 = SO_2$ i.e. 1 molecule of sulphur S added to 1 molecule of oxygen O_2 causes a reaction and produces 1 molecule of sulphur dioxide SO_2
(b) $Zn + H_2SO_4 = ZnSO_4 + H_2$ i.e. 1 molecule of zinc Zn added to 1 molecule of sulphuric acid H_2SO_4 causes a reaction and produces 1 molecule of zinc sulphate $ZnSO_4$ and 1 molecule of hydrogen H_2

In a chemical equation:

(i) each element must have the same total number of atoms on each side of the equation; for example, in chemical equation (b) above each side of the equation has 1 zinc atom, 2 hydrogen atoms, 1 sulphur atom and 4 oxygen atoms
(ii) a number written in front of a molecule multiplies all the atoms in that molecule

---

Acids and Alkalis

An **acid** is a compound containing hydrogen in which the hydrogen can be easily replaced by a metal. For example, in equation (b) above, it is shown that zinc reacts with sulphuric acid to give zinc sulphate and hydrogen.

An acid produces hydrogen ions H^+ in solution (an **ion** being a charged particle formed when atoms or molecules lose or gain electrons). Examples of acids include: sulphuric acid, H_2SO_4, hydrochloric acid, HCl and nitric acid HNO_3

A **base** is a substance that can neutralise an acid (i.e. remove the acidic properties of acids). An **alkali** is a soluble base. When in solution an alkali produces hydroxyl ions, OH^-. Examples of alkalis include: sodium hydroxide, NaOH (i.e. caustic soda), calcium hydroxide, $Ca(OH)_2$, ammonium hydroxide, NH_4OH and potassium hydroxide, KOH (i.e. caustic potash).

A **salt** is the product of the neutralisation between an acid and a base, i.e.

acid + base = salt + water

For example: $HCl + NaOH = NaCl + H_2O$

$$H_2SO_4 + 2KOH = K_2SO_4 + 2H_2O$$

$$H_2SO_4 + CuO = CuSO_4 + H_2O$$

Examples of salts include: sodium chloride, NaCl (i.e. common salt), potassium sulphate, K_2SO_4, copper sulphate, $CuSO_4$ and calcium carbonate, $CaCO_3$ (i.e. limestone).

An **indicator** is a chemical substance, which when added to a solution, indicates the acidity or alkalinity of the solution by changing colour. Litmus is a simple two-colour indicator which turns red in the presence of acids and blue in the presence of alkalis. Two other examples of indicators are ethyl orange (red for acids, yellow for alkalis) and phenolphthalein (colourless for acids, pink for alkalis).

The **pH scale** (pH meaning 'the potency of hydrogen') represents, on a scale from 0 to 14, degrees of acidity and alkalinity. 0 is strongly acidic, 7 is neutral and 14 is strongly alkaline. Some average pH values include: concentrated hydrochloric acid, HCl 1.0, lemon juice 3.0, milk 6.6, pure water 7.0, sea water 8.2, concentrated sodium hydroxide, NaOH 13.0

Acids have the following properties:

(i) Almost all acids react with carbonates and bicarbonates, (a carbonate being a compound containing carbon and oxygen — an example being sodium carbonate, i.e. washing soda)

(ii) Dilute acids have a sour taste; examples include citric acid (lemons), acetic acid (vinegar) and lactic acid (sour milk).

(iii) Acid solutions turn litmus paper red, methyl orange red and phenolphthalein colourless, as mentioned above.

(iv) Most acids react with higher elements in the electrochemical series (see chapter 42) and hydrogen is released.

Alkalis have the following properties:

(i) Alkalis neutralise acids to form a salt and water only.

(ii) Alkalis have little effect on metals.

(iii) Alkalis turn litmus paper blue, methyl orange yellow and phenolphthalein pink, as mentioned above.

(iv) Alkalis are slippery when handled; strong alkalis are good solvents for certain oils and greases.

6 Standard Quantity Symbols and their Units

Quantity	Quantity symbol	Unit	Unit symbol
Acceleration,			
gravitational	g	metres per second squared	m/s^2 or $m\ s^{-2}$
linear	a	metres per second squared	m/s^2 or $m\ s^{-2}$
Angular acceleration	α	radians per second squared	rad/s^2
Angular velocity	ω	radians per second	rad/s
Area	A	square metres	m^2
Area, second moment of	I	$(metre)^4$	m^4
Capacitance	C	farads	F
Capacity	V	litres	l
Coefficient of friction	μ	no unit	
Coefficient of linear			
expansion	α	per degree Celsius	$/^{\circ}C$
Conductance	G	siemens	S
Cubic expansion, coefficient of	γ	per degree Celsius	$/^{\circ}C$
Current	I	ampere	A
Density	ρ	kilogram per cubic metre	kg/m^3
Density, relative	d	no unit	
Dryness fraction	x	no unit	
Efficiency	η	no unit	
Elasticity, modulus of	E	Pascal ($1\ Pa = 1\ N/m^2$)	Pa
Electric field strength	E	volts per metre	V/m
Electric flux density	D	coulomb per square metre	C/m^2
Energy	W	joules	J
Energy, internal	U, E	joules	J
Energy, specific internal	u, e	kilojoules per kilogram	kJ/kg
Enthalpy	H	joules	J
Enthalpy, specific	h	kilojoules per kilogram	kJ/kg
Entropy	S	kilojoules per kelvin	kJ/K
Expansion:			
coefficient of cubic	γ	per degree Celsius	$/^{\circ}C$
coefficient of linear	α	per degree Celsius	$/^{\circ}C$
coefficient of superficial	β	per degree Celsius	$/^{\circ}C$
Field strength: electric	E	volts per metre	V/m
magnetic	H	ampere per metre	A/m
Flux density: electric	D	coulomb per square metre	C/m^2
magnetic	B	tesla ($1\ T = 1\ Wb/m^2$)	T
Flux: electric	ψ	coulomb	C
magnetic	Φ	weber	Wb
Force	F	newton	N
Frequency	f	hertz	Hz
Heat capacity, specific	c	kilojoules per kilogram kelvin	kJ/(kg K)

Quantity	Quantity symbol	Unit	Unit symbol
Impedance	Z	ohms	Ω
Inductance: self	L	henry	H
mutual	M	henry	H
Internal energy	U, E	joules	J
specific	u, e	kilojoules per kilogram	kJ/kg
Inertia, moment of	I, J	kilogram metre squared	kg m^2
Length	l	metre	m
Luminous intensity	l	candela	cd
Magnetic field strength	H	ampere per metre	A/m
Magnetic flux	Φ	weber	Wb
density	B	tesla	T
Magnetomotive force	F	ampere	A
Mass	m	kilogram	kg
Mass, rate of flow	V	cubic metre per second	m^3/s
Modulus of elasticity	E	Pascal	Pa
rigidity	G	Pascal	Pa
Moment of force	M	newton metre	Nm
Moment of inertia	I, J	kilogram metre squared	kg m^2
Mutual inductance	M	henry	H
Number of turns in a winding	N	no unit	
Periodic time	T	second	s
Permeability: absolute	μ	henry per metre	H/m
of free space	μ_0	henry per metre	H/m
relative	μ_r	no unit	
Permittivity: absolute	ε	farad per metre	F/m
of free space	ε_0	farad per metre	F/m
relative	ε_r	no unit	
Polar moment of area	J	(metre)^4	m^4
Power: active	P	watt	W
apparent	S	volt ampere	VA
reactive	Q	volt ampere reactive (vars)	var
Pressure	P	Pascal (1 Pa = 1 N/m^2)	Pa
Quantity of heat	Q	joule	J
Quantity of electricity	Q	coulomb	C
Reactance	X	ohm	Ω
Reluctance	S	per henry or ampere per weber	/H or A/Wb
Resistance	R	ohm	Ω
Resistivity	ρ	ohm metre	Ωm
Second moment of area	I	(metre)^4	m^4
Shear strain	γ	no unit	
stress	τ	Pascal	Pa
Specific gas constant	R	kilojoules per kilogram kelvin	kJ/(kg K)
Specific heat capacity	c	kilojoules per kilogram kelvin	kJ/(kg K)
Specific volume	v	cubic metres per kilogram	m^3/kg
Strain, direct	ε	no unit	
Stress, direct	σ	Pascal	Pa
Shear modulus of rigidity	G	Pascal	Pa
Temperature coefficient of resistance	α	per degree Celsius	/°C
Temperature, thermodynamic	T	kelvin	K

Quantity	Quantity symbol	Unit	Unit symbol
Time	t	second	s
Torque	T	newton metre	Nm
Velocity	v	metre per second	m/s
angular	ω	radian per second	rad/s
Voltage	V	volt	V
Volume	V	cubic metre	m^3
Volume, rate of flow	V	cubic metre per second	m^3/s
Wavelength	λ	metre	m
Work	W	joule	J
Young's modulus of elasticity	E	Pascal	Pa

Part Two **Mechanical Engineering and Physical Science**

7 Speed and Velocity

Speed

Speed is the rate of covering distance and is given by:

$$\text{speed} = \frac{\text{distance travelled}}{\text{time taken}}$$

The usual units for speed are metres per second, (m/s or m s^{-1}), or kilometres per hour, (km/h or km h^{-1}). Thus if a person walks 5 kilometres in 1 hour, the speed of the person is $\frac{5}{1}$, that is, 5 kilometres per hour.

The symbol for the SI unit of speed (and velocity) is written as m s^{-1}, called the 'index notation'. However, engineers usually use the symbol m/s, called the 'oblique notation', and it is this notation that is largely used in this chapter and other chapters on mechanics. One of the exceptions is when labelling the axes of graphs, when two obliques occur, and in this case the index notation is used. Thus for speed or velocity, the axis markings are speed/m s^{-1} or velocity/m s^{-1}.

For example, if a man walks 600 metres in 5 minutes

$$\text{then speed} = \frac{\text{distance travelled}}{\text{time taken}} = \frac{600 \text{ m}}{5 \text{ min}}$$

$$= \frac{600 \text{ m}}{5 \text{ min}} \times \frac{1 \text{ min}}{60 \text{ s}} = \mathbf{2 \text{ m/s}}$$

$$\text{and } 2 \text{ m/s} = \frac{2 \text{ m}}{1 \text{ s}} \times \frac{1 \text{ km}}{1000 \text{ m}} \times \frac{3600 \text{ s}}{1 \text{ h}} = 2 \times 3.6 = \mathbf{7.2 \text{ km/h}}$$

(Note: to change from m/s to km/h, multiply by 3.6)

Distance/time Graph

One way of giving data on the motion of an object is graphically. A graph of distance travelled (the scale on the vertical axis of the graph) against time (the scale on the horizontal axis of the graph) is called a **distance/time graph**. Thus if an aeroplane travels 500 kilometres in its first hour of flight and 750 kilometres in its second hour of flight, then after 2 hours, the total distance travelled is (500 + 750) kilometres, that is, 1250 kilometres. The distance/time graph for this flight is shown in Figure 7.1.

The **average speed** is given by:

$$\frac{\text{total distance travelled}}{\text{total time taken}}$$

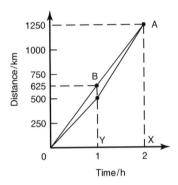

Figure 7.1

Thus, the average speed of the aeroplane is

$$\frac{(500 + 750)\ \text{km}}{(1 + 1)\ \text{h}} = \frac{1250}{2} = 625\ \text{km/h}.$$

If points O and A are joined in Figure 7.1, the slope of line OA is defined as

$$\frac{\text{change in distance (vertical)}}{\text{change in time (horizontal)}}$$

for any two points on line OA.

For point A, the change in distance is AX, that is, 1250 kilometres, and the change in time is OX, that is, 2 hours. Hence the average speed is $\dfrac{1250}{2}$, i.e. 625 kilometres per hour.

Alternatively, for point B on line OA, the change in distance is BY, that is, 625 kilometres, and the change in time is OY, that is 1 hour, hence the average speed is $\dfrac{625}{1}$, i.e. 625 kilometres per hour.

In general, the slope of line, say, MN on the distance/time graph gives the average speed of an object travelling between points M and N.

Speed/time Graph

If a graph is plotted of speed against time, the area under the graph gives the distance travelled. Thus the distance covered by the object when moving from 0 to B in Figure 7.2, is given by the area beneath the speed/time graph, shown shaded.

Area of triangle OAC $= \frac{1}{2} \times$ base $\times$ perpendicular height

$$= \frac{1}{2} \times 5\ \text{s} \times 10\ \frac{\text{m}}{\text{s}} = 25\ \text{m}$$

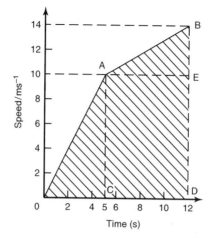

Figure 7.2

Area of rectangle AEDC $=$ base $\times$ height

$$= (12 - 5) \text{ s} \times (10 - 0) \frac{\text{m}}{\text{s}} = 70 \text{ m}$$

Area of triangle ABE $= \frac{1}{2} \times$ base $\times$ perpendicular height

$$= \frac{1}{2} \times (12 - 5) \text{ s} \times (14 - 10) \frac{\text{m}}{\text{s}}$$

$$= \frac{1}{2} \times 7 \text{ s} \times 4 \frac{\text{m}}{\text{s}} = 14 \text{ m}$$

Hence **the distance covered by the object moving from O to B** is:

$$(25 + 70 + 14) \text{ m} = \mathbf{109 \text{ m}}$$

Velocity

The **velocity** of an object is the speed of the object **in a specified direction**. Thus, if a plane is flying due south at 500 kilometres per hour, its speed is 500 kilometres per hour, but its velocity is 500 kilometres per hour due south. It follows that if the plane had flown in a circular path for one hour at a speed of 500 kilometres per hour, so that one hour after taking off it is again over the airport, its average velocity in the first hour of flight is zero.

The average velocity is given by

$$\frac{\text{distance travelled in a specified direction}}{\text{time taken}}$$

If a plane flies from place O to place A, a distance of 300 kilometres in one hour, A being due north of O, then OA in Figure 7.3 represents the first

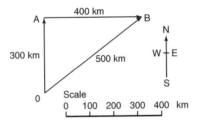

Figure 7.3

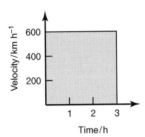

Figure 7.4

hour of flight. It then flies from A to B, a distance of 400 kilometres during the second hour of flight, B being due east of A, thus AB in Figure 7.3 represents its second hour of flight.

Its average velocity for the two hour flight is:

$$\frac{\text{distance OB}}{2 \text{ hours}} = \frac{500 \text{ km}}{2 \text{ h}} = 250 \text{ km/h in direction OB}$$

A graph of velocity (scale on the vertical axis) against time (scale on the horizontal axis) is called a velocity/time graph. The graph shown in Figure 7.4 represents a plane flying for 3 hours at a constant speed of 600 kilometres per hour in a specified direction. The shaded area represents velocity (vertically) multiplied by time (horizontally), and has units of $\frac{\text{kilometres}}{\text{hours}} \times \text{hours}$, i.e. kilometres, and represents the distance travelled in a specific direction. In this case, distance $= 600 \frac{\text{km}}{\text{h}} \times 3 \text{ h} = 1800 \text{ km}$.

Another method of determining the distance travelled is from:

distance travelled = average velocity × time

Thus if a plane travels due south at 600 kilometres per hour for 20 minutes, the distance covered is

$$\frac{600 \text{ km}}{1 \text{ h}} \times \frac{20}{60} \text{ h, i.e. } 200 \text{ km}$$

8 Acceleration

Introduction to Acceleration

Acceleration is the rate of change of velocity with time. The average acceleration, a, is given by:

$$a = \frac{\text{change in velocity}}{\text{time taken}}$$

The usual units are metres per second squared (m/s² or m s⁻²). If u is the initial velocity of an object in metres per second, v is the final velocity in metres per second and t is the time in seconds elapsing between the velocities of u and v, then:

$$\boxed{\text{average acceleration, } a = \frac{v-u}{t}\,\text{m/s}^2}$$

Velocity/time Graph

A graph of velocity (scale on the vertical axis) against time (scale on the horizontal axis) is called a velocity/time graph, as introduced in Chapter 7. For the velocity/time graph shown in Figure 8.1, the slope of line OA is given by (AX/OX). AX is the change in velocity from an initial velocity, u, of zero to a final velocity, v, of 4 metres per second. OX is the time taken for this

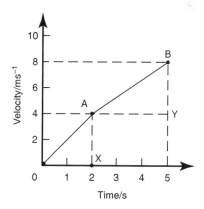

Figure 8.1

change in velocity, thus

$$\frac{AX}{OX} = \frac{\text{change in velocity}}{\text{time taken}}$$

$$= \text{the acceleration in the first two seconds}$$

From the graph:

$$\frac{AX}{OX} = \frac{4 \text{ m/s}}{2 \text{ s}} = 2 \text{ m/s}^2$$

i.e. the acceleration is 2 m/s^2.

Similarly, the slope of line AB in Figure 8.1 is given by (BY/AY), i.e. the acceleration between 2 s and 5 s is

$$\frac{8-4}{5-2} = \frac{4}{3} = 1\frac{1}{3} \text{ m/s}^2.$$

In general, the slope of a line on a velocity/time graph gives the acceleration. The words 'velocity' and 'speed' are commonly interchanged in everyday language. Acceleration is a vector quantity and is correctly defined as the rate of change of velocity with respect to time. However, acceleration is also the rate of change of speed with respect to time in a certain specified direction.

Free-fall and Equation of Motion

If a dense object such as a stone is dropped from a height, called **free-fall**, it has a constant acceleration of approximately 9.8 m/s^2. In a vacuum, all objects have this same constant acceleration, vertically downwards, that is, a feather has the same acceleration as a stone. However, if free-fall takes place in air, dense objects have the constant acceleration of 9.8 m/s^2 over short distances, but objects that have a low density, such as feathers, have little or no acceleration.

For bodies moving with a constant acceleration, the average acceleration is the constant value of the acceleration, and since from earlier:

$$a = \frac{v - u}{t}$$

then $a \times t = v - u$ from which **v = u + at**

where u is the initial velocity in m/s,
        v is the final velocity in m/s
        a is the constant acceleration in m/s^2 and
        t is the time in s

When symbol 'a' has a negative value, it is called **deceleration** or **retardation**. The equation $v = u + at$ is called an **equation of motion**.

For example, if a stone is dropped from an aeroplane the stone is free falling and thus has an acceleration, a, of approximately 9.8 m/s^2 (taking downward motion as positive). The initial downward velocity of the stone, u, is zero. The velocity v after $2s$ is given by: $v = u + at = 0 + 9.8 \times 2 = 19.6$ m/s, i.e. **the velocity of the stone after 2 s is approximately 19.6 m/s.**

In another example, if a train is travelling at, say, 30 km/h and it accelerates uniformly to 50 km/h in 2 minutes, then $u = 30$ km/h $= \dfrac{30}{3.6}$ m/s, $v = 50$ km/h $= \dfrac{50}{3.6}$ m/s and t $= 2$ min $= 2 \times 60 = 120$ s

Thus, $\qquad\qquad\qquad v = u + at$, i.e. $\dfrac{50}{3.6} = \dfrac{30}{3.6} + \text{a} \times 120$

Transposing, gives $\qquad\qquad\qquad 120 \times a = \dfrac{50 - 30}{3.6}$

and $\qquad\qquad\qquad a = \dfrac{20}{3.6 \times 120} = 0.0463$ m/s^2

i.e. **the uniform acceleration of the train is 0.0463 m/s^2**

9 Force, Mass and Acceleration

Introduction

When an object is pushed or pulled, a **force** is applied to the object. This force is measured in **newtons (N)**. The effects of pushing or pulling an object are:

 (i) to cause a change in the motion of the object, and
(ii) to cause a change in the shape of the object.

If a change occurs in the motion of the object, that is, its velocity changes from u to v, then the object accelerates. Thus, it follows that acceleration results from a force being applied to an object. If a force is applied to an object and it does not move, then the object changes shape, that is, deformation of the object takes place. Usually the change in shape is so small that it cannot be detected by just watching the object. However, when very sensitive measuring instruments are used, very small changes in dimensions can be detected.

A force of attraction exists between all objects. The factors governing the size of this force F are the masses of the objects and the distances between their centres:

$$F \propto \frac{m_1 m_2}{d^2}$$

Thus, if a person is taken as one object and the earth as a second object, a force of attraction exists between the person and the earth. This force is called the **gravitational force** and is the force that gives a person a certain weight when standing on the earth's surface. It is also this force that gives freely falling objects a constant acceleration in the absence of other forces.

Newton's Laws of Motion

To make a stationary object move or to change the direction in which the object is moving requires a force to be applied externally to the object. This concept is known as **Newton's first law of motion** and may be stated as:

An object remains in a state of rest, or continues in a state of uniform motion in a straight line, unless it is acted on by an externally applied force

Since a force is necessary to produce a change of motion, an object must have some resistance to a change in its motion. The force necessary to give a stationary pram a given acceleration is far less than the force necessary to give a stationary car the same acceleration. The resistance to a change in motion is called the **inertia** of an object and the amount of inertia depends on the mass of the object. Since a car has a much larger mass than a pram, the inertia of a car is much larger than that of a pram.

Newton's second law of motion may be stated as:

The acceleration of an object acted upon by an external force is proportional to the force and is in the same direction as the force

Thus, force α acceleration, or force = a constant × acceleration, this constant of proportionality being the mass of the object, i.e.

force = mass × acceleration

The unit of force is the newton (N) and is defined in terms of mass and acceleration. One newton is the force required to give a mass of 1 kilogram an acceleration of 1 metre per second squared. Thus

$$\boxed{F = ma}$$

where F is the force in newtons (N), m is the mass in kilograms (kg) and a is the acceleration in metres per second squared (m/s^2), i.e.

$$1 \text{ N} = \frac{1 \text{ kg m}}{\text{s}^2}$$

It follows that 1 m/s^2 = 1 N/kg. Hence a gravitational acceleration of 9.8 m/s^2 is the same as a gravitational field of 9.8 N/kg

Newton's third law of motion may be stated as:

For every force, there is an equal and opposite reacting force

Thus, an object on, say, a table, exerts a downward force on the table and the table exerts an equal upward force on the object, known as a **reaction force** or just a **reaction**.

For example, to calculate the force needed to accelerate a boat of mass 20 tonne uniformly from rest to a speed of 21.6 km/h in 10 minutes:

final velocity, $\quad v = 21.6$ km/h $= \dfrac{21.6}{3.6} = 6$ m/s, and the time,

$\quad t = 10$ min $= 600$ s

Thus $\quad v = u + at$, i.e. $6 = 0 + a \times 600$, from which,

$\quad a = \dfrac{6}{600} = 0.01$ m/s^2

From Newton's second law, $F = ma$

i.e. $\qquad\qquad$ **force** $= 20\,000 \times 0.01$ N $=$ **200 N**

In another example, if the moving head of a machine tool requires a force of 1.2 N to bring it to rest in 0.8 s from a cutting speed of 30 m/min, then from Newton's second law, $F = ma$, from which, the mass of the moving head, $m = \dfrac{F}{a}$.

32

The law of motion $v = u + at$ can be used to find acceleration a, where $v = 0$, $u = 30$ m/min $= \dfrac{30}{60}$ m/s $= 0.5$ m/s, and $t = 0.8$ s.

Thus,
$$0 = 0.5 + a \times 0.8$$

from which,
$$a = -\frac{0.5}{0.8} = -0.625 \text{ m/s}^2$$

or a retardation of 0.625 m/s^2

Thus the **mass**,
$$m = \frac{F}{a} = \frac{1.2}{0.625} = \textbf{1.92 kg}$$

Centripetal Acceleration

When an object moves in a circular path at constant speed, its direction of motion is continually changing and hence its velocity (which depends on both magnitude and direction) is also continually changing. Since acceleration is the (change in velocity)/(time taken) the object has an acceleration.

Let the object be moving with a constant angular velocity of ω and a tangential velocity of magnitude v and let the change of velocity for a small change of angle of $\theta(= \omega t)$ be V (see Figure 9.1(a)). Then, $v_2 - v_1 = V$.

The vector diagram is shown in Figure 9.1(b) and since the magnitudes of v_1 and v_2 are the same, i.e. v, the vector diagram is also an isosceles triangle.

Bisecting the angle between v_2 and v_1 gives:

$$\sin \frac{\theta}{2} = \frac{V/2}{v_2} = \frac{V}{2v}$$

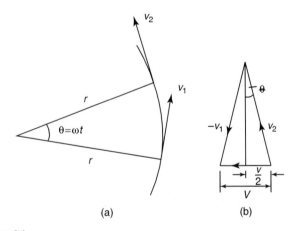

(a)   (b)

Figure 9.1

i.e.
$$V = 2v \sin \frac{\theta}{2} \qquad (1)$$

Since $\theta = \omega t$, then
$$t = \frac{\theta}{\omega} \qquad (2)$$

Dividing (1) by (2) gives:

$$\frac{V}{t} = \frac{2v \sin \dfrac{\theta}{2}}{\dfrac{\theta}{\omega}} = \frac{v\omega \sin \dfrac{\theta}{2}}{\dfrac{\theta}{2}}$$

For small angles, $\dfrac{\sin \dfrac{\theta}{2}}{\dfrac{\theta}{2}}$ is very nearly equal to unity,

hence, $\dfrac{V}{t} = \dfrac{\text{change of velocity}}{\text{change of time}} = $ acceleration, $a = v\omega$

But $\omega = v/r$, thus $v\omega = v \times \dfrac{v}{r} = \dfrac{v^2}{r}$

That is, **the acceleration a is** $\dfrac{v^2}{r}$ and is towards the centre of the circle of motion (along V). It is called the **centripetal acceleration**. If the mass of the rotating object is m, then by Newton's second law, the **centripetal force** is $\dfrac{m v^2}{r}$, and its direction is towards the centre of the circle of motion.

For example, if a vehicle of mass 750 kg travels round a bend of radius 150 m, at 50.4 km/h, then the **centripetal force** $= \dfrac{mv^2}{r} = \dfrac{750 \times \left(\dfrac{50.4}{3.6}\right)^2}{150} = $ **980 N**

10 Centre of Gravity and Equilibrium

Centre of Gravity

The **centre of gravity** of an object is a point where the resultant gravitational force acting on the body may be taken to act. For objects of uniform thickness lying in a horizontal plane, the centre of gravity is vertically in line with the point of balance of the object. For a thin uniform rod the point of balance and hence the centre of gravity is halfway along the rod as shown in Figure 10.1(a).

A thin flat sheet of a material of uniform thickness is called a **lamina** and the centre of gravity of a rectangular lamina lies at the point of intersection of its diagonals, as shown in Figure 10.1(b). The centre of gravity of a circular lamina is at the centre of the circle, as shown in Figure 10.1(c).

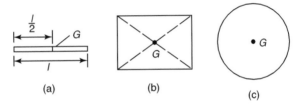

(a) (b) (c)

Figure 10.1

Equilibrium

An object is in **equilibrium** when the forces acting on the object are such that there is no tendency for the object to move. The state of equilibrium of an object can be divided into three groups.

(i) If an object is in **stable equilibrium** and it is slightly disturbed by pushing or pulling (i.e. a disturbing force is applied), the centre of gravity is raised and when the disturbing force is removed, the object returns to its original position. Thus a ball bearing in a hemispherical cup is in stable equilibrium, as shown in Figure 10.2(a).

(ii) An object is in **unstable equilibrium** if, when a disturbing force is applied, the centre of gravity is lowered and the object moves away from its original position. Thus, a ball bearing balanced on top of a hemispherical cup is in unstable equilibrium, as shown in Figure 10.2(b).

(iii) When an object in **neutral equilibrium** has a disturbing force applied, the centre of gravity remains at the same height and the object does not move when the disturbing force is removed. Thus, a ball bearing on a flat horizontal surface is in neutral equilibrium, as shown in Figure 10.2(c).

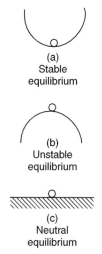

(a)
Stable
equilibrium

(b)
Unstable
equilibrium

(c)
Neutral
equilibrium

Figure 10.2

11 Forces Acting at a Point

Forces

When forces are all acting in the same plane, they are called **coplanar**. When forces act at the same time and at the same point, they are called **concurrent forces.**

Force is a **vector quantity** and thus has both a magnitude and a direction. A vector can be represented graphically by a line drawn to scale in the direction of the line of action of the force.

To distinguish between vector and scalar quantities, various ways are used, as detailed in chapter 3. The method adopted in this chapter is to denote vector quantities in **bold print**.

Thus, *ab* in Figure 11.1 represents a force of 5 newtons acting in a direction due east.

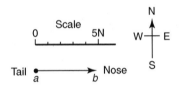

Figure 11.1

The Resultant of Two Coplanar Forces

For two forces acting at a point, there are three possibilities.

(a) For forces acting in the same direction and having the same line of action, the single force having the same effect as both of the forces, called the **resultant force** or just the **resultant**, is the arithmetic sum of the separate forces. Forces of F_1 and F_2 acting at point P, as shown in Figure 11.2(a), have exactly the same effect on point P as force F shown in Figure 11.2(b),

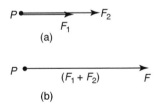

Figure 11.2

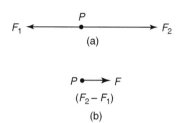

Figure 11.3

where $F = F_1 + F_2$ and acts in the same direction as F_1 and F_2. Thus F is the resultant of F_1 and F_2

(b) For forces acting in opposite directions along the same line of action, the resultant force is the arithmetic difference between the two forces. Forces of F_1 and F_2 acting at point P as shown in Figure 11.3(a) have exactly the same effect on point P as force F shown in Figure 11.3(b), where $F = F_2 - F_1$ and acts in the direction of F_2, since F_2 is greater than F_1. Thus F is the resultant of F_1 and F_2

(c) When two forces do not have the same line of action, the magnitude and direction of the resultant force may be found by a procedure called vector addition of forces. There are two graphical methods of performing **vector addition**, known as the **triangle of forces** method and the **parallelogram of forces** method.

Triangle of Forces Method

A simple procedure for the triangle of forces method of vector addition is as follows:

(i) Draw a vector representing one of the forces, using an appropriate scale and in the direction of its line of action.

(ii) From the **nose** of this vector and using the same scale, draw a vector representing the second force in the direction of its line of action.

(iii) The resultant vector is represented in both magnitude and direction by the vector drawn from the tail of the first vector to the nose of the second vector.

For example, to determine the magnitude and direction of the resultant of a force of 15 N acting horizontally to the right and a force of 20 N inclined at an angle of 60° to the 15 N force using the triangle of forces method:

Using the above procedure and with reference to Figure 11.4

(i) **ab** is drawn 15 units long horizontally

(ii) from b, **bc** is drawn 20 units long, inclined at an angle of 60° to **ab**
(Note, in angular measure, an angle of 60° from **ab** means 60° in an anticlockwise direction)

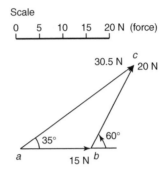

Figure 11.4

(iii) by measurement, the resultant **ac** is 30.5 units long inclined at an angle of 35° to *ab*.

Hence, the resultant force is **30.5 N**, inclined at an angle of **35°** to the 15 N force.

The Parallelogram of Forces Method

A simple procedure for the parallelogram of forces method of vector addition is as follows:

 (i) Draw a vector representing one of the forces, using an appropriate scale and in the direction of its line of action.
 (ii) From the **tail** of this vector and using the same scale draw a vector representing the second force in the direction of its line of action.
(iii) Complete the parallelogram using the two vectors drawn in (i) and (ii) as two sides of the parallelogram.
(iv) The resultant force is represented in both magnitude and direction by the vector corresponding to the diagonal of the parallelogram drawn from the tail of the vectors in (i) and (ii)

For example, to find the magnitude and direction of the resultant of a force of 250 N acting at an angle of 135° and a force of 400 N acting at an angle of −120° using the parallelogram of forces method:

From the above procedure and with reference to Figure 11.5:

 (i) **ab** is drawn at an angle of 135° and 250 units in length
 (ii) **ac** is drawn at an angle of −120° and 400 units in length
(iii) **bd** and **cd** are drawn to complete the parallelogram
(iv) **ad** is drawn. By measurement, **ad** is 413 units long at an angle of −156°.

Hence, the resultant force is **413 N** at an angle of **−156°**

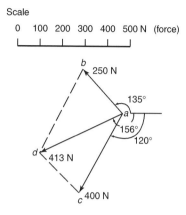

Figure 11.5

Resultant of Coplanar Forces by Calculation

An alternative to the graphical methods of determining the resultant of two coplanar forces is by **calculation**. This can be achieved by **trigonometry** using the **cosine rule** and the **sine rule**, or by **resolution of forces** (see later).

For example, to determine the magnitude and direction of the resultant of a force of 8 kN acting at an angle of 50° to the horizontal and a force of 5 kN acting at an angle of −30° to the horizontal using the cosine and sine rules:

The space diagram is shown in Figure 11.6(a). A sketch is made of the vector diagram, *oa* representing the 8 kN force in magnitude and direction and *ab* representing the 5 kN force in magnitude and direction. The resultant is given by length *ob*. By the cosine rule,

$$ob^2 = oa^2 + ab^2 - 2(oa)(ab)\cos \angle oab$$

$$= 8^2 + 5^2 - 2(8)(5)\cos 100°$$

$$(\text{since } \angle oab = 180° - 50° - 30° = 100°)$$

$$= 64 + 25 - (-13.892) = 102.892$$

Hence $ob = \sqrt{102.892} = 10.14$ kN

By the sine rule,

$$\frac{5}{\sin \angle aob} = \frac{10.14}{\sin 100°}$$

from which, $\sin \angle aob = \dfrac{5 \sin 100°}{10.14} = 0.4856$

Hence $\angle aob = \sin^{-1}(0.4856) = 29.05°$.

Thus angle ϕ in Figure 11.6(b) is $50° - 29.05° = 20.95°$

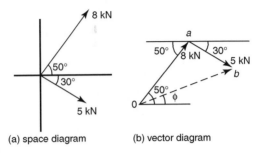

| (a) space diagram | (b) vector diagram |

Figure 11.6

Hence the resultant of the two forces is 10.14 kN acting at an angle of 20.95° to the horizontal

---

Resultant of more than Two Coplanar Forces

For the three coplanar forces F_1, F_2 and F_3 acting at a point as shown in Figure 11.7, the vector diagram is drawn using the nose-to-tail method.

The procedure is:

(i) Draw **oa** to scale to represent force F_1 in both magnitude and direction (see Figure 11.8)

(ii) From the nose of **oa**, draw **ab** to represent force F_2

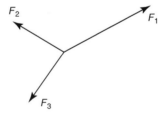

Figure 11.7

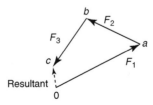

Figure 11.8

(iii) From the nose of **ab**, draw **bc** to represent force F_3

(iv) The resultant vector is given by length **oc** in Figure 11.8. The direction of resultant **oc** is from where we started, i.e. point o, to where we finished, i.e. point c. When acting by itself, the resultant force, given by **oc**, has the same effect on the point as forces F_1, F_2 and F_3 have when acting together. The resulting vector diagram of Figure 11.8 is called the **polygon of forces**.

For example, three coplanar forces acting at a point are: force A, 12 N acting horizontally to the right, force B, 7 N inclined at 60° to force A, and force C, 15 N inclined at 150° to force A. The magnitude and direction of their resultant is determined graphically as follows:

The space diagram is shown in Figure 11.9. The vector diagram, shown in Figure 11.10, is produced as follows:

 (i) **oa** represents the 12 N force in magnitude and direction
 (ii) From the nose of **oa**, **ab** is drawn inclined at 60° to **oa** and 7 units long
(iii) From the nose of **ab**, **bc** is drawn 15 units long inclined at 150° to **oa** (i.e. 150° to the horizontal)
(iv) **oc** represents the resultant; by measurement, the resultant is 13.8 N inclined at $\phi = 80°$ to the horizontal

Thus the resultant of the three forces, F_A, F_B and F_C is a force of 13.8 N at 80° to the horizontal

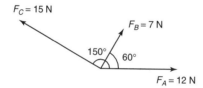

Figure 11.9

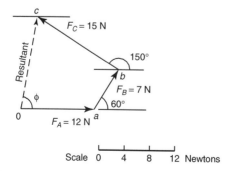

Figure 11.10

42

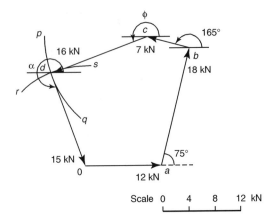

Figure 11.11

Coplanar Forces in Equilibrium

When three or more coplanar forces are acting at a point and the vector diagram closes, there is no resultant. The forces acting at the point are in **equilibrium**.

For example, five coplanar forces are acting on a body and the body is in equilibrium. The forces are: 12 kN acting horizontally to the right, 18 kN acting at an angle of 75°, 7 kN acting at an angle of 165°, 16 kN acting from the nose of the 7 kN force, and 15 kN acting from the nose of the 16 kN force. To determine the directions of the 16 kN and 15 kN forces relative to the 12 kN force:

With reference to Figure 11.11, *oa* is drawn 12 units long horizontally to the right. From point *a*, *ab* is drawn 18 units long at an angle of 75°. From *b*, *bc* is drawn 7 units long at an angle of 165°. The direction of the 16 kN force is not known, thus arc *pq* is drawn with a compass, with centre at *c*, radius 16 units. Since the forces are at equilibrium, the polygon of forces must close. Using a compass with centre at 0, arc rs is drawn having a radius 15 units. The point where the arcs intersect is at *d*.

By measurement, angle

$$\phi = 198° \text{ and } \alpha = 291°$$

Thus the 16 kN force acts at an angle of 198° (or −162°) to the 12 kN force, and the 15 kN force acts at an angle of 291° (or −69°) to the 12 kN force.

Resolution of Forces

A vector quantity may be expressed in terms of its **horizontal** and **vertical components**.

For example, a vector representing a force of 10 N at an angle of 60°
to the horizontal is shown in Figure 11.12. If the horizontal line *oa* and the
vertical line *ab* are constructed as shown, then *oa* is called the horizontal
component of the 10 N force, and *ab* the vertical component of the 10 N
force.

By trigonometry, $\cos 60° = \dfrac{oa}{ob}$, hence the horizontal component,

$$oa = 10 \ \cos 60°$$

Also, $\sin 60° = \dfrac{ab}{ob}$, hence the vertical component, $ab = 10 \sin 60°$

This process is called **finding the horizontal and vertical components
of a vector** or **the resolution of a vector**, and can be used as an alternative to
graphical methods for calculating the resultant of two or more coplanar forces
acting at a point.

For example, to calculate the resultant of a 10 N force acting at 60° to the
horizontal and a 20 N force acting at −30° to the horizontal (see Figure 11.13)
the procedure is as follows:

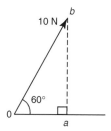

Figure 11.12

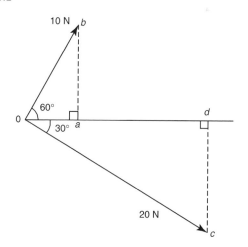

Figure 11.13

(i) Determine the horizontal and vertical components of the 10 N force, i.e.

horizontal component, $\quad oa = 10\cos 60° = 5.0$ N, and

vertical component, $\quad ab = 10\sin 60° = 8.66$ N

(ii) Determine the horizontal and vertical components of the 20 N force, i.e.

horizontal component, $\quad od = 20\cos(-30°) = 17.32$ N, and

vertical component, $\quad cd = 20\sin(-30°) = -10.0$ N

(iii) Determine the total horizontal component, i.e.

$$oa + od = 5.0 + 17.32 = 22.32 \text{ N}$$

(iv) Determine the total vertical component, i.e.

$$ab + cd = 8.66 + (-10.0) = -1.34 \text{ N}$$

(v) Sketch the total horizontal and vertical components as shown in Figure 11.14. The resultant of the two components is given by length **or** and, by

Pythagoras' theorem, $\qquad \mathbf{or} = \sqrt{22.32^2 + 1.34^2}$

$$= 22.36 \text{ N}$$

and using trigonometry, angle $\quad \phi = \tan^{-1}\dfrac{1.34}{22.32} = 3.44°$

Hence the resultant of the 10 N and 20 N forces shown in Figure 11.13 is **22.36 N at an angle of −3.44° to the horizontal.**

The above example demonstrates the use of resolution of forces for calculating the resultant of two coplanar forces acting at a point. However, the method may also be used for more than two forces acting at a point.

For example, three coplanar forces acting at a point are: 200 N acting at 20° to the horizontal, 400 N acting at 165° to the horizontal, and 500 N acting at 250° to the horizontal. To determine by resolution of forces the resultant of the forces:

A tabular approach using a calculator may be made as shown below.

	Horizontal component		
Force 1	$200\cos 20°$	=	187.94
Force 2	$400\cos 165°$	=	−386.37
Force 3	$500\cos 250°$	=	−171.01
Total horizontal component		=	−369.44

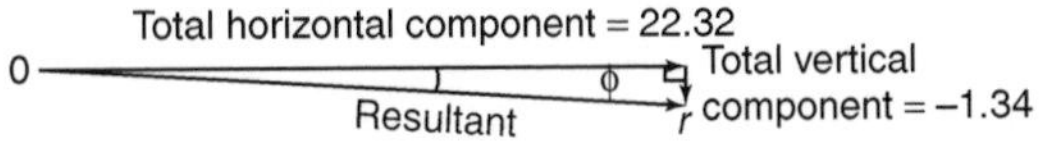

Figure 11.14

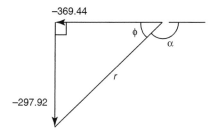

Figure 11.15

	Vertical component	
Force 1	$200 \sin 20°$	$= \quad 68.40$
Force 2	$400 \sin 165°$	$= -103.53$
Force 3	$500 \sin 250°$	$= -469.85$
Total vertical component		$= -297.92$

The total horizontal and vertical components are shown in Figure 11.15.

Resultant $\quad r = \sqrt{369.44^2 + 297.92^2} = 474.60$, and

angle $\quad \phi = \tan^{-1} \dfrac{297.92}{369.44} = 38.88°$,

from which, $\quad \alpha = 180° - 38.88° = 141.12°$

Thus the resultant of the three forces given is 474.6 N acting at an angle of −141.12° (or +218.88°) to the horizontal.

Summary

(a) To determine the **resultant of two coplanar forces** acting at a point, four methods are commonly used. They are:

    by drawing: (1) triangle of forces method, and

                       (2) parallelogram of forces method, and

    by calculation: (3) use of cosine and sine rules, and

                       (4) resolution of forces

(b) To determine the **resultant of more than two coplanar forces** acting at a point, two methods are commonly used. They are:

    by drawing: (1) polygon of forces method, and

    by calculation: (2) resolution of forces

12 Simply Supported Beams

The Moment of a Force

When using a spanner to tighten a nut, a force tends to turn the nut in a clockwise direction. This turning effect of a force is called the **moment of a force** or more briefly, a **moment**. The size of the moment acting on the nut depends on two factors:

(a) the size of the force acting at right angles to the shank of the spanner, and
(b) the perpendicular distance between the point of application of the force and the centre of the nut.

In general, with reference to Figure 12.1, the moment M of a force acting about a point P is force × perpendicular distance between the line of action of the force and P, i.e.

$$\boxed{M = F \times d}$$

The unit of a moment is the **newton metre (Nm)**. Thus, if force F in Figure 12.1 is 7 N and distance d is 3 m, then the moment M is 7 N × 3 m, i.e. 21 Nm.

Equilibrium and the Principle of Moments

If more than one force is acting on an object and the forces do not act at a single point, then the turning effect of the forces, that is, the moment of the forces, must be considered.

Figure 12.2 shows a beam with its support (known as its pivot or fulcrum) at P, acting vertically upwards, and forces F_1 and F_2 acting vertically downwards at distances a and b, respectively, from the fulcrum.

A beam is said to be in **equilibrium** when there is no tendency for it to move. There are two conditions for equilibrium:

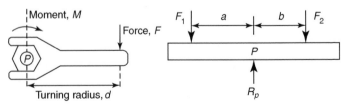

Figure 12.1   **Figure 12.2**

(i) The sum of the forces acting vertically downwards must be equal to the sum of the forces acting vertically upwards, i.e. for Figure 12.2,

$$R_p = F_1 + F_2$$

(ii) The total moment of the forces acting on a beam must be zero; for the total moment to be zero:

> *the sum of the clockwise moments about any point must be equal to the sum of the anticlockwise moments about that point*

This statement is known as the **principle of moments**.
Hence, taking moments about P in Figure 12.2,

$F_2 \times b =$ the clockwise moment,

and $\quad F_1 \times a =$ the anticlockwise moment

Thus for equilibrium: $\boxed{F_1 a = F_2 b}$

For example, for the centrally supported uniform beam shown in Figure 12.3, to determine the values of forces F_1 and F_2 when the beam is in equilibrium:

At equilibrium: (i) $R = F_1 + F_2$ i.e. $5 = F_1 + F_2$ (1)

and (ii) $F_1 \times 3 = F_2 \times 7$ (2)

From equation (1), $F_2 = 5 - F_1$

Substituting for F_2 in equation (2) gives:

$$F_1 \times 3 = (5 - F_1) \times 7$$

i.e. $\quad 3F_1 = 35 - 7F_1$

$\quad 10F_1 = 35$

from which, $\quad F_1 = 3.5$ kN

Since $\quad F_2 = 5 - F_1, F_2 = 1.5$ kN

Thus at equilibrium, force $F_1 = 3.5$ kN and force $F_2 = 1.5$ kN

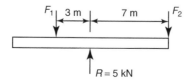

Figure 12.3

Simply Supported Beams Having Point Loads

A **simply supported beam** is one that rests on two supports and is free to move horizontally.

Two typical simply supported beams having loads acting at given points on the beam, called **point loading**, are shown in Figure 12.4

A man whose mass exerts a force F vertically downwards, standing on a wooden plank which is simply supported at its ends, may, for example, be represented by the beam diagram of Figure 12.4(a) if the mass of the plank is neglected. The forces exerted by the supports on the plank, R_A and R_B, act vertically upwards, and are called **reactions**.

When the forces acting are all in one plane, the algebraic sum of the moments can be taken about **any** point.

For the beam in Figure 12.4(a) at equilibrium:

(i) $R_A + R_B = F$, and
(ii) taking moments about A, $F(a) = R_B(a + b)$
(Alternatively, taking moments about C, $R_A a = R_B b$)

For the beam in Figure 12.4(b), at equilibrium

(i) $R_A + R_B = F_1 + F_2$, and
(ii) taking moments about B, $R_A(a + b) + F_2 c = F_1 b$

Typical **practical applications** of simply supported beams with point loadings include bridges, beams in buildings, and beds of machine tools.

For example, for the beam shown in Figure 12.5, the force acting on support A, R_A, and distance d, neglecting any forces arising from the mass of the beam, are calculated as follows:

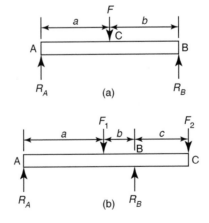

Figure 12.4

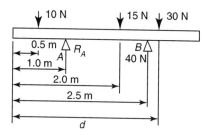

Figure 12.5

$$\text{(the forces acting in an upward direction)} = \text{(the forces acting in a downward direction)}$$

Hence

$$(R_A + 40) \text{ N} = (10 + 15 + 30) \text{ N}$$

$$R_A = 10 + 15 + 30 - 40 = \mathbf{15 \text{ N}}$$

Taking moments about the left-hand end of the beam and applying the principle of moments gives:

$$\text{clockwise moments} = \text{anticlockwise moments}$$

$$(10 \times 0.5) + (15 \times 2.0) \text{ Nm} + 30 \text{ N} \times d$$

$$= (15 \times 1.0) + (40 \times 2.5) \text{ Nm}$$

i.e.

$$35 \text{ Nm} + 30 \text{ N} \times d = 115 \text{ Nm}$$

from which,

$$\text{\textbf{distance, d}} = \frac{(115 - 35) \text{ Nm}}{30 \text{ N}} = \mathbf{2.67 \text{ m}}$$

13 Shearing Force and Bending Moments

Shearing Force

For equilibrium of a beam, the forces to the left of any section such as X in Figure 13.1, must balance the forces to the right. Also, the moment about X of the forces to the left must balance the moments about X of the forces to the right.

Although for equilibrium the forces and moments cancel, the magnitude and nature of these forces and moments are important as they determine both the stresses at X and the beam curvature and deflection. The resultant force to the left of X and the resultant force to the right of X (forces or components of forces transverse to the beam) constitute a pair of forces tending to shear the beam at this section. **Shearing force** is defined as the force transverse to the beam at a given section tending to cause it to shear at that section.

By convention, if the tendency is to shear as shown in Figure 13.2(a), the shearing force is regarded as positive, i.e. $+F$; if the tendency to shear is as shown in Figure 13.2(b), it is regarded as negative, i.e. $-F$.

Bending Moment

The **bending moment** at a given section of a beam is defined as the resultant moment about that section of either all of the forces to its left or of all

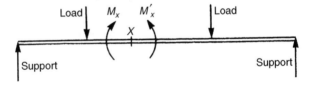

Figure 13.1

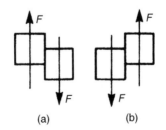

(a)   (b)

Figure 13.2

of the forces to its right. In Figure 13.1 it is M_X or M'_X. These moments, clockwise to the left and anticlockwise to the right, will cause the beam to bend concave upwards, called 'sagging'. By convention this is regarded as positive bending (i.e. bending moments is a positive bending moment). Where the curvature produced is concave downwards, called 'hogging', the bending moment is regarded as negative.

The values of shearing force and bending moment will usually vary along a beam. Diagrams showing the shearing force and bending moment for all sections of a beam are called shearing force and bending moment diagrams respectively.

Shearing forces and shearing force diagrams are less important than bending moments, but can be very useful in giving pointers to the more important bending moment diagrams. For example, wherever the shearing force is zero, the bending moment will be a maximum or minimum.

For example, the shearing force and bending moment diagrams for the beam shown in Figure 13.3 are obtained as follows:

It is first necessary to calculate the reactions at A and B. The beam is simply supported at A and B, which means that it rests on supports at these points giving vertical reactions. The general conditions for equilibrium require that the resultant moment about any point must be zero, and total upward force must equal total downward force. Therefore, taking moments about A, the moment R_B must balance the moment of the load at C:

$$R_B \times 8 \text{ m} = 24 \text{ kN} \times 5 \text{ m} = 120 \text{ kN}$$

from which
$$R_B = \frac{120 \text{ kNm}}{8 \text{ m}} = 15 \text{ kN}$$

and
$$R_A = 24 \text{ kN} - 15 \text{ kN} = 9 \text{ kN}$$

Immediately to the right of A the shearing force is due to R_A and is therefore 9 kN. As this force to the left of the section considered is upwards, the shearing force is positive. The shearing force is the same for all points between A and C as no other forces come on the beam between these points.

When a point to the right of C is considered, the load at C as well as R_A must be considered, or alternatively, R_B on its own. The shearing force is 15 kN, either obtained from $R_B = 15$ kN, or from load at $C - R_A = 15$ kN. For any point between C and B the force to the right is upwards and the shearing force is therefore negative. It should be noted that the shearing force changes suddenly at C.

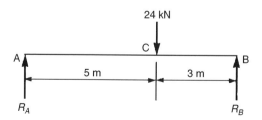

Figure 13.3

The bending moment at A is zero, as there are no forces to the left. At a point 1 m to the right of A the moment of the only force R_A to the left of the point is $R_A \times 1$ m = 9 kNm. At this moment to the left is clockwise the bending moment is positive, i.e. it is +9 kNm. At points 2 m, 3 m, 4 m and 5 m to the right of A the bending moments are respectively:

$$R_A \times 2 \text{ m} = 9 \text{ kN} \times 2 \text{ m} = 18 \text{ kNm}$$

$$R_A \times 3 \text{ m} = 9 \text{ kN} \times 3 \text{ m} = 27 \text{ kNm}$$

$$R_A \times 4 \text{ m} = 9 \text{ kN} \times 4 \text{ m} = 36 \text{ kNm}$$

$$R_A \times 5 \text{ m} = 9 \text{ kN} \times 5 \text{ m} = 45 \text{ kNm}$$

All are positive bending moments.

For points to the right of C, the load at C as well as R_A must be considered or, more simply, R_B alone can be used. At points 5 m, 6 m and 7 m from A, the bending moments are respectively:

$$R_B \times 3 \text{ m} = 15 \text{ kN} \times 3 \text{ m} = 45 \text{ kNm}$$

$$R_B \times 2 \text{ m} = 15 \text{ kN} \times 2 \text{ m} = 30 \text{ kNm}$$

$$R_B \times 1 \text{ m} = 15 \text{ kN} \times 1 \text{ m} = 15 \text{ kNm}$$

As these moments to the right of the points considered are anticlockwise they are all positive bending moments. At B the bending moment is zero as there is no force to its right. The results are summarised in the table below:

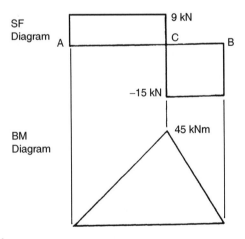

Figure 13.4

Distance from A (m)	0	1	2	3	4	5	6	7	8
Shearing force (kN)	+9	+9	+9	+9	+9	+9			
							−15	−15	−15	−15
Bending moment (kNm)	0	+9	+18	+27	+36	+45	+30	+15	+15

Making use of the above values, the diagrams are as shown in Figure 13.4. A stepped shearing force diagram, with horizontal and vertical lines only, is always obtained when the beam carries concentrated loads only. A sudden change in shearing force occurs where the concentrated loads, including the reactions at supports, occur. For this type of simple loading the bending moment diagram always consists of straight lines, usually sloping. Sudden changes of bending moment cannot occur except in the unusual circumstances of a moment being supplied to a beam as distinct from a load.

14 Bending Stress

A fundamental equation for the bending of beams is:

$$\frac{\sigma}{y} = \frac{M}{I} \left(= \frac{E}{R} \right)$$

where σ = stress due to bending at distance y from the neutral axis

M = bending moment

I = second moment of area of section of beam about its neutral axis

E = modulus of elasticity

R = radius of curvature

Section modulus $Z = \dfrac{I}{y_{max}}$

The second moments of area of the beam sections most commonly met with are (about the central axis XX):

(a) **Solid rectangle** (see Figure 14.1)

$$I = \frac{BD^3}{12}$$

(b) **Symmetrical hollow rectangle or I-section** (See Figure 14.2)

$$I = \frac{BD^3 - bd^3}{12}$$

(a) **Solid rod** (see Figure 14.3)

$$I = \frac{\pi D^4}{64}$$

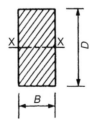

Figure 14.1

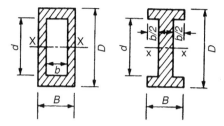

Figure 14.2

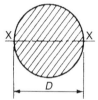

Figure 14.3

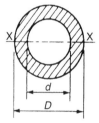

Figure 14.4

(b) **Tube** (see Figure 14.4)

$$I = \frac{\pi(D^4 - d^4)}{64}$$

The neutral axis of any section, where bending produces no strain and therefore no stress, always passes through the centroid of the section. For the symmetrical sections listed above this means that for vertical loading the neutral axis is the horizontal axis of symmetry.

For example, let the maximum bending moment on a beam be 120 Nm. If the beam section is rectangular 18 mm wide and 36 mm deep, the maximum bending stress is calculated as follows:

Second moment of area of section about the neutral axis,

$$I = \frac{bD^3}{12} = \frac{(18)(36)^3}{12} = 6.9984 \times 10^4 \text{ mm}^4$$

Maximum distance from neutral axis, $y = \dfrac{36}{2} = 18$ mm

Since $\dfrac{\sigma}{y} = \dfrac{M}{I}$ then the maximum bending stress σ will occur where M and y have their maximum values,

i.e. $\sigma = \dfrac{My}{I} = \dfrac{120\ \text{Nm} \times 18\ \text{mm}}{6.9984 \times 10^4\ \text{mm}^4} = \dfrac{120\ \text{Nm} \times 18 \times 10^{-3}\ \text{m}}{6.9984 \times 10^4 \times 10^{-12}\ \text{m}^4}$

$$= 30.86\ \text{MN/m}^2$$

$$= \mathbf{30.86\ MPa}$$

In another example, a cantilever is of tubular construction with internal and external diameters of 100 mm and 120 mm. If the length of the cantilever is 800 mm, the maximum load which it can carry at its free end if the maximum stress is not to exceed 50 MPa (assuming the weight of the beam is ignored) is determined as follows:

The second moment of area of the section is

$$I = \frac{\pi}{64}(D^4 - d^4) = \frac{\pi}{64}(120^4 - 100^4) = 5.27 \times 10^6\ \text{mm}^4$$

From $\dfrac{\sigma}{y} = \dfrac{M}{I}$ the maximum permissible bending moment M will give the maximum permissible stress at the maximum distance y from the neutral axis,

i.e. $M = \dfrac{\sigma I}{y} = \dfrac{50\ \text{MPa} \times 5.27 \times 10^6\ \text{mm}^4}{60\ \text{mm}}$

$$= \dfrac{50 \times 10^6\ \dfrac{\text{N}}{\text{m}^2} \times 5.27 \times 10^{-6}\ \text{m}^4}{60 \times 10^{-3}\ \text{m}}$$

$$= \dfrac{50 \times 5.27 \times 10^3}{60}\ \text{Nm}$$

$$= 4.392\ \text{kNm}$$

If W is the load (in kN) at the free end of the cantilever, the bending moment at a point distance x from the free end is Wx with a maximum value where the cantilever is built into the wall, given by:

$$W\ \text{kN} \times 0.8\ \text{m} = 0.8\ W\ \text{kNm}$$

Equating this to the calculated maximum permissible bending moment gives:

$$0.8\ W = 4.392 \text{ from which, } \textbf{maximum load at free end,}$$

$$W = \dfrac{4.392\ \text{kNm}}{0.8\ \text{m}} = \mathbf{5.49\ kN}$$

15 Linear and Angular Motion

The Radian

The unit of angular displacement is the radian, where one radian is the angle subtended at the centre of a circle by an arc equal in length to the radius, as shown in Figure 15.1

The relationship between angle in radians (θ), arc length (s) and radius of a circle (r) is:

$$s = r\theta \qquad (1)$$

Since the arc length of a complete circle is $2\pi r$ and the angle subtended at the centre is $360°$, then from equation (1), for a complete circle,

$$2\pi r = r\theta \quad \text{or} \quad \theta = 2\pi \text{ radians}$$

Thus,

$$2\pi \text{ radians corresponds to } 360° \qquad (2)$$

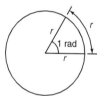

Figure 15.1

Linear and Angular Velocity

Linear velocity

Linear velocity v is defined as the rate of change of linear displacement s with respect to time t, and for motion in a straight line:

$$\text{Linear velocity} = \frac{\text{change of displacement}}{\text{change of time}}$$

i.e.

$$v = \frac{s}{t} \qquad (3)$$

The unit of linear velocity is metres per second (m/s)

Angular velocity

The speed of revolution of a wheel or a shaft is usually measured in revolutions per minute or revolutions per second but these units do not form part of a coherent system of units. The basis used in SI units is the angle turned through in one second.

Angular velocity ω is defined as the rate of change of angular displacement θ, with respect to time t, and for an object rotating about a fixed axis at a constant speed:

$$\text{angular velocity} = \frac{\text{angle turned through}}{\text{time taken}}$$

i.e.
$$\boxed{\omega = \frac{\theta}{t}} \tag{4}$$

The unit of angular velocity is radians per second (rad/s)

An object rotating at a constant speed of n revolutions per second subtends an angle of $2\pi n$ radians in one second, that is, its angular velocity,

$$\boxed{\omega = 2\pi n \ \textbf{rad/s}} \tag{5}$$

From equation (1), $s = r\theta$, and from equation (4), $\theta = \omega t$, hence

$$s = r\omega t \quad \text{or} \quad \frac{s}{t} = \omega r$$

However, from equation (3), $v = \dfrac{s}{t}$,

hence
$$\boxed{v = \omega r} \tag{6}$$

Equation (6) gives the relationship between linear velocity, v, and angular velocity, ω.

For example, if a wheel of diameter 540 mm is rotating at $(1500/\pi)$ rev/min, the angular velocity of the wheel and the linear velocity of a point on the rim of the wheel is calculated as follows:

From equation (5), angular velocity $\omega = 2\pi n$, where n is the speed of revolution in revolutions per second, i.e.

$$n = \frac{1500}{60\pi} \text{ revolutions per second.}$$

Thus, **angular velocity**,

$$\omega = 2\pi \left(\frac{1500}{60\pi} \right) = \textbf{50 rad/s}$$

The linear velocity of a point on the rim, $v = \omega r$, where r is the radius of the wheel, i.e. 0.54/2 or 0.27 m

Thus, **linear velocity**, $v = 50 \times 0.27 = \textbf{13.5 m/s}$

Linear and Angular Acceleration

Linear acceleration, a, is defined as the rate of change of linear velocity with respect to time (as introduced in Chapter 8). For an object whose linear velocity is increasing uniformly:

$$\text{linear acceleration} = \frac{\text{change of linear velocity}}{\text{time taken}}$$

i.e.

$$a = \frac{v_2 - v_1}{t} \tag{7}$$

The unit of linear acceleration is metres per second squared (m/s^2). Rewriting equation (7) with v_2 as the subject of the formula gives:

$$v_2 = v_1 + at \tag{8}$$

Angular acceleration, α, is defined as the rate of change of angular velocity with respect to time. For an object whose angular velocity is increasing uniformly:

$$\text{Angular acceleration} = \frac{\text{change of angular velocity}}{\text{time taken}}$$

i.e.

$$\alpha = \frac{\omega_2 - \omega_1}{t} \tag{9}$$

The unit of angular acceleration is radians per second squared (rad/s^2). Rewriting equation (9) with ω_2 as the subject of the formula gives:

$$\omega_2 = \omega_1 + \alpha t \tag{10}$$

From equation (6), $v = \omega r$. For motion in a circle having a constant radius r, $v_2 = \omega_2 r$ and $v_1 = \omega_1 r$, hence equation (7) can be rewritten as

$$a = \frac{\omega_2 r - \omega_1 r}{t} = \frac{r(\omega_2 - \omega_1)}{t}$$

But from equation (9), $\dfrac{\omega_2 - \omega_1}{t} = \alpha$

Hence $\boxed{a = r\alpha}$ $\hspace{2cm}$ (11)

For example, if the speed of a shaft increases uniformly from 300 revolutions per minute to 800 revolutions per minute in 10s, the angular acceleration is determined as follows:

Initial angular velocity,

$$\omega_1 = 300 \text{ rev/min} = 300/60 \text{ rev/s} = \frac{300 \times 2\pi}{60} \text{ rad/s},$$

final angular velocity,

$$\omega_2 = \frac{800 \times 2\pi}{60} \text{ rad/s and time, } t = 10 \text{ s}$$

Hence, from equation (9),

$$\text{angular acceleration, } \alpha = \frac{\dfrac{800 \times 2\pi}{60} - \dfrac{300 \times 2\pi}{60}}{10} \text{ rad/s}^2$$

$$= \frac{500 \times 2\pi}{60 \times 10} = \mathbf{5.24 \text{ rad/s}^2}$$

If the diameter of the shaft is 50 mm, then from equation (11), $a = r\alpha$, thus the

$$\text{linear acceleration, } a = r\alpha = \frac{50}{2} \times 10^{-3} \times 5.24 = \mathbf{0.131 \text{ m/s}^2}$$

Further Equations of Motion

From equation (3), $s = vt$, and if the linear velocity is changing uniformly from v_1 to v_2, then $s =$ mean linear velocity $\times$ time

i.e. $\boxed{s = \left(\dfrac{v_1 + v_2}{2}\right) t}$ (12)

From equation (4), $\theta = \omega t$, and if the angular velocity is changing uniformly from ω_1 to ω_2, then $\theta =$ mean angular velocity $\times$ time

i.e. $\boxed{\theta = \left(\dfrac{\omega_1 + \omega_2}{2}\right) t}$ (13)

Two further equations of linear motion may be derived from equations (8) and (12):

$\boxed{s = v_1 t + \tfrac{1}{2} a t^2}$ (14)

and $\boxed{v_2^2 = v_1^2 + 2as}$ (15)

Two further equations of angular motion may be derived from equations (10) and (13):

$\boxed{\theta = \omega_1 t + \tfrac{1}{2} \alpha t^2}$ (16)

and $\boxed{\omega_2^2 = \omega_1^2 + 2\alpha\theta}$ (17)

Table 15.1 summarises the principal equations of linear and angular motion for uniform changes in velocities and constant accelerations and also gives the relationships between linear and angular quantities.

For example, the shaft of an electric motor, initially at rest, accelerates uniformly for 0.4 s at 15 rad/s^2. To determine the angle (in radians) turned through by the shaft in this time:

Table 15.1

s = arc length (m)	r = radius of circle (m)
t = time (s)	θ = angle (rad)
v = linear velocity (m/s)	ω = angular velocity (rad/s)
v_1 = initial linear velocity (m/s)	ω_1 = initial angular velocity (rad/s)
v_2 = final linear velocity (m/s)	ω_2 = final angular velocity (rad/s)
a = linear acceleration (m/s^2)	α = angular acceleration (rad/s^2)
n = speed of revolution (rev/s)	

Equation number	Linear motion	Angular motion
(1)	$s = r\theta$ m	
(2)		2π rad = 360°
(3) and (4)	$v = \dfrac{s}{t}$ m/s	$\omega = \dfrac{\theta}{t}$ rad/s
(5)		$\omega = 2\pi n$ rad/s
(6)	$v = \omega r$ m/s^2	
(8) and (10)	$v_2 = (v_1 + at)$ m/s	$\omega_2 = (\omega_1 + \alpha t)$ rad/s
(11)	$a = r\alpha$ m/s^2	
(12) and (13)	$s = \left(\dfrac{v_1 + v_2}{2}\right) t$	$\theta = \left(\dfrac{\omega_1 + \omega_2}{2}\right) t$
(14) and (16)	$s = v_1 t + \frac{1}{2}at^2$	$\theta = \omega_1 t + \frac{1}{2}\alpha t^2$
(15) and (17)	$v_2^2 = v_1^2 + 2as$	$\omega_2^2 = \omega_1^2 + 2\alpha\theta$

From equation (16), $\theta = \omega_1 t + \frac{1}{2}\alpha t^2$

Since the shaft is initially at rest, $\omega_1 = 0$ and $\theta = \frac{1}{2}\alpha t^2$;
the angular acceleration, $\alpha = 15$ rad/s^2 and time $t = 0.4$ s
Hence, **angle turned through**, $\theta = \frac{1}{2} \times 15 \times 0.4^2 = \mathbf{1.2}$ **rad**

Relative Velocity

A vector quantity is represented by a straight line lying along the line of action of the quantity and having a length that is proportional to the size of the quantity, as shown in chapter 3. Thus **ab** in Figure 15.2 represents a velocity of 20 m/s, whose line of action is due west. The bold letters, **ab**, indicate a vector quantity and the order of the letters indicate that the time of action is from a to b.

For example, consider two aircraft A and B flying at a constant altitude, A travelling due north at 200 m/s and B travelling 30° east of north, written $N\ 30°E$, at 300 m/s, as shown in Figure 15.3.

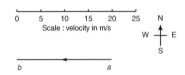

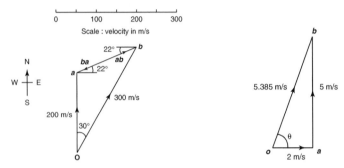

Figure 15.2

Figure 15.3   **Figure 15.4**

Relative to a fixed point *o*, *oa* represents the velocity of A and *ob* the velocity of B.

The **velocity of B relative to A**, that is, the velocity at which B seems to be travelling to an observer on A, is given by *ab*, and by measurement is **160 m/s in a direction E 22° N**.

The **velocity of A relative to B**, that is, the velocity at which A seems to be travelling to an observer on B, is given by *ba* and by measurement is **160 m/s in a direction W 22° S**.

In another example, a crane is moving in a straight line with a constant horizontal velocity of 2 m/s. At the same time it is lifting a load at a vertical velocity of 5 m/s. The velocity of the load relative to a fixed point on the earth's surface is calculated as follows:

A vector diagram depicting the motion of the crane and load is shown in Figure 15.4. *oa* represents the velocity of the crane relative to a fixed point on the earth's surface and *ab* represents the velocity of the load relative to the crane. The velocity of the load relative to the fixed point on the earth's surface is *ob*. By Pythagoras' theorem:

$$ob^2 = oa^2 + ab^2 = 4 + 25 = 29$$

Hence $ob = \sqrt{29} = 5.385$ m/s

Tan $\theta = \frac{5}{2} = 2.5$, hence $\theta = \tan^{-1} 2.5 = 68.20°$

i.e. the velocity of the load relative to a fixed point on the earth's surface is **5.385 m/s in a direction 68.20° to the motion of the crane**

16 Friction

Introduction to Friction

When an object, such as a block of wood, is placed on a floor and sufficient force is applied to the block, the force being parallel to the floor, the block slides across the floor. When the force is removed, motion of the block stops; thus there is a force which resists sliding. This force is called **dynamic** or **sliding friction**. A force may be applied to the block, which is insufficient to move it. In this case, the force resisting motion is called the **static friction** or **striction**. Thus there are two categories into which a frictional force may be split:

(i) dynamic or sliding friction force which occurs when motion is taking place, and
(ii) static friction force which occurs before motion takes place

There are three factors that affect the size and direction of frictional forces:

(i) The size of the frictional force depends on the type of surface (a block of wood slides more easily on a polished metal surface than on a rough concrete surface).
(ii) The size of the frictional force depends on the size of the force acting at right angles to the surfaces in contact, called the **normal force**; thus, if the weight of a block of wood is doubled, the frictional force is doubled when it is sliding on the same surface.
(iii) The direction of the frictional force is always opposite to the direction of motion. Thus the frictional force opposes motion, as shown in Figure 16.1.

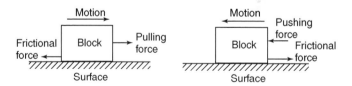

Figure 16.1

Coefficient of Friction

The **coefficient of friction, μ**, is a measure of the amount of friction existing between two surfaces. A low value of coefficient of friction indicates that the force required for sliding to occur is less than the force required when the coefficient of friction is high. The value of the coefficient of friction is

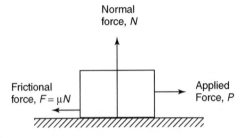

Figure 16.2

given by:

$$\mu = \frac{\text{frictional force } (F)}{\text{normal force } (N)}$$

Transposing gives: frictional force $= \mu \times$ normal force, i.e.

$$\boxed{F = \mu N}$$

The direction of the forces given in this equation is as shown in Figure 16.2

The coefficient of friction is the ratio of a force to a force, and hence has no units. Typical values for the coefficient of friction when sliding is occurring, i.e. the dynamic coefficient of friction, are:

for polished oiled metal surfaces	less than 0.1
for glass on glass	0.4
for rubber on tarmac	close to 1.0

For example, the material of a brake is being tested and it is found that the dynamic coefficient of friction between the material and steel is 0.91. The normal force, when the frictional force is 0.728 kN, is calculated as follows:

Since $\quad F = \mu N$, then normal force, $N = \dfrac{F}{\mu} = \dfrac{728}{0.91} = 800$ N

i.e. **the normal force is 800 N**

Applications of Friction

In some applications, a low coefficient of friction is desirable, for example, in bearings, pistons moving within cylinders, on ski runs, and so on. However, for such applications as force being transmitted by belt drives and braking systems, a high value of coefficient is necessary.

Advantages and Disadvantages of Frictional Forces

Instances where frictional forces are an **advantage** include:

(i) Almost all fastening devices rely on frictional forces to keep them in place once secured, examples being screws, nails, nuts, clips and clamps.
(ii) Satisfactory operation of brakes and clutches rely on frictional forces being present.
(iii) In the absence of frictional forces, most accelerations along a horizontal surface are impossible; for example, a person's shoes just slip when walking is attempted and the tyres of a car just rotate with no forward motion of the car being experienced.

Disadvantages of frictional forces include:

(i) Energy is wasted in the bearings associated with shafts, axles and gears due to heat being generated.
(ii) Wear is caused by friction, for example, in shoes, brake lining materials and bearings.
(iii) Energy is wasted when motion through air occurs (it is much easier to cycle with the wind rather than against it).

Design Implications

Two examples of design implications, which arise due to frictional forces and how lubrication may or may not help, are:

(i) Bearings are made of an alloy called white metal, which has a relatively low melting point. When the rotating shaft rubs on the white metal bearing, heat is generated by friction, often in one spot and the white metal may melt in this area, rendering the bearing useless. Adequate lubrication (oil or grease) separates the shaft from the white metal, keeps the coefficient of friction small and prevents damage to the bearing. For very large bearings, oil is pumped under pressure into the bearing and the oil is used to remove the heat generated, often passing through oil coolers before being re-circulated. Designers should ensure that the heat generated by friction can be dissipated.
(ii) Wheels driving belts, to transmit force from one place to another, are used in many workshops. The coefficient of friction between the wheel and the belt must be high, and dressing the belt with a tar-like substance may increase it. Since frictional force is proportional to the normal force, a slipping belt is made more efficient by tightening it, thus increasing the normal and hence the frictional force. Designers should incorporate some belt tension mechanism into the design of such a system.

17 Waves

Introduction to Waves

Wave motion is a travelling disturbance through a medium or through space, in which energy is transferred from one point to another without movement of matter.

Examples where wave motion occurs include:

  (i) water waves, such as are produced when a stone is thrown into a still pool of water
 (ii) waves on strings
(iii) waves on stretched springs
 (iv) sound waves
  (v) light waves (see page 75)
 (vi) radio waves
(vii) infra-red waves, which are emitted by hot bodies
(viii) ultra-violet waves, which are emitted by very hot bodies and some gas discharge lamps
 (ix) x-ray waves, which are emitted by metals when they are bombarded by high speed electrons
  (x) gamma-rays which are emitted by radioactive elements.

Examples (i) to (iv) are **mechanical waves** and they require a medium (such as air or water) in order to move. Examples (v) to (x) are **electromagnetic waves** and do not require any medium — they can pass through a vacuum.

Wave Types

There are two types of waves, these being transverse and longitudinal waves:

  (i) **Transverse waves** are where the particles of the medium move perpendicular to the direction of movement. For example, when a stone is thrown onto a pool of still water, the ripple moves radially outwards but the movement of a floating object shows that the water at a particular point merely moves up and down. Light and radio waves are other examples of transverse waves.
 (ii) **Longitudinal waves** are where the particles of the medium vibrate back and forth parallel to the direction of the wave travel. Examples include sound waves and waves in springs.

Figure 17.1 shows a cross section of a typical wave.

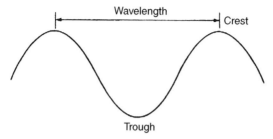

Figure 17.1

Wavelength, Frequency and Velocity

Wavelength is the distance between two successive identical parts of a wave (for example, between two crests as shown in Figure 17.1). The symbol for wavelength is λ (Greek lambda) and its unit is metres.

Frequency is the number of complete waves (or cycles) passing a fixed point in one second. The symbol for frequency is f and its unit is the hertz, Hz.

The **velocity**, v of a wave is given by:

Velocity = frequency × wavelength

i.e. $\boxed{v = f\lambda}$

The unit of velocity is metres per second.

For example, if BBC radio 4 is transmitted at a frequency of 198 kHz and a wavelength of 1500 m, the velocity of the radio wave v is given by:

$$v = f\lambda = (198 \times 10^3)(1500) = \mathbf{2.97 \times 10^8 \ m/s}$$

Reflection and Refraction

Reflection is a change in direction of a wave while the wave remains in the same medium. There is no change in the speed of a reflected wave. All waves are reflected when they meet a surface through which they cannot pass. For example,

(i) light rays are reflected by mirrors,
(ii) water waves are reflected at the end of a bath or by a sea wall,
(iii) sound waves are reflected at a wall (which can produce an echo),
(iv) a wave reaching the end of a spring or string is reflected, and
(v) television waves are reflected by satellites above the Earth.

Experimentally, waves produced in an open tank of water may readily be observed to reflect off a sheet of glass placed at right angles to the surface of the water.

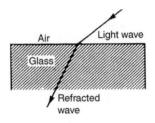

Figure 17.2

Refraction is a change in direction of a wave as it passes from one medium to another. All waves refract, and examples include:

(i) a light wave changing its direction at the boundary between air and glass, as shown in Figure 17.2,
(ii) sea waves refracting when reaching more shallow water, and
(iii) sound waves refracting when entering air of different temperature (see below).

Experimentally, if one end of a water tank is made shallow the waves may be observed to travel more slowly in these regions and are seen to change direction as the wave strikes the boundary of the shallow area. The greater the change of velocity the greater is the bending or refraction.

Sound Waves and their Characteristics

A **sound wave** is a series of alternate layers of air, one layer at a pressure slightly higher than atmospheric, called compressions, and the other slightly lower, called rarefactions. In other words, **sound is a pressure wave**. Figure 17.3(a) represents layers of undisturbed air; Figure 17.3(b) shows what happens to the air when a sound wave passes.

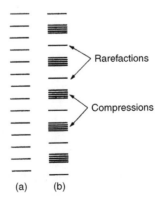

Figure 17.3

Sound waves exhibit the following characteristics:

(i) Sound waves can travel through solids, liquids and gases, but not through a vacuum.

(ii) Sound has a finite (i.e. fixed) velocity, the value of which depends on the medium through which it is travelling. The velocity of sound is also affected by temperature. Some typical values for the velocity of sound are: air 331 m/s at 0°C, and 342 m/s at 18°C, water 1410 m/s at 20°C and iron 5100 m/s at 20°C.

(iii) Sound waves can be reflected, the most common example being an echo. Echo-sounding is used for charting the depth of the sea.

(iv) Sound waves can be refracted. This occurs, for example, when sound waves meet layers of air at different temperatures. If a sound wave enters a region of higher temperature the medium has different properties and the wave is bent as shown in Figure 17.4, which is typical of conditions that occur at night.

Sound waves are produced as a result of **vibrations**.

(i) In brass instruments, such as trumpets and trombones, or wind instruments, such as clarinets and oboes, sound is due to the vibration of columns of air.

(ii) In stringed instruments, such as guitars and violins, sound is produced by vibrating strings causing air to vibrate. Similarly, the vibration of vocal chords produces speech.

(iii) Sound is produced by a tuning fork due to the vibration of the metal prongs.

(iv) Sound is produced in a loudspeaker due to vibrations in the cone.

The pitch of a sound depends on the frequency of the vibrations; the higher the frequency, the higher is the pitch. The frequency of sound depends on the form of the vibrator. The valves of a trumpet or the slide of a trombone lengthen or shorten the air column and the fingers alter the length of strings on a guitar or violin. The shorter the air column or vibrating string the higher the frequency and hence pitch. Similarly, a short tuning fork will produce a higher pitch note than a long tuning fork.

The human ear can perceive frequencies between about 20 Hz and 20 kHz.

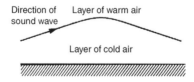

Direction of sound wave   Layer of warm air

Layer of cold air

Figure 17.4

18 Interference and Diffraction

Interference

At the point where two waves cross, the total displacement is the vector sum of the individual displacements due to each wave at that point. This is the **principle of superposition**. If these two waves are either both transverse or both longitudinal, **interference** effects may be observed. It is not necessary for the two waves to have the same frequencies or amplitudes for the above statements to be true, although these are the waves considered in this chapter.

Consider two transverse waves of the same frequency and amplitude travelling in opposite directions superimposed on one another. Interference takes place between the two waves and a **standing** or **stationary wave** is produced. The standing wave is shown in Figure 18.1

The wave does not progress to the left or right and certain parts of the wave called **nodes**, labelled N in the diagram, do not oscillate. Those positions on the wave that undergo maximum disturbance are called **antinodes**, labelled A. The distance between adjacent nodes or adjacent antinodes is $\frac{\lambda}{2}$, where λ is the wavelength. Standing waves may be set up in a string, for example, when a wave is reflected at the end of the string and is superimposed on the incoming wave. Under these circumstances standing waves are produced only for certain frequencies. Also, the nodes may not be perfect because the reflected wave may have a slightly reduced amplitude.

Two sound (longitudinal) waves of the same amplitude and frequency travelling in opposite directions and superimposed on each other also produce a standing wave. In this case there are displacement nodes where the medium does not oscillate and displacement antinodes where the displacement is a maximum.

The interference effects mentioned above are not always restricted to the line between the two sources of waves. Two dimensional interference patterns are produced on the surface of water in a ripple tank, for example. In this case, two dippers, usually oscillating in phase and with the same frequency, produce circular ripples on the surface of the water and interference takes place where the circular ripples overlap. The resulting interference pattern is shown in Figure 18.2. The sources of the waves are S_1 and S_2.

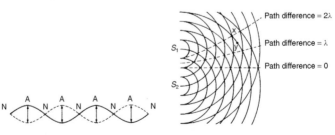

Figure 18.1   Figure 18.2

Consider a point X on the surface. $(S_2X - S_1X)$ is called the path difference. If $(S_2X - S_1X) = n\lambda$, where n is an integer, and λ is the wavelength, the waves arriving at X from S_1 and S_2 must be in phase. (In Figure 18.2, $S_2X - S_1X = 2\lambda$). At X, constructive interference takes place and the resulting amplitude is a maximum. At a second point, Y, positioned such that $(S_2Y - S_1Y) = (n + \frac{1}{2})\lambda$, the waves arriving at Y from S_1 and S_2 are out of phase. The resultant amplitude at Y is a minimum and destructive interference has taken place. (In Figure 18.2, $(S_2Y - S_1Y) = \lambda$, that is, $n = 1$).

Diffraction

When sea waves are incident on a barrier that is parallel to them a disturbance is observed beyond the barrier in that region where it might be thought that the water would remain undisturbed. This is because waves may spread round obstacles into regions which would be in shadow if the energy travelled exactly in straight lines. This phenomenon is called **diffraction.** All waves whether transverse or longitudinal exhibit this property. If light, for example, is incident on a narrow slit, diffraction takes place. The diffraction pattern on the screen placed beyond the slit is not perfectly sharp. The intensity of the image varies as shown in Figure 18.3.

A consequence of diffraction is that if light from two sources that are close together pass through a slit or small circular aperture, the diffraction patterns of the two sources may overlap to such an extent that they appear to be one source. If they are to be distinguished as two separate sources, the angular separation, θ, in radians, of the two sources, must be greater than $\dfrac{\lambda}{b}$, where λ is the wavelength of the light and b is the width of the slit (see Figure 18.4).

For a circular aperture the condition becomes:

$$\theta > \frac{1.22\lambda}{b}.$$

This is known as the **Rayleigh criterion**.

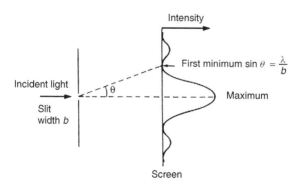

Figure 18.3

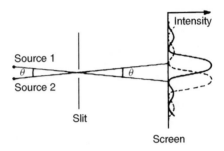

Figure 18.4

If light falls on two narrow parallel slits with a small separation, light passes through both slits and because of diffraction there is an overlapping of the light and interference takes place. This is shown in Figure 18.5. The interference effects are similar to those described for water ripples above.

Suppose the light from the two slits meet at a point on a distant screen. Since the distance between the slits is much less that the slit to screen distance the two light beams will be very nearly parallel. See Figure 18.6. If the path difference is $n\lambda$, where n is an integer and λ is the wavelength there will be constructive interference and a maximum intensity occurs on the screen.

But from Figure 18.6, the path difference is BC, that is, $d \sin \theta$. Thus for a maximum intensity on the screen, $n\lambda = d \sin \theta$, that is:

$$\sin \theta = \frac{n\lambda}{d}$$

Thus maximum values occur where:

$$\sin \theta = \frac{\lambda}{d}, \quad \frac{2\lambda}{d}, \quad \frac{3\lambda}{d}, \quad \text{and so on.}$$

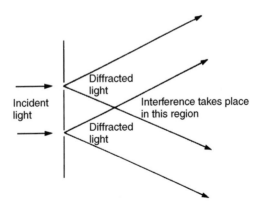

Figure 18.5

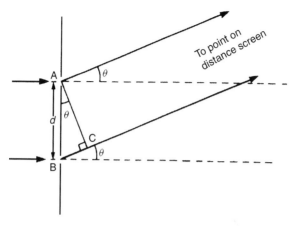

Figure 18.6

The intensity of the interference pattern on the screen at various distances from the polar axis is shown in Figure 18.7. The pattern is modified by the type of diffraction pattern produced by a single slit.

A diffraction grating is similar to the two-slit arrangement, but with a very large number of slits. Very sharp values of maximum intensity are produced in this case. If the slit separation is d and light is incident along the normal to the grating, the condition for a maximum is:

$$n\lambda = d \sin\theta$$

When $n = 0$, then $\sin\theta = 0$, and the light travels straight through the grating to give the **zero order** maximum (marked 0 in Figure 18.7).

If $n = 1$, then $\sin\theta = \dfrac{\lambda}{d}$. This gives the direction of the **first order** maximum and the path difference from adjacent slits to the first order maximum is λ.

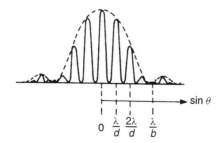

Figure 18.7

If $n = 2$, then $\sin \theta = \dfrac{2\lambda}{d}$. This gives the direction of the **second order** maximum and the path difference is 2λ.

If white light is incident on a diffraction grating a continuous spectrum is produced because the angle at which the first order emerges from the grating depends on the wavelength. Thus the diffraction grating may be used to determine the wavelengths present in a source of light.

X-ray diffraction

Atoms in a crystal diffract X-rays that are incident upon them and information may be gained about crystal structure from the analysis of the diffraction pattern obtained. When X-rays strike atoms in a crystal, each atom scatters the X-rays in all directions. However, in certain directions constructive interference takes place. In Figure 18.8 a lattice of atoms is shown, in which X-rays strike atoms and are scattered. The X-rays emerging in a particular direction are considered.

Three planes of atoms are shown. The X-rays 'reflected' from the top and middle planes (and any other pair of adjacent planes) will be in phase if their path difference is $n\lambda$, where n is an integer and λ is the wavelength of the X-rays.

$$\begin{aligned}
\text{The path difference} \quad &= AB + BC \\
&= d \sin \theta + d \sin \theta \\
&= 2d \sin \theta
\end{aligned}$$

where d is the separation between planes.

Thus the condition for constructive interference is:

$$n\lambda = 2d \sin \theta$$

The angle θ is called the **glancing angle** and the equation is known as **Bragg's law**.

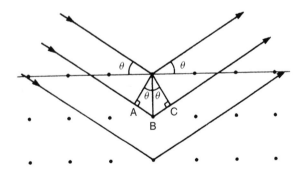

Figure 18.8

19 Light Rays

Introduction

Light is an electromagnetic wave (see page 66) and the straight line paths followed by very narrow beams of light, along which light energy travels, are called rays.

The behaviour of light rays may be investigated by using a ray-box. This consists merely of a lamp in a box containing a narrow slit which emits rays of light.

Light always travels in straight lines although its direction can be changed by reflection or refraction.

Reflection of Light

Figure 19.1 shows a ray of light, called the incident ray, striking a plane mirror at O, and making an angle i with the normal, which is a line drawn at right angles to the mirror at O.

i is called the **angle of incidence**. r is called the **angle of reflection**. There are **two laws of reflection**:

(i) The angle of incidence is equal to the angle of reflection (i.e. $i = r$ in Figure 19.1)
(ii) The incident ray, the normal at the point of incidence and the reflected ray all lie in the same plane.

A Simple Periscope

A simple periscope arrangement is shown in Figure 19.2. A ray of light from O strikes a plane mirror at an angle of 45° at point P. Since from the laws

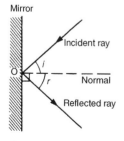

Figure 19.1

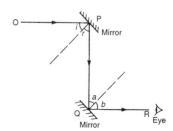

Figure 19.2

76

of reflection the angle of incidence i is equal to the angle of reflection r then $i = r = 45°$. Thus angle OPQ $= 90°$ and the light is reflected through $90°$. The ray then strikes another mirror at $45°$ at point Q. Thus $a = b = 45°$, angle PQR $= 90°$ and the light ray is again reflected through $90°$. Thus the light from O finally travels in the direction QR, which is parallel to OP, but displaced by the distance PQ. The arrangement thus acts as a **periscope**.

Refraction of Light

When a ray of light passes from one medium to another the light undergoes a change in direction. This displacement of light rays is called **refraction**.

Figure 19.3 shows the path of a ray of light as it passes through a parallel-sided glass block. The incident ray AB that has an angle of incidence i enters the glass block at B. The direction of the ray changes to BC such that the angle r is less than angle i. r is called the angle of refraction. When the ray emerges from the glass at C the direction changes to CD, angle r' being greater than i'. The final emerging ray CD is parallel to the incident ray AB.

Lenses

In general, when entering a more dense medium from a less dense medium, light is refracted towards the normal and when it passes from a dense to a less dense medium it is refracted away from the normal. **Lenses** are pieces of glass or other transparent material with a spherical surface on one or both sides. When light is passed through a lens it is refracted.

Lenses are used in spectacles, magnifying glasses and microscopes, telescopes, cameras and projectors.

There are a number of different shaped lenses and two of the most common are shown in Figure 19.4.

Figure 19.4(a) shows a **bi-convex lens**, so called since both its surfaces curve outwards. Figure 19.4(b) shows a **bi-concave lens**, so called since both

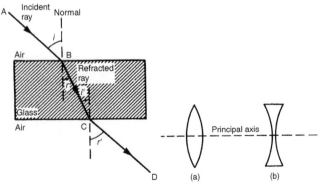

Figure 19.3   **Figure 19.4**

of its surfaces curve inwards. The line passing through the centre of curvature of the lens surface is called the **principal axis**.

Figure 19.5 shows a number of parallel rays of light passing through a bi-convex lens. They are seen to converge at a point F on the principal axis.

Figure 19.6 shows parallel rays of light passing through a bi-concave lens. They are seen to diverge such that they appear to come from a point F, which lies between the source of light and the lens, on the principal axis.

In both Figure 19.5 and Figure 19.6, F is called the **principal focus** or the **focal point**, and the distance from F to the centre of the lens is called the **focal length** of the lens.

An image is the point from which reflected rays of light entering the eye appear to have originated. If the rays actually pass through the point then a real image is formed. Such images can be formed on a screen. Figure 19.7 illustrates how the eye collects rays from an object after reflection from a plane mirror. To the eye, the rays appear to come from behind the mirror and the eye sees what seems to be an image of the object as far behind the mirror as the object is in front. Such an image is called a virtual image and this type cannot be shown on a screen.

Lenses are important since they form images when an object is placed at an appropriate distance from the lens.

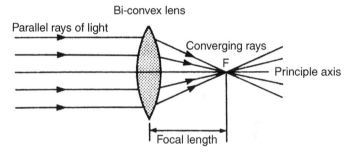

Figure 19.5

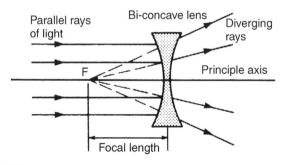

Figure 19.6

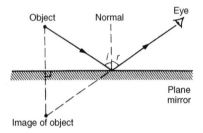

Figure 19.7

Bi-convex Lenses and their Applications

(i) Figure 19.8 shows an object O (a source of light) at a distance of more than twice the focal length from the lens. To determine the position and size of the image, two rays only are drawn, one parallel with the principal axis and the other passing through the centre of the lens. The image, I, produced is real, inverted (i.e. upside down), smaller than the object (i.e. diminished) and at a distance between one and two times the focal length from the lens. This arrangement is used in a **camera**.

(ii) Figure 19.9 shows an object O at a distance of twice the focal length from the lens. This arrangement is used in a **photocopier**.

(iii) Figure 19.10 shows an object O at a distance of between one and two focal lengths from the lens. The image I is real, inverted, magnified (i.e. greater than the object) and at a distance of more than twice the focal length from the lens. This arrangement is used in a **projector**

(iv) Figure 19.11 shows an object O at the focal length of the lens. After passing through the lens the rays are parallel. Thus the image I can be considered as being found at infinity and being real, inverted and very much magnified. This arrangement is used in a **spotlight**.

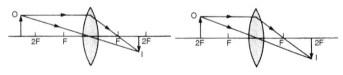

Figure 19.8   **Figure 19.9**

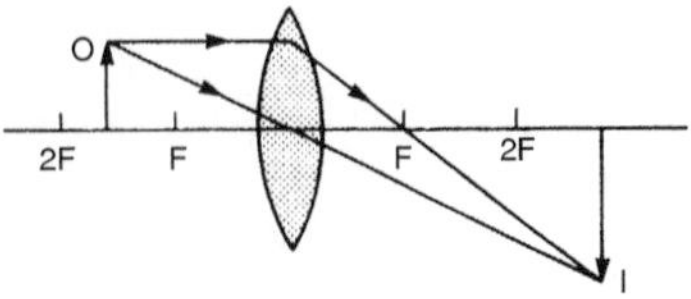

Figure 19.10 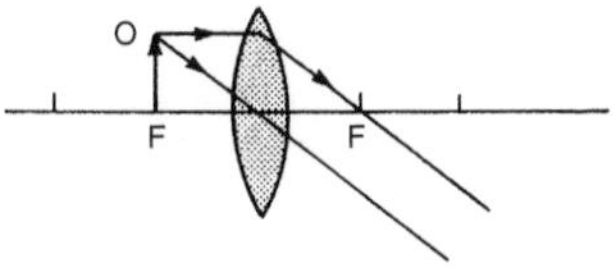   **Figure 19.11**

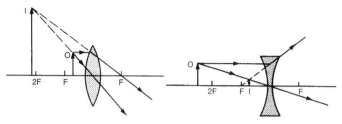

Figure 19.12   **Figure 19.13**

(v) Figure 19.12 shows an object O lying inside the focal length of the lens. The image I is virtual, since the rays of light only appear to come from it, is on the same side of the lens as the object, is upright and magnified. This arrangement is used in a **magnifying glass**.

Bi-concave Lenses

For a bi-concave lens, as shown in Figure 19.13, the object O can be any distance from the lens and the image I formed is virtual, upright, diminished and is found on the same side of the lens as the object. This arrangement is used in some types of **spectacles**.

A Compound Microscope

A **compound microscope** is able to give large magnification by the use of two (or more) lenses. An object O is placed outside the focal length F_o of a bi-convex lens, called the objective lens (since it is near to the object), as shown in Figure 19.14. This produces a real, inverted, magnified image I_1.

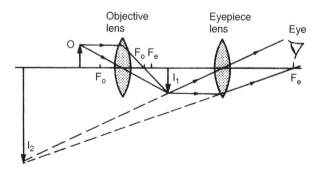

Figure 19.14

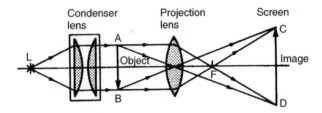

Figure 19.15

This image then acts as the object for the eyepiece lens (i.e. the lens nearest the eye), and falls inside the focal length F_o of the lens. The eyepiece lens then produces a magnified, virtual, inverted image I_2 as shown in Figure 19.14.

A Simple Projector

A **simple projector arrangement** is shown in Figure 19.15 and consists of a source of light and two-lens system. L is a brilliant source of light, such as a tungsten filament. One lens system, called the condenser (usually consisting of two converging lenses as shown), is used to produce an intense illumination of the object AB, which is a slide transparency or film. The second lens, called the projection lens, is used to form a magnified, real, upright image of the illuminated object on a distant screen CD.

20 Work, Energy and Power

Work

If a body moves as a result of a force being applied to it, the force is said to do work on the body. The amount of work done is the product of the applied force and the distance, i.e.

$$\text{work done} = \text{force} \times \frac{\text{distance moved in the}}{\text{direction of the force}}$$

The unit of work is the **joule**, **J**, which is defined as the amount of work done when a force of 1 newton acts for a distance of 1 m in the direction of the force. Thus, **1 J = 1 Nm**

If a graph is plotted of experimental values of force (on the vertical axis) against distance moved (on the horizontal axis) a force/distance graph or work diagram is produced. **The area under the graph represents the work done**.

For example, a constant force of 20 N used to raise a load a height of 8 m may be represented on a force/distance graph as shown in Figure 20.1. The area under the graph, shown shaded, represents the work done. Hence

$$\text{work done} = 20 \text{ N} \times 8 \text{ m} = \textbf{160 J}$$

In another example, a spring extended by 20 mm by a force of 500 N may be represented by the work diagram shown in Figure 20.2, where

$$\textbf{work done} = \text{shaded area} = \tfrac{1}{2} \times \text{ base } \times \text{ height}$$

$$= \tfrac{1}{2} \times (20 \times 10^{-3} \text{ m}) \times 500 \text{ N} = \textbf{5 J}$$

It is shown in chapter 9 that force = mass × acceleration, and that if an object is dropped from a height it has a constant acceleration of around 9.81 m/s².

For example, if a mass of 8 kg is lifted vertically 4 m, the work done is given by:

$$\textbf{work done} = \text{force} \times \text{distance}$$

$$= (\text{mass} \times \text{acceleration}) \times \text{distance}$$

$$= (8 \times 9.81) \times 4 = \textbf{313.92 N}$$

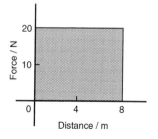

Figure 20.1

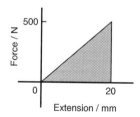

Figure 20.2

The work done by a variable force may be found by determining the area enclosed by the force/distance graph using an approximate method (such as the mid-ordinate rule)

Energy

Energy is the capacity, or ability, to do work. The unit of energy is the joule, the same as for work. Energy is expended when work is done. There are several **forms of energy** and these include:

   (i) Mechanical energy   (ii) Heat or thermal energy
 (iii) Electrical energy   (iv) Chemical energy
  (v) Nuclear energy   (vi) Light energy
(vii) Sound energy

Energy may be converted from one form to another. **The principle of conservation of energy** states that the total amount of energy remains the same in such conversions, i.e. energy cannot be created or destroyed.

Some examples of energy conversions include:

    (i) Mechanical energy is converted to electrical energy by a generator
   (ii) Electrical energy is converted to mechanical energy by a motor
  (iii) Heat energy is converted to mechanical energy by a steam engine
  (iv) Mechanical energy is converted to heat energy by friction
   (v) Heat energy is converted to electrical energy by a solar cell
  (vi) Electrical energy is converted to heat energy by an electric fire
 (vii) Heat energy is converted to chemical energy by living plants
(viii) Chemical energy is converted to heat energy by burning fuels
  (ix) Heat energy is converted to electrical energy by a thermocouple
   (x) Chemical energy is converted to electrical energy by batteries
  (xi) Electrical energy is converted to light energy by a light bulb
 (xii) Sound energy is converted to electrical energy by a microphone.
(xiii) Electrical energy is converted to chemical energy by electrolysis.

Efficiency is defined as the ratio of the useful output energy to the input energy. The symbol for efficiency is η (Greek letter eta). Hence

$$\boxed{\text{efficiency, } \eta = \frac{\textbf{useful output energy}}{\textbf{input energy}}}$$

Efficiency has no units and is often stated as a percentage. A perfect machine would have an efficiency of 100%. However, all machines have an efficiency lower than this due to friction and other losses.

    For example, if the input energy to a motor is 1000 J and the output energy is 800 J then the efficiency is:

$$\frac{800}{1000} \times 100\% = \textbf{80\%}$$

In another example, if a machine exerts a force of 200 N in lifting a mass through a height of 6 m, the efficiency of the machine if 2 kJ of energy are supplied to it is calculated as follows:

Work done in lifting mass = force × distance moved

$$= \text{weight of body} \times \text{distance moved}$$

$$= 200 \text{ N} \times 6 \text{ m} = 1200 \text{ J}$$

$$= \text{useful energy output}$$

Energy input = 2 kJ = 2000 J

Efficiency, η $= \dfrac{\text{useful output energy}}{\text{input energy}} = \dfrac{1200}{2000}$

$$= \textbf{0.6} \text{ or } \textbf{60\%}$$

Power

Power is a measure of the rate at which work is done or at which energy is converted from one form to another.

$$\boxed{\textbf{Power } P = \frac{\textbf{energy used}}{\textbf{time taken}}} \text{ or } \boxed{P = \frac{\textbf{work done}}{\textbf{time taken}}}$$

The unit of power is the **watt, W**, where 1 watt is equal to 1 joule per second. The watt is a small unit for many purposes and a larger unit called the kilowatt, kW, is used, where 1 kW = 1000 W

The power output of a motor that does 120 kJ of work in 30 s is thus given by

$$P = \frac{120 \text{ kJ}}{30 \text{ s}} = 4 \text{ kW}$$

(For electrical power, see Chapter 40)

Since work done = force × distance, then

$$\text{power} = \frac{\text{work done}}{\text{time taken}} = \frac{\text{force} \times \text{distance}}{\text{time taken}} = \text{force} \times \frac{\text{distance}}{\text{time taken}}$$

However, $\dfrac{\text{distance}}{\text{time taken}} = \text{velocity}$

Hence $\quad \boxed{\textbf{power} = \textbf{force} \times \textbf{velocity}}$

For example, if a lorry is travelling at a constant velocity of 72 km/h and the force resisting motion is 800 N, then the tractive power necessary to keep the lorry moving at this speed is given by:

$$\textbf{power} = \text{force} \times \text{velocity} = (800 \text{ N}) \left(\frac{72}{3.6} \text{ m/s} \right)$$

$$= 16000 \text{ Nm/s} = 16000 \text{ J/s} = 16000 \text{ W or } \textbf{16 kW}$$

21 Potential and Kinetic Energy

Introduction

Mechanical engineering is concerned principally with two kinds of energy, potential energy and kinetic energy.

Potential Energy

Potential energy is energy due to the position of the body. The force exerted on a mass of m kg is mg N (where $g = 9.81$ m/s^2, the acceleration due to gravity). When the mass is lifted vertically through a height h m above some datum level, the work done is given by: force × distance = $(mg)(h)$ J. This work done is stored as potential energy in the mass.

Hence, | **potential energy = mgh joules**

(the potential energy at the datum level being taken as zero).

For example, if a car of mass 800 kg is climbing an incline at 10° to the horizontal, the increase in potential energy of the car as it moves a distance of 50 m up the incline is determined as follows:

With reference to Figure 21.1, $\sin 10° = \dfrac{\text{opposite}}{\text{hypotenuse}} = \dfrac{h}{50}$, from which, $h = 50 \sin 10° = 8.682$ m

Hence, increase in potential energy = mgh

$$= 800 \, \text{kg} \times 9.81 \, \text{m/s}^2 \times 8.682 \, \text{m}$$

$$= \mathbf{68\,140 \, J} \text{ or } \mathbf{68.14 \, kJ}$$

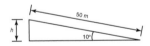

Figure 21.1

Kinetic Energy

Kinetic energy is the energy due to the motion of a body. Suppose a force F acts on an object of mass m originally at rest (i.e. $u = 0$) and accelerates it to a velocity v in a distance s:

$$\text{work done} = \text{force} \times \text{distance}$$

$$= Fs = (ma)(s) \quad \text{(if no energy is lost)}$$

where a is the acceleration. Since $v^2 = u^2 + 2as$ (see chapter 15) and $u = 0$, $v^2 = 2as$, from which $a = \dfrac{v^2}{2s}$, hence, work done $= (ma)(s) = (m)\left(\dfrac{v^2}{2s}\right)(s) = \frac{1}{2}mv^2$

This energy is called the kinetic energy of the mass m, i.e.

$$\boxed{\text{kinetic energy} = \tfrac{1}{2}mv^2 \text{ joules}}$$

For example, at the instant of striking, a hammer of mass 30 kg has a velocity of 15 m/s. The kinetic energy in the hammer is given by:

$$\text{kinetic energy} = \tfrac{1}{2}mv^2 = \tfrac{1}{2}(30\ \text{kg})(15\ \text{m/s})^2 = \textbf{3375 J}$$

Principle of Conservation of Energy

Energy may be converted from one form to another. The **principle of conservation of energy** states that the total amount of energy remains the same in such conversions, i.e. energy cannot be created or destroyed.

In mechanics, the potential energy possessed by a body is frequently converted into kinetic energy, and vice versa. When a mass is falling freely, its potential energy decreases as it loses height, and its kinetic energy increases as its velocity increases. Ignoring air frictional losses, at all times:

potential energy + kinetic energy = a constant

If friction is present, then work is done overcoming the resistance due to friction and this is dissipated as heat. Then,

initial energy = final energy + work done overcoming frictional resistance

Kinetic energy is not always conserved in collisions. Collisions in which kinetic energy is conserved (i.e. stays the same) are called **elastic collisions**, and those in which it is not conserved are termed **inelastic collisions.**

Kinetic Energy of Rotation

The tangential velocity v of a particle of mass m moving at an angular velocity ω rad/s at a radius r metres (see Figure 21.2) is given by:

$$v = \omega r \text{ rad/s}$$

The kinetic energy of a particle of mass m is given by:

$$\text{kinetic energy} = \tfrac{1}{2}mv^2 = \tfrac{1}{2}m(\omega r)^2 = \tfrac{1}{2}\ \mathbf{m\omega^2 r^2} \text{ joules}$$

The total kinetic energy of a system of masses rotating at different radii about a fixed axis but with the same angular velocity ω, as shown in Figure 21.3, is given by: total kinetic energy $= \tfrac{1}{2}m_1\omega^2 r_1^2 + \tfrac{1}{2}m_2\omega^2 r_2^2 + \tfrac{1}{2}m_3\omega^2 r_3^2$

$$= (m_1 r_1^2 + m_2 r_2^2 + m_3 r_3^2)\dfrac{\omega^2}{2}$$

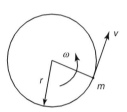

Figure 21.2   **Figure 21.3**

In general, this may be written as:

$$\textbf{total kinetic energy} = \left(\sum mr^2\right)\frac{\omega^2}{2} = I\frac{\omega^2}{2}$$

where $I(=\sum mr^2)$ is called the **moment of inertia** of the system about the axis of rotation.

The moment of inertia of a system is a measure of the amount of work done to give a system an angular velocity of ω rad/s, or the amount of work that can be done by a system turning at ω rad/s.

In general, total kinetic energy $= I\dfrac{\omega^2}{2} = Mk^2\dfrac{\omega^2}{2}$

where $M(=\sum m)$ is the total mass and k is called **the radius of gyration** of the system for the given axis.

If all the masses were concentrated at the radius of gyration it would give the same moment of inertia as the actual system.

For example, a system consists of three small masses rotating at the same speed about a fixed axis; the masses and their radii of rotation are: 16 g at 256 mm, 23 g at 192 mm and 31 g at 176 mm.

The **moment of inertia of the system** $= \sum mr^2$

$$= 16 \text{ g} \times (0.256 \text{ m})^2 + 23 \text{ g} \times (0.192 \text{ m})^2 + 31 \text{ g} \times (0.176 \text{ m})^2$$

$$= 1.0486 \text{ g m}^2 + 0.8479 \text{ g m}^2 + 0.9603 \text{ g m}^2$$

$$= \textbf{2.857 g m}^2 \text{ or } \textbf{2857 kg mm}^2$$

This is also Mk^2 where $M = 16 + 23 + 31 = 70$ g and k is the overall radius of gyration. Thus 70 g $\times k^2 = 2.857$ g m^2 from which, **radius of gyration,**

$$k = \sqrt{\left(\frac{2.857}{70}\right)} = 0.202 \text{ m} = \textbf{202 mm}$$

If the speed of rotation is 1250 rev/min, then the **kinetic energy in the system**

$$= I\frac{\omega^2}{2} = (2.857 \times 10^{-3} \text{kg m}^2)\left(\frac{\left(\dfrac{1250 \times 2\pi}{60}\text{rad/s}\right)^2}{2}\right) = \textbf{24.48 J}$$

Flywheels

The function of a flywheel is to restrict fluctuations of speed by absorbing and releasing large quantities of kinetic energy for small speed variations.

To do this they require large moments of inertia and to avoid excessive mass they need to have radii of gyration as large as possible. Most of the mass of a flywheel is usually in its rim.

For example, a cast iron flywheel is required to release 2.10 kJ of kinetic energy when its speed falls from 3020 rev/min to 3010 rev/min. The moment of inertia of the flywheel is assumed to be concentrated in its rim which is to be of rectangular section, the external and internal diameters being 670 mm and 600 mm. The radius of gyration of the rim may be assumed to be its mean radius. Taking the density of cast iron as 7800 kg/m^3, the required width for the flywheel is determined as follows:

The change of kinetic energy is given by $\dfrac{I(\omega_2^2 - \omega_1^2)}{2}$ hence the moment of inertia I required for the flywheel is given by

$$I = \frac{(\text{change of kinetic energy}) \times 2}{(\omega_2^2 - \omega_1^2)}$$

$$= \frac{2100 \text{ J} \times 2}{\left(\dfrac{3020 \times 2\pi}{60} \text{rad/s}\right)^2 - \left(\dfrac{3010 \times 2\pi}{60} \text{rad/s}\right)^2} = 6.3515 \text{ kg m}^2$$

This is also given by Mk^2, where M is the mass of the rim and k is the radius of gyration, assumed here to be the mean radius of the rim, given by

$$\frac{\dfrac{670 \text{ mm}}{2} + \dfrac{600 \text{ mm}}{2}}{2} = 317.5 \text{ mm or } 0.3175 \text{ m}$$

Since $I = Mk^2$ then $M = \dfrac{I}{k^2} = \dfrac{6.3515 \text{ kg m}^2}{(0.3175 \text{ m})^2} = 63.007 \text{ kg}$

This is the product of the volume V and density ρ of the cast iron.

Since $M = V\rho$ then $V = \dfrac{M}{\rho} = \dfrac{63.007 \text{ kg}}{7800 \text{ kg/m}^3} = 0.0080778 \text{ m}^3$

The volume V is given by $V = \dfrac{\pi b}{4}(D^2 - d^2)$ where b is the width of the rim and D and d are the external and internal diameters respectively.

Hence, **width of rim,** $b = \dfrac{4V}{\pi(D^2 - d^2)} = \dfrac{4 \times 0.0080778 \text{ m}^3}{\pi(0.67^2 - 0.60^2)}$

$$= 0.1157 \text{ m} = \textbf{115.7 mm}$$

22 Simple Machines

Machines

A machine is a device that can change the magnitude or line of action, or both magnitude and line of action of a force. A simple machine usually amplifies an input force, called the **effort**, to give a larger output force, called the **load**. Some typical examples of simple machines include pulley systems, screw-jacks, gear systems and lever systems.

Force Ratio, Movement Ratio and Efficiency

The **force ratio** or **mechanical advantage** is defined as the ratio of load to effort, i.e.

$$\text{Force ratio} = \frac{\text{load}}{\text{effort}} \qquad (1)$$

Since both load and effort are measured in newtons, force ratio is a ratio of the same units and thus is a dimension-less quantity.

The **movement ratio** or **velocity ratio** is defined as the ratio of the distance moved by the effort to the distance moved by the load, i.e.

$$\text{Movement ratio} = \frac{\text{distance moved by the effort}}{\text{distance moved by the load}} \qquad (2)$$

Since the numerator and denominator are both measured in metres, movement ratio is a ratio of the same units and thus is a dimension-less quantity.

The **efficiency of a simple machine** is defined as the ratio of the force ratio to the movement ratio, i.e.

$$\text{Efficiency} = \frac{\text{force ratio}}{\text{movement ratio}}$$

Since the numerator and denominator are both dimension-less quantities, efficiency is a dimension-less quantity. It is usually expressed as a percentage, thus:

$$\text{Efficiency} = \frac{\text{force ratio}}{\text{movement ratio}} \times 100\% \qquad (3)$$

Due to the effects of friction and inertia associated with the movement of any object, some of the input energy to a machine is converted into heat and losses occur. Since losses occur, the energy output of a machine is less than the energy input, thus the mechanical efficiency of any machine cannot reach 100%.

For example, a simple machine raises a load of 160 kg through a distance of 1.6 m. The effort applied to the machine is 200 N and moves through a distance of 16 m.

From equation (1),

$$\textbf{force ratio} = \frac{\text{load}}{\text{effort}} = \frac{160 \text{ kg}}{200 \text{ N}} = \frac{160 \times 9.8 \text{ N}}{200 \text{ N}} = \textbf{7.84}$$

From equation (2),

$$\textbf{movement ratio} = \frac{\text{distance moved by the effort}}{\text{distance moved by the load}} = \frac{16 \text{ m}}{1.6 \text{ m}} = \textbf{10}$$

From equation (3),

$$\textbf{efficiency} = \frac{\text{force ratio}}{\text{movement ratio}} \times 100\% = \frac{7.84}{10} \times 100 = \textbf{78.4\%}$$

For simple machines, the relationship between effort and load is of the form: $F_e = aF_l + b$, where F_e is the effort, F_l is the load and a and b are constants.

From equation (1),

$$\text{force ratio} = \frac{\text{load}}{\text{effort}} = \frac{F_l}{F_e} = \frac{F_l}{aF_l + b}$$

Dividing both numerator and denominator by F_l gives:

$$\frac{F_l}{aF_l + b} = \frac{1}{a + \dfrac{b}{F_l}}$$

When the load is large, F_l is large and $\dfrac{b}{F_l}$ is small compared with a. The force ratio then becomes approximately equal to $\dfrac{1}{a}$ and is called the **limiting force ratio**, i.e.

$$\boxed{\textbf{limiting ratio} = \frac{1}{a}}$$

The limiting efficiency of a simple machine is defined as the ratio of the limiting force ratio to the movement ratio, i.e.

$$\boxed{\textbf{Limiting efficiency} = \frac{1}{a \times \textbf{movement ratio}} \times \textbf{100\%}}$$

where a is the constant for the law of the machine: $F_e = aF_l + b$

Due to friction and inertia, the limiting efficiency of simple machines is usually well below 100%.

For example, in a test on a simple machine, the effort/load graph was a straight line of the form $F_e = aF_l + b$. Two values lying on the graph were at $F_e = 10$ N, $F_l = 30$ N, and at $F_e = 74$ N, $F_l = 350$ N. The movement ratio of the machine was 17.

The equation $F_e = aF_l + b$ is of the form $y = mx + c$, where m is the gradient of the graph. The slope of the line passing through points (x_1, y_1) and (x_2, y_2) of the graph $y = mx + c$ is given by: $m = \dfrac{y_2 - y_1}{x_2 - x_1}$

Thus for $F_e = aF_l + b$, the slope a is given by: $a = \dfrac{74-10}{350-30} = \dfrac{64}{320} = 0.2$

The **limiting force ratio** is $\dfrac{1}{a}$, that is $\dfrac{1}{0.2} = \mathbf{5}$

The **limiting efficiency** $= \dfrac{1}{a \times \text{movement ratio}} \times 100$

$$= \dfrac{1}{0.2 \times 17} \times 100 = \mathbf{29.4\%}$$

Pulleys

A **pulley system** is a simple machine. A single-pulley system, shown in Figure 22.1(a), changes the line of action of the effort, but does not change the magnitude of the force.

A two-pulley system, shown in Figure 22.1(b), changes both the line of action and the magnitude of the force. Theoretically, each of the ropes marked (i) and (ii) share the load equally, thus the theoretical effort is only half of the load, i.e. the theoretical force ratio is 2. In practice the actual force ratio is less than 2 due to losses.

A three-pulley system is shown in Figure 22.1(c). Each of the ropes marked (i), (ii) and (iii) carry one-third of the load, thus the theoretical force ratio is 3. In general, for a multiple pulley system having a total of n pulleys, the theoretical force ratio is n. Since the theoretical efficiency of a pulley system (neglecting losses) is 100 and since from equation (3),

$$\text{efficiency} = \dfrac{\text{force ratio}}{\text{movement ratio}} \times 100\%$$

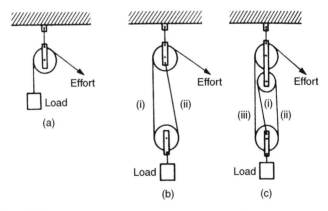

Figure 22.1

it follows that when the force ratio is n,

$$100 = \frac{n}{\text{movement ratio}} \times 100$$

that is, the movement ratio is also n.

For example, a load of 80 kg is lifted by a three-pulley system and the applied effort is 392 N.

From equation (1), the **force ratio** $= \dfrac{\text{load}}{\text{effort}} = \dfrac{80 \times 9.8}{392} = \mathbf{2}$

The **movement ratio = 3** (since it is a three-pulley system)

From equation (3),

$$\textbf{efficiency} = \frac{\text{force ratio}}{\text{movement ratio}} \times 100\%$$

$$= \frac{2}{3} \times 100 = \mathbf{66.67\%}$$

The Screw-jack

A **simple screw-jack** is shown in Figure 22.2 and is a simple machine since it changes both the magnitude and the line of action of a force.

The screw of the table of the jack is located in a fixed nut in the body of the jack. As the table is rotated by means of a bar, it raises or lowers a load placed on the table. For a single-start thread, as shown, for one complete revolution of the table, the effort moves through a distance $2\pi r$, and the load

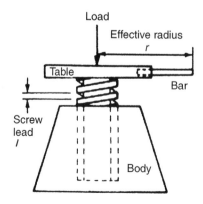

Figure 22.2

moves through a distance equal to the lead of the screw, say, l

$$\text{Movement ratio} = \frac{2\pi r}{l} \qquad (4)$$

For example, a screw-jack is used to support the axle of a car, the load on it being 2.4 kN. The screw jack has an effort of effective radius 200 mm and a single-start square thread, having a lead of 5 mm. If an effort of 60 N is required to raise the car axle:

$$\text{Force ratio} = \frac{\text{load}}{\text{effort}} = \frac{2400 \text{ N}}{60 \text{ N}} = \textbf{40}$$

From equation (4),

$$\text{movement ratio} = \frac{2\pi r}{l} = \frac{2\pi(200) \text{ mm}}{5 \text{ mm}} = \textbf{251.3}$$

Hence,

$$\text{efficiency} = \frac{\text{force ratio}}{\text{movement ratio}} \times 100 = \frac{40}{251.3} \times 100 = \textbf{15.9\%}$$

Gear Trains

A **simple gear train** is used to transmit rotary motion and can change both the magnitude and the line of action of a force, hence is a simple machine. The gear train shown in Figure 22.3 consists of **spur gears** and has an effort applied to one gear, called the driver, and a load applied to the other gear, called the **follower**.

In such a system, the teeth on the wheels are so spaced that they exactly fill the circumference with a whole number of identical teeth, and the teeth on the driver and follower mesh without interference. Under these conditions, the number of teeth on the driver and follower are in direct proportion to the circumference of these wheels, i.e.

$$\frac{\textbf{number of teeth on driver}}{\textbf{number of teeth on follower}} = \frac{\textbf{circumference of driver}}{\textbf{circumference of follower}} \qquad (5)$$

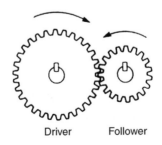

Driver   Follower

Figure 22.3

If there are, say, 40 teeth on the driver and 20 teeth on the follower then the follower makes two revolutions for each revolution of the driver. In general:

$$\frac{\text{number of revolutions made by driver}}{\text{number of revolutions made by the follower}} = \frac{\text{number of teeth on follower}}{\text{number of teeth on driver}} \quad (6)$$

It follows from equation (6) that the speeds of the wheels in a gear train are inversely proportional to the number of teeth. The ratio of the speed of the driver wheel to that of the follower is the movement ratio, i.e.

$$\textbf{Movement ratio} = \frac{\textbf{speed of driver}}{\textbf{speed of follower}} = \frac{\textbf{teeth on follower}}{\textbf{teeth on driver}} \quad (7)$$

When the same direction of rotation is required on both the driver and the follower an **idler wheel** is used as shown in Figure 22.4

Let the driver, idler, and follower be A, B and C, respectively, and let N be the speed of rotation and T be the number of teeth. Then from equation (7),

$$\frac{N_B}{N_A} = \frac{T_A}{T_B} \quad \text{and} \quad \frac{N_C}{N_B} = \frac{T_B}{T_C}$$

Thus $\dfrac{\text{speed of A}}{\text{speed of C}} = \dfrac{N_A}{N_C} = \dfrac{N_B \dfrac{T_B}{T_A}}{N_B \dfrac{T_B}{T_C}} = \dfrac{T_B}{T_A} \times \dfrac{T_C}{T_B} = \dfrac{T_C}{T_A}$

This shows that the movement ratio is independent of the idler, only the direction of the follower being altered.

A **compound gear train** is shown in Figure 22.5, in which gear wheels B and C are fixed to the same shaft and hence $N_B = N_C$

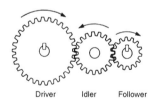

Driver   Idler   Follower

Figure 22.4

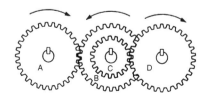

Figure 22.5

From equation (7), $\dfrac{N_A}{N_B} = \dfrac{T_B}{T_A}$ i.e. $N_B = N_A \times \dfrac{T_A}{T_B}$

Also, $\dfrac{N_D}{N_C} = \dfrac{T_C}{T_D}$ i.e. $N_D = N_C \times \dfrac{T_C}{T_D}$

But $N_B = N_C$, hence $N_D = N_A \times \dfrac{T_A}{T_B} \times \dfrac{T_C}{T_D}$ $\qquad$ (8)

For compound gear trains having, say, P gear wheels,

$$N_P = N_A \times \dfrac{T_A}{T_B} \times \dfrac{T_C}{T_D} \times \dfrac{T_E}{T_F} \ldots \times \dfrac{T_O}{T_P}$$

from which,

$$\boxed{\text{movement ratio} = \dfrac{N_A}{N_P} = \dfrac{T_B}{T_A} \times \dfrac{T_D}{T_C} \ldots \times \dfrac{T_P}{T_O}} \qquad (9)$$

For example, a compound gear train consists of a driver gear A, having 40 teeth, engaging with gear B, having 160 teeth. Attached to the same shaft as B, gear C has 48 teeth and meshes with gear D on the output shaft, having 96 teeth.

From equation (9), **movement ratio** $= \dfrac{\text{speed of A}}{\text{speed of D}} = \dfrac{T_B}{T_A} \times \dfrac{T_D}{T_C}$

$$= \dfrac{160}{40} \times \dfrac{96}{48} = \mathbf{8}$$

If the force ratio is, say, 6, then the **efficiency** $= \dfrac{6}{8} \times 100 = \mathbf{75\%}$

Levers

A **lever** can alter both the magnitude and the line of action of a force and is thus classed as a simple machine. There are three types or orders of levers, as shown in Figure 22.6

A lever of the first order has the fulcrum placed between the effort and the load, as shown in Figure 22.6(a).

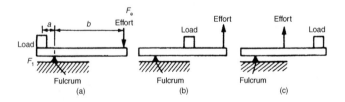

Figure 22.6

A lever of the second order has the load placed between the effort and the fulcrum, as shown in Figure 22.6(b).

A lever of the third order has the effort applied between the load and the fulcrum, as shown in Figure 22.6(c).

Problems on levers can largely be solved by applying the principle of moments (see Chapter 12). Thus for the lever shown in Figure 22.6(a), when the lever is in equilibrium,

anticlockwise moment = clockwise moment

i.e. $$a \times F_l = b \times F_e$$

Thus, **force ratio** $= \dfrac{F_l}{F_e} = \dfrac{b}{a} = \dfrac{\textbf{distance of effort from fulcrum}}{\textbf{distance of load from fulcrum}}$

For example, the load on a first-order lever is 1.2 kN, the distance between the fulcrum and load is 0.5 m and the distance between the fulcrum and effort is 1.5 m.

Applying the principle of moments, for equilibrium:

anticlockwise moment = clockwise moment

i.e. $$1200 \text{ N} \times 0.5 \text{ m} = \text{effort} \times 1.5 \text{ m}$$

from which, $$\textbf{effort} = \frac{1200 \times 0.5}{1.5} = \textbf{400 N}$$

$$\textbf{force ratio} = \frac{F_l}{F_e} = \frac{1200}{400} = \textbf{3}$$

Alternatively, $$\textbf{force ratio} = \frac{b}{a} = \frac{1.5}{0.5} = \textbf{3}$$

This result shows that to lift a load of, say, 300 N, an effort of 100 N is required.

Since, efficiency $= \dfrac{\text{force ratio}}{\text{movement ratio}} \times 100\%$

then **movement ratio** $= \dfrac{\text{force ratio}}{\text{efficiency}} \times 100 = \dfrac{3}{100} \times 100 = \textbf{3}$

This result shows that to raise the load by, say, 100 mm, the effort has to move 300 mm.

23 The Effects of Forces on Materials

Introduction

A **force** exerted on a body can cause a change in either the shape or the motion of the body. The unit of force is the **newton, N**.

No solid body is perfectly rigid and when forces are applied to it, changes in dimensions occur. Such changes are not always perceptible to the human eye since they are so small. For example, the span of a bridge will sag under the weight of a vehicle and a spanner will bend slightly when tightening a nut. It is important for engineers and designers to appreciate the effects of forces on materials, together with their mechanical properties.

The three main types of mechanical force that can act on a body are:

(i) tensile, (ii) compressive, and (iii) shear

Tensile Force

Tension is a force that tends to stretch a material, as shown in Figure 23.1. Examples include:

  (i) the rope or cable of a crane carrying a load is in tension
 (ii) rubber bands, when stretched, are in tension
(iii) a bolt; when a nut is tightened, a bolt is under tension

A tensile force, i.e. one producing tension, increases the length of the material on which it acts.

Compressive Force

Compression is a force that tends to squeeze or crush a material, as shown in Figure 23.2. Examples include:

  (i) a pillar supporting a bridge is in compression
 (ii) the sole of a shoe is in compression
(iii) the jib of a crane is in compression

Figure 23.1   **Figure 23.2**

A compressive force, i.e. one producing compression, will decrease the length of the material on which it acts.

Shear Force

Shear is a force that tends to slide one face of the material over an adjacent face. Examples include:

(i) a rivet holding two plates together is in shear if a tensile force is applied between the plates as shown in Figure 23.3
(ii) a guillotine cutting sheet metal, or garden shears, each provide a shear force
(iii) a horizontal beam is subject to shear force
(iv) transmission joints on cars are subject to shear forces

A shear force can cause a material to bend, slide or twist.

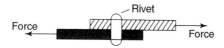

Figure 23.3

Stress

Forces acting on a material cause a change in dimensions and the material is said to be in a state of **stress**. Stress is the ratio of the applied force F to cross-sectional area A of the material. The symbol used for tensile and compressive stress is σ (Greek letter sigma). The unit of stress is the **Pascal, Pa**, where 1 Pa = 1 N/m^2. Hence $\boxed{\sigma = \dfrac{F}{A} \ \textbf{Pa}}$

where F is the force in newtons and A is the cross-sectional area in square metres.

For tensile and compressive forces, the cross-sectional area is that which is at right angles to the direction of the force.

For example, a rectangular bar having a cross-sectional area of 75 mm^2 has a tensile force of 15 kN applied to it. Then

$$\textbf{the stress in bar, } \sigma = \frac{F}{A} = \frac{15 \times 10^3 \text{ N}}{75 \times 10^{-6} \text{ m}^2}$$

$$= 0.2 \times 10^9 \text{ Pa} = \textbf{200 MPa}$$

Strain

The fractional change in a dimension of a material produced by a force is called the **strain**. For a tensile or compressive force, strain is the ratio of the change of length to the original length. The symbol used for strain is ε (Greek epsilon). For a material of length l metres which changes in length by an amount x metres when subjected to stress,

$$\boxed{\varepsilon = \frac{x}{l}}$$

Strain is dimension-less and is often expressed as a percentage, i.e.

$$\text{percentage strain} = \frac{x}{l} \times 100$$

For example, a bar 1.60 m long, contracts by 0.1 mm when a compressive load is applied to it.

Hence,   **strain** $\varepsilon = \dfrac{\text{contraction}}{\text{original length}} = \dfrac{0.1 \text{ mm}}{1.60 \times 10^3 \text{ mm}}$

$$= \frac{0.1}{1600} = \mathbf{0.0000625}$$

and **percentage strain** $= 0.0000625 \times 100 = \mathbf{0.00625\%}$

Shear Stress and Strain

For a shear force, the shear strain is equal to $\dfrac{\text{force}}{\text{area}}$, where the area is that which is parallel to the direction of the force. The symbol for shear stress is the Greek letter tau, τ

Hence, from Figure 23.4, **shear stress**, $\tau = \dfrac{F}{bd}$

Shear strain is denoted by the Greek letter gamma, γ, and with reference to Figure 23.4, **shear strain,** $\gamma = \dfrac{\delta}{l}$

Modulus of rigidity, $G = \dfrac{\text{shear stress}}{\text{shear strain}}$, i.e. $\boldsymbol{G = \dfrac{\tau}{\gamma}}$

For any metal the modulus of rigidity G is approximately 0.4 of the modulus of elasticity E.

Torsional Stress and Strain

With reference to Figure 23.5: $\boxed{\dfrac{\tau}{r} = \dfrac{T}{J} = \dfrac{G\theta}{l}}$

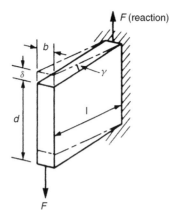

Figure 23.4

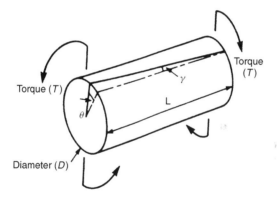

Figure 23.5

where τ = shear stress at radius r, T = torque on shaft, J = polar second moment of area of section of shaft, G = modulus of rigidity, θ = angle of twist (in radians) in a length l of shaft.

The polar second moment of area of a solid shaft is:

$$J = \frac{\pi D^4}{32} \text{ where D is the diameter}$$

The polar second moment of area of a hollow shaft is:

$$J = \frac{\pi}{32}(D^4 - d^4) \text{ where } D = \text{external diameter and}$$

d = internal diameter

For example, a hollow shaft of length 2.4 m, has external and internal diameters of 100 mm and 80 mm. The torque the shaft can transmit if the maximum permissible shear stress is 45 **MPa** is given by:

$$\frac{\tau}{\gamma} = \frac{T}{J} = \frac{G\theta}{l} \text{ from which,}$$

$$\textbf{torque } T = \frac{\tau J}{r} = \frac{(45 \times 10^6 \text{ Pa}) \left(\frac{\pi}{32}(100^4 - 80^4) \times 10^{-12} \text{ m}^4 \right)}{50 \times 10^{-3} \text{ m}}$$

$$= \textbf{5.217 kNm}$$

If the modulus of rigidity is 80 GPa, then the angle of twist between the ends is determined from: $\dfrac{\tau}{r} = \dfrac{G\theta}{l}$ from which,

$$\textbf{angle of twist } \theta = \frac{\tau l}{Gr} = \frac{(45 \times 10^6)(2.4 \text{ m})}{(80 \times 10^9)(50 \times 10^{-3})} = 0.027 \text{ rad}$$

$$= 0.027 \times \frac{180}{\pi} = \textbf{1.547}°$$

Elasticity and Elastic Limit

Elasticity is the ability of a material to return to its original shape and size on the removal of external forces.
Plasticity is the property of a material of being permanently deformed by a force without breaking. Thus if a material does not return to the original shape, it is said to be plastic.

Within certain load limits, mild steel, copper, polythene and rubber are examples of elastic materials; lead and plasticine are examples of plastic materials.

If a tensile force applied to a uniform bar of mild steel is gradually increased and the corresponding extension of the bar is measured, then provided the applied force is not too large, a graph depicting these results is likely to be as shown in Figure 23.6. Since the graph is a straight line, **extension is directly proportional to the applied force**.

If the applied force is large, it is found that the material no longer returns to its original length when the force is removed. The material is then said to have passed its **elastic limit** and the resulting graph of force/extension is no longer a straight line. Stress, $\sigma = \dfrac{F}{A}$, from above, and since, for a particular bar, area A can be considered as a constant, then $F \propto \sigma$.

Strain $\varepsilon = \dfrac{x}{l}$, from above, and since for a particular bar l is constant, then $x \propto \varepsilon$. Hence for stress applied to a material below the elastic limit a graph of stress/strain will be as shown in Figure 23.7, and is a similar shape to the force/extension graph of Figure 23.6

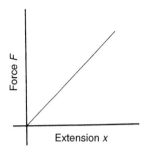

Figure 23.6

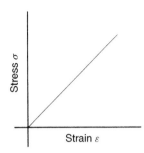

Figure 23.7

Hooke's Law

Hooke's law states:

> *Within the elastic limit, the extension of a material is proportional to the applied force*

It follows, from above, that:

> *Within the elastic limit of a material, the strain produced is directly proportional to the stress producing it*

Young's modulus of elasticity

Within the elastic limit, stress α strain, hence

$$\text{stress} = (\text{a constant}) \times \text{strain}$$

This constant of proportionality is called **Young's modulus of elasticity** and is given the symbol E. The value of E may be determined from the gradient of the straight line portion of the stress/strain graph. The dimensions of E are pascals (the same as for stress, since strain is dimension-less).

$$\boxed{E = \frac{\sigma}{\varepsilon} \ \text{Pa}}$$

Some **typical values** for Young's modulus of elasticity, E, include:
Aluminium 70 GPa (i.e. 70×10^9 Pa), brass 90 GPa, copper 96 GPa, diamond 1200 GPa, mild steel 210 GPa, lead 18 GPa, tungsten 410 GPa, cast iron 110 GPa, zinc 85 GPa

Stiffness

A material having a large value of Young's modulus is said to have a high value of stiffness, where stiffness is defined as:

$$\text{Stiffness} = \frac{\text{force } F}{\text{extension } x}$$

For example, mild steel is much stiffer than lead.

Since $E = \dfrac{\sigma}{\varepsilon}$, $\sigma = \dfrac{F}{A}$ and $\varepsilon = \dfrac{x}{l}$, then

$$E = \frac{\dfrac{F}{A}}{\dfrac{x}{l}} = \frac{Fl}{Ax} = \left(\frac{F}{x}\right)\left(\frac{l}{A}\right)$$

i.e.

$$E = (\text{stiffness}) \times \left(\frac{l}{A}\right)$$

Stiffness $\left(= \dfrac{F}{x}\right)$ is also the gradient of the force/extension graph, hence

$$E = (\text{gradient of force/extension graph})\left(\frac{l}{A}\right)$$

Since l and A for a particular specimen are constant, the greater Young's modulus the greater the stiffness.

For example, in an experiment to determine the modulus of elasticity of a sample of mild steel, a wire is loaded and the corresponding extension noted. The results of the experiment are as shown.

Load (N)	0	40	110	160	200	250	290	340
Extension (mm)	0	1.2	3.3	4.8	6.0	7.5	10.0	16.2

The mean diameter of the wire is 1.3 mm and its length is 8.0 m.

A graph of load/extension is shown in Figure 23.8

$$E = \frac{\sigma}{\varepsilon} = \frac{\dfrac{F}{A}}{\dfrac{x}{l}} = \left(\frac{F}{x}\right)\left(\frac{l}{A}\right)$$

$\dfrac{F}{x}$ is the gradient of the straight line part of the load/extension graph.

Gradient, $\dfrac{F}{x} = \dfrac{BC}{AC} = \dfrac{200 \text{ N}}{6 \times 10^{-3} \text{ m}} = 33.33 \times 10^3$ N/m

Modulus of elasticity $= (\text{gradient of graph})\left(\dfrac{l}{A}\right)$

Length of specimen, $l = 8.0$ m

and cross-sectional area $A = \dfrac{\pi d^2}{4} = \dfrac{\pi (0.0013)^2}{4}$

$$= 1.327 \times 10^{-6} \text{ m}^2$$

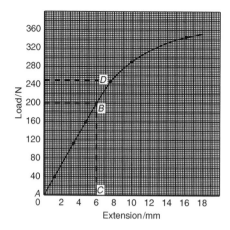

Figure 23.8

Hence, **modulus of elasticity,**

$$E = (33.33 \times 10^3) \left(\frac{8.0}{1.327 \times 10^{-6}} \right) = \textbf{201 GPa}$$

The elastic limit is at point D in Figure 20.6 where the graph no longer follows a straight line. This point corresponds to a load of 250 N as shown.

$$\textbf{Stress at elastic limit} = \frac{\text{force}}{\text{area}} = \frac{250}{1.327 \times 10^{-6}} = 188.4 \times 10^6 \text{ Pa}$$

$$= \textbf{188.4 MPa}$$

Ductility, Brittleness and Malleability

Ductility is the ability of a material to be plastically deformed by elongation, without fracture. This is a property that enables a material to be drawn out into wires. For ductile materials such as mild steel, copper and gold, large extensions can result before fracture occurs with increasing tensile force. Ductile materials usually have a percentage elongation value of about 15% or more.

Brittleness is the property of a material manifested by fracture without appreciable prior plastic deformation. Brittleness is a lack of ductility, and brittle materials such as cast iron, glass, concrete, brick and ceramics, have virtually no plastic stage, the elastic stage being followed by immediate fracture. Little or no 'waist' occurs before fracture in a brittle material undergoing a tensile test.

Malleability is the property of a material whereby it can be shaped when cold by hammering or rolling. A malleable material is capable of undergoing plastic deformation without fracture. Examples of malleable materials include lead, gold, putty and mild steel.

24 Tensile Testing

The Tensile Test

A **tensile test** is one in which a force is applied to a specimen of a material in increments and the corresponding extension of the specimen noted. The process may be continued until the specimen breaks into two parts and this is called **testing to destruction**. The testing is usually carried out using a universal testing machine that can apply either tensile or compressive forces to a specimen in small, accurately measured steps. **British Standard 18** gives the standard procedure for such a test. Test specimens of a material are made to standard shapes and sizes and two typical test pieces are shown in Figure 24.1. The results of a tensile test may be plotted on a load/extension graph and a typical graph for a mild steel specimen is shown in Figure 24.2.

(i) Between A and B is the region in which Hooke's law applies and stress is directly proportional to strain. The gradient of AB is used when determining Young's modulus of elasticity (see Chapter 23).

(ii) Point B is the **limit of proportionality** and is the point at which stress is no longer proportional to strain when a further load is applied.

(iii) Point C is the **elastic limit** and a specimen loaded to this point will effectively return to its original length when the load is removed, i.e. there is negligible permanent extension.

(iv) Point D is called the **yield point** and at this point there is a sudden extension with no increase in load. The yield stress of the material is given by:

$$\text{yield stress} = \frac{\text{load where yield begins to take place}}{\text{original cross-sectional area}}$$

The yield stress gives an indication of the ductility of the material (see Chapter 23).

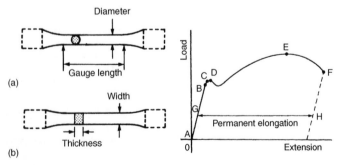

Figure 24.1   **Figure 24.2**

(v) Between points D and E extension takes place over the whole gauge length of the specimen.

(vi) Point E gives the maximum load which can be applied to the specimen and is used to determine the **ultimate tensile strength (UTS)** of the specimen (often just called the tensile strength)

$$\textbf{UTS} = \frac{\textbf{maximum load}}{\textbf{original cross-sectional area}}$$

(vii) Between points E and F the cross-sectional area of the specimen decreases, usually about half way between the ends, and a **waist** or **neck** is formed before fracture.

$$\textbf{Percentage reduction in area} = \frac{\begin{array}{c}\textbf{original c.s.a.}\\\textbf{−final c.s.a.}\end{array}}{\textbf{original c.s.a.}} \times \textbf{100\%}$$

The percentage reduction in area provides information about the malleability of the material (see Chapter 23).

The value of stress at point F is greater than at point E since although the load on the specimen is decreasing as the extension increases, the cross-sectional area is also reducing.

(viii) At point F the specimen fractures.

(ix) Distance GH is called the **permanent elongation** and

$$\begin{array}{c}\textbf{permanent}\\\textbf{elongation}\end{array}\Big\} = \frac{\begin{array}{c}\textbf{increase in length during}\\\textbf{test to destruction}\end{array}}{\textbf{original length}} \times \textbf{100\%}$$

For example, a rectangular zinc specimen is subjected to a tensile test and the data from the test is shown below. Width of specimen 40 mm; breadth of specimen 2.5 mm; gauge length 120 mm.

Load (kN)	10	17	25	30	35	37.5	38.5	37	34	32
Extension (mm)	0.15	0.25	0.35	0.55	1.00	1.50	2.50	3.50	4.50	5.00

Fracture occurs when the extension is 5.0 mm and the maximum load recorded is 38.5 kN.

A load/extension graph is shown in Figure 24.3

The limit of proportionality occurs at point P on the graph, where the initial gradient of the graph starts to change. This point has a load value of 26.5 kN.

Stress at the limit of proportionality is given by:

$$\sigma = \frac{\text{force}}{\text{area}} = \frac{26.5 \times 10^3 \text{ N}}{40 \times 2.5 \times 10^{-6} \text{ m}^2} = 265 \times 10^6 \text{ Pa} = \textbf{265 MPa}$$

The gradient of the straight line portion of the graph is given by:

$$\frac{\text{BC}}{\text{AC}} = \frac{25000 \text{ N}}{0.35 \times 10^{-3} \text{ m}} = 71.43 \times 10^6 \text{ N/m}$$

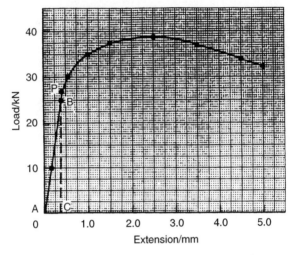

Figure 24.3

$$\text{Young's modulus of elasticity} = (\text{gradient of graph}) \left(\frac{l}{A} \right)$$

$$= \left(71.43 \times 10^6 \right) \left(\frac{120 \times 10^{-3}}{100 \times 10^{-6}} \right)$$

$$= 85.72 \times 10^9 \text{ Pa} = \textbf{85.72 GPa}$$

$$\text{Ultimate tensile strength} = \frac{\text{maximum load}}{\text{original c.s.a.}}$$

$$= \frac{38.5 \times 10^3 \text{ N}}{100 \times 10^{-6} \text{ m}^2}$$

$$= 385 \times 10^6 \text{ Pa} = \textbf{385 MPa}$$

$$\text{Percentage elongation} = \frac{\text{extension at fracture point}}{\text{original length}} \times 100$$

$$= \frac{5.0 \text{ mm}}{120 \text{ mm}} \times 100 = \textbf{4.17\%}$$

25 Hardness and Impact Tests

Hardness

The hardness of a material may be defined in the following ways:

 (i) the ability to scratch other materials
 (ii) the ability to resist scratching
(iii) the ability to resist plastic spreading under indentation
(iv) the ability to resist elastic deformation under indentation
 (v) the ability to resist deformation by rolling

Hardness Tests

Hardness tests are based on pressing a hard substance, such as a diamond or a steel sphere having known dimensions, into the material under test. The hardness can be determined from the size of indentation made for a known load. The three principal hardness tests are:

(a) the Brinell test,
(b) the Vickers test, and
(c) the Rockwell test.

The Brinell Test

In a standard **Brinell test**, a hardened steel ball having a diameter of 10 mm is squeezed into the material under a load of 3000 kg. The diameter of the indentation produced is measured under a microscope. The Brinell hardness number, H_B, is given by:

$$H_B = \frac{\text{load}}{\text{spherical area of indent}} = \frac{F}{\frac{\pi D}{2}[D - \sqrt{D^2 - d^2}]}$$

where F is the load in kilograms (usually 3000 kg),
         D is the diameter of the steel ball in millimetres
              (usually 10 mm), and
         d is the diameter of the indentation in millimetres.

Variations on the standard test include smaller loads used for soft materials, balls of different diameters (usually restricted to 1, 2 and 5 mm) and the steel ball being replaced by one made of tungsten carbide for use with very hard materials. Values of Brinell hardness number vary from about 900 for very hard materials having an equivalent tensile strength for steel of 3000 MPa, down to about 100 for materials having an equivalent tensile strength for steel of about 350 MPa.

The following precautions are required when carrying out a Brinell test.

(i) The material must be sufficiently wide and thick. The impression must have its centre not less than two and a half times its diameter from any edge. The thickness must be at least ten times the depth of the impression, this depth being given by: $\dfrac{F}{\pi DH}$ where H is the estimated Brinell hardness number.

(ii) The surface of the material should, if possible, be ground flat and polished.

(iii) The load should be held for 5 s.

(iv) Two diameters of the impression at right angles should be read and their mean used in the calculation.

(v) When stating the result of the test the ball number and load should be stated; for example, $H\,10/3000 = 410$

Special machines are made but a Brinell test can be carried out on most universal testing machines.

For materials of the same quality and for families of materials an approximate direct proportional relationship seems to exist between tensile strength and Brinell hardness number. For example, a nickel-chrome steel which is hardened and then tempered to various temperatures has tensile strengths varying from 1900 MPa to 1070 MPa as the Brinell hardness number varies from 530 to 300. A constant of proportionality k for: tensile strength $= (k \times$ hardness) in this case is 3.57 for all tempering temperatures. Similarly, for a family of carbon steels, the tensile strength varies from 380 MPa to 790 MPa as the carbon content increases. The Brinell hardness number varies from 115 to 230 over the same range of carbon values and the constant of proportionality in this case is 3.35. Because of the general approximate relationship between tensile strength and hardness, tables exist relating these quantities, the tables usually based on a constant of proportionality of about 3.35.

Vickers Test

In a **Vickers diamond pyramid hardness test**, a square-based diamond pyramid is pressed into the material under test. The angle between opposite faces of the diamond is 136° and the load applied is one of the values 5, 10, 30, 50 or 120 kg, depending on the hardness of the material. The Vickers diamond hardness number, H_V, is given by:

$$H_V = \frac{\text{load}}{\text{surface area of indentation}} = \frac{F}{\dfrac{d^2}{1.854}}$$

where F is the load in kilograms and d is the length of the diagonal of the square of indentation in millimetres.

The Rockwell Test

The **Rockwell hardness test** is mainly used for rapid routine testing of finished material, the hardness number being indicated directly on a dial. The value of hardness is based directly on the depth of indentation of either a steel ball or a cone shaped diamond with a spherically rounded tip, called a 'brale'. Whether the steel ball or brale is selected for use depends on the hardness of the material under test, the steel ball being used for materials having a hardness up to that of medium carbon steels.

Several different scales are shown on the dial, and can include Rockwell *A* to *H* scales together with Rockwell *K*, *N* and *T* scales. Examples of the scale used are:

Scale A: using a brale and a 60 kg load
Scale B: using a brale and a 150 kg load
Scale C: using a 1/16th inch steel ball and 100 kg load, and so on.

The big advantage of the Rockwell test over Brinell and Vickers tests is the speed with which it can be made. As it is also independent of surface condition it is well suited to production line testing. British Standards, however, require hardness numbers to be based on the surface area of any indentation.

Other Hardness Tests

The Brinell, Vickers and Rockwell tests are examples of static hardness tests. Another example is the **Firth Hardometer test**, which is very similar to the Vickers test.

Examples of dynamic hardness tests are those using the **Herbert pendulum Hardness Tester** and the **Shore Scleroscope**. The former uses an arched rocker resting on a steel or diamond pivot; hardness can be indicated by the time taken for ten swings or by the difference between an initial angular displacement and the first swing. The Scleroscope is a portable apparatus in which a diamond-tipped hammer falls on to the material under test. The height of the rebound gives the hardness number.

Other Non-destructive Tests

Flaws inside a casting can be revealed by X-ray methods. Surface flaws can be revealed by electro-magnetic methods and by those using ultra-violet light. The former are applicable only to ferrous metals but the latter can be used for other metals and for other materials, such as plastics and ceramics.

Impact Tests

To give an indication of the toughness of a material, that is, the energy needed to fracture it, impact tests are carried out. Two such tests are the **Izod test**, principally used in Great Britain, and the **Charpy test** that is widely used in other parts of Europe.

Izod Test

In an Izod test, a square test piece of side 10 mm and having a vee-notch of angle 45° machined along one side, is clamped firmly in a vice in the base of the Izod test machine. A heavy pendulum swings down to strike the specimen and fractures it. The difference between the release angle of the pendulum measured to the vertical and the overswing angle after fracturing the specimen is proportional to the energy expended in fracturing the specimen, and can be read from a scale on the testing machine. An Izod test is basically an acceptance test, that is, the value of impact energy absorbed is either acceptable or is not acceptable. The results of an Izod test cannot be used to determine impact strength under other conditions.

Charpy Test

A Charpy test is similar to an Izod test, the only difference being the method of mounting the test specimen and a capability of varying the mass of the pendulum. In the Izod test, the specimen is gripped at one end and is supported as a cantilever, compared with the specimen being supported at each end as a beam in the Charpy test. One other difference is that the notch is at the centre of the supported beam and faces away from the striker.

26 Measurement of Strain

Introduction

An essential requirement of engineering design is the accurate determination of stresses and strains in components under working conditions. 'Strength of materials' is a subject relating to the physical nature of substances which are acted upon by external forces. No solid body is perfectly rigid, and when forces are applied to it changes in dimensions occur. Such changes are not always perceptible to the human eye since they are so small. For example, a spanner will bend slightly when tightening a nut, and the span of a bridge will sag under the weight of a car.

Strain

The change in the value of a linear dimension of a body, say x, divided by the original value of the dimension, say l, gives a great deal of information about what is happening to the material itself. This ratio is called strain, ε, and is dimensionless, i.e. $\varepsilon = \dfrac{x}{l}$

Stress

The force (F) acting on an area (A) of a body is called the stress, σ, and is measured in pascals (Pa) or newtons per square metre (N/m^2), i.e. $\sigma = \dfrac{F}{A}$

Young's Modulus of Elasticity

If a solid body is subjected to a gradually increasing stress, and if both the stress and the resulting strain are measured, a graph of stress against strain may be drawn. Up to a certain value of stress the graph is a straight line. That particular value is known as the limit of proportionality and its value varies for different materials. The gradient of the straight line is a constant known as **Young's modulus of elasticity**, E

$$E = \frac{\text{stress}}{\text{strain}} \text{ up to the limit of proportionality} = \frac{\dfrac{F}{A}}{\dfrac{x}{l}} = \frac{Fl}{xA} \text{ Pa} \qquad (1)$$

Young's modulus of elasticity is a constant for a given material. As an example, mild steel has a value of E of about 210×10^9 Pa (i.e. 210 GPa).

Elastic Limit

If on removal of external forces a body recovers its original shape and size, the material is said to be **elastic**. If it does not return to its original shape, it is said to be **plastic**. Copper, steel and rubber are examples of elastic materials while lead and plasticine are plastic materials. However, even for elastic materials there is a limit to the amount of strain from which it can recover its original dimensions. This limit is called the **elastic limit** of the material. The elastic limit and the limit of proportionality for all engineering materials are virtually the same. If a body is strained beyond the elastic limit permanent deformation will occur.

The Need for Strain Measurement

In designing a structure, such as an electricity transmission tower carrying overhead power lines or support pillars and spans of new designs of bridges, the engineer is greatly concerned about the mechanical properties of the materials he is going to use. Many laboratory tests have been designed to provide important information about materials. Such tests include tensile, compression, torsion, impact, creep and fatigue tests and each attempt to provide information about the behaviour of materials under working conditions. (A typical tensile test is described in Chapter 24)

It is possible to design a structure that is strong enough to withstand the forces encountered in service, but is, nonetheless, useless because of the amount of elastic deformation. Hence, tests made on materials up to the elastic limit are of great importance. A material that has a relatively high value of Young's modulus is said to have a high value of **stiffness**, stiffness being the ratio of force to extension, i.e. $\dfrac{F}{x}$

From equation (1) $E = \dfrac{\text{stress}}{\text{strain}} = \left(\dfrac{F}{x} \right) \left(\dfrac{l}{A} \right) = \text{stiffness} \left(\dfrac{l}{A} \right)$

Thus, when the determination of Young's modulus of elasticity, E, of a material is required, an accurate stress/strain or load/extension graph must be obtained. The actual strain is very small and this means that very small extensions must be measured with a high degree of accuracy.

The measurement of extension, and thus strain, is achieved in the laboratory with an instrument called an **extensometer**. Although some extensometers can be used in such practical situations as a crane under load, it is more usual to use in these situations an electrical device called a **strain gauge**.

A knowledge of stress and strain is the foundation of economy and safety in design.

Extensometers

An extensometer is an instrument used in engineering and metallurgical design to measure accurately the minute elastic extensions of materials, in order to forecast their behaviour during use. There are several different designs of extensometer including the Lindley, the Huggenburger and the Hounsfield.

The Lindley extensometer

This is probably the most common type of extensometer used for measuring tensile strains. This instrument consists of two arms, A and B, connected by a strip of spring steel that acts as a hinge. The unstressed specimen of the material is clamped at points C and D by pointed screws, the distance between C and D usually being 50 mm. Thus 50 mm is termed the 'gauge length'. A dial test indicator is placed between the arms A and B as shown in the typical arrangement of the Lindley extensometer in Figure 26.1

The point D is halfway between the hinge and the indicator; hence the movement of the pointer on the test indicator will record twice the extension of the specimen. However, the indicator is normally calibrated so that it indicates extension directly, each graduation representing an extension of 1 micron (i.e. 10^{-6} m or 0.001 mm). Extensions may be measured to an accuracy of 0.0001 mm using the Lindley extensometer.

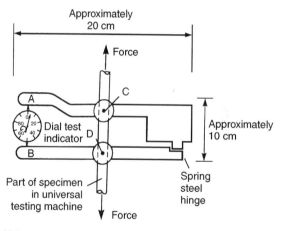

Figure 26.1

The Huggenburger extensometer

This is a simple, rugged and accurate instrument that may be used to measure tensile or compressive strains. Its construction is based on a lever multiplying system capable of obtaining magnifications in the order of 2000. Figure 26.2 shows a simplified schematic arrangement of a front view of the Huggenburger

114

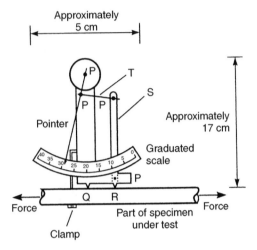

Approximately
5 cm

P

T

S

Pointer

Approximately
17 cm

40 35 30 25 20 15 10 5 0

Graduated
scale

P

Force

Q R

Force

Clamp

Part of specimen
under test

Figure 26.2

extensometer clamped to a specimen, where Q and R are two knife-edges, usually either 10 mm or 20 mm apart. Any strain encountered by the specimen
under test will alter the gauge length QR. In Figure 26.2, the specimen is
shown in tension, thus QR will increase in length. This change is transmitted
by pivots (labelled P) and levers S and T to the pointer, and is indicated on the
scale according to the multiplication factor. The supplier who calibrates each
device after manufacture supplies this factor, of approximately 2000, to the
instrument user. This type of extensometer enables extensions to be recorded
to an accuracy comparable with the Lindley extensometer and may be used
in the laboratory or in the field.

The Hounsfield extensometer

This may be used in conjunction with a Hounsfield Tensometer (which is a
universal portable testing machine capable of applying tensile or compressive
forces to metals, plastics, textiles, timber, paper and so on), or with any other
testing machine. The extensometer is a precision instrument for measuring
the extension of a test specimen over a 50 mm gauge length, while the test
specimen is loaded in the testing machine. The instrument can be attached
to round specimens of material of up to 25 mm in diameter or rectangular
sections of material of up to 25 mm square at precisely 50 mm gauge length
without prior marking of the specimen. Figure 26.3 shows a typical Hounsfield
extensometer viewed from two different elevations.

The gauge length rod is screwed into position, making the fixed centres
exactly 50 mm apart. The extensometer is then clamped to the test piece
before the gauge length rod is unscrewed. With the test piece still unloaded
the micrometer is wound in until the platinum contacts meet, thus completing
the circuit (shown by the lamp lighting up). The micrometer reading is then
taken and the micrometer head unwound. After the load is placed on the

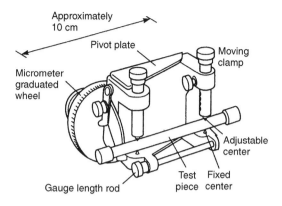

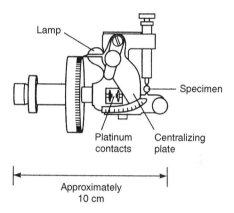

Figure 26.3

specimen the micrometer head is again wound in and a new reading taken when the lamp lights. The difference between the two micrometer readings is an indication of the extension of the test piece for the particular load applied. Each division of the micrometer wheel is equal to 0.002 mm. The accuracy of the Hounsfield extensometer compares favourably with other extensometers, and an advantage, in certain circumstances, of this instrument is its small overall size.

Strain Gauges

A strain gauge is an electrical device used for measuring mechanical strain, i.e. the change in length accompanying the application of a stress. The strain gauge consists essentially of a very fine piece of wire that is cemented, or glued strongly, to the part where the strain is to be measured. When the length of a piece of wire is changed, a change in its electrical resistance occurs, this change in resistance being proportional to the change in length of the wire. Thus, when the wire is securely cemented to the part that is being strained, a change of electrical resistance of the wire occurs due to the change in length. By measuring this change of resistance the strain can be determined. The strain gauge was first introduced in the USA in 1939 and since that time it has come into widespread use, particularly in the aircraft industry, and is now the basis of one of the most useful of stress analysis techniques. A typical simple strain gauge is shown in Figure 26.4.

Rolling out a thin foil of the resistive material, and then cutting away parts of the foil by a photo-etching process to create the required grid pattern form a modern strain gauge. Such a device is called a **foil strain gauge** and a typical arrangement is shown in Figure 26.5. A foil-strain gauge has many advantages over the earlier method and these include:

(i) better adhesion between conductor and backing material,
(ii) better heat dissipation,
(iii) a more robust construction,
(iv) easier to attach leads to,
(v) accurate reproducibility of readings, and
(vi) smaller sizes are possible.

In order to obtain a deflection on a galvanometer, G, proportional to the strain occurring in the gauge, it must be connected into one arm of a Wheatstone

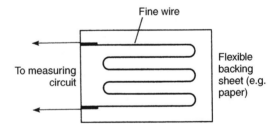

Figure 26.4

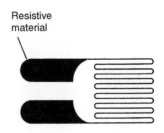

Figure 26.5

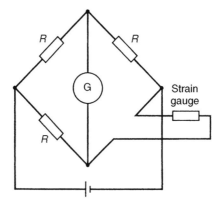

Figure 26.6

bridge, as shown in Figure 26.6. A Wheatstone bridge circuit having four equal resistances in the arms has zero deflection on the galvanometer, but when the resistance of one or more of the arms changes, then the bridge galvanometer deflects from zero, the amount of deflection being a measure of the change in resistance. If the resistance change occurs in a strain gauge as a result of applied strain, then the bridge galvanometer deflection is a measure of the amount of strain. For very accurate measurements of strain there are a number of possible sophistications. These are not described in detail in this chapter but include:

  (i) the use of a temperature-compensating dummy gauge to make the bridge output independent of temperature, since the resistance of a gauge varies with temperature and such a resistance change may be misread as strain in the material,
 (ii) an additional bridge balancing circuit to obtain zero galvanometer deflection for zero strain, and
(iii) the addition of an amplifier to amplify the signal output from the bridge in applications where the level of strain is such that the bridge deflection is too small to readily detect on a galvanometer.

A typical selection of **practical situations** where strain gauges are used include:

  (i) the airframe and skin of an aircraft in flight,
 (ii) electricity pylons, cranes and support pillars and spans of new designs of bridges, where strain must be tested to validate the design, and
(iii) applications in harsh environments and remote positions, such as inside nuclear boilers, on turbine blades, in vehicle engines, on helicopter blades in flight and under water on oil rig platforms, where a knowledge of strain is required.

27 Linear Momentum and Impulse

Linear Momentum

The **momentum** of a body is defined as the product of its mass and its velocity, i.e. **momentum = mu**, where m = mass (in kg) and u = velocity (in m/s). The unit of momentum is kg m/s.

Since velocity is a vector quantity, **momentum is a vector quantity**, i.e. it has both magnitude and direction.

For example, the momentum of a pile driver of mass 400 kg when it is moving downwards with a speed of 12 m/s is given by:

$$\textbf{momentum} = \text{mass} \times \text{velocity} = 400 \text{ kg} \times 12 \text{ m/s}$$

$$= \textbf{4800 kg m/s downwards}$$

Newton's first law of motion states:

a body continues in a state of rest or in a state of uniform motion in a straight line unless acted on by some external force

Hence the momentum of a body remains the same provided no external forces act on it.

The principle of conservation of momentum for a closed system (i.e. one on which no external forces act) may be stated as:

the total linear momentum of a system is a constant

The total momentum of a system before collision in a given direction is equal to the total momentum of the system after collision in the same direction. In Figure 27.1, masses m_1 and m_2 are travelling in the same direction with velocity $u_1 > u_2$. A collision will occur, and applying the principle of conservation of momentum:

total momentum before impact = total momentum after impact

i.e. $$m_1 u_1 + m_2 u_2 = m_1 v_1 + m_2 v_2$$

where v_1 and v_2 are the velocities of m_1 and m_2 after impact.

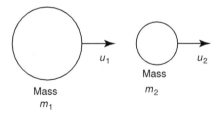

Figure 27.1

For example, a wagon of mass 10 t is moving at a speed of 6 m/s and collides with another wagon of mass 15 t, which is stationary. After impact, the wagons are coupled together. To determine the common velocity of the wagons after impact:

$$\text{Total momentum before impact} = m_1 u_1 + m_2 u_2$$

$$= (10000 \times 6) + (15000 \times 0)$$

$$= 60000 \text{ kg m/s}$$

Let the common velocity of the wagons after impact be v m/s

$$\left.\begin{array}{c}\text{Since total momentum}\\\text{before impact}\end{array}\right\} = \text{total momentum after impact:}$$

$$60000 = m_1 v + m_2 v$$

$$= v(m_1 + m_2) = v(25000)$$

Hence $$v = \frac{60000}{25000} = 2.4 \text{ m/s}$$

i.e. **the common velocity after impact is 2.4 m/s in the direction in which the 10 t wagon is initially travelling.**

Impulse and Impulsive Forces

Newton's second law of motion states:

the rate of change of momentum is directly proportional to the applied force producing the change, and takes place in the direction of this force

In the SI system, the units are such that:

the applied force = rate of change of momentum

$$= \frac{\text{change of momentum}}{\text{time taken}} \tag{1}$$

When a force is suddenly applied to a body due to either a collision with another body or being hit by an object such as a hammer, the time taken in equation (1) is very small and difficult to measure. In such cases, the total effect of the force is measured by the change of momentum it produces.

Forces that act for very short periods of time are called **impulsive forces**. The product of the impulsive force and the time during which it acts is called the **impulse** of the force and is equal to the change of momentum produced by the impulsive force, i.e.

impulse = applied force × time = change in linear momentum

For example, the average force exerted on the work-piece of a press-tool operation is 150 kN, and the tool is in contact with the work-piece for 50 ms.

From above,

$$\text{change of linear momentum} = \text{applied force} \times \text{time} \; (= \text{impulse})$$

Hence,

$$\left.\begin{array}{r}\textbf{change in momentum} \\ \textbf{of work-piece}\end{array}\right\} = 150 \times 10^3 \text{ N} \times 50 \times 10^{-3} \text{ s}$$

$$= \textbf{7500 kg m/s} \; (\text{since } 1 \text{ N} = 1 \text{ kg m/s}^2)$$

Examples where impulsive forces occur include when a gun recoils and when a free-falling mass hits the ground. Solving problems associated with such occurrences often requires the use of the equation of motion: $v^2 = u^2 + 2as$, from Chapter 15.

For example, the hammer of a pile-driver of mass 1 t falls a distance of 1.5 m on to a pile. The blow takes place in 25 ms and the hammer does not rebound.

Using the equation of motion: $v^2 = u^2 + 2gs$

then

$$v^2 = 0^2 + 2(9.81)(1.5)$$

from which,   impact velocity, $v = \sqrt{(2)(9.81)(1.5)} = 5.425$ m/s

Neglecting the small distance moved by the pile and hammer after impact,

$$\text{momentum lost by hammer} = \text{the change of momentum}$$

$$= mv = 1000 \text{ kg} \times 5.425 \text{ m/s}$$

$$\text{Rate of change of momentum} = \frac{\text{change of momentum}}{\text{change of time}}$$

$$= \frac{1000 \times 5.425}{25 \times 10^{-3}} = 217000 \text{ N}$$

Since the impulsive force is the rate of change of momentum, **the average force exerted on the pile is 217 kN**

When a pile is being hammered into the ground, the ground resists the movement of the pile and this resistance is called a **resistive force**.

Newton's third law of motion may be stated as:

for every force there is an equal and opposite force

The force applied to the pile is the resistive force; the pile exerts an equal and opposite force on the ground.

In practice, when impulsive forces occur, energy is not entirely conserved and some energy is changed into heat, noise, and so on.

28 Torque

Couple and Torque

When two equal forces act on a body as shown in Figure 28.1, they cause the body to rotate, and the system of forces is called a **couple**.

The turning moment of a couple is called a **torque**, T. In Figure 28.1,

$$\text{torque} = \text{magnitude of either force} \times \begin{array}{l}\text{perpendicular distance}\\ \text{between the forces}\end{array}$$

i.e. $$\boxed{T = Fd}$$

The unit of torque is the **newton metre, Nm**

When a force F newtons is applied at a radius r metres from the axis of, say, a nut to be turned by a spanner, as shown in Figure 28.2, the torque T applied to the nut is given by: $T = Fr$ **Nm**

For example, the torque when a pulley wheel of diameter 300 mm has a force of 80 N applied at the rim, is given by:

$$\textbf{torque } T = Fr = (80)\left(\frac{300}{2}\times 10^{-3}\right) = \textbf{12 Nm}$$

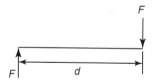

Figure 28.1

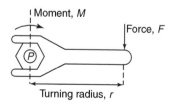

Figure 28.2

Work Done and Power Transmitted by a Constant Torque

Figure 28.3(a) shows a pulley wheel of radius r metres attached to a shaft and a force F newtons applied to the rim at point P

Figure 28.3(b) shows the pulley wheel having turned through an angle θ radians as a result of the force F being applied. The force moves through a distance s, where arc length $s = r\theta$

$$\text{Work done} = \text{force} \times \text{distance moved by force}$$

$$= F \times r\theta = Fr\theta \text{ Nm} = Fr\theta \text{ J}$$

However, Fr is the torque T, hence,

$$\boxed{\textbf{work done} = T\theta \textbf{ joules}}$$

$$\text{Average power} = \frac{\text{work done}}{\text{time taken}} = \frac{T\theta}{\text{time taken}} \text{ for a constant torque } T$$

However, (angle θ)/(time taken) = angular velocity, ω rad/s

Hence,

$$\boxed{\textbf{power, } P = T\omega \textbf{ watts}}$$

Angular velocity, $\omega = 2\pi n$ rad/s where n is the speed in rev/s

Hence,

$$\boxed{\textbf{power, } P = 2\pi nT \textbf{ watts}}$$

For example, the torque developed by a motor whose spindle is rotating at 1000 rev/min and developing a power of 2.50 kW is given by:

power $P = 2\pi nT$, from which,

$$\text{torque, } T = \frac{P}{2\pi n} \text{ Nm}$$

$$= \frac{2500}{2\pi \left(\dfrac{1000}{60} \right)} = \frac{2500 \times 60}{2\pi \times 1000} = \textbf{23.87 Nm}$$

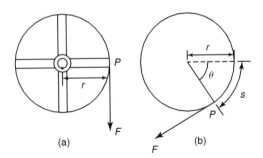

(a)　　　　(b)

Figure 28.3

From above, work done $= T\theta$, and if this work is available to increase the kinetic energy of a rotating body of moment of inertia I, then:

$T\theta = I \left(\dfrac{\omega_2^2 - \omega_1^2}{2} \right)$ where ω_1 and ω_2 are the initial and final angular

velocities, i.e. $T\theta = I \left(\dfrac{\omega_2 + \omega_1}{2} \right) (\omega_2 - \omega_1)$

However, $\left(\dfrac{\omega_2 + \omega_1}{2} \right)$ is the mean angular velocity, i.e. $\dfrac{\theta}{t}$, where t is the time, and $(\omega_2 - \omega_1)$ is the change in angular velocity, i.e. αt, where a is the angular acceleration

Hence, $T\theta = I \left(\dfrac{\theta}{t} \right) (\alpha t)$ from which, $\boxed{\textbf{torque } T = I\alpha}$

where I is the moment of inertia in kg m^2, α is the angular acceleration in rad/s^2 and T is the torque in Nm.

For example, if a shaft system has a moment of inertia of 37.5 kg m^2, the torque required to give it an angular acceleration of 5.0 rad/s^2 is given by:

torque, $T = I\alpha = (37.5)(5.0) = \textbf{187.5 Nm}$

Power Transmission by Belt Drives

A common and simple method of **transmitting power** from one shaft to another is by means of a **belt** passing over pulley wheels which are keyed to the shafts, as shown in Figure 28.4. Typical applications include an electric motor driving a lathe or a drill, and an engine driving a pump or generator.

For a belt to transmit power between two pulleys there must be a difference in tensions in the belt on either side of the driving and driven pulleys. For the direction of rotation shown in Figure 28.4, $F_2 > F_1$

The torque T available at the driving wheel to do work is given by:

$$\boxed{T = (F_2 - F_1)r_x \text{ Nm}}$$

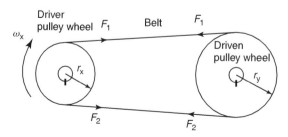

Figure 28.4

and the available power P is given by:

$$\boxed{P = T\omega = (F_2 - F_1)r_x\omega_x \text{ watts}}$$

From chapter 15, the linear velocity of a point on the driver wheel, $v_x = r_x\omega_x$
Similarly, the linear velocity of a point on the driven wheel, $v_y = r_y\omega_y$.

Assuming no slipping, $\quad v_x = v_y$ i.e. $r_x\omega_x = r_y\omega_y$

Hence $\hspace{4cm} r_x(2\pi n_x) = r_y(2\pi n_y)$

from which, $\hspace{3cm} \boxed{\dfrac{r_x}{r_y} = \dfrac{n_y}{n_x}}$

Percentage efficiency $= \dfrac{\text{useful work output}}{\text{energy output}} \times 100$

or $\hspace{1cm} \boxed{\textbf{efficiency} = \dfrac{\textbf{power output}}{\textbf{power input}} \times 100\%}$

For example, a 15 kW motor is driving a shaft at 1150 rev/min by means of pulley wheels and a belt. The tensions in the belt on each side of the driver pulley wheel are 400 N and 50 N. The diameters of the driver and driven pulley wheels are 500 mm and 750 mm respectively. The power output from the motor is given by:

$$\text{power output} = (F_2 - F_1)r_x\omega_x$$

$$= (400-50)\left(\frac{500}{2} \times 10^{-3}\right)\left(\frac{1150 \times 2\pi}{60}\right)$$

$$= 10.54 \text{ kW}$$

Hence, the **efficiency of the motor** $= \dfrac{\text{power output}}{\text{power input}}$

$$= \frac{10.54}{15} \times 100 = \textbf{70.27\%}$$

The speed of the driven pulley is obtained from:

$$\frac{r_x}{r_y} = \frac{n_y}{n_x} \text{ from which,}$$

$$\textbf{speed of driven pulley wheel, } n_y = \frac{n_x r_x}{r_y} = \frac{1150 \times 0.25}{\dfrac{0.750}{2}}$$

$$= \textbf{767 rev/min}$$

Flat and V-belts

The ratio of the tensions for a flat belt **when the belt is on the point of slipping** is: $\dfrac{T_1}{T_2} = e^{\mu\theta}$

where μ is the coefficient of friction between belt and pulley,

θ is the angle of lap, in radians (see Figure 28.5), and

e is the exponent ≈ 2.718

For a vee belt as in Figure 28.6, the ratio is: $\dfrac{T_1}{T_2} = e^{\frac{\mu\theta}{\sin\alpha}}$

where α is the half angle of the groove and of the belt.

This gives a much larger ratio than for the flat belt. The V-belt is jammed into its groove and is less likely to slip. Referring to Figure 28.6, the force of friction on each side is μR_N where R_N is the normal (perpendicular) reaction on each side. The triangle of forces shows that $R_N = \dfrac{\frac{R}{2}}{\sin\alpha}$ where R is the resultant reaction. The force of friction giving rise to the difference between T_1 and T_2 is therefore $\mu R_N = \left(\dfrac{\mu}{\sin\alpha}\right) R$

The corresponding force of friction for a flat belt is μR. Comparing the forces of friction for flat and V-belts it can be said that the V-belt is equivalent to a flat belt with a coefficient of friction given by $\dfrac{\mu}{sin\alpha}$

The net force exerted on the belt by a driving pulley, or the belt on a driven pulley is $T_1 - T_2$. The power transmitted by a belt is therefore $(T_1 - T_2)v$, where v is the speed of the belt. If $(T_1 - T_2)$ is measured in newtons and v in metres per second, the power will be Nm/s, i.e. watts (W).

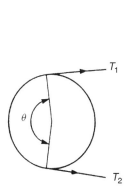

Figure 28.5

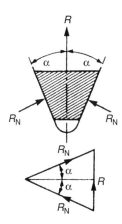

Figure 28.6

The belt speed is given in m/s by ωr, where ω is the angular speed of the pulley (rad/s) and r is its radius (m).

Also, $\omega = \dfrac{2\pi N}{60}$ where N is the pulley speed in rev/min.

It should be noted that:

(a) slipping would be expected to occur first on the pulley having the smaller angle of lap
(b) in practice, care would be taken to ensure that slipping is not likely to occur.

For example, a pulley has vee grooves to take six belts which are required to transmit 125 kW from the pulley when its speed is 1920 rev/min. The angle of the grooves is 45°, the angle of lap is 175° and the effective diameter of the pulley is 265 mm. If the coefficient of friction is 0.32, the lowest possible value for the tension on the tight side of each belt is calculated as follows:

The power transmitted by the six belts is given by: $6(T_1 - T_2)v$ where T_1 and T_2 are the tensions on the tight and slack sides respectively of one belt, and $v \ (= \omega r)$ is the belt speed,

i.e. $\quad 125000 = 6(T_1 - T_2)\left(\dfrac{1920 \times 2\pi}{60}\right)\left(\dfrac{0.265}{2}\right)$

and $\quad T_1 - T_2 = \dfrac{125000 \text{ W}}{6 \times \dfrac{1920 \times 2\pi}{60} \text{ rad/s} \times 0.1325 \text{ m/s}} = 782.01 \text{ N}$

The tension T_1 will have its lowest possible value when the belts are on the point of slipping. For this condition $\dfrac{T_1}{T_2} = e^{\frac{\mu\theta}{\sin\alpha}} = e^{\frac{(0.32)\left(\frac{175\pi}{180}\right)}{\sin 22.5°}} = 12.859$

i.e. $T_1 = 12.859 T_2$ or $T_2 = \dfrac{T_1}{12.859}$

Hence $T_1 - T_2 = T_1\left(1 - \dfrac{1}{12.859}\right) = 0.92223 \, T_1$

The value of $T_1 - T_2$ has been calculated as 782.01 N

The **lowest possible value for** T_1 is therefore $\dfrac{782.01}{0.92223} = \textbf{848.0 N}$

29 Heat Energy

Introduction

Heat is a form of energy and is measured in joules.

Temperature is the degree of hotness or coldness of a substance. Heat and temperature are thus not the same thing. For example, twice the heat energy is needed to boil a full container of water than half a container — that is, different amounts of heat energy are needed to cause an equal rise in the temperature of different amounts of the same substance.

Temperature is measured either (i) on the **Celsius (°C)** scale (formerly Centigrade), where the temperature at which ice melts, i.e. the freezing point of water, is taken as 0°C and the point at which water boils under normal atmospheric pressure is taken as 100°C, or (ii) on the **thermodynamic scale**, in which the unit of temperature is the **kelvin (K).** The kelvin scale uses the same temperature interval as the Celsius scale but as its zero takes the 'absolute zero of temperature' which is at about −273°C. Hence,

$$\text{kelvin temperature} = \text{degree Celsius} + 273$$

i.e.

$$\boxed{\text{K} = (\text{°C}) + 273}$$

For example, 0°C = 273 K, 25°C = 298 K and 100°C = 373 K

The Measurement of Temperature

A **thermometer** is an instrument that measures temperature. Any substance that possesses one or more properties that vary with temperature can be used to measure temperature. These properties include changes in length, area or volume, electrical resistance or in colour. Examples of temperature measuring devices include:

(i) **liquid-in-glass thermometer**, which uses the expansion of a liquid with increase in temperature as its principle of operation

(ii) **thermocouples**, which use the e.m.f. set up when the junction of two dissimilar metals is heated

(iii) **resistance thermometer**, which uses the change in electrical resistance caused by temperature change, and

(iv) **pyrometers**, which are devices for measuring very high temperatures, using the principle that all substances emit radiant energy when hot, the rate of emission depending on their temperature.

Each of these temperature measuring devices, together with others, are described in Chapter 31.

Specific Heat Capacity

The **specific heat capacity** of a substance is the quantity of heat energy required to raise the temperature of 1 kg of the substance by 1°C.

The symbol used for specific heat capacity is c and the units are J/(kg °C) or J/(kg K). (Note that these units may also be written as $J\,kg^{-1}°C^{-1}$ or $J\,kg^{-1}\,K^{-1}$)

Some typical values of specific heat capacity for the range of temperature 0°C to 100°C include:

Water	4190 J/(kg °C),	Ice	2100 J/(kg °C)
Aluminium	950 J/(kg °C),	Copper	390 J/(kg °C)
Iron	500 J/(kg °C),	Lead	130 J/(kg °C)

Hence, to raise the temperature of 1 kg of iron by 1°C requires 500 J of energy, to raise the temperature of 5 kg of iron by 1°C requires (500×5) J of energy, and to raise the temperature of 5 kg of iron by 40°C requires $(500 \times 5 \times 40)$ J of energy, i.e. 100 kJ

In general, the quantity of heat energy, Q, required to raise a mass m kg of a substance with a specific heat capacity c J/(kg °C) from temperature t_1°C to t_2°C is given by:

$$\boxed{Q = mc(t_2 - t_1) \text{ joules}}$$

For example, the quantity of heat required to raise the temperature of 5 kg of water from 0°C to 100°C, assuming the specific heat capacity of water is 4200 J/(kg °C), is given by:

$$\text{quantity of heat energy, } Q = mc(t_2 - t_1)$$

$$= 5 \text{ kg} \times 4200 \text{ J/(kg°C)} \times (100-0)°C$$

$$= 5 \times 4200 \times 100$$

$$= \mathbf{2100000 \ J} \text{ or } \mathbf{2100 \ kJ} \text{ or } \mathbf{2.1 \ MJ}$$

Change of State

A material may exist in any one of three states — solid, liquid or gas. If heat is supplied at a constant rate to some ice initially at, say, −30°C, its temperature rises as shown in Figure 29.1. Initially the temperature increases from −30°C to 0°C as shown by the line AB. It then remains constant at 0°C for the time BC required for the ice to melt into water.

When melting commences the energy gained by continual heating is offset by the energy required for the change of state and the temperature remains constant even though heating is continued. When the ice is completely melted to water, continual heating raises the temperature to 100°C, as shown by CD in Figure 29.1. The water then begins to boil and the temperature again remains constant at 100°C, shown as DE, until all the water has vaporised.

Continual heating raises the temperature of the steam as shown by EF in the region where the steam is termed superheated.

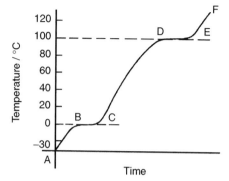

Figure 29.1

Changes of state from solid to liquid or liquid to gas occur without change of temperature and such changes are reversible processes. When heat energy flows to or from a substance and causes a change of temperature, such as between A and B, between C and D and between E and F in Figure 29.1, it is called **sensible heat** (since it can be 'sensed' by a thermometer).

Heat energy which flows to or from a substance while the temperature remains constant, such as between B and C and between D and E in Figure 29.1, is called **latent heat** (latent means concealed or hidden).

Latent Heats of Fusion and Vaporisation

The **specific latent heat of fusion** is the heat required to change 1 kg of a substance from the solid state to the liquid state (or vice versa) at constant temperature.

The **specific latent heat of vaporisation** is the heat required to change 1 kg of a substance from a liquid to a gaseous state (or vice versa) at constant temperature.

The units of the specific latent heats of fusion and vaporisation are J/kg, or more often kJ/kg, and some typical values are shown in Table 29.1

The quantity of heat Q supplied or given out during a change of state is given by:

$$Q = mL$$

where m is the mass in kilograms and L is the specific latent heat.

For example, the heat required to convert 10 kg of ice at 0°C to water at 0°C is given by 10 kg × 335 kJ/kg = 3350 kJ or 3.35 MJ.

Besides changing temperature, the effects of supplying heat to a material can involve changes in dimensions, as well as in colour, state and electrical resistance. Most substances expand when heated and contract when cooled, and there are many practical applications and design implications of thermal movement (see Chapter 30).

Table 29.1

	Latent heat of fusion (kJ/kg)	Melting point (°C)
Mercury	11.8	−39
Lead	22	327
Silver	100	957
Ice	335	0
Aluminium	387	660

	Latent heat of vaporisation (kJ/kg)	Boiling point (°C)
Oxygen	214	−183
Mercury	286	357
Ethyl alcohol	857	79
Water	2257	100

Principle of Operation of a Simple Refrigerator

The boiling point of most liquids may be lowered if the pressure is lowered. In a simple refrigerator a working fluid, such as ammonia or freon, has the pressure acting on it reduced. The resulting lowering of the boiling point causes the liquid to vaporise. In vaporising, the liquid takes in the necessary latent heat from its surroundings, i.e. the freezer, which thus becomes cooled. The vapour is immediately removed by a pump to a condenser that is outside of the cabinet, where it is compressed and changed back into a liquid, giving out latent heat. The cycle is repeated when the liquid is pumped back to the freezer to be vaporised.

Conduction, Convection and Radiation

Heat may be **transferred** from a hot body to a cooler body by one or more of three methods, these being: (a) by **conduction**, (b) by **convection**, or (c) by **radiation**.

Conduction

Conduction is the transfer of heat energy from one part of a body to another (or from one body to another) without the particles of the body moving.

Conduction is associated with solids. For example, if one end of a metal bar is heated, the other end will become hot by conduction. Metals and metallic alloys are good conductors of heat, whereas air, wood, plastic, cork, glass and gases are examples of poor conductors (i.e. they are heat insulators).

Practical applications of conduction include:

(i) A domestic saucepan or dish conducts heat from the source to the contents. Also, since wood and plastic are poor conductors of heat they are used for saucepan handles.

(ii) The metal of a radiator of a central heating system conducts heat from the hot water inside to the air outside.

Convection

Convection is the transfer of heat energy through a substance by the actual movement of the substance itself. Convection occurs in liquids and gases, but not in solids. When heated, a liquid or gas becomes less dense. It then rises and is replaced by a colder liquid or gas and the process repeats. For example, electric kettles and central heating radiators always heat up at the top first.

Examples of convection are:

(i) Natural circulation hot water heating systems depend on the hot water rising by convection to the top of the house and then falling back to the bottom of the house as it cools, releasing the heat energy to warm the house as it does so.

(ii) Convection currents cause air to move and therefore affect climate.

(iii) When a radiator heats the air around it, the hot air rises by convection and cold air moves in to take its place.

(iv) A cooling system in a car radiator relies on convection.

(v) Large electrical transformers dissipate waste heat to an oil tank. The heated oil rises by convection to the top, then sinks through cooling fins, losing heat as it does so.

(vi) In a refrigerator, the cooling unit is situated near the top. The air surrounding the cold pipes become heavier as it contracts and sinks towards the bottom. Warmer, less dense air is pushed upwards and in turn is cooled. A cold convection current is thus created.

Radiation

Radiation is the transfer of heat energy from a hot body to a cooler one by electromagnetic waves. Heat radiation is similar in character to light waves (see Chapter 19) — it travels at the same speed and can pass through a vacuum — except that the frequency of the waves are different. Waves are emitted by a hot body, are transmitted through space (even a vacuum) and are not detected until they fall on another body. Radiation is reflected from shining, polished surfaces but absorbed by dull, black surfaces.

Practical applications of radiation include:

(i) heat from the sun reaching earth
(ii) heat felt by a flame
(iii) cooker grills
(iv) industrial furnaces
(v) infra-red space heaters

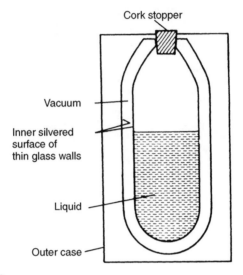

Figure 29.2

Vacuum Flask

A cross-section of a typical vacuum flask is shown in Figure 29.2 and is seen to be a double-walled bottle with a vacuum space between them, the whole supported in a protective outer case.

Very little heat can be transferred by conduction because of the vacuum space and the cork stopper (cork is a bad conductor of heat). Also, because of the vacuum space, no convection is possible. Radiation is minimised by silvering the two glass surfaces (radiation is reflected off shining surfaces).

Thus a vacuum flask is an example of prevention of all three types of heat transfer and is therefore able to keep hot liquids hot and cold liquids cold.

Use of Insulation in Conserving Fuel

Fuel used for heating a building is becoming increasingly expensive. By the careful use of insulation, heat can be retained in a building for longer periods and the cost of heating thus minimised.

(i) Since convection causes hot air to rise it is important to insulate the roof space, which is probably the greatest source of heat loss in the home. This can be achieved by laying fibre-glass between the wooden joists in the roof space.

(ii) Glass is a poor conductor of heat. However, large losses can occur through thin panes of glass and such losses can be reduced by using double-glazing. Two sheets of glass, separated by air, are used. Air is a very good insulator but the air space must not be too large otherwise convection currents can occur which would carry heat across the space.

(iii) Hot water tanks should be lagged to prevent conduction and convection of heat to the surrounding air.

(iv) Brick, concrete, plaster and wood are all poor conductors of heat. A house is made from two walls with an air gap between them. Air is a poor conductor and trapped air minimises losses through the wall. Heat losses through the walls can be prevented almost completely by using cavity wall insulation, i.e. plastic-foam.

Besides changing temperature, the effects of supplying heat to a material can involve changes in dimensions, as well as in colour, state and electrical resistance.

Most substances expand when heated and contract when cooled, and there are many practical applications and design implications of thermal movement as explained in Chapter 30 following.

30 Thermal Expansion

Introduction

When heat is applied to most materials, **expansion** occurs in all directions. Conversely, if heat energy is removed from a material (i.e. the material is cooled) **contraction** occurs in all directions. The effects of expansion and contraction each depend on the **change of temperature** of the material.

Practical Applications of Thermal Expansion

Some practical applications where expansion and contraction of solid materials must be allowed for include:

(i) Overhead electrical transmission lines are hung so that they are slack in summer, otherwise their contraction in winter may snap the conductors or bring down pylons.

(ii) Gaps need to be left in lengths of railway lines to prevent buckling in hot weather (except where these are continuously welded).

(iii) Ends of large bridges are often supported on rollers to allow them to expand and contract freely.

(iv) Fitting a metal collar to a shaft or a steel tyre to a wheel is often achieved by first heating them so that they expand, fitting them in position, and then cooling them so that the contraction holds them firmly in place; this is known as a 'shrink-fit'. By a similar method hot rivets are used for joining metal sheets.

(v) The amount of expansion varies with different materials. Figure 30.1(a) shows a bimetallic strip at room temperature (i.e. two different strips of metal riveted together). When heated, brass expands more than steel, and since the two metals are riveted together the bimetallic strip is forced into an arc as shown in Figure 30.1(b). Such a movement can be arranged to make or break an electric circuit and bimetallic strips are used, in particular, in thermostats (which are temperature-operated switches) used to control central heating systems, cookers, refrigerators, toasters, irons, hot water and alarm systems.

(vi) Motor engines use the rapid expansion of heated gases to force a piston to move.

(vii) Designers must predict, and allow for, the expansion of steel pipes in a steam-raising plant so as to avoid damage and consequent danger to health.

Expansion and Contraction of Water

Water is a liquid that at low temperature displays an unusual effect. If cooled, contraction occurs until, at about 4°C, the volume is at a minimum. As the

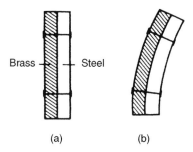

Brass   Steel

(a)   (b)

Figure 30.1

temperature is further decreased from 4°C to 0°C expansion occurs, i.e. the volume increases. When ice is formed, considerable expansion occurs and it is this expansion that often causes frozen water pipes to burst.

A practical application of the expansion of a liquid is with thermometers, where the expansion of a liquid, such as mercury or alcohol, is used to measure temperature.

Coefficient of Linear Expansion

The amount by which unit length of a material expands when the temperature is raised one degree is called the **coefficient of linear expansion** of the material and is represented by α (Greek alpha).

The units of the coefficient of linear expansion are m/(mK), although it is usually quoted as just /K or K^{-1}. For example, copper has a coefficient of linear expansion value of 17×10^{-6} K^{-1}, which means that a 1 m long bar of copper expands by 0.000017 m if its temperature is increased by 1 K (or 1°C). If a 6 m long bar of copper is subjected to a temperature rise of 25 K then the bar will expand by $(6 \times 0.000017 \times 25)$ m, i.e. 0.00255 m or 2.55 mm. (Since the kelvin scale uses the same temperature interval as the Celsius scale, a **change** of temperature of, say, 50°C, is the same as a change of temperature of 50 K).

If a material, initially of length l_1 and at a temperature of t_1 and having a coefficient of linear expansion α, has its temperature increased to t_2, then the new length l_2 of the material is given by:

$$l_2 = l_1 [1 + \alpha(t_2 - t_1)]$$ (1)

Some typical values for the coefficient of linear expansion include:

Aluminium	23×10^{-6} K^{-1}	Brass	18×10^{-6} K^{-1}
Concrete	12×10^{-6} K^{-1}	Copper	17×10^{-6} K^{-1}
Gold	14×10^{-6} K^{-1}	Invar (nickel-	
Iron	$11–12 \times 10^{-6}$ K^{-1}	steel alloy)	0.9×10^{-6} K^{-1}
Steel	$15–16 \times 10^{-6}$ K^{-1}	Nylon	100×10^{-6} K^{-1}
Zinc	31×10^{-6} K^{-1}	Tungsten	4.5×10^{-6} K^{-1}

For example, the copper tubes in a boiler are 4.20 m long at a temperature of 20°C. Then, when surrounded only by feed water at 10°C, the final length of the tubes, l_2, is given by:

$$l_2 = l_1[1 + \alpha(t_2 - t_1)] = 4.20[1 + (17 \times 10^{-6})(10 - 20)]$$

$$= 4.20[1 - 0.00017] = \mathbf{4.1993 \ m}$$

i.e. **the tube contracts by 0.7 mm** when the temperature decreases from 20°C to 10°C

When the boiler is operating and the mean temperature of the tubes is, say, 320°C, the final length of the tubes, l_2, is given by:

$$l_2 = l_1[1 + \alpha(t_2 - t_1)] = 4.20[1 + (17 \times 10^{-6})(320 - 20)]$$

$$= 4.20[1 + 0.0051] = \mathbf{4.2214 \ m}$$

i.e. **the tube extends by 21.4 mm** when the temperature rises from 20°C to 320°C.

Coefficient of Superficial Expansion

The amount by which unit area of a material increases when the temperature is raised by one degree is called the **coefficient of superficial (i.e. area) expansion** and is represented by β (Greek beta).

If a material having an initial surface area A_1 at temperature t_1 and having a coefficient of superficial expansion β, has its temperature increased to t_2, then the new surface area A_2 of the material is given by:

$$\boxed{A_2 = A_1 \ [1 + \beta(t_2 - t_1)]} \tag{2}$$

It may be shown that the coefficient of superficial expansion is twice the coefficient of linear expansion, i.e. $\beta = 2\alpha$, to a very close approximation.

Coefficient of Cubic Expansion

The amount by which unit volume of a material increases for a one degree rise of temperature is called the **coefficient of cubic (or volumetric) expansion** and is represented by γ (Greek gamma).

If a material having an initial volume V_1 at temperature t_1 and having a coefficient of cubic expansion γ, has its temperature raised to t_2, then the new volume V_2 of the material is given by:

$$\boxed{V_2 = V_1[1 + \gamma(t_2 - t_1)]} \tag{3}$$

It may be shown that the coefficient of cubic expansion is three times the coefficient of linear expansion, i.e. $\gamma = 3\alpha$, to a very close approximation.

A liquid has no definite shape and only its cubic or volumetric expansion need be considered. Thus with expansions in liquids, equation (3) is used.

Some **typical values** for the coefficient of cubic expansion measured at 20°C (i.e. 293 K) include:

| Ethyl alcohol | 1.1×10^{-3} K^{-1} | Mercury | 1.82×10^{-4} K^{-1} |
| Paraffin oil | 9×10^{-2} K^{-1} | Water | 2.1×10^{-4} K^{-1} |

The coefficient of cubic expansion γ is only constant over a limited range of temperature.

For example, mercury contained in a thermometer has a volume of 476 mm^3 at 15°C. When the volume is, say, 478 mm^3 then:

  final volume, $V_2 = V_1[1 + \gamma(t_2 - t_1)]$, from equation (3)

i.e. $V_2 = V_1 + V_1\gamma(t_2 - t_1)$, from which

$$(t_2 - t_1) = \frac{V_2 - V_1}{V_1\gamma} = \frac{478 - 476}{(476)(1.82 \times 10^{-4})}) = 23.09°C$$

Hence $t_2 = 23.09 + 15 = 38.09°C$

Hence the temperature at which the volume of mercury is 478 mm^3 is **38.09°C**

31 The Measurement of Temperature

Introduction

A change in temperature of a substance can often result in a change in one or more of its physical properties. Thus, although temperature cannot be measured directly, its effects can be measured. Some properties of substances used to determine changes in temperature include changes in dimensions, electrical resistance, state, type and volume of radiation, and colour.

Temperature measuring devices available are many and varied. Those described in the following paragraphs are those most often used in science and industry.

Liquid-in-glass Thermometer

A **liquid-in-glass thermometer** uses the expansion of a liquid with increase in temperature as its principle of operation.

Construction

A typical liquid-in-glass thermometer is shown in Figure 31.1 and consists of a sealed stem of uniform small-bore tubing, called a capillary tube, made of glass, with a cylindrical glass bulb formed at one end. The bulb and part of the stem are filled with a liquid such as mercury or alcohol and the remaining part of the tube is evacuated. Etching graduations on the stem forms a temperature scale. A safety reservoir is usually provided, into which the liquid can expand without bursting the glass if the temperature is raised beyond the upper limit of the scale.

Principle of operation

The operation of a liquid-in-glass thermometer depends on the liquid expanding with increase in temperature and contracting with decrease in temperature. The position of the end of the column of liquid in the tube is a measure of the temperature of the liquid in the bulb shown as 15 °C in Figure 31.1, which is about room temperature. Two fixed points are needed to calibrate

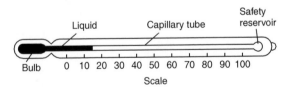

Figure 31.1

the thermometer, with the interval between these points being divided into 'degrees'. In the first thermometer, made by Celsius, the fixed points chosen were the temperature of melting ice (0°C) and that of boiling water at standard atmospheric pressure (100°C), in each case the blank stem being marked at the liquid level. The distance between these two points, called the fundamental interval, was divided into 100 equal parts, each equivalent to 1°C, thus forming the scale.

The **clinical thermometer**, with a limited scale around body temperature, the **maximum and/or minimum thermometer**, recording the maximum day temperature and minimum night temperature, and the **Beckman thermometer**, which is used only in accurate measurement of temperature change and has no fixed points, are particular types of liquid-in-glass thermometer which all operate on the same principle.

Advantages

The liquid-in-glass thermometer is simple in construction, relatively inexpensive, easy to use and portable, and is the most widely used method of temperature measurement having industrial, chemical, clinical and meteorological applications.

Disadvantages

Liquid-in-glass thermometers tend to be fragile and hence easily broken, can only be used where the liquid column is visible, cannot be used for surface temperature measurements, cannot be read from a distance and are unsuitable for high temperature measurements.

Advantages of mercury

The use of mercury in a thermometer has many advantages, for mercury:

(i) is clearly visible,
(ii) has a fairly uniform rate of expansion,
(iii) is readily obtainable in the pure state,
(iv) does not 'wet' the glass,
(v) is a good conductor of heat.

Mercury has a freezing point of −39°C and cannot be used in a thermometer below this temperature. Its boiling point is 357°C but before this temperature is reached some distillation of the mercury occurs if the space above the mercury is a vacuum. To prevent this, and to extend the upper temperature limits to over 500°C, an inert gas such as nitrogen under pressure is used to fill the remainder of the capillary tube. Alcohol, often dyed red to be seen in the capillary tube, is considerably cheaper than mercury and has a freezing point of −113°C, which is considerably lower than for mercury. However it has a low boiling point at about 79°C.

Errors

Typical errors in liquid-in-glass thermometers may occur due to:

(i) the slow cooling rate of glass,
(ii) incorrect positioning of the thermometer,

(iii) a delay in the thermometer becoming steady (i.e. slow response time),
(iv) non-uniformity of the bore of the capillary tube, which means that equal intervals marked on the stem do not correspond to equal temperature intervals.

Thermocouples

Thermocouples use the e.m.f. set up when the junction of two dissimilar metals is heated.

Principle of operation

At the junction between two different metals, say, copper and constantan, there exists a difference in electrical potential, which varies with the temperature of the junction. This is known as the 'thermo-electric effect'. If the circuit is completed with a second junction at a different temperature, a current will flow round the circuit. This principle is used in the thermocouple. Two different metal conductors having their ends twisted together are shown in Figure 31.2. If the two junctions are at different temperatures, a current I flows round the circuit.

The deflection on the galvanometer G depends on the difference in temperature between junctions X and Y and is caused by the difference between voltages V_x and V_y. The higher temperature junction is usually called the 'hot junction' and the lower temperature junction the 'cold junction'. If the cold junction is kept at a constant known temperature, the galvanometer can be calibrated to indicate the temperature of the hot junction directly. The cold junction is then known as the reference junction.

In many instrumentation situations, the measuring instrument needs to be located far from the point at which the measurements are to be made. Extension leads are then used, usually made of the same material as the thermocouple but of smaller gauge. The reference junction is then effectively moved to their ends. The thermocouple is used by positioning the hot junction where the temperature is required. The meter will indicate the temperature of the hot junction only if the reference junction is at 0°C for:

(temperature of hot junction) = (temperature of the cold junction)

+ (temperature difference)

In a laboratory the reference junction is often placed in melting ice, but in industry it is often positioned in a thermostatically controlled oven or buried underground where the temperature is constant.

Construction

Thermocouple junctions are made by twisting together two wires of dissimilar metals before welding them. The construction of a typical copper-constantan thermocouple for industrial use is shown in Figure 31.3. Apart from the actual

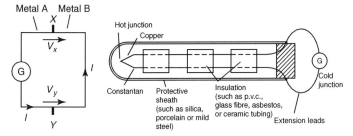

Figure 31.2   **Figure 31.3**

junction the two conductors used must be insulated electrically from each other with appropriate insulation and is shown in Figure 31.3 as twin-holed tubing. The wires and insulation are usually inserted into a sheath for protection from environments in which they might be damaged or corroded.

Applications

A copper-constantan thermocouple can measure temperature from $-250°C$ up to about $400°C$, and is used typically with boiler flue gases, food processing and with sub-zero temperature measurement. An iron-constantan thermocouple can measure temperature from $-200°C$ to about $850°C$, and is used typically in paper and pulp mills, re-heat and annealing furnaces and in chemical reactors. A chromel-alumel thermocouple can measure temperatures from $-200°C$ to about $1100°C$ and is used typically with blast furnace gases, brick kilns and in glass manufacture. For the measurement of temperatures above $1100°C$ radiation pyrometers are normally used. However, thermocouples are available made of platinum-platinum/rhodium, capable of measuring temperatures up to $1400°C$, or tungsten-molybdenum which can measure up to $2600°C$.

Advantages

A thermocouple:

  (i) has a very simple, relatively inexpensive construction,
 (ii) can be made very small and compact,
(iii) is robust,
 (iv) is easily replaced if damaged,
  (v) has a small response time,
 (vi) can be used at a distance from the actual measuring instrument and is thus ideal for use with automatic and remote-control systems.

Sources of error

Sources of error in the thermocouple that are difficult to overcome include:

  (i) voltage drops in leads and junctions,
 (ii) possible variations in the temperature of the cold junction,

(iii) stray thermoelectric effects, which are caused by the addition of further metals into the 'ideal' two-metal thermocouple circuit. Additional leads are frequently necessary for extension leads or voltmeter terminal connections.

A thermocouple may be used with a battery- or mains-operated electronic thermometer instead of a millivoltmeter. These devices amplify the small e.m.f.'s from the thermocouple before feeding them to a multi-range voltmeter calibrated directly with temperature scales. These devices have great accuracy and are almost unaffected by voltage drops in the leads and junctions.

Resistance Thermometers

Resistance thermometers use the change in electrical resistance caused by temperature change.

Construction

Resistance thermometers are made in a variety of sizes, shapes and forms depending on the application for which they are designed. A typical resistance thermometer is shown diagrammatically in Figure 31.4. The most common metal used for the coil in such thermometers is platinum even though its sensitivity is not as high as other metals such as copper and nickel. However, platinum is a very stable metal and provides reproducible results in a resistance thermometer. A platinum resistance thermometer is often used as a calibrating device. Since platinum is expensive, connecting leads of another metal, usually copper, are used with the thermometer to connect it to a measuring circuit. The platinum and the connecting leads are shown joined at A and B in Figure 31.4, although sometimes this junction may be made outside of the sheath. However, these leads often come into close contact with the heat source that can introduce errors into the measurements. Including a pair of identical leads, called dummy leads, which experience the same temperature change as the extension leads, may eliminate these.

Principle of operation

With most metals a rise in temperature causes an increase in electrical resistance, and since resistance can be measured accurately this property can be

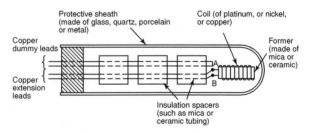

Figure 31.4

used to measure temperature. If the resistance of a length of wire at 0°C is R_0, and its resistance at θ°C is R_θ, then $R_\theta = R_0(1 + \alpha\theta)$, where α is the temperature coefficient of resistance of the material (see Chapter 41).

Rearranging gives: $$\text{temperature, } \theta = \frac{R_\theta - R_0}{\alpha R_0}$$

Values of R_0 and α may be determined experimentally or obtained from existing data. Thus, if R_θ can be measured, temperature θ can be calculated. This is the principle of operation of a resistance thermometer. Although a sensitive ohmmeter can be used to measure R_θ, for more accurate determinations a **Wheatstone bridge** circuit is used as shown in Figure 31.5 (see also chapter 50). This circuit compares an unknown resistance R_θ with others of known values, R_1 and R_2 being fixed values and R_3 being variable. Galvanometer G is a sensitive centre-zero microammeter. R_3 is varied until zero deflection is obtained on the galvanometer, i.e. no current flows through G and the bridge is said to be 'balanced'.

At balance: $R_2 R_\theta = R_1 R_3$

from which, $$R_\theta = \frac{R_1 R_3}{R_2}$$

and if R_1 and R_2 are of equal value, then $R_\theta = R_3$

A resistance thermometer may be connected between points A and B in Figure 31.5 and its resistance R_θ at any temperature θ accurately measured. Dummy leads included in arm BC help to eliminate errors caused by the extension leads which are normally necessary in such a thermometer.

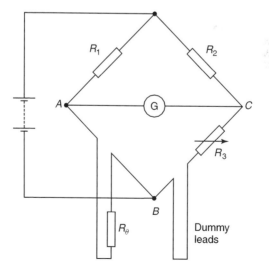

Figure 31.5

Limitations

Resistance thermometers using a nickel coil are used mainly in the range $-100°C$ to $300°C$, whereas platinum resistance thermometers are capable of measuring with greater accuracy temperatures in the range $-200°C$ to about $800°C$. This upper range may be extended to about $1500°C$ if high melting point materials are used for the sheath and coil construction.

Advantages and disadvantages of a platinum coil

Platinum is commonly used in resistance thermometers since it is chemically inert, i.e. unreactive, resists corrosion and oxidation and has a high melting point of $1769°C$. A disadvantage of platinum is its slow response to temperature variation.

Applications

Platinum resistance thermometers may be used as calibrating devices or in applications such as heat-treating and annealing processes and can be adapted easily for use with automatic recording or control systems. Resistance thermometers tend to be fragile and easily damaged especially when subjected to excessive vibration or shock.

Thermistors

A thermistor is a semi-conducting material such as mixtures of oxides of copper, manganese, cobalt, etc. in the form of a fused bead connected to two leads. As its temperature is increased its resistance rapidly decreases. Typical resistance/temperature curves for a thermistor and common metals are shown in Figure 31.6. The resistance of a typical thermistor can vary from 400 Ω at 0°C to 100 Ω at 140°C

Advantages

The main advantages of a thermistor are its high sensitivity and small size. It provides an inexpensive method of measuring and detecting small changes in temperature.

Pyrometers

A pyrometer is a device for measuring very high temperatures and uses the principle that all substances emit radiant energy when hot, the rate of emission depending on their temperature. The measurement of thermal radiation is therefore a convenient method of determining the temperature of hot sources and is particularly useful in industrial processes. There are two main types of pyrometer, namely the total radiation pyrometer and the optical pyrometer.

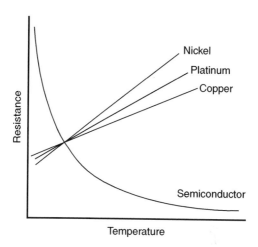

Figure 31.6

Pyrometers are very convenient instruments since they can be used at a safe and comfortable distance from the hot source. Thus applications of pyrometers are found in measuring the temperature of molten metals, the interiors of furnaces or the interiors of volcanoes. Total radiation pyrometers can also be used in conjunction with devices which record and control temperature continuously.

Total radiation pyrometer

A typical arrangement of a total radiation pyrometer is shown in Figure 31.7. Radiant energy from a hot source, such as a furnace, is focused on to the hot junction of a thermocouple after reflection from a concave mirror. The temperature rise recorded by the thermocouple depends on the amount of radiant energy received, which in turn depends on the temperature of the hot source. The galvanometer G shown connected to the thermocouple records the

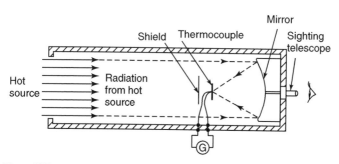

Figure 31.7

current that results from the e.m.f. developed and may be calibrated to give a direct reading of the temperature of the hot source. The thermocouple is protected from direct radiation by a shield as shown and the hot source may be viewed through the sighting telescope. For greater sensitivity, a **thermopile** may be used, a thermopile being a number of thermocouples connected in series. Total radiation pyrometers are used to measure temperature in the range 700°C to 2000°C

Optical pyrometers

When the temperature of an object is raised sufficiently two visual effects occur; the object appears brighter and there is a change in colour of the light emitted. These effects are used in the optical pyrometer where a comparison or matching is made between the brightness of the glowing hot source and the light from a filament of known temperature.

The most frequently used optical pyrometer is the **disappearing filament pyrometer** and a typical arrangement is shown in Figure 31.8. A filament lamp is built into a telescope arrangement that receives radiation from a hot source, an image of which is seen through an eyepiece. A red filter is incorporated as a protection to the eye.

A variable resistor controls the current flowing through the lamp. As the current is increased the temperature of the filament increases and its colour changes. When viewed through the eyepiece the filament of the lamp appears superimposed on the image of the radiant energy from the hot source. The current is varied until the filament glows as brightly as the background. It will then merge into the background and seem to disappear. The current required to achieve this is a measure of the temperature of the hot source and the ammeter can be calibrated to read the temperature directly. Optical pyrometers may be used to measure temperatures up to, and even in excess of, 3000°C

Advantages of pyrometers

(i) There is no practical limit to the temperature that a pyrometer can measure.

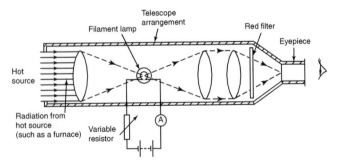

Figure 31.8

(ii) A pyrometer need not be brought directly into the hot zone and so is free from the effects of heat and chemical attack that can often cause other measuring devices to deteriorate in use.

(iii) Very fast rates of change of temperature can be followed by a pyrometer.

(iv) The temperature of moving bodies can be measured.

(v) The lens system makes the pyrometer virtually independent of its distance from the source.

Disadvantages of pyrometers

(i) A pyrometer is often more expensive than other temperature measuring devices.

(ii) A direct view of the heat process is necessary.

(iii) Manual adjustment is necessary.

(iv) A reasonable amount of skill and care is required in calibrating and using a pyrometer. For each new measuring situation the pyrometer must be re-calibrated.

(v) The temperature of the surroundings may affect the reading of the pyrometer and such errors are difficult to eliminate.

Temperature Indicating Paints and Crayons

Temperature indicating paints contain substances that change their colour when heated to certain temperatures. This change is usually due to chemical decomposition, such as loss of water, in which the change in colour of the paint after having reached the particular temperature will be a permanent one. However, in some types the original colour returns after cooling. Temperature indicating paints are used where the temperature of inaccessible parts of apparatus and machines is required. They are particularly useful in heat-treatment processes where the temperature of the component needs to be known before a quenching operation. There are several such paints available and most have only a small temperature range so that different paints have to be used for different temperatures. The usual range of temperatures covered by these paints is from about 30°C to 700°C.

Temperature sensitive crayons consist of fusible solids compressed into the form of a stick. The melting point of such crayons is used to determine when a given temperature has been reached. The crayons are simple to use but indicate a single temperature only, i.e. its melting point temperature. There are over 100 different crayons available, each covering a particular range of temperature. Crayons are available for temperatures within the range of 50°C to 1400°C. Such crayons are used in metallurgical applications such as preheating before welding, hardening, annealing or tempering, or in monitoring the temperature of critical parts of machines or for checking mould temperatures in the rubber and plastics industries.

Bimetallic thermometers

Bimetallic thermometers depend on the expansion of metal strips that operate an indicating pointer. Two thin metal strips of differing thermal expansion are welded or riveted together and the curvature of the bimetallic strip changes with temperature change. For greater sensitivity the strips may be coiled into a flat spiral or helix, one end being fixed and the other being made to rotate a pointer over a scale. Bimetallic thermometers are useful for alarm and over-temperature applications where extreme accuracy is not essential. If the whole is placed in a sheath, protection from corrosive environments is achieved but with a reduction in response characteristics. The normal upper limit of temperature measurement by this thermometer is about 200°C, although with special metals the range can be extended to about 400°C.

Mercury-in-steel Thermometer

The **mercury-in-steel thermometer** is an extension of the principle of the mercury-in-glass thermometer. Mercury in a steel bulb expands via a small bore capillary tube into a pressure indicating device, say a Bourdon gauge, the position of the pointer indicating the amount of expansion and thus the temperature. The advantages of this instrument are that it is robust and, by increasing the length of the capillary tube, the gauge can be placed some distance from the bulb and can thus be used to monitor temperatures in positions that are inaccessible to the liquid-in-glass thermometer. Such thermometers may be used to measure temperatures up to 600°C.

Gas Thermometers

The gas thermometer consists of a flexible U-tube of mercury connected by a capillary tube to a vessel containing gas. The change in the volume of a fixed mass of gas at constant pressure, or the change in pressure of a fixed mass of gas at constant volume, may be used to measure temperature. This thermometer is cumbersome and rarely used to measure temperature directly, but it is often used as a standard with which to calibrate other types of thermometer. With pure hydrogen the range of the instrument extends from −240°C to 1500°C and measurements can be made with extreme accuracy.

32 Pressure in Fluids

Pressure

The **pressure** acting on a surface is defined as the perpendicular force per unit area of surface. The unit of pressure is the **pascal, Pa**, where 1 pascal is equal to 1 newton per square metre. Thus

$$\text{pressure, } p = \frac{F}{A} \text{ pascals}$$

where F is the force in newtons acting at right angles to a surface of area A square metres.

For example, when a force of 20 N acts uniformly over, and perpendicular to, an area of 4 m², then the pressure on the area, p, is given by:

$$p = \frac{20 \text{ N}}{4 \text{ m}^2} = 5 \text{ Pa}$$

Fluid Pressure

A fluid is either a liquid or a gas and there are four basic factors governing the pressure within fluids.

(a) The pressure, at a given depth, in a fluid is equal in all directions, as shown in Figure 32.1(a).
(b) The pressure at a given depth in a fluid is independent of the shape of the container in which the fluid is held. In Figure 32.1(b), the pressure at X is the same as the pressure at Y.
(c) Pressure acts at right angles to the surface containing the fluid. In Figure 32.1(c), the pressures at points A to F all act at right angles to the container.
(d) When a pressure is applied to a fluid, this pressure is transmitted equally in all directions. In Figure 32.1(d), if the mass of the fluid is neglected, the pressures at points A to D are all the same.

The pressure, p, at any point in a fluid depends on three factors:

(i) the density of the fluid, ρ, in kg/m³
(ii) the gravitational acceleration, g, taken as approximately 9.81 m/s² (or the gravitational field force in N/kg), and
(iii) the height of fluid vertically above the point, h metres.

The relationship connecting these quantities is: $\boxed{p = \rho g h \text{ pascals}}$

150

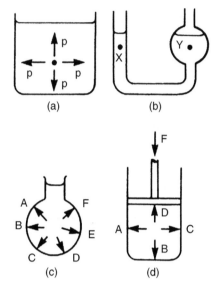

Figure 32.1

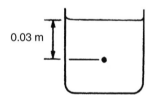

Figure 32.2

For example, when the container shown in Figure 32.2 is filled with water of density 1000 kg/m³, the pressure due to the water at a depth of 0.03 m below the surface is given by: $p = \rho g h = (1000 \times 9.81 \times 0.03) = \mathbf{294.3\ Pa}$

Atmospheric Pressure

The air above the earth's surface is a fluid, having a density, ρ, which varies from approximately 1.225 kg/m³ at sea level to zero in outer space. Since $p = \rho g h$, where height h is several thousands of metres, the air exerts a pressure on all points on the earth's surface. This pressure, called **atmospheric pressure**, has a value of approximately 101 kilopascals. Two terms are commonly used when measuring pressures:

(a) **absolute pressure**, meaning the pressure above that of an absolute vacuum (which is zero pressure), and

(b) **gauge pressure**, meaning the pressure above that normally present due to the atmosphere.

Thus: **absolute pressure = atmospheric pressure + gauge pressure**

Thus, a gauge pressure of 50 kPa is equivalent to an absolute pressure of $(100 + 50)$ kPa, i.e. 150 kPa, since the atmospheric pressure is approximately 100 kPa.

For example, the absolute pressure at a point on a submarine, at a depth of 30 m below the surface of the sea, when the atmospheric pressure is 101 kPa (taking the density of sea water as 1030 kg/m^3 and the gravitational acceleration as 9.81 m/s^2) is calculated as follows:

The pressure due to the sea, that is, the gauge pressure (p_g) is given by:

$$p_g = \rho g h = 1030 \times 9.81 \times 30 = 303\,129 \text{ Pa} = 303.13 \text{ kPa}$$

Absolute pressure = atmospheric pressure + gauge pressure

$$= (101 + 303.13) \text{ kPa} = 404.13 \text{ kPa}$$

i.e. **the absolute pressure at a depth of 30 m is 404.13 kPa**

Another unit of pressure, used in particular for atmospheric pressure, is the bar, where **1 bar $= 10^5$ N/m^2 $= 100$ kPa**

For more detail on atmospheric, absolute and gauge pressure, see Chapter 33 on the measurement of pressure.

Archimedes' Principle

Archimedes' principle states that:

If a solid body floats, or is submerged, in a liquid, the liquid exerts an upthrust on the body equal to the gravitational force on the liquid displaced by the body

In other words, if a solid body is immersed in a liquid, the apparent loss of weight is equal to the weight of liquid displaced. If V is the volume of the body below the surface of the liquid, then the apparent loss of weight W is given by:

$$W = V\omega = V\rho g$$

where ω is the specific weight (i.e. weight per unit volume) and ρ is the density. If a body floats on the surface of a liquid all of its weight appears to have been lost. The weight of liquid displaced is equal to the weight of the floating body.

For example, a body weighs 2.760 N in air and 1.925 N when completely immersed in water of density 1000 kg/m^3.

The apparent loss of weight is: 2.760 N − 1.925 N = 0.835 N

This is the weight of water displaced, i.e. $V\rho g$, where V is the volume of the body and ρ is the density of water.

Thus, $0.835 \text{ N} = (V)(1000 \text{ kg/m}^3)(9.81 \text{ m/s}^2)$

from which, $V = \dfrac{0.835}{9810} = \mathbf{8.512 \times 10^{-5} \ m^3}$

The **density of the body** $= \dfrac{\text{mass}}{\text{volume}} = \dfrac{\text{weight}}{gV}$

$$= \dfrac{2.760 \text{ N}}{(9.81 \text{ m/s}^2)(8.512 \times 10^{-5} \text{ m}^3)}$$

$$= \mathbf{3305 \ kg/m^3 = 3.305 \ tonne/m^3}$$

The **relative density of the body** $= \dfrac{\text{density}}{\text{density of water}}$

$$= \dfrac{3305 \text{ kg/m}^3}{1000 \text{ kg/m}^3} = \mathbf{3.305}$$

Measurement of Pressure

There are various ways of measuring pressure, and these include by:

(a) barometers,
(b) manometers,
(c) pressure gauges, and
(d) vacuum gauges.

The construction and principle of operation of each of these devices are described in Chapter 33.

Let us look briefly at just one of these instruments the manometer. A manometer is a device used for measuring relatively small pressures, either above or below atmospheric pressure. A simple U-tube manometer is shown in Figure 32.3. Pressure p acting in, say, a gas main, pushes the liquid in the

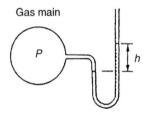

Figure 32.3

U-tube until equilibrium is obtained. At equilibrium:

pressure in gas main, p = (atmospheric pressure, p_a) + (pressure due to the column of liquid, ρgh)

i.e. $p = p_a + \rho gh$

For example, if the atmospheric pressure, p_a is 101 kPa, the liquid in the U-tube is water of density 1000 kg/m³ and height, h is 300 mm, then:

$$\text{absolute gas pressure} = (101\,000 + 1000 \times 9.81 \times 0.3)\ \text{Pa}$$

$$= (101\,000 + 2943)\ \text{Pa}$$

$$= 103\,943\ \text{Pa} = 103.94\ \text{kPa}$$

The gauge pressure of the gas is 2.94 kPa.

By filling the U-tube with a more dense liquid, say, mercury having a density of 13 600 kg/m³, for a given height of U-tube, the pressure which can be measured is increased by a factor of 13.6.

More on Hydrostatic Pressure

The pressure p at the base of the tank shown in Figure 32.4(a) is:

$$\boxed{p = \rho gh = wh}$$

where w is the specific weight, i.e. the weight per unit volume, its unit being N/m³.

The pressure increases to this value uniformly from zero at the free surface. The pressure variation is shown in Figure 32.4(b).

At any intermediate depth x the pressure is: $p_x = \rho gx = wx$

It may be shown that the average pressure on any wetted plane surface is the pressure at the centroid, the centre of area. The sloping sides of the tank

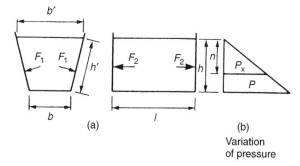

(a)   (b)
Variation
of pressure

Figure 32.4

of Figure 32.4 are rectangular and therefore the average pressure on them is the pressure at half depth: $\dfrac{p}{2} = \dfrac{\rho gh}{2} = \dfrac{wh}{2}$

The force on a sloping side is the product of this average pressure and the area of the sloping side: $F_1 = \dfrac{\rho gh}{2} \times lh' = \dfrac{whlh'}{2}$ where h' is the slant height. The pressure and consequently the force F_1 are at right angles to the sloping side as shown in Figure 32.4.

The average pressure on the vertical trapezoidal ends of the tank is not the pressure at half depth ~~it~~ is rather higher. The depth of the centroid is given by: $\dfrac{h}{3}\left(\dfrac{2b + b'}{b + b'}\right)$ (see Figure 32.4)

The average pressure on the end is therefore:

$$\frac{\rho gh}{3}\left(\frac{2b + b'}{b + b'}\right) = \frac{wh}{3}\left(\frac{2b + b'}{b + b'}\right)$$

The forces F_2 on the vertical trapezoidal ends of the tank are horizontal forces given by the product of this average pressure and the area of the trapezium, $\left(\dfrac{b + b'}{2}\right)h$

The force on the base of the tank is $(\rho gh) \times$ (area of base) $= \rho ghlb = whlb$

For a tank with vertical sides this is the weight of liquid in the tank.

In any vessel containing homogeneous liquid at rest and in continuous contact, the pressure must be the same at all points at the same level. In a U-tube, as shown in Figure 32.5, with the liquid in the lower part at rest, the pressure must be the same on both sides for all levels up to X_1X_2. The pressure at X_1, however, is greater than the pressure at Y_1 by the amount $p_1 = w_1h = \rho_1gh$, where w_1, ρ_1 are the specific weight and density respectively of the liquid, or gas, between X_1 and Y_1.

Similarly, the pressure at X_2 is greater than the pressure at Y_2 by an amount given by $p_2 = w_2h = \rho_2gh$, where w_2 and ρ_2 are the specific weight and density respectively of the liquid in the bottom of the U-tube.

For practical reasons, ρ_2 must be greater than ρ_1 and the pressure at Y_1 will exceed that at Y_2 by: $p_2 - p_1 = (w_2 - w_1)h = (\rho_2 - \rho_1)gh$

If the upper limits of the U-tube contain air or any other gas or gas mixture, w_1 and ρ_1 can reasonably be ignored, giving: $p_2 - p_1 = w_2h = \rho_2\,gh$

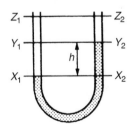

Figure 32.5

If the upper limbs contain a lighter liquid, then the pressure difference may be expressed as: $p_2 - p_1 = \rho_1(d-1)gh = (d-1)w_1h$ where $d = \dfrac{\rho_2}{\rho_1}$

A common arrangement is mercury and water, in which case d is the relative density of mercury, approximately 13.6. This gives: $p_2 - p_1 = 12.6\rho gh = 12.6wh$, ρ and w being respectively the density and specific weight of water. The pressure difference at Z_1Z_2 will be the same as Y_1Y_2 if both limbs contain the same liquid between these levels. This follows from the fact that the pressure increase from Z_1 to Y_1 is the same as the increase from Z_2 to Y_2.

For example, the pressure difference between two points on a horizontal pipe carrying water and running full is to be indicated on a U-tube below the pipe. The bottom of the U-tube contains mercury; the rest is filled with water. Taking the density of water as 1000 kg/m^3 and the relative density of mercury as 13.6, the pressure difference indicated when the difference in the mercury levels in the U-tube is 310 mm is determined as follows:

The pressure difference at level XX' in Figure 32.6 is the same as the pressure difference between the two points in the pipe.

At level YY', however, there is no pressure difference, as below this there is a U-tube of mercury at rest. The difference in pressure at XX' and hence between the points on the pipe is therefore the difference in pressure between that due to a column of mercury 310 mm high (i.e. $X'Y'$) and that due to a column of water 310 mm (i.e. XY). The pressure difference is therefore:

$$310 \text{ mm} \times w_m - 310 \text{ mm} \times w_w = 310 \text{ mm} \ (w_m - w_w)$$

$$= 310 \text{ mm} \ (\rho_m g - \rho_w g)$$

$$= 310 \text{ mm} \ g\rho_w(d_m - 1)$$

where w_m = specific weight of mercury,
w_w = specific weight of water,
ρ_m = density of mercury,
ρ_w = density of water,
d_m = relative density of mercury

Hence,

pressure difference = 0.31 m $\times$ 1000 kg/m^3 $\times$ 9.81 m/s^2(13.6$-$1)

$$= 0.31 \times 1000 \times 9.81 \times 12.6 = \textbf{38.32 kPa}$$

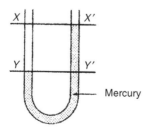

Figure 32.6

33 Measurement of Pressure

Introduction

As stated in Chapter 32, pressure is the force exerted by a fluid per unit area. A fluid (i.e. liquid, vapour or gas) has a negligible resistance to a shear force, so that the force it exerts always acts at right angles to its containing surface. The SI unit of pressure is the **pascal, Pa**, which is unit force per unit area, i.e. **1 Pa = 1 N/m²**. The pascal is a very small unit and a commonly used larger unit is the bar, where **1 bar = 10^5 Pa**

Atmospheric pressure is due to the mass of the air above the earth's surface. Atmospheric pressure changes continuously. A standard value of atmospheric pressure, called 'standard atmospheric pressure', is often used, having a value of 101 325 Pa or 1.01325 bars or 1013.25 millibars. This latter unit, the millibar, is usually used in the measurement of meteorological pressures. (Note that when atmospheric pressure varies from 101 325 Pa it is no longer standard.)

Pressure indicating instruments are made in a wide variety of forms because of their many different applications. Apart from the obvious criteria such as pressure range, accuracy and response, many measurements also require special attention to material, sealing and temperature effects. The fluid whose pressure is being measured may be corrosive or may be at high temperatures. Pressure indicating devices used in science and industry include:

(i) barometers
(ii) manometers
(iii) Bourdon pressure gauge
(iv) McLeod and Pirani gauges

Barometers

A barometer is an instrument for measuring atmospheric pressure. It is affected by seasonal changes of temperature. Barometers are therefore also used for the measurement of altitude and also as one of the aids in weather forecasting. The value of atmospheric pressure will thus vary with climatic conditions, although not usually by more than about 10% of standard atmospheric pressure.

Construction and principle of operation

A simple barometer consists of a glass tube, just under 1 m in length, sealed at one end, filled with mercury and then inverted into a trough containing more mercury. Care must be taken to ensure that no air enters the tube during this latter process. Such a barometer is shown in Figure 33.1(a) and it is seen that the level of the mercury column falls, leaving an empty space, called a vacuum. Atmospheric pressure acts on the surface of the mercury in the trough

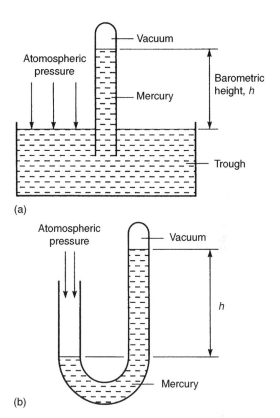

Figure 33.1

as shown and this pressure is equal to the pressure at the base of the column of mercury in the inverted tube, i.e. the pressure of the atmosphere is supporting the column of mercury. If the atmospheric pressure falls the barometer height h decreases. Similarly, if the atmospheric pressure rises then h increases. Thus atmospheric pressure can be measured in terms of the height of the mercury column. It may be shown that for mercury the height h is 760 mm at standard atmospheric pressure, i.e. a vertical column of mercury 760 mm high exerts a pressure equal to the standard value of atmospheric pressure.

There are thus several ways in which atmospheric pressure can be expressed:

Standard atmospheric pressure $= 101\,325$ Pa or 101.325 kPa

$$= 101\,325 \text{ N/m}^2 \text{ or } 101.325 \text{ kN/m}^2$$

$$= 1.01325 \text{ bars or } 1013.25 \text{ mbars}$$

$$= 760 \text{ mm of mercury}$$

Another arrangement of a typical barometer is shown in Figure 33.1(b) where a U-tube is used instead of an inverted tube and trough, the principle being similar. If, instead of mercury, water was used as the liquid in a barometer, then the barometric height h at standard atmospheric pressure would be 13.6 times more than for mercury, i.e. about 10.4 m high, which is not very practicable. This is because the relative density of mercury is 13.6

Types of barometer

The **Fortin barometer** is an example of a mercury barometer that enables barometric heights to be measured to a high degree of accuracy (in the order of one-tenth of a millimetre or less). Its construction is merely a more sophisticated arrangement of the inverted tube and trough shown in Figure 33.1(a), with the addition of a vernier scale to measure the barometric height with great accuracy. A disadvantage of this type of barometer is that it is not portable.

A Fortin barometer is shown in Figure 33.2. Mercury is contained in a leather bag at the base of the mercury reservoir, and height, H, of the mercury in the reservoir can be adjusted using the screw at the base of the barometer to depress or release the leather bag. To measure the atmospheric pressure the screw is adjusted until the pointer at H is just touching the surface of the mercury and the height of the mercury column is then read using the main and vernier scales. The measurement of atmospheric pressure using a Fortin barometer is achieved much more accurately than by using a simple barometer.

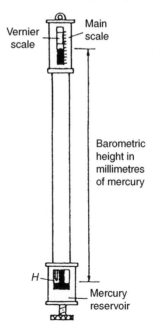

Figure 33.2

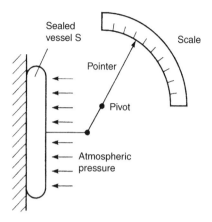

Figure 33.3

A portable type often used is the **aneroid barometer**. Such a barometer consists basically of a circular, hollow, sealed vessel, *S*, and usually made from thin flexible metal. The air pressure in the vessel is reduced to nearly zero before sealing, so that a change in atmospheric pressure will cause the shape of the vessel to expand or contract. These small changes can be magnified by means of a lever and be made to move a pointer over a calibrated scale. Figure 33.3 shows a typical arrangement of an aneroid barometer. The scale is usually circular and calibrated in millimetres of mercury. These instruments require frequent calibration.

Absolute and Gauge Pressure

A barometer measures the true or absolute pressure of the atmosphere. The term **absolute pressure** means the pressure above that of an absolute vacuum (which is zero pressure). In Figure 33.4 a pressure scale is shown with the line AB representing absolute zero pressure (i.e. a vacuum) and line CD representing atmospheric pressure. With most practical pressure-measuring instruments the part of the instrument that is subjected to the pressure being measured is also subjected to atmospheric pressure. Thus practical instruments actually determine the difference between the pressure being measured and atmospheric pressure. The pressure that the instrument is measuring is then termed the **gauge pressure**. In Figure 33.4, the line EF represents an absolute pressure that has a value greater than atmospheric pressure, i.e. the 'gauge' pressure is positive.

Thus,

absolute pressure = gauge pressure + atmospheric pressure

Hence a gauge pressure of, say, 60 kPa recorded on an indicating instrument when the atmospheric pressure is 101 kPa is equivalent to an absolute pressure of 60 kPa + 101 kPa, or 161 kPa.

160

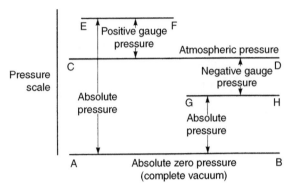

Figure 33.4

   Pressure-measuring indicating instruments are referred to generally as **pressure gauges** (which acts as a reminder that they measure 'gauge' pressure).

   It is possible, of course, for the pressure indicated on a pressure gauge to be below atmospheric pressure, i.e. the gauge pressure is negative. Such a gauge pressure is often referred to as a **vacuum**, even though it does not necessarily represent a complete vacuum at absolute zero pressure. The line GH in Figure 30.4 shows such a pressure. An indicating instrument used for measuring such pressures is called a **vacuum gauge**.

   A vacuum gauge indication of, say, 0.4 bar means that the pressure is 0.4 bar less than atmospheric pressure. If atmospheric pressure is 1 bar, then the absolute pressure is $1 - 0.4$ or 0.6 bar.

The Manometer

A manometer is a device for measuring or comparing fluid pressures, and is the simplest method of indicating such pressures.

U-tube manometer

A U-tube manometer consists of a glass tube bent into a U shape and containing a liquid such as mercury. A U-tube manometer is shown in Figure 33.5(a). If limb A is connected to a container of gas whose pressure is above atmospheric, then the pressure of the gas will cause the levels of mercury to move as shown in Figure 33.5(b), such that the difference in height is h_1. The measuring scale can be calibrated to give the gauge pressure of the gas as h_1 mm of mercury.

   If limb A is connected to a container of gas whose pressure is below atmospheric then the levels of mercury will move as shown in Figure 33.5(c), such that their pressure difference is h_2 mm of mercury.

   It is also possible merely to compare two pressures, say, P_A and P_B, using a U-tube manometer. Figure 33.5(d) shows such an arrangement with

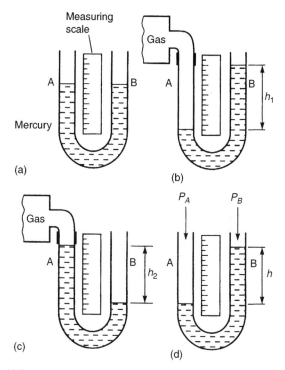

Figure 33.5

$(P_B - P_A)$ equivalent to h mm of mercury. One application of this differential pressure-measuring device is in determining the velocity of fluid flow in pipes (see Chapter 38).

For the measurement of lower pressures, water or paraffin may be used instead of mercury in the U-tube to give larger values of h and thus greater sensitivity.

Inclined manometers

For the measurement of very low pressures, greater sensitivity is achieved by using an inclined manometer, a typical arrangement of which is shown in Figure 33.6. With the inclined manometer the liquid used is water and the scale attached to the inclined tube is calibrated in terms of the vertical height h. Thus when a vessel containing gas under pressure is connected to the reservoir, movement of the liquid levels of the manometer occurs. Since small-bore tubing is used the movement of the liquid in the reservoir is very small compared with the movement in the inclined tube and is thus neglected. Hence the scale on the manometer is usually used in the range 0.2 mbar to 2 mbar.

162

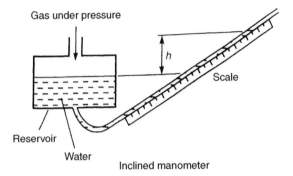

Gas under pressure

h

Scale

Reservoir

Water

Inclined manometer

Figure 33.6

The length of tube used naturally limits the pressure of a gas that a manometer is capable of measuring. Most manometer tubes are less than 2 m in length and this restricts measurement to a maximum pressure of about 2.5 bar (or 250 kPa) when mercury is used.

The Bourdon Pressure Gauge

Pressures many times greater than atmospheric can be measured by the Bourdon pressure gauge, which is the most extensively used of all pressure-indicating instruments. It is a robust instrument. Its main component is a piece of metal tube (called the Bourdon tube), usually made of phosphor bronze or alloy steel, of oval or elliptical cross-section, sealed at one end and bent into an arc. In some forms the tube is bent into a spiral for greater sensitivity. A typical arrangement is shown in Figure 33.7(a). One end, E, of the Bourdon tube is fixed and the fluid whose pressure is to be measured is connected to this end. The pressure acts at right angles to the metal tube wall as shown in the cross-section of the tube in Figure 33.7(b). Because of its elliptical shape it is clear that the sum of the pressure components, i.e. the total force acting on the sides A and C, exceeds the sum of the pressure components acting on ends B and D. The result is that sides A and C tend to move outwards and B and D inwards tending to form a circular cross-section. As the pressure in the tube is increased the tube tends to uncurl, or if the pressure is reduced the tube curls up further. The movement of the free end of the tube is, for practical purposes, proportional to the pressure applied to the tube, this pressure, of course, being the gauge pressure (i.e. the difference between atmospheric pressure acting on the outside of the tube and the applied pressure acting on the inside of the tube). By using a link, a pivot and a toothed segment as shown in Figure 33.7(a), the movement can be converted into the rotation of a pointer over a graduated calibrated scale.

The Bourdon tube pressure gauge is capable of measuring high pressures up to 10^4 bar (i.e. 7600 m of mercury) with the addition of special safety features. A pressure gauge must be calibrated, and this is done either

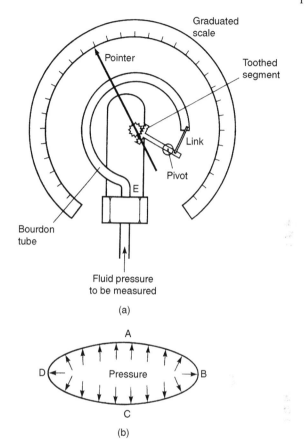

Figure 33.7

by a manometer, for low pressures, or by a piece of equipment called a **'dead weight tester'**. This tester consists of a piston operating in an oil-filled cylinder of known bore, and carrying accurately known weights as shown in Figure 33.8. The gauge under test is attached to the tester and a screwed piston or ram applies the required pressure until the weights are just lifted. While the gauge is being read, the weights are turned to reduce friction effects.

Vacuum Gauges

Vacuum gauges are instruments for giving a visual indication, by means of a pointer, of the amount by which the pressure of a fluid applied to the gauge

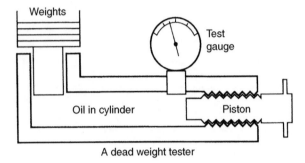

A dead weight tester

Figure 33.8

is less than the pressure of the surrounding atmosphere. Two examples of vacuum gauges are the McLeod gauge and the Pirani gauge.

McLeod gauge

The McLeod gauge is normally regarded as a standard and is used to calibrate other forms of vacuum gauges. The basic principle of this gauge is that it takes a known volume of gas at a pressure so low that it cannot be measured, and then compresses the gas in a known ratio until the pressure becomes large enough to be measured by an ordinary manometer. This device is used to measure low pressures, often in the range 10^{-6} to 1.0 mm of mercury. A disadvantage of the McLeod gauge is that it does not give a continuous reading of pressure and is not suitable for registering rapid variations in pressure.

Pirani gauge

The Pirani gauge measures the resistance and thus the temperature of a wire through which current is flowing. The thermal conductivity decreases with the pressure in the range 10^{-1} to 10^{-4} mm of mercury so that the increase in resistance can be used to measure pressure in this region. The Pirani gauge is calibrated by comparison with a McLeod gauge.

34 Ideal gas Laws

Introduction

The relationships that exist between pressure, volume and temperature in a gas are given in a set of laws called **the gas laws**.

Boyle's Law

Boyle's law states:

> *the volume V of a fixed mass of gas is inversely proportional to its absolute pressure p at constant temperature*

i.e. $p \propto \dfrac{1}{V}$ or $p = \dfrac{k}{V}$ or $pV = k$, at constant temperature,

where p = absolute pressure in pascals (Pa), V = volume in m^3, and k = a constant.

Changes that occur at constant temperature are called **isothermal changes**.

When a fixed mass of gas at constant temperature changes from pressure p_1 and volume V_1 to pressure p_2 and volume V_2 then:

$$\boxed{p_1 V_1 = p_2 V_2}$$

For example, a gas occupies a volume of 0.10 m^3 at a pressure of 1.8 MPa. The pressure, if the volume is changed to 0.06 m^3 at constant temperature, is given by: $p_1 V_1 = p_2 V_2$

Hence $(1.8)(0.10) = p_2(0.06)$ from which,

$$\textbf{pressure } p_2 = \frac{1.8 \times 0.10}{0.06} = \textbf{3 MPa}$$

Charles' Law

Charles' law states:

> *for a given mass of gas at constant pressure, the volume V is directly proportional to its thermodynamic temperature T*

i.e. $V \propto T$ or $V = kT$ or $\dfrac{V}{T} = k$, at constant pressure,

where T = thermodynamic temperature in kelvin (K).

A process that takes place at constant pressure is called an **isobaric process**.

The relationship between the Celsius scale of temperature and the thermodynamic or absolute scale is given by:

$$\text{kelvin} = \text{degrees Celsius} + 273$$

i.e. $\boxed{K = {}^\circ C + 273}$ or $\boxed{{}^\circ C = K - 273}$

(as stated in Chapter 29)

If a given mass of gas at a constant pressure occupies a volume V_1 at a temperature T_1 and a volume V_2 at temperature T_2, then:

$$\boxed{\frac{V_1}{T_1} = \frac{V_2}{T_2}}$$

For example, a gas occupies a volume of 1.2 litres at 20°C. If the pressure is kept constant, the volume it occupies at 130°C is determined from:

$$\frac{V_1}{T_1} = \frac{V_2}{T_2} \text{ from which, } V_2 = V_1\left(\frac{T_2}{T_1}\right) = (1.2)\left(\frac{130 + 273}{20 + 273}\right) = (1.2)\left(\frac{403}{293}\right)$$

$$= \mathbf{1.65 \text{ litres}}$$

The Pressure Law

The pressure law states:

> *the pressure p of a fixed mass of gas is directly proportional to its thermodynamic temperature T at constant volume*

i.e. $p \propto T$ or $p = kT$ or $\dfrac{p}{T} = k$

When a fixed mass of gas at constant volume changes from pressure p_1 and temperature T_1, to pressure p_2 and temperature T_2 then:

$$\boxed{\frac{p_1}{T_1} = \frac{p_2}{T_2}}$$

For example, gas initially at a temperature of 17°C and pressure 150 kPa is heated at constant volume until its temperature is 124°C. Assuming no loss of gas, the final pressure of the gas is determined from:

$$\frac{p_1}{T_1} = \frac{p_2}{T_2} \text{ from which, } p_2 = \frac{p_1 T_2}{T_1} = \frac{(150)(397)}{290} = \mathbf{205.3 \text{ kPa}}$$

Dalton's Law of Partial Pressure

Dalton's law of partial pressure states:

> *the total pressure of a mixture of gases occupying a given volume is equal to the sum of the pressures of each gas, considered separately, at constant temperature*

The pressure of each constituent gas when occupying a fixed volume alone is known as the **partial pressure** of that gas.

An **ideal gas** is one that completely obeys the gas laws given above. In practice no gas is an ideal gas, although air is very close to being one. For calculation purposes the difference between an ideal and an actual gas is very small.

For example, a gas R in a container exerts a pressure of 200 kPa at a temperature of 18°C. Gas Q is added to the container and the pressure increases to 320 kPa at the same temperature. The pressure that gas Q alone exerts at the same temperature, is given by Dalton's law of partial pressure:

the pressure of gas Q alone, $p_Q = p - p_R = 320 - 200$

$$= 120 \text{ kPa}$$

Characteristic Gas Equation

Frequently, when a gas is undergoing some change, the pressure, temperature and volume all vary simultaneously. Provided there is no change in the mass of a gas, the above gas laws can be combined, giving:

$$\boxed{\frac{p_1 V_1}{T_1} = \frac{p_2 V_2}{T_2} = k}$$ where k is a constant

For an ideal gas, constant $k = mR$, where m is the mass of the gas in kg, and R is the **characteristic gas constant**, i.e.

$$\frac{pV}{T} = mR \quad \text{or} \quad \boxed{pV = mRT}$$

This is called the **characteristic gas equation**. In this equation, $p =$ absolute pressure in pascals, $V =$ volume in m^3, $m =$ mass in kg, $R =$ characteristic gas constant in J/(kg K), and $T =$ thermodynamic temperature in kelvin.

Some **typical values** of the characteristic gas constant R include: air, 287 J/(kg K), hydrogen 4160 J/(kg K), oxygen 260 J/(kg K) and carbon dioxide 184 J/(kg K).

For example, some air at a temperature of 40°C and pressure 4 bar occupies a volume of 0.05 m^3. The mass of the air, assuming the characteristic gas constant for air to be 287 J/(kg K), is determined from: $pV = mRT$, from which, mass of air,

$$m = \frac{pV}{RT} = \frac{(4 \times 10^5 \text{ Pa})(0.05 \text{ m}^3)}{(287 \text{ J/kg K})(40 + 273)\text{K}} = \textbf{0.223 kg or 223 g}$$

Standard temperature and pressure (i.e. **STP**) refers to a temperature of 0°C, i.e. 273 K, and normal atmospheric pressure of 101.325 kPa

---

Kinetic Theory of Gases

The kinetic theory of gases suggests that gases are composed of particles in motion. The continual bombardment of any surface by the gas causes a pressure to be exerted; the greater the density of a gas, the more frequent the number of collisions between molecules and the surface and the greater the pressure exerted. Hence the pressure increases either when the volume of a certain mass of gas is reduced, or when more gas is pumped into a vessel. When the temperature of a gas is increased, the speed of the molecules increases, causing an increase in both the number and the momentum imparted by each collision. This accounts for the increase in pressure of a gas with increase in temperature.

Maxwell (in 1860) explained some of the properties of a gas by assuming that the molecules of a gas make elastic collisions, spend negligible time actually in collision, and themselves occupy a negligible part of the volume of the gas. Also, the attractive forces between molecules are assumed negligible.

It may be shown that for gas occupying a volume V at a pressure p and containing n molecules each of mass m moving at an average velocity of c,

$$\boxed{pV = \tfrac{1}{2}mnc^2}$$

Also, the kinetic energy of the molecules of a gas is proportional to its thermodynamic temperature.

When a liquid evaporates, molecules with sufficient kinetic energy escape from the liquid's surface. The higher the temperature of the liquid the greater the average kinetic energy of the molecules and the greater the number of molecules that are able to escape. Since it is the molecules with the highest kinetic energy that escape, the average kinetic energy of the remaining molecules decreases and thus the liquid cools.

If a liquid evaporates a **vapour** is formed. When a vapour exists in the presence of its own liquid a **saturated vapour** is formed. If all the liquid evaporates an **unsaturated vapour** is produced. The higher the temperature the greater the number of molecules that escapes to form the vapour. These molecules bombard the walls of the container and thus exert a pressure.

The **saturated vapour pressure** depends only on the temperature of the vapour. The saturated vapour pressure of water at various temperatures is shown in Table 34.1. A liquid boils at a temperature when its saturated vapour pressure is equal to the atmospheric pressure. Thus water will boil at a temperature greater than 100°C if the atmospheric pressure is increased. This is the principle of the **pressure cooker**.

A saturated vapour does not obey the gas laws since its pressure depends only on temperature. An unsaturated vapour will obey the gas laws fairly

Table 34.1

Temperature (°C)	Saturated vapour pressure of water (10^3 Pa)
0	0.61
10	1.23
20	2.33
30	4.23
40	7.35
50	12.3
60	19.9
70	31.2
80	47.4
90	70.2
100	101
150	476
200	1550

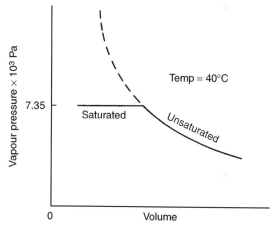

Figure 34.1

closely as long as it remains unsaturated. If an unsaturated vapour at a particular temperature is decreased in volume its pressure will rise in accordance with Boyle's law until it reaches the saturated vapour pressure at that particular temperature (see Figure 34.1). When the vapour pressure at 40°C reaches 7.35×10^3 Pa the vapour becomes saturated as it starts to liquefy.

35 Properties of Water and Steam

Principle of Conservation of Energy

When two systems are at different temperatures, the transfer of energy from one system to the other is called **heat transfer**. For a block of hot metal cooling in air, heat is transferred from the hot metal to the cool air.

The **principle of conservation of energy** may be stated as

energy cannot be created nor can it be destroyed

and since heat is a form of energy, this law applies to heat transfer problems.

A more convenient way of expressing this law when referring to heat transfer problems is:

$$\left(\begin{array}{c} \text{initial energy of the system} \\ + \text{ energy entering the system} \end{array} \right) = \left(\begin{array}{c} \text{final energy} \\ \text{of the system} \\ + \\ \text{energy leaving} \\ \text{the system} \end{array} \right)$$

$$\text{or} \quad \text{energy entering the system} = \left(\begin{array}{c} \text{change of energy} \\ \text{within the system} \\ + \\ \text{energy leaving} \\ \text{the system} \end{array} \right)$$

Internal Energy

Fluids consist of a very large number of molecules moving in random directions within the fluid. When the fluid is heated, the speeds of the molecules are increased, increasing the kinetic energy of the molecules. There is also an increase in volume due to an increase in the average distance between molecules, causing the potential energy of the fluid to increase. The **internal energy**, U, of a fluid is the sum of the internal kinetic and potential energies of the molecules of a fluid, measured in joules. It is not usual to state the internal energy of a fluid as a particular value in heat transfer problems, since it is normally only the **change** in internal energy that is required.

The amount of internal energy of a fluid depends on:

(a) the type of fluid; in gases the molecules are well separated and move with high velocities, thus a gaseous fluid has higher internal energy than the same mass of a liquid
(b) the mass of a fluid; the greater the mass, the greater the number of molecules and hence the greater the internal energy

(c) the temperature; the higher the temperature the greater the velocity of the molecules

---

Enthalpy

The sum of the internal energy and the pressure energy of a fluid is called the **enthalpy** of the fluid, denoted by the symbol H and measured in joules. The product of pressure p and volume V gives the pressure energy, or work done, i.e.

$$\text{pressure energy} = pV \text{ joules}$$

Thus, enthalpy = internal energy + pressure energy (or work done)

i.e. $\boxed{H = U + pV}$

As for internal energy, the actual value of enthalpy is usually unimportant and it is the **change** in enthalpy that is usually required. In heat transfer problems involving steam and water, water is considered to have zero enthalpy at a standard pressure of 101 kPa and a temperature of 0°C. The word 'specific' associated with quantities indicates 'per unit mass'. Thus the **specific enthalpy** is obtained by dividing the enthalpy by the mass and is denoted by the symbol h. Thus:

$$\text{specific enthalpy } h = \frac{\textbf{enthalpy}}{\textbf{mass}} = \frac{H}{m}$$

The units of specific enthalpy are joules per kilogram (J/kg)

For example, in a closed system, that is, a system in which the mass of fluid remains a constant, the internal energy changes from 25 kJ to 50 kJ and the work done by the system is 55 kJ. The heat transferred to the system to effect this change is given by:

$$H = U + pV = [(50 - 25) + 55] \text{ kJ}$$

$$= (25 + 55) \text{ kJ} = 80 \text{ kJ}$$

That is, **the heat transferred to the system is 80 kJ**

Sensible Heat

The specific enthalpy of water, h_f, at temperature θ°C is the quantity of heat needed to raise 1 kg of water from 0°C to θ°C, and is called the **sensible heat** of the water. Its value is given by:

$$\text{specific heat capacity of water } (c) \times \text{temperature change } (\theta)$$

i.e. $\boxed{h_f = c\theta}$

The specific heat capacity of water varies with temperature and pressure but is normally taken as 4.2 kJ/kg, thus

$$h_f = 4.2\theta \text{ kJ/kg}$$

Saturated Steam

When water is heated at a uniform rate, a stage is reached (at 100°C at standard atmospheric pressure) where the addition of more heat does not result in a corresponding increase in temperature. The temperature at which this occurs is called the **saturation temperature**, t_{SAT}, and the water is called **saturated water**. As heat is added to saturated water, it is turned into **saturated steam**. The amount of heat required to turn 1 kg of saturated water into saturated steam is called the **specific latent heat of vaporisation**, and is given the symbol, h_{fg}. The total specific enthalpy of steam at saturation temperature, h_g, is given by:

the specific sensible heat + the specific latent heat of vaporization

i.e. $$\boxed{h_g = h_f + h_{fg}}$$

Dryness Factor

If the amount of heat added to saturated water is insufficient to turn all the water into steam, then the ratio:

$$\frac{\text{mass of saturated steam}}{\text{total mass of steam and water}}$$

is called the **dryness fraction** of the steam, denoted by the symbol q.
The steam is called **wet steam** and its total enthalpy h is given by:

enthalpy of saturated water + (dryness fraction) (enthalpy of latent heat

of vaporisation)

i.e. $$\boxed{h = h_f + qh_{fg}}$$

Superheated Steam

When the amount of heat added to water at saturation temperature is sufficient to turn all the water into steam, it is called either saturated vapour or **dry saturated steam**. The addition of further heat results in the temperature of the steam rising and it is then called **superheated steam**. The specific enthalpy of superheated steam above that of dry saturated steam is given by:

$c(t_{SUP} - t_{SAT})$, where c is the specific heat capacity of the steam and t_{SUP} is the temperature of the superheated steam. The total specific enthalpy of the superheated steam is given by:

$$\boxed{h_f + h_{fg} + c(t_{SUP} - t_{SAT})} \quad \text{or} \quad \boxed{h_g + c(t_{SUP} - t_{SAT})}$$

Temperature/Specific Enthalpy Graph

The relationship between temperature and specific enthalpy can be shown graphically and a typical temperature/specific enthalpy diagram is shown in Figure 35.1. In this figure, AB represents the sensible heat region where any increase in enthalpy results in a corresponding increase in temperature. BC is called the **evaporation line** and points between B and C represent the wet steam region (or latent region), point C representing dry saturated steam. Points to the right of C represent the superheated steam region.

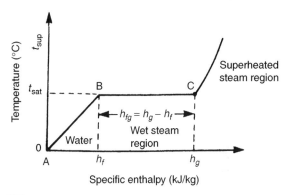

Figure 35.1

Steam Tables

The boiling point of water, t_{SAT} and the various specific enthalpies associated with water and steam [h_f, h_{fg} and $c(t_{SUP} - t_{SAT})$] all vary with pressure. These values at various pressures have been tabulated in **steam tables**, extracts from these being shown in Tables 35.1 and 35.2

In Table 35.1, the pressure in both bar and kilopascals, and saturated water temperature, are shown in columns on the left. The columns on the right give the corresponding specific enthalpies of water (h_f) and dry saturated steam (h_g), together with the specific enthalpy of the latent heat of vaporization (h_{fg}). The columns on the right of Table 35.2 give the specific

Table 35.1

Pressure		Saturation temperature t_{SAT} (°C)	Specific enthalpy (kJ/kg)		
(bar)	(kPa)		Saturated water h_f	Latent heat h_{fg}	Saturated vapour h_g
1	100	99.6	417	2258	2675
1.5	150	111.4	467	2226	2693
2	200	120.2	505	2202	2707
3	300	133.5	561	2164	2725
4	400	143.6	605	2134	2739
5	500	151.8	640	2109	2749
6	600	158.8	670	2087	2757
7	700	165.0	697	2067	2764
8	800	170.4	721	2048	2769
9	900	175.4	743	2031	2774
10	1000	179.9	763	2015	2778
15	1500	198.3	845	1947	2792
20	2000	212.4	909	1890	2799
30	3000	233.8	1008	1795	2803
40	4000	250.3	1087	1714	2801

Table 35.2

Pressure		Saturation Temperature t_{SAT}(°C)	Saturated vapour h_g	Specific enthalpy (kJ/kg) Superheated steam at				
(bar)	(kPa)			200°C	250°C	300°C	350°C	400°C
1	100	99.6	2675	2876	2975	3075	3176	3278
1.5	150	111.4	2693	2873	2973	3073	3175	3277
2	200	120.2	2707	2871	2971	3072	3174	3277
3	300	133.5	2725	2866	2968	3070	3172	3275
4	400	143.6	2739	2862	2965	3067	3170	3274
5	500	151.8	2749	2857	2962	3065	3168	3272
6	600	158.8	2757	2851	2958	3062	3166	3270
7	700	165.0	2764	2846	2955	3060	3164	3269
8	800	170.4	2769	2840	2951	3057	3162	3267
9	900	175.4	2774	2835	2948	3055	3160	3266
10	1000	179.9	2778	2829	2944	3052	3158	3264
15	1500	198.3	2792	2796	2925	3039	3148	3256
20	2000	212.4	2799		2904	3025	3138	3248
30	3000	233.8	2803		2858	2995	3117	3231
40	4000	250.3	2801			2963	3094	3214

enthalpies of dry saturated steam (h_g) and superheated steam at various temperatures. The values stated refer to zero enthalpy. However, if the degree of superheat is given, this refers to the saturation temperature. Thus at a pressure of 100 kPa, the column headed, say, 250°C has a degree of superheat of (250 − 99.6)°C, that is 150.4°C.

For example, let some dry saturated steam at a pressure of 1.0 MPa be cooled at constant pressure until it has a dryness fraction of 0.6. The change in the specific enthalpy of the steam is determined as follows:

From Table 35.1, the specific enthalpy of dry saturated steam h_g, at a pressure of 1.0 MPa (i.e. 1000 kPa) is 2778 kJ/kg. From earlier, the specific enthalpy of wet steam is $h_f + q h_{fg}$. At a pressure of 1.0 MPa, h_f is 763 kJ/kg and h_{fg} is 2015 kJ/kg. Thus, the specific enthalpy of the wet steam is given by:

$$763 + (0.6 \times 2015) = 1972 \text{ kJ/kg}$$

The **change in specific enthalpy** $= 2778 - 1972 = $ **806 kJ/kg**

In another example, let steam leave a boiler at a pressure of 3.0 MPa and a temperature of 400°C. The degree of superheat may be determined from Table 35.2. At a pressure of 3.0 MPa, i.e. 3000 kPa, the saturation temperature is 233.8°C, hence **the degree of superheat** is $(400 - 233.8) = $ **166.2°C**

The specific enthalpy of superheated steam at 3.0 MPa and 400°C is given in Table 32.2 as **3231 kJ/kg**.

Superheated steam behaves very nearly as if it is an ideal gas and the gas laws introduced in Chapter 34 may be used to determine the relationship between pressure, volume and temperature.

36 Surface Tension and Viscosity

Surface Tension

The force of attraction between molecules in a liquid gives rise to what is termed surface tension.

The **surface tension** γ of a liquid is the force per unit length acting in the surface perpendicular to one side of a line in the surface.

The **free surface energy** σ is the energy required to create an additional unit area against the attractive forces of the molecules. The surface tension γ and the free surface energy σ are numerically the same, as shown below.

Consider a wire frame as shown in Figure 36.1 on which there is a soap film. XY is a sliding wire. The length of soap film in contact with the sliding wire is l. The force F due to surface tension on the wire XY is $2\gamma l$, the factor 2 occurring because there are two surfaces to the soap film. γ is the surface tension of the soap film. If the wire is moved a distance x to the right, the work done against the force of surface tension is $2\gamma l x$.

Thus the increase in the surface tension is $2lx$. Hence the energy required to create an additional unit area of film is: $\dfrac{2\gamma l x}{2lx} = \gamma$

But this is the definition of free surface energy σ. Thus $\sigma = \gamma$ numerically.

Because of differences between the cohesive force between molecules of liquid and the adhesive force between molecules of liquids and molecules of solids, a liquid surface is usually curved where it makes contact with a solid. For example, the surface of water in a glass tube is concave and the surface of mercury in a glass tube is convex, as shown in Figure 36.2

The angle of contact θ is defined as the angle between the solid surface and the tangent to the liquid surface. θ is measured through the liquid as shown in Figure 36.3. If $\theta < 90°$ the liquid is said to 'wet' the solid surface.

Liquids for which $\theta < 90°$ rise in a tube with a small internal diameter (such as a capillary tube). Figure 36.4 shows a liquid that has risen a height h up a capillary tube of radius r.

The force due to the surface tension acting on the meniscus depends upon the circumference of the meniscus and the surface tension γ.

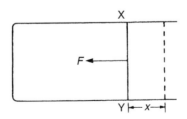

Figure 36.1

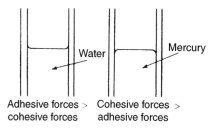

Figure 36.2

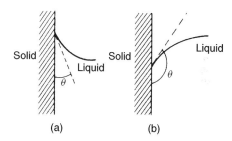

Figure 36.3

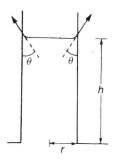

Figure 36.4

The upward vertical component of the force due to surface tension is:

$$\gamma \times \text{circumference} \times \cos\theta = \gamma(2\pi r)\cos\theta$$

The downward vertical force on the column of liquid is due to:

weight of liquid = volume × density × earth's gravitational field

$$= (\pi r^2 h)\rho g$$

These two forces are equal, i.e. $\gamma(2\pi r)\cos\theta = (\pi r^2 h)\rho g$

from which,

$$\text{height } h = \frac{2\gamma\cos\theta}{r\rho g}$$

For example, if the surface tension of mercury at 20 °C is 0.465 Nm^{-1} and its angle of contact with glass is 140°, the capillary rise h of the mercury in a capillary tube of internal radius 2 mm is given by:

$$h = \frac{2\gamma\cos\theta}{r\rho g} = \frac{2(0.465)(-0.766)}{(2\times 10^{-3})(13.6\times 10^3)(9.81)}$$

$$= -2.67\times 10^{-3} = -2.67 \text{ mm}$$

(The negative sign indicates that the mercury level in the capillary tube falls).

Excess Pressure

It may be shown that there is a pressure inside a spherical drop of liquid that exceeds the surrounding air pressure by an amount equal to $\frac{2\gamma}{R}$ where R is the radius of the drop. This is called the **excess pressure**.

(i) For a spherical drop of liquid in air the excess pressure is $\frac{2\gamma}{R}$

(ii) For a bubble of gas in a liquid the excess pressure is $\frac{2\gamma}{R}$

(iii) For a soap bubble in air the excess pressure is $\frac{4\gamma}{R}$ since a soap bubble has two surfaces.

Energy Needed to Break Intermolecular Bond

From a knowledge of the free surface energy of a liquid an approximate value of the energy needed to break an intermolecular bond may be found. A molecule moving to the surface of a liquid as a new surface is created has its number of near neighbours decreased from ten to five. If n is the number of molecules per unit area of surface, then $\frac{5n}{2}$ bonds are broken for each unit area of surface produced and $\frac{5n\varepsilon}{2}$ is the energy needed if ε is the energy required to break one bond. Thus, as σ, the force surface energy, is the energy required to produce unit area of surface, we have: $\sigma = \frac{5n\varepsilon}{2}$, from which, $\varepsilon = \frac{2\sigma}{5n}$

Molar Latent Heat of Vaporisation

The molar latent heat of vaporisation of a substance is the energy required to evaporate 1 mole of the substance at standard pressure. If a solid that has a hexagonal close packed crystal structure is considered, each atom has twelve near neighbours. In the liquid form each atom has about ten near neighbours and bonds must be broken as the solid turns into a liquid and loses two near neighbours per atom. A vapour has no near neighbours and thus the liquid loses ten near neighbours per atom in evaporating.

In one mole there are 6×10^{23} atoms and thus if each atom has ten near neighbours the number of bonds that need to be broken is:

$$\frac{10 \times 6 \times 10^{23}}{2}$$

The divisor of 2 is present because each bond connects two atoms. Thus, the molar latent heat is $5N_A\varepsilon$, where N_A is the number of atoms in a mole, i.e.

$$\text{molar latent heat } = 5N_A \left(\frac{2\sigma}{5n} \right) = \frac{2N_A\sigma}{n}$$

Therefore liquids with high values of σ should have high values of molar latent heat of vaporisation. This is reasonably confirmed by experiment.

Viscosity

Liquids (and gases) in contact with a solid surface stick to that surface. If a liquid flows on a solid surface we can consider the liquid to consist of layers. The bottom layer remains in contact with the solid and at rest. The other layers slide on one another and travel with velocities that increase the further the layer is from the solid, as shown in Figure 36.5. This is a description of streamline flow. If the velocity increases to beyond a critical value the flow becomes **turbulent** and the description in terms of layers no longer applies. In Figure 36.5, the arrows indicate the velocities of different layers. This condition will exist when the liquid is subjected to a **shear** force. The opposition to this is called the **viscosity** of the liquid.

Consider two parallel layers of liquid separated by a distance Δy travelling at velocities v and $v + \Delta y$. The lower layer tends to impede the flow of the upper layer and exerts a retarding force F on it, whereas the lower layer itself experiences an accelerating force F exerted on it by the upper layer.

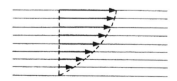

Figure 36.5

The tangential stress between the two layers is $\dfrac{F}{A}$ where A is the area of contact between the layers. The ratio $\dfrac{\Delta v}{\Delta y}$ is called the velocity gradient.

Newton realised that for some fluids:

Tangential stress α velocity gradient, and thus, $\dfrac{F}{A} = \eta\dfrac{\Delta v}{\Delta y}$ where η is a constant called the **coefficient of viscosity**. Thus,

the coefficient of viscosity, $\boxed{\eta = \dfrac{\textbf{tangential stress}}{\textbf{velocity gradient}}}$

η usually decreases with increasing temperature although 'viscostatic' oils are almost temperature independent. The units of the coefficient of viscosity are N s m^{-2} or, alternatively, $\text{kg m}^{-1}\text{s}^{-1}$ (since $1\,\text{N} = 1\,\text{kg m s}^{-2}$)

Poiseulle's Formula

Poiseulle's formula for streamline flow through a circular pipe gives an expression for the volume V of liquid passing per second:

$$V = \frac{\pi p r^4}{8\eta l}$$

where r is the radius of the pipe, p is the pressure difference between the ends of the pipe, l is the length of the pipe and η is the coefficient of viscosity of the liquid.

For example, in Figure 36.6 water flows from a tank through a tube of length 1 m and internal radius 2 mm. If the viscosity of water is $1 \times 10^{-3}\ \text{kg m}^{-1}\text{s}^{-1}$, the rate at which water is collected in the small container is determined as follows:

From Poiseulle's formula the volume collected per second, V, is given by:

$$V = \frac{\pi p r^4}{8\eta l}$$

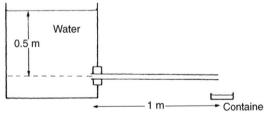

Figure 36.6

where the pressure difference between the ends of the tube,

$$p = h\rho g = (0.5 \times 10^3)(9.81)$$

Hence
$$V = \frac{\pi(0.5 \times 10^3)(9.81)(2 \times 10^{-3})^4}{8(1 \times 10^{-3})(1)}$$

$$= \mathbf{3.08 \times 10^{-5} \ m^3 s^{-1}}$$

This rate will not be maintained because the water level in the tank will fall and pressure p will decrease.

Stoke's Law

Stoke's law gives an expression for the force F due to viscosity acting on a sphere moving with streamline flow through a liquid, i.e.

$$\boxed{\text{force } F = 6\pi\eta r v}$$

where r is the radius of the sphere, and v its velocity.

For example, if a steel ball-bearing of radius 1 mm falls through water, its terminal velocity v is determined as follows, assuming the density of steel is 7.8×10^3 kg m^{-3}, the density of water is 1×10^3 kg m^{-3}, the viscosity of water is 1×10^{-3} kg m^{-1} s^{-1} and that streamline flow is maintained:

The volume of the sphere is $\frac{4}{3}\pi r^3$, thus the weight of the sphere is $\frac{4}{3}\pi r^3 \rho g$ where ρ is the density of the sphere and g is the earth's gravitational field.

The volume of the liquid displaced is $\frac{4}{3}\pi r^3$, thus the upthrust (i.e. weight of liquid displaced) is $\frac{4}{3}\pi r^3 \rho_0 g$ where ρ_0 is the density of the liquid.

When the terminal velocity is reached there is no resultant force on the sphere. Thus the weight of the sphere = viscous drag + upthrust

i.e.
$$\tfrac{4}{3}\pi r^3 \rho g = 6\pi\eta r v + \tfrac{4}{3}\pi r^3 \rho_0 g$$

hence
$$6\pi\eta r v = \tfrac{4}{3}\pi r^3 (\rho - \rho_0)g$$

and terminal velocity

$$\mathbf{v} = \frac{\frac{4}{3}\pi r^3 (\rho - \rho_0)g}{6\pi\eta r} = \frac{2r^2 g(\rho - \rho_0)}{9\eta}$$

$$= \frac{2(1 \times 10^{-3})^2(9.81)(7.8 \times 10^3 - 1 \times 10^3)}{9(1 \times 10^{-3})} = \mathbf{14.8 \ ms^{-1}}$$

37 Fluids in Motion

Bernoulli's Equation

Bernoulli's equation is the principle of the conservation of energy applied to fluids in motion:

$$\frac{p}{w} + \frac{v^2}{2g} + Z = \text{constant} \tag{1}$$

or

$$\frac{p}{\rho} + \frac{v^2}{2} + Zg = \text{constant} \tag{2}$$

All of the quantities on the left of each equation apply to a specified fixed point in the moving fluid:

p = pressure (gauge pressure unless otherwise specified)
w = specific weight (weight per unit volume)
v = velocity
g = acceleration due to gravity
Z = height above some specified datum
ρ = density

The two significant differences between this application of the principle of the conservation of energy and the application of the principle of solids in motion are:

(i) this application is to a steady process mass (or weight) flowing per second which has to be considered, instead of a given fixed mass or weight
(ii) a third form of energy, that is, pressure energy, must be considered; the corresponding form of energy in dealing with solids, strain energy, is only occasionally met.

Each of the terms in equation (1) represents energy per unit weight of fluid. The basic unit of each term is the metre, i.e. $\frac{\text{Nm}}{\text{N}}$ or $\frac{\text{J}}{\text{N}}$

The basic unit for pressure is the same as for stress, N/m² or pascal, Pa. Each term in the equation is called a **head**.

$\frac{p}{w}$ is the **pressure head** $\left(\frac{\text{N/m}^2}{\text{N/m}^3} = \text{m}\right)$

$\frac{v^2}{2g}$ is the **velocity** or **kinetic head** $\left(\frac{(\text{m/s})^2}{\text{m/s}^2} = \text{m}\right)$

Z is the **potential head** (m)

The sum of the three heads is called **total head (H)**
  Equation (2) gives energies per unit mass.

In practise, as with solids, some energy is lost, converted into heat. The elimination of this loss is an important aspect of the mechanics of fluids.

If a pipe is filled by moving liquid the volumetric rate of flow, V, i.e. the volume passing per second, must be the same at every section: $V = A_1v_1 = A_2v_2$ where A_1 and v_1 are the cross-sectional area and velocity at one selected section and A_2 and v_2 are area and velocity at a second section. The equation

$A_1v_1 = A_2v_2$ is called the **equation of continuity**.

The basic unit of volumetric rate of flow $\dot{V}$ is m³/s which is a large unit. The smaller unit, litres per second, is often preferred.

$$1 \text{ litre} = 1000 \text{ cm}^3 = 1 \text{ m}^3 \times 10^{-3}$$

(For very low rates of flow, litres per minute units may be preferred).

For example, let the rate of flow of water through a pipe of 32 mm diameter be 2.8 litre/s. The total head at a point where the pressure is 28.4 kPa with reference to a datum 1.84 m below is determined as follows:

$$\text{velocity of flow } v = \frac{\dot{V}}{A} = \frac{2.8 \times 10^{-3} \text{ m}^3/\text{s}}{\frac{\pi}{4}(32)^2 \times 10^{-6} \text{ m}^2} = 3.4815 \text{ m/s}$$

$$\text{the velocity head } \frac{v^2}{2g} = \frac{(3.4815)^2}{2(9.81)} = 0.6178 \text{ m}$$

$$\text{the specific weight } w = \rho g = (1000 \text{ kg/m}^3)(9.81 \text{ m/s}^2)$$

$$= 9.81 \text{ kN/m}^3$$

$$\text{the pressure head } \frac{p}{w} = \frac{28.4 \text{ kPa}}{9.81 \text{ kN/m}^2} = 2.895 \text{ m}$$

$$\text{the potential head } Z = 1.84 \text{ m}$$

Hence the **total head**,

$$H = \frac{p}{w} + \frac{v^2}{2g} + Z = 2.895 + 0.6178 + 1.84 = \textbf{5.353 m}$$

Flow through Orifices

Water issuing from a tank as a horizontal jet, as shown in Figure 37.1, has a velocity head only, if the datum is taken at the level of orifice. Water, which will eventually form the jet, starts at the top of the tank with a potential head only, h. Equating initial potential and final velocity heads:

$$\frac{v^2}{2g} = h \text{ from which, } v = \sqrt{2gh}$$

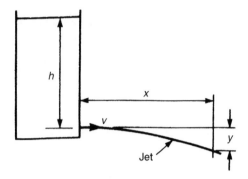

Figure 37.1

In practise, some energy loss occurs.

The ratio $\dfrac{\text{actual velocity of jet}}{\text{theoretical velocity of jet}}$ is called the **coefficient of velocity,** C_v, of the orifice. The actual velocity can be obtained from accurate observation of coordinates x and y of points on the jet trajectory.

The diameter of the jet is also found in practise to be less than the diameter of the orifice. The ratio $\dfrac{\text{cross-sectional area of jet}}{\text{area of orifice}}$ is called the **coefficient of contraction, C_c,** of the orifice.

The ratio $\dfrac{\text{actual rate of discharge}}{\text{theoretical rate of discharge}}$ is called the **coefficient of discharge, C_d,** of the orifice.

But $\dfrac{\text{actual rate of discharge}}{\text{theoretical rate of discharge}} = \dfrac{\text{actual velocity of jet} \times \text{c.s.a. of jet}}{\text{theoretical velocity of jet} \times \text{area of orifice}}$

i.e.
$$\boxed{C_d = C_v \times C_c}$$

For example, an orifice in the bottom of a water tank has a diameter of 12.3 mm. Assuming coefficients of contraction and velocity of 0.64 and 0.96 respectively, the depth of water required in the tank to give a rate of discharge through the orifice of 0.25 litres per second is determined as follows:

The theoretical rate of discharge from the tank,

$$\dot{V} = Av = \left(\frac{\pi d^2}{4}\right)\sqrt{2gh} = \frac{\pi(0.0125)^2}{4}\sqrt{(2)(9.81)h}$$

$$= 0.00054357\sqrt{h} \ \text{m}^3/\text{s}$$

$$= 0.54357\sqrt{h} \ \text{litres/s, where } h \text{ is in metres.}$$

Coefficient of discharge, $C_d = C_v \times C_c = 0.96 \times 0.64 = 0.6144$

$$C_d = \frac{\text{actual rate of discharge}}{\text{theoretical rate of discharge}}$$

thus, (actual rate of discharge) $= (C_d)$ (theoretical rate of

discharge)

$$= (0.6144)(0.54357\sqrt{h})$$

$$= 0.33397\sqrt{h} \text{ litres/s}$$

hence $\quad 0.25 = 0.33397\sqrt{h}$

from which, **depth of water, $h = \left(\dfrac{0.25}{0.33397} \right)^2 = 0.5604$ m**

$$= \mathbf{560.4 \ mm}$$

Impact of a Jet

The force exerted by a jet of water on a plate is, from Newton's third law of motion, equal and opposite to the force exerted by the plate on the water. From Newton's second law, this is equal to the rate of change of momentum of water.

$$\text{Rate of change of momentum} = \frac{\text{mass} \times \text{change of velocity}}{\text{time}}$$

In dealing with solids this is interpreted as:

$$\text{mass} \times \frac{\text{change of velocity}}{\text{time}} = \text{mass} \times \text{acceleration}$$

In dealing with the continuous process of fluid flow it must be interpreted as:

$$\frac{\text{mass}}{\text{time}} \times \text{change of velocity}$$

$$= \text{mass rate of flow} \times \text{change of velocity} = \dot{M}v$$

In the case of the jet striking a flat plate at right angles, as in Figure 37.2, the final velocity in the original direction is zero, so that v is the change of velocity in this direction. Also, if d is the diameter of the jet:

$$\dot{M} = \dot{V}p = Av\rho = \frac{\pi d^2}{4} v\rho$$

Force on plate, $\boldsymbol{F = \dot{M}v = \dfrac{\pi d^2}{4} v^2 \rho}$

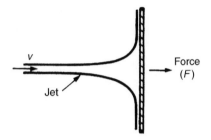

Figure 37.2

The force will be in newtons if the jet diameter is in metres, the jet velocity in metres per second and density in kilograms per metre cubed and mass flow rate M in kg/s.

For example, let a jet of water with a diameter of 12.5 mm and a velocity of 40 m/s strike a stationery flat plate at right angles.

Mass rate of flow,

$$\dot{M} = A v \rho = \frac{\pi d^2}{4} v \rho = \frac{\pi (0.0125)^2}{4} (40)(1000) = 4.9087 \text{ kg/s}$$

Force on plate,

$$F = \dot{M} v = (4.9087 \text{ kg/s})(40 \text{ m/s}) = 196.3 \text{ kg m/s}^2 = \mathbf{196.3 \ N}$$

38 Measurement of Fluid Flow

Introduction

The measurement of fluid flow is of great importance in many industrial processes, some examples including air flow in the ventilating ducts of a coal mine, the flow rate of water in a condenser at a power station, the flow rate of liquids in chemical processes, the control and monitoring of the fuel, lubricating and cooling fluids of ships and aircraft engines, and so on. Fluid flow is one of the most difficult of industrial measurements to carry out, since flow behaviour depends on a great many variables concerning the physical properties of a fluid.

There are available a large number of fluid flow measuring instruments generally called **flowmeters**, which can measure the flow rate of liquids (in m^3/s) or the mass flow rate of gaseous fluids (in kg/s). The two main categories of flowmeters are differential pressure flowmeters and mechanical flowmeters.

Differential Pressure Flowmeters

When certain flowmeters are installed in pipelines they often cause an obstruction to the fluid flowing in the pipe by reducing the cross-sectional area of the pipeline. This causes a change in the velocity of the fluid, with a related change in pressure. Figure 38.1 shows a section through a pipeline into which a flowmeter has been inserted. The flow rate of the fluid may be determined from a measurement of the difference between the pressures on the walls of the pipe at specified distances upstream and downstream of the flowmeter. Such devices are known as **differential pressure flowmeters**.

The pressure difference in Figure 38.1 is measured using a manometer connected to appropriate pressure tapping points. The pressure is seen to be greater upstream of the flowmeter than downstream, the pressure difference being shown as h.

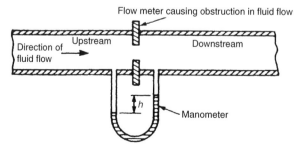

Figure 38.1

Calibration of the manometer depends on the shape of the obstruction, the positions of the pressure tapping points and the physical properties of the fluid.

In industrial applications the pressure difference is detected by a differential pressure cell, the output from which is either an amplified pressure signal or an electrical signal.

Examples of differential pressure flowmeters commonly used include:

(a) Orifice plate
(b) Venturi tube
(c) Flow nozzles
(d) Pitot-static tube

British Standard reference BS 1042: Part 1: 1964 and Part 2A: 1973 'Methods for the measurement of fluid flow in pipes' gives specifications for measurement, manufacture, tolerances, accuracy, sizes, choice, and so on, of differential flowmeters.

Orifice Plate

Construction

An orifice plate consists of a circular, thin, flat plate with a hole (or orifice) machined through its centre to fine limits of accuracy. The orifice has a diameter less than the pipeline into which the plate is installed and a typical section of an installation is shown in Figure 38.2(a). Orifice plates are manufactured in stainless steel, monel metal, polyester glass fibre, and for large pipes, such as sewers or hot gas mains, in brick and concrete.

Principles of operation

When a fluid moves through a restriction in a pipe, the fluid accelerates and a reduction in pressure occurs, the magnitude of which is related to the flow rate of the fluid. The variation of pressure near an orifice plate is shown in Figure 38.2(b). The position of minimum pressure is located downstream from the orifice plate where the flow stream is narrowest. This point of minimum cross-sectional area of the jet is called the '**vena contracta**'. Beyond this point the pressure rises but does not return to the original upstream value and there is a permanent pressure loss. This loss depends on the size and type of orifice plate, the positions of the upstream and downstream pressure tappings and the change in fluid velocity between the pressure tappings that depends on the flow rate and the dimensions of the orifice plate.

In Figure 38.2(a) corner pressure tappings are shown at A and B. Alternatively, with an orifice plate inserted into a pipeline of diameter d, pressure tappings are often located at distances of d and $d/2$ from the plate respectively upstream and downstream. At distance d upstream the flow pattern is not influenced by the presence of the orifice plate and distance $d/2$ coincides with the vena contracta

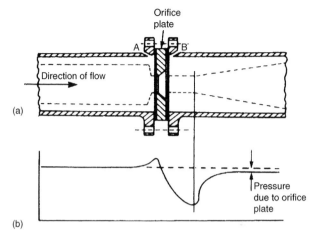

Figure 38.2

Advantages of orifice plates

(i) They are relatively inexpensive
(ii) They are usually thin enough to fit between an existing pair of pipe flanges

Disadvantages of orifice plates

(i) The sharpness of the edge of the orifice can become worn with use, causing calibration errors
(ii) The possible build-up of matter against the plate
(iii) A considerable loss in the pumping efficiency due to the pressure loss downstream of the plate

Applications

Orifice plates are usually used in medium and large pipes and are best suited to the indication and control of essentially constant flow rates. Several applications are found in the general process industries.

Venturi Tube

Construction

The Venturi tube or venturimeter is an instrument for measuring with accuracy the flow rate of fluids in pipes. A typical arrangement of a section through such a device is shown in Figure 38.3, and consists of a short converging conical tube called the inlet or upstream cone, leading to a cylindrical portion called

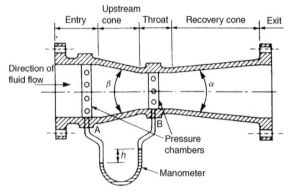

Figure 38.3

the throat. A diverging section called the outlet or recovery cone follows this. The entrance and exit diameter is the same as that of the pipeline into which it is installed. Angle β is usually a maximum of 21°, giving a taper of $\beta/2$ of 10.5°. The length of the throat is made equal to the diameter of the throat. Angle α is about 5° to 7° to ensure a minimum loss of energy but where this is unimportant α can be as large as 14° to 15°.

Pressure tappings are made at the entry (at A) and at the throat (at B) and the pressure difference h which is measured using a manometer, a differential pressure cell or similar gauge, is dependent on the flow rate through the meter. Usually pressure chambers are fitted around the entrance pipe and the throat circumference with a series of tapping holes made in the chamber to which the manometer is connected. This ensures that an average pressure is recorded. The loss of energy due to turbulence that occurs just downstream with an orifice plate is largely avoided in the venturimeter due to the gradual divergence beyond the throat. Venturimeters are usually made a permanent installation in a pipeline and are manufactured usually from stainless steel, cast iron, monel metal or polyester glass fibre.

Advantages of venturimeters

(i) High accuracy results are possible
(ii) There is a low-pressure loss in the tube (typically only 2% to 3% in a well proportioned tube)
(iii) Venturimeters are unlikely to trap any matter from the fluid being metered

Disadvantages of venturimeters

(i) High manufacturing costs
(ii) The installation tends to be rather long (typically 120 mm for a pipe of internal diameter 50 mm)

Flow Nozzle

The flow nozzle lies between an orifice plate and the venturimeter both in performance and cost. A typical section through a flow nozzle is shown in Figure 38.4 where pressure tappings are located immediately adjacent to the upstream and downstream faces of the nozzle (i.e. at points A and B). The fluid flow does not contract any further as it leaves the nozzle and the pressure loss created is considerably less than that occurring with orifice plates. Flow nozzles are suitable for use with high velocity flows for they do not suffer the wear that occurs in orifice plate edges during such flows.

Pitot-static Tube

A Pitot-static tube is a device for measuring the velocity of moving fluids or of the velocity of bodies moving through fluids. It consists of one tube, called the Pitot tube, with an open end facing the direction of the fluid motion, shown as pipe R in Figure 38.5, and a second tube, called the piezometer tube, with the opening at 90° to the fluid flow, shown as T in Figure 38.5. Pressure recorded by a pressure gauge moving with the flow, i.e. static or stationary relative to the fluid, is called free stream pressure and connecting a pressure

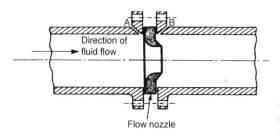

Figure 38.4

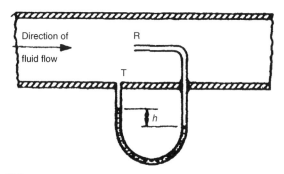

Figure 38.5

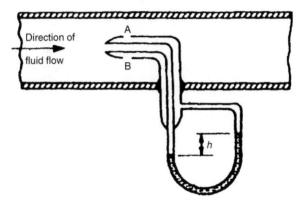

Figure 38.6

gauge to a small hole in the wall of a pipe, such as point T in Figure 38.5, is the easiest method of recording this pressure. The difference in pressure $(p_R - p_T)$, shown as h in the manometer of Figure 35.5, is an indication of the speed of the fluid in the pipe.

Figure 38.6 shows a practical Pitot-static tube consisting of a pair of concentric tubes. The centre tube is the impact probe that has an open end which faces 'head-on' into the flow. The outer tube has a series of holes around its circumference located at right angles to the flow, as shown by A and B in Figure 38.6. The manometer, showing a pressure difference of h, may be calibrated to indicate the velocity of flow directly.

Applications

A Pitot-static tube may be used for both turbulent and non-turbulent flow. The tubes can be made very small compared with the size of the pipeline and the monitoring of flow velocity at particular points in the cross-section of a duct can be achieved. The device is generally unsuitable for routine measurements and in industry is often used for making preliminary tests of flow rate in order to specify permanent flow measuring equipment for a pipeline. The main use of Pitot tubes is to measure the velocity of solid bodies moving through fluids, such as the velocity of ships. In these cases, the tube is connected to a Bourdon pressure gauge that can be calibrated to read velocity directly. A development of the Pitot tube, a **pitometer**, tests the flow of water in water mains and detects leakages.

Advantages of Pitot-static tubes

  (i) They are inexpensive devices
 (ii) They are easy to install
(iii) They produce only a small pressure loss in the tube
 (iv) They do not interrupt the flow

Disadvantages of Pitot-static tubes

(i) Due to the small pressure difference, they are only suitable for high velocity fluids
(ii) They can measure the flow rate only at a particular position in the cross-section of the pipe
(iii) They easily become blocked when used with fluids carrying particles

Mechanical Flowmeters

With mechanical flowmeters, the fluid flowing past it displaces a sensing element situated in a pipeline.

Examples of mechanical flowmeters commonly used include:

(a) Deflecting vane flowmeter
(b) Turbine type meters

Deflecting Vane Flowmeter

The deflecting vane flowmeter consists basically of a pivoted vane suspended in the fluid flow stream as shown in Figure 38.7

When a jet of fluid impinges on the vane it deflects from its normal position by an amount proportional to the flow rate. The movement of the vane is indicated on a scale that may be calibrated in flow units. This type of meter is normally used for measuring liquid flow rates in open channels or for measuring the velocity of air in ventilation ducts. The main disadvantages of this device are that it restricts the flow rate and it needs to be recalibrated for fluids of differing densities.

Turbine Type Meters

Turbine type flowmeters are those that use some form of multi-vane rotor and are driven by the fluid being investigated. Three such devices are the

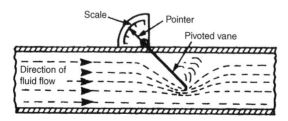

Figure 38.7

cup anemometer, the rotary vane positive displacement meter and the turbine flowmeter.

(a) **Cup anemometer**. An anemometer is an instrument that measures the velocity of moving gases and is most often used for the measurement of wind speed. The cup anemometer has three or four cups of hemispherical shape mounted at the end of arms radiating horizontally from a fixed point. The cup system spins round the vertical axis with a speed approximately proportional to the velocity of the wind. With the aid of a mechanical and/or electrical counter the wind speed can be determined and the device is easily adapted for automatic recording.

(b) **Rotary vane positive displacement meters** measure the flow rate by indicating the quantity of liquid flowing through the meter in a given time. A typical device is shown in section in Figure 38.8 and consists of a cylindrical chamber into which is placed a rotor containing a number of vanes (six in this case). Liquid entering the chamber turns the rotor and a known amount of liquid is trapped and carried round to the outlet. If x is the volume displaced by one blade then for each revolution of the rotor in Figure 35.8 the total volume displaced is $6x$. The rotor shaft may be coupled to a mechanical counter and electrical devices which may be calibrated to give flow volume. This type of meter in its various forms is used widely for the measurement of domestic and industrial water consumption, for the accurate measurement of petrol in petrol pumps and for the consumption and batch control measurements in the general process and food industries for measuring flows as varied as solvents, tar and molasses (i.e. thickish treacle).

(c) A **turbine flowmeter** contains in its construction a rotor to which blades are attached which spin at a velocity proportional to the velocity of the fluid which flows through the meter. A typical section through such a meter is shown in Figure 38.9. The number of revolutions made by the turbine blades may be determined by a mechanical or electrical device enabling the flow rate or total flow to be determined. Advantages

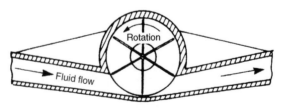

Figure 38.8

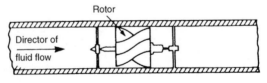

Figure 38.9

of turbine flowmeters include a compact durable form, high accuracy, wide temperature and pressure capability and good response characteristics. Applications include the volumetric measurement of both crude and refined petroleum products in pipelines up to 600 mm bore, and in the water, power, aerospace, process and food industries, and with modification may be used for natural, industrial and liquid gas measurements. Turbine flowmeters require periodic inspection and cleaning of the working parts.

Float and Tapered-tube Meter

Principle of operation

With orifice plates and venturimeters the area of the opening in the obstruction is fixed and any change in the flow rate produces a corresponding change in pressure. With the float and tapered-tube meter the area of the restriction may be varied so as to maintain a steady pressure differential. A typical meter of this type is shown diagrammatically in Figure 38.10 where a vertical tapered tube contains a 'float' that has a density greater than the fluid.

The float in the tapered tube produces a restriction to the fluid flow. The fluid can only pass in the annular area between the float and the walls of the tube. This reduction in area produces an increase in velocity and hence a pressure difference, which causes the float to rise. The greater the flow rate,

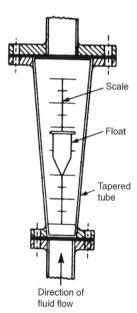

Direction of
fluid flow

Figure 38.10

the greater is the rise in the float position, and vice versa. The position of the float is a measure of the flow rate of the fluid and this is shown on a vertical scale engraved on a transparent tube of plastic or glass. For air, a small sphere is used for the float but for liquids there is a tendency to instability and the float is then designed with vanes that cause it to spin and thus stabilize itself as the liquid flows past. Such meters are often called '**rotameters**'. Calibration of float and tapered tube flowmeters can be achieved using a Pitot-static tube or, more often, by using a weighing meter in an instrument repair workshop.

Advantages of float and tapered-tube flowmeters

(i) They have a very simple design
(ii) They can be made direct reading
(iii) They can measure very low flow rates

Disadvantages of float and tapered-tube flowmeters

(i) They are prone to errors, such as those caused by temperature fluctuations
(ii) They can only be installed vertically in a pipeline
(iii) They cannot be used with liquids containing large amounts of solids in suspension
(iv) They need to be recalibrated for fluids of different densities

Practical applications of float and tapered-tube meters are found in the medical field, in instrument purging, in mechanical engineering test rigs and in simple process applications, in particular for very low flow rates. Many corrosive fluids can be handled with this device without complications.

Electromagnetic Flowmeter

The flow rate of fluids that conduct electricity, such as water or molten metal, can be measured using an electromagnetic flowmeter whose principle of operation is based on the laws of electromagnetic induction. When a conductor of length l moves at right angles to a magnetic field of density B at a velocity v, an induced e.m.f. e is generated, given by $e = Blv$ (see Chapter 48, page 271).

With the electromagnetic flowmeter arrangement shown in Figure 38.11, the fluid is the conductor and the e.m.f. is detected by two electrodes placed across the diameter of the non-magnetic tube.

Rearranging $e = Blv$ gives: **velocity,** $v = \dfrac{e}{Bl}$

Thus with B and l known, when e is measured, the velocity of the fluid can be calculated.

Main advantages of electromagnetic flowmeters

(i) Unlike other methods, there is nothing directly to impede the fluid flow
(ii) There is a linear relationship between the fluid flow and the induced e.m.f.

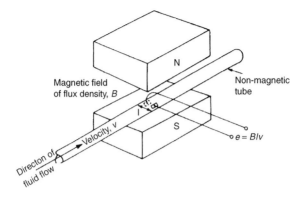

Figure 38.11

(iii) Flow can be metered in either direction by using a centre-zero measuring instrument

Applications of electromagnetic flowmeters are found in the measurement of speeds of slurries, pastes and viscous liquids, and they are also widely used in the water production, supply and treatment industry.

Hot-wire Anemometer

A simple hot-wire anemometer consists of a small piece of wire which is heated by an electric current and positioned in the air or gas stream whose velocity is to be measured. The stream passing the wire cools it, the rate of cooling being dependent on the flow velocity. In practice there are various ways in which this is achieved:

(i) If a constant current is passed through the wire, variation in flow results in a change of temperature of the wire and hence a change in resistance which may be measured by a Wheatstone bridge arrangement. The change in resistance may be related to fluid flow.
(ii) If the wire's resistance, and hence temperature, is kept constant, a change in fluid flow results in a corresponding change in current which can be calibrated as an indication of the flow rate.
(iii) A thermocouple may be incorporated in the assembly, monitoring the hot wire and recording the temperature which is an indication of the air or gas velocity.

Advantages of the hot-wire anemometer

(a) Its size is small
(b) It has great sensitivity

39 Simple Harmonic Motion and Natural Vibrations

Simple Harmonic Motion

Simple harmonic motion is defined as a periodic motion of a point along a straight line, such that its acceleration is always towards a fixed point in that line and is proportional to its distance from that point.

Simple harmonic motion (SHM) may be considered as the projection on a diameter of a movement at uniform speed around the circumference of a circle.

In Figure 39.1, P moves with uniform speed $v(= \omega r)$ around a circle of radius r; the point X projected from P on diameter AB moves with SHM. The acceleration of P is the centripetal acceleration, $\omega^2 r$. The displacement (measured from the mean position O), the velocity and acceleration of X are respectively:

$$\text{displacement, } x = OX = r \cos \theta = r \cos \omega t$$

where t is the time measured from the instant when P and X are at A and $\theta = 0$

$$\text{velocity, } v = \omega r \sin \theta = -\omega r \sin \omega t$$

$$\text{acceleration, } a = -\omega^2 r \cos \theta = -\omega^2 r \cos \omega t = -\omega^2 x$$

The expressions for velocity and acceleration can be derived from that for displacement by differentiating with respect to time. The negative signs in the expressions for velocity and acceleration show that for the position X in Figure 36.1, both velocity and acceleration are in the opposite direction from the displacement. Displacement and acceleration are always in opposite directions. The periodic time T of the motion is the time taken for one complete

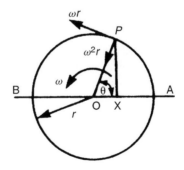

Figure 39.1

oscillation of X. In this time, OP makes one complete revolution and therefore:

$$\text{periodic time } T = \frac{2\pi}{\omega}$$

Since $a = \omega^2 x$, $\omega = \sqrt{\dfrac{a}{x}}$ and therefore,

$$T = 2\pi\sqrt{\frac{x}{a}} = 2\pi\sqrt{\frac{\text{displacement}}{\text{corresponding acceleration}}}$$

The frequency n is the number of complete oscillations in one second.

$$n = \frac{\omega}{2\pi} = \frac{1}{T} = \frac{1}{2\pi}\sqrt{\frac{a}{x}}$$

The unit for n of one oscillation per second is called the hertz, Hz. The maximum velocity of X occurs at the mid-point, where it equals the velocity of P,

i.e. $$\mathbf{v}_{max} = \boldsymbol{\omega}\mathbf{r}$$

The maximum acceleration of X occurs at the extreme positions A and B where it equals the acceleration of P,

i.e. $$\mathbf{a}_{max} = \boldsymbol{\omega}^2\mathbf{r}$$

The velocity of X is zero at A and B; its acceleration is zero at O. The amplitude of the oscillation is r. The distance AB, i.e. $2r$, is sometimes called the stroke or travel of the motion.

For example, a body moves with SHM of amplitude 45 mm and frequency 2.5 Hz.

Hence, frequency $= \dfrac{\omega}{2\pi} = 2.5$ from which, $\omega = (2\pi)(2.5) = 5\pi = 15.708$ rad/s

The maximum velocity $= \omega r = 15.708 \times 45 = \mathbf{706.9}$ **mm/s** and this occurs at the mean position.

The maximum acceleration $= \omega^2 r = (15.708)^2 \times 45 = \mathbf{11.10}$ **m/s^2**, which occurs at each extreme position.

For a displacement of, say, $x = 25$ mm, $\cos\theta = \dfrac{25}{45}$ (see Figure 39.1), from which, $\theta = 56.25°$

The velocity for a displacement of 25 mm from the mean position is:

$$\omega r \sin\theta = (15.708)(45)\sin 56.25° = \mathbf{587.7}\ \mathbf{mm/s}$$

The acceleration for a displacement of 25 mm from the mean position is:

$$\omega^2 r \cos\theta = \omega^2 x = (15.708)^2(25) = \mathbf{6169}\ \mathbf{mm/s^2}\ \text{or}\ \mathbf{6.169}\ \mathbf{m/s^2}$$

Natural Vibration

Motion closely approximating to SHM occurs in a number of natural or free vibrations. Many examples are met where a body oscillates under a control that

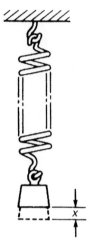

Figure 39.2

obeys Hooke's law, for example, a spring or a beam. Consider, for example, the helical spring shown in Figure 39.2. If, from its position of rest, the mass M is pulled down a distance r and then released, the mass will oscillate in a vertical line. In the rest position, the force in the spring will exactly balance the force of gravity on the mass.

If s is the stiffness of the spring, that is, force per unit change of length, then for a displacement x from the rest position, the change in the force in the spring is sx. This change of force is the unbalanced or accelerating force F acting on the mass M, i.e. $F = sx$

$$\text{Acceleration } a = \frac{F}{M} = \frac{sx}{M}$$

This shows that the acceleration is directly proportional to displacement from its rest position. The motion is therefore SHM. The periodic time is given by:

$$T = 2\pi\sqrt{\frac{x}{a}} = 2\pi\sqrt{\frac{x}{\frac{sx}{M}}} = 2\pi\sqrt{\frac{M}{s}}$$

For example, a load of 10 kg is hung from a vertical helical spring and it causes an extension of 15 mm. The load is pulled down a further distance of 18 mm and then released.

Thus, the weight of the load $= Mg = 10 \times 9.81 = 98.1$ N

Stiffness of spring,

$$s = \frac{\text{force}}{\text{extension}} = \frac{98.1 \text{ N}}{15 \text{ mm}} = 6.54 \text{ N/mm} = 6.54 \text{ kN/m}$$

Frequency of the vibration,

$$n = \frac{1}{T} = \frac{1}{2\pi\sqrt{\dfrac{M}{s}}} = \frac{\sqrt{s}}{2\pi\sqrt{M}} = \frac{\sqrt{6.54 \text{ kN/m}}}{2\pi\sqrt{10 \text{ kg}}}$$

$$= \frac{\sqrt{654\dfrac{\text{kg m/s}^2}{\text{kg m}}}}{2\pi} = \textbf{4.07Hz}$$

The amplitude r of the vibration is 18 mm.

The maximum velocity of the load is: $\omega r = (2\pi n)(r) = (2\pi)(4.07)(18) = $ **460.3 mm/s**

The maximum acceleration of the load is: $\omega^2 r = [(2\pi)(4.07)]^2(18) = $ **11771 mm/s^2 or 11.77 m/s^2**

$$\left.\begin{array}{c}\text{The maximum force} \\ \text{in the spring}\end{array}\right\} = \text{maximum extension} \times \text{stiffness}$$

$$= (15 \text{ mm} + 18 \text{ mm})(6.54 \text{ N/mm})$$

$$= \textbf{215.8 N}$$

Simple Pendulum

Another common example of a vibration giving a close approximation to SHM is the movement of a simple pendulum. This is defined as a mass of negligible dimensions on the end of a cord or rod of negligible mass. For a small displacement x of the bob A from its mean position C in Figure 39.3, the accelerating force F on the bob, weight W, is $W\sin\theta$, which very nearly equals $W\theta$ if θ is a small angle and measured in radians.

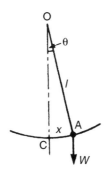

Figure 39.3

Therefore, acceleration of the bob, $a = \dfrac{F}{M} = \dfrac{W\theta}{M} = \dfrac{Mg\theta}{M} = g\theta$ where $M\left(=\dfrac{W}{g}\right)$ is the mass of the bob.

However, $x = l\theta$ when θ is measured in radians, hence, $\theta = \dfrac{x}{l}$ and $a = g\theta = \dfrac{gx}{l}$

The acceleration is therefore shown to be proportional to displacement, satisfying the definition of SHM, and the periodic time is:

$$T = 2\pi\sqrt{\frac{x}{a}} = 2\pi\sqrt{\frac{x}{\frac{gx}{l}}} = 2\pi\sqrt{\frac{l}{g}}$$

For example, a simple pendulum has a length of 780 mm. Periodic time of the pendulum, $T = 2\pi\sqrt{\dfrac{l}{g}} = 2\pi\sqrt{\dfrac{0.78 \text{ m}}{9.81 \text{ m/s}^2}} = \textbf{1.772 s}$

If the amplitude of movement of the bob is 80 mm, then the maximum velocity of the bob is: $\omega r = \left(\dfrac{2\pi}{T}\right) r$ where r is the amplitude.

Hence maximum velocity $= \left(\dfrac{2\pi}{1.772}\right)(80) = \textbf{283.7 mm/s}$

and maximum acceleration $= \omega^2 r = \left(\dfrac{2\pi}{T}\right)^2 (r) = \left(\dfrac{2\pi}{1.772}\right)^2 (80)$

$$= \textbf{1006 mm/s}^2 \text{ or } \textbf{1.006 m/s}^2$$

The angular motion of the pendulum must not be confused with the angular motion of an imaginary line used in the analysis of simple harmonic motion. The imaginary line (OP in Figure 39.1) rotates at a constant speed. The angular velocity of the pendulum is variable, having its maximum value in the vertical position. For a velocity v of the bob, the angular velocity of the pendulum is:

$$\omega_p = \frac{v}{l}$$

The angular acceleration of the pendulum is greatest in the extreme positions. For an acceleration a of the bob, the angular acceleration of the pendulum is:

$$\alpha_p = \frac{a}{l}$$

Part Three Electrical Engineering Science

40 An Introduction to Electric Circuits

Electrical/Electronic System Block Diagrams

An electrical/electronic **system** is a group of components connected together to perform a desired function. Figure 40.1 shows a simple public address system, where a microphone is used to collect acoustic energy in the form of sound pressure waves and converts this to electrical energy in the form of small voltages and currents; the signal from the microphone is then amplified by means of an electronic circuit containing transistors/integrated circuits before it is applied to the loudspeaker.

A **sub-system** is a part of a system that performs an identified function within the whole system; the amplifier in Figure 40.1 is an example of a sub-system.

A **component** or **element** is usually the simplest part of a system that has a specific and well-defined function — for example, the microphone in Figure 40.1.

The illustration in Figure 40.1 is called a **block diagram** and electrical/electronic systems, which can often be quite complicated, can be better understood when broken down in this way. It is not always necessary to know precisely what is inside each sub-system in order to know how the whole system functions.

As another example of an engineering system, Figure 40.2 illustrates a temperature control system containing a heat source (such as a gas boiler), a fuel controller (such as an electrical solenoid valve), a thermostat and a source of electrical energy. The system of Figure 40.2 can be shown in block diagram form as in Figure 40.3; the thermostat compares the actual room temperature with the desired temperature and switches the heating on or off.

There are many types of engineering systems. A **communications system** is an example, where a local area network could comprise a file server, coaxial cable, network adapters, several computers and a laser printer;an **electromechanical system** is another example, where a car electrical system could comprise a battery, a starter motor, an ignition coil, a contact breaker and a distributor. Block diagrams may represent all such systems as these.

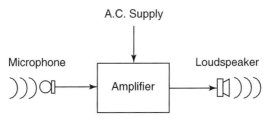

Figure 40.1

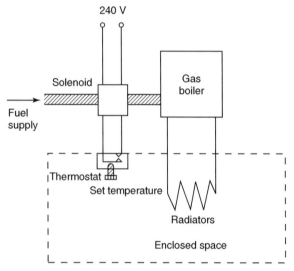

Figure 40.2

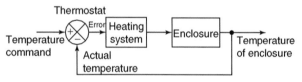

Figure 40.3

Standard Symbols for Electrical Components

Symbols are used for components in electrical circuit diagrams and some of the more common ones are shown in Figure 40.4.

Electric Current and Quantity of Electricity

All **atoms** consist of **protons, neutrons** and **electrons**. The protons, which have positive electrical charges, and the neutrons, which have no electrical charge, are contained within the **nucleus**. Removed from the nucleus are minute negatively charged particles called electrons. Atoms of different materials differ from one another by having different numbers of protons, neutrons and electrons. An equal number of protons and electrons exist within an atom

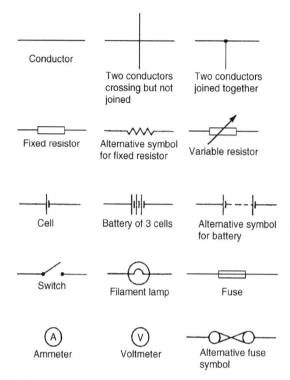

Figure 40.4

and it is said to be electrically balanced, as the positive and negative charges cancel each other out. When there are more than two electrons in an atom the electrons are arranged into **shells** at various distances from the nucleus.

All atoms are bound together by powerful forces of attraction existing between the nucleus and its electrons. Electrons in the outer shell of an atom, however, are attracted to their nucleus less powerfully than are electrons whose shells are nearer the nucleus.

It is possible for an atom to lose an electron; the atom, which is now called an **ion**, is not now electrically balanced, but is positively charged and is thus able to attract an electron to itself from another atom. Electrons that move from one atom to another are called free electrons and such random motion can continue indefinitely. However, if an electric pressure or **voltage** is applied across any material there is a tendency for electrons to move in a particular direction. This movement of free electrons, known as **drift**, constitutes an electric current flow. **Thus current is the rate of movement of charge.**

Conductors are materials that contain electrons that are loosely connected to the nucleus and can easily move through the material from one atom to another. **Insulators** are materials whose electrons are held firmly to their nucleus.

The unit used to measure the **quantity of electrical charge** Q is called the **coulomb** C (where 1 coulomb $= 6.24 \times 10^{18}$ electrons)

If the drift of electrons in a conductor takes place at the rate of one coulomb per second the resulting current is said to be a current of one ampere.

Thus 1 ampere = 1 coulomb per second or 1 A = 1 C/s

Hence 1 coulomb = 1 ampere second or 1 C = 1 As

Generally, if I is the current in amperes and t the time in seconds during which the current flows, then $I \times t$ represents the quantity of electrical charge in coulombs, i.e. quantity of electrical charge transferred,

$$\boxed{Q = I \times t \text{ coulombs}}$$

For example, if a current of 10 A flows for four minutes, the quantity of electricity transferred, $Q = It = 10 \times 240 = \mathbf{2400\ C}$

Electrical Potential and e.m.f.

The **unit of electric potential** is the volt (V), where one volt is one joule per coulomb. One volt is defined as the difference in potential between two points in a conductor which, when carrying a current of one ampere, dissipates a power of one watt, i.e.

$$\text{volts} = \frac{\text{watts}}{\text{amperes}} = \frac{\text{joules/second}}{\text{amperes}} = \frac{\text{joules}}{\text{ampere seconds}} = \frac{\text{joules}}{\text{coulombs}}$$

A change in electric potential between two points in an electric circuit is called a **potential difference (p.d.)**. The **electromotive force (e.m.f.)** provided by a source of energy such as a battery or a generator is measured in volts.

For a continuous current to flow between two points in a circuit a potential difference or **voltage, V**, is required between them; a complete conducting path is necessary to and from the source of electrical energy.

Figure 40.5 shows a cell connected across a filament lamp. Current flow, by convention, is considered as flowing from the positive terminal of the cell, around the circuit to the negative terminal.

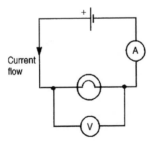

Figure 40.5

Resistance and Conductance

The flow of electric current is subject to friction. This friction, or opposition, is called **resistance R** and is the property of a conductor that limits current. The **unit of electric resistance** is the **ohm (Ω)**; 1 ohm is defined as the resistance that will have a current of 1 ampere flowing through it when 1 volt is connected across it.

$$\text{resistance } R = \frac{\text{Potential difference}}{\text{current}} \quad \text{i.e.} \quad \boxed{R = \frac{V}{I} \text{ ohms}}$$

where V is the potential difference across the two points, in volts, and I is the current flowing between the two points, in amperes.

The reciprocal of resistance is called **conductance** and is measured in siemens (S). Thus,

$$\text{conductance} \quad \boxed{G = \frac{1}{R} \text{ siemens}}$$

Basic Electrical Measuring Instruments

An **ammeter** is an instrument used to measure current and must be connected **in series** with the circuit. Figure 40.5 shows an ammeter connected in series with the lamp to measure the current flowing through it. Since all the current in the circuit passes through the ammeter it must have a very **low resistance**.

A **voltmeter** is an instrument used to measure p.d. and must be connected **in parallel** with the part of the circuit whose p.d. is required. In Figure 40.5, a voltmeter is connected in parallel with the lamp to measure the p.d. across it. To avoid a significant current flowing through it a voltmeter must have a very **high resistance**.

An **ohmmeter** is an instrument for measuring resistance.

A **multimeter**, or universal instrument, may be used to measure voltage, current and resistance. An '**Avometer**' is a typical example.

The **cathode ray oscilloscope (CRO)** may be used to observe waveforms and to measure voltages and currents. The display of a CRO involves a spot of light moving across a screen. The amount by which the spot is deflected from its initial position depends on the p.d. applied to the terminals of the CRO and the range selected. The displacement is calibrated in 'volts per cm'. For example, if the spot is deflected 3 cm and the volts/cm switch is on 10 V/cm then the magnitude of the p.d. is 3 cm × 10 V/cm, i.e. 30 V.

(See Chapter 50 for more detail about electrical measuring instruments and measurements)

Linear and Non-linear Devices

Figure 40.6 shows a circuit in which current I can be varied by the variable resistor R_2. For various settings of R_2, the current flowing in resistor R_1,

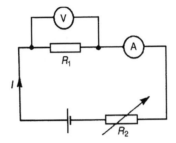

Figure 40.6

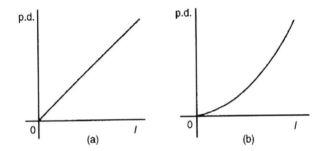

Figure 40.7

displayed on the ammeter, and the p.d. across R_1, displayed on the voltmeter, are noted and a graph is plotted of p.d. against current. The result is shown in Figure 40.7(a) where the straight-line graph passing through the origin indicates that current is directly proportional to the p.d. Since the gradient, i.e. $\dfrac{\text{p.d.}}{\text{current}}$ is constant, resistance R_1 is constant. A resistor is thus an example of a **linear device**.

If the resistor R_1 in Figure 40.6 is replaced by a component such as a lamp then the graph shown in Figure 40.7(b) results when values of p.d. are noted for various current readings. Since the gradient is changing, the lamp is an example of a **non-linear device**.

Ohm's Law

Ohm's law states that the current I flowing in a circuit is directly proportional to the applied voltage V and inversely proportional to the resistance R, provided the temperature remains constant. Thus,

$$I = \frac{V}{R} \quad \text{or} \quad V = IR \quad \text{or} \quad R = \frac{V}{I}$$

Table 40.1

Prefix	Name	Meaning	Example
M	mega	multiply by 1 000 000 (i.e. $\times 10^6$)	2 MΩ = 2 000 000 ohms
k	kilo	multiply by 1000 (i.e. $\times 10^3$)	10 kV = 10 000 volts
m	milli	divide by 1000 (i.e. $\times 10^{-3}$)	25 mA = $\dfrac{25}{1000}$ A = 0.025 amperes
μ	micro	divide by 1 000 000 (i.e. $\times 10^{-6}$)	50 μV = $\dfrac{50}{1\,000\,000}$ V = 0.00005 volts

For example, the current flowing through a resistor is 0.8 A when a p.d. of 20 V is applied. From Ohm's law,

$$\text{resistance } R = \frac{V}{I} = \frac{20}{0.8} = \frac{200}{8} = \mathbf{25\ \Omega}$$

Multiples and Sub-multiples

Currents, voltages and resistances can often be very large or very small. Thus **multiples and sub-multiples** of units are often used, as stated in chapter 1. The most common ones, with an example of each, are listed in Table 40.1.

For example, a 100 V battery is connected across a resistor and causes a current of 5 mA to flow. The resistance of the resistor is given by:

$$R = \frac{V}{I} = \frac{100}{5 \times 10^{-3}} = \frac{100 \times 10^3}{5} = 20 \times 10^3 = \mathbf{20\ k\Omega}$$

If the voltage is reduced to, say, 25 V, the current flowing,

$$I = \frac{V}{R} = \frac{25}{20 \times 10^3} = \frac{25}{20} \times 10^{-3} = \mathbf{1.25\ mA}$$

Conductors and Insulators

A **conductor** is a material having a low resistance that allows electric current to flow in it. All metals are conductors and some examples include copper, aluminium, brass, platinum, silver, gold and carbon.

An **insulator** is a material having a high resistance that does not allow electric current to flow in it. Some examples of insulators include plastic, rubber, glass, porcelain, air, paper, cork, mica, ceramics and certain oils.

Electrical Power

The product of potential difference V and current I gives power P in an electrical circuit. The unit of power is the **watt, W.**

Hence $\boxed{P = V \times I \text{ watts}}$

Using Ohm's, it may be shown that power is also given by:

$$\boxed{P = I^2R = \frac{V^2}{R} \text{ watts}}$$

There are thus three possible formulae that may be used for calculating power.

For example, the power dissipated when a current of 4 mA flows through a resistance of 5 kΩ is given by:

power $P = I^2R = (4 \times 10^{-3})^2(5 \times 10^3) = 16 \times 10^{-6} \times 5 \times 10^3$

$$= 80 \times 10^{-3} = \textbf{0.08 W or 80 mW}$$

Alternatively, since $I = 4 \times 10^{-3}$ and $R = 5 \times 10^3$ then from Ohm's law, voltage $V = IR = 4 \times 10^{-3} \times 5 \times 10^3 = 20$ V

Hence, **power** $P = V \times I = 20 \times 4 \times 10^{-3} = \textbf{80 mW}$

Electrical Energy

$$\boxed{\textbf{Electrical energy = power} \times \textbf{time}}$$

If the power is measured in watts and the time in seconds then the unit of energy is watt-seconds or **joules**. If the power is measured in kilowatts and the time in hours then the unit of energy is **kilowatt-hours**, often called the '**unit of electricity**'. The 'electricity meter' in the home records the number of kilowatt-hours used and is thus an energy meter.

(1 kWh = 1000 watt hour = $1000 \times 60 \times 60 = 3\,600\,000$ watt seconds or joules).

For example, electrical equipment in an office takes a current of 13 A from a 240 V supply. If the equipment is used for 30 hours each week and 1 kWh of energy costs 7p, the cost per week of electricity is calculated as follows:

Power $= VI$ watts $= 240 \times 13 = 3120$ W $= 3.12$ kW

Energy used per week $=$ power $\times$ time $= (3.12$ kW$) \times (30$ h$)$

$$= 93.6 \text{ kWh}$$

213

Cost at 7 p per kWh = 93.6 × 7 = 655.2 p

Hence the **weekly cost of electricity = £6.55**

Main Effects of Electric Current

The three main effects of an electric current are: (a) magnetic effect (b) chemical effect (c) heating effect

Some practical applications of the effects of an electric current include:

Magnetic effect: bells, relays, motors, generators, transformers, telephones, car-ignition and lifting magnets (see chapter 47)

Chemical effect: primary and secondary cells and electroplating (see chapter 42)

Heating effect: cookers, water heaters, electric fires, irons, furnaces, kettles and soldering irons

---

Fuses

A **fuse** is used to prevent overloading of electrical circuits. The fuse, which is made of material having a low melting point, utilises the heating effect of an electric current. A fuse is placed in an electrical circuit and if the current becomes too large the fuse wire melts and so breaks the circuit. A circuit diagram symbol for a fuse is shown in Figure 40.1, on page 207.

For example, an electric toaster has a power rating of 1 kW, and is connected to a 240 V supply.

Power taken, $P = VI$, from which,

$$\text{current } I = \frac{P}{V} = \frac{1000}{240} = \frac{100}{24} = 4.17 \text{ A}$$

Hence a **5 A fuse** is most appropriate for the toaster.

41 Resistance Variation

Resistance and Resistivity

The resistance of an electrical conductor depends on four factors, these being: (a) the length of the conductor, (b) the cross-sectional area of the conductor, (c) the type of material and (d) the temperature of the material.

Resistance, R, is directly proportional to length, l, of a conductor, i.e. $R \propto l$. Thus, for example, if the length of a piece of wire is doubled, then the resistance is doubled.

Resistance, R, is inversely proportional to cross-sectional area, a, of a conductor, i.e. $R \propto \dfrac{1}{a}$. Thus, for example, if the cross-sectional area of a piece of wire is doubled then the resistance is halved.

Since $R \propto l$ and $R \propto \dfrac{1}{a}$ then $R \propto \dfrac{l}{a}$. By inserting a constant of proportionality into this relationship the type of material used may be taken into account.

The constant of proportionality is known as the **resistivity** of the material and is given the symbol ρ (Greek rho).

Thus,   resistance $\boxed{R = \dfrac{\rho l}{a} \text{ ohms}}$

ρ is measured in ohm metres (Ωm)

The value of the resistivity is that resistance of a unit cube of the material measured between opposite faces of the cube.

Resistivity varies with temperature and some typical values of resistivities measured at about room temperature are given below:

Copper 1.7×10^{-8} Ωm (or 0.017 $\mu\Omega$m), Aluminium 2.6×10^{-8} Ωm (or 0.026 $\mu\Omega$m), Carbon (graphite) 10×10^{-8} Ωm (0.10 $\mu\Omega$m), Glass 1×10^{10} Ωm (or 10^4 $\mu\Omega$m), Mica 1×10^{13} Ωm (or 10^7 $\mu\Omega$m)

Note that good conductors of electricity have a low value of resistivity and good insulators have a high value of resistivity.

For example, the resistance of a 2 km length of aluminium overhead power cable, if the cross-sectional area of the cable is 100 mm², is given by:

$$R = \frac{\rho l}{a} = \frac{(2.6 \times 10^{-8} \ \Omega\text{m})(2000 \ \text{m})}{(100 \times 10^{-6} \ \text{m}^2)} = \frac{0.026 \times 2000}{100} \ \Omega = \mathbf{0.52 \ \Omega}$$

Temperature Coefficient of Resistance

In general, as the temperature of a material increases, most conductors increase in resistance, insulators decrease in resistance, whilst the resistance of some special alloys remain almost constant.

The **temperature coefficient of resistance** of a material is the increase in the resistance of a 1 Ω resistor of that material when it is subjected to a rise of temperature of 1°C. The symbol used for the temperature coefficient of resistance is α (Greek alpha). Thus, if some copper wire of resistance 1 Ω is heated through 1°C and its resistance is then measured as 1.0043 Ω then $\alpha = 0.0043 \ \Omega/\Omega°C$ for copper. The units are usually expressed only as 'per °C', i.e. $\alpha = 0.0043/°C$ for copper. If the 1 Ω resistor of copper is heated through 100°C then the resistance at 100°C would be $1 + 100 \times 0.0043 = 1.43 \ \Omega$

Some typical values of temperature coefficient of resistance measured at 0°C are given below:

Copper	0.0043/°C	Aluminium	0.0038/°C
Nickel	0.0062/°C	Carbon	−0.00048/°C
Constantan	0	Eureka	0.00001/°C

(Note that the negative sign for carbon indicates that its resistance falls with increase of temperature.)

If the resistance of a material at 0°C is known the resistance at any other temperature can be determined from:

$$R_\theta = R_0(1 + \alpha_0\theta)$$

where R_0 = resistance at 0°C,

R_θ = resistance at temperature θ°C and

α_0 = temperature coefficient of resistance at 0°C

For example, a coil of copper wire has a resistance of 100 Ω when its temperature is 0°C. Its resistance at 70°C, if the temperature coefficient of resistance of copper at 0°C is 0.0043/°C, is given by:

resistance at 100 °C, $R_{100} = R_0(1 + \alpha_0\theta) = 100[1 + (0.0043)(70)]$

$$= 100[1 + 0.301] = \mathbf{130.1 \ \Omega}$$

If the resistance of a material at room temperature (approximately 20°C), R_{20}, and the temperature coefficient of resistance at 20°C, α_{20}, are known, then the resistance R_θ at temperature θ°C is given by:

$$R_\theta = R_{20}[1 + \alpha_{20}(\theta - 20)]$$

For example, a coil of copper wire has a resistance of 10 Ω at 20°C. If the temperature coefficient of resistance of copper at 20°C is 0.004/°C the resistance of the coil when the temperature rises to 100°C is given by:

resistance at 100°C, $R_{100} = 10[1 + (0.004)(100 - 20)]$

$$= 10[1 + 0.32] = 10(1.32) = \mathbf{13.2 \ \Omega}$$

If the resistance at 0°C is not known, but is known at some other temperature θ_1, then the resistance at any temperature can be found as follows:

$$R_1 = R_0(1 + \alpha_0\theta_1) \text{ and } R_2 = R_0(1 + \alpha_0\theta_2)$$

Dividing one equation by the other gives:

$$\boxed{\frac{R_1}{R_2} = \frac{1 + \alpha_0\theta_1}{1 + \alpha_0\theta_2}} \text{ where } R_2 = \text{resistance at temperature } \theta_2$$

For example, some copper wire has a resistance of 200 Ω at 20°C. A current is passed through the wire and the temperature rises to 90°C. The resistance of the wire at 90°C, assuming that the temperature coefficient of resistance is 0.004/°C at 0°C, is given by:

$$R_{90} = \frac{R_{20}[1 + 90\alpha_0]}{[1 + 20\alpha_0]} = \frac{200[1 + 90(0.004)]}{[1 + 20(0.004)]} = \frac{200[1 + 0.36]}{[1 + 0.08]}$$

$$= \frac{200(1.36)}{(1.08)} = \textbf{251.85 } \Omega$$

Resistor Colour Coding and Ohmic Values

(a) Colour code for fixed resistors

The colour code for fixed resistors is given in Table 41.1

(i) For a **four-band fixed resistor** (i.e. resistance values with two significant figures):

Table 41.1

Colour	Significant Figures	Multiplier	Tolerance
Silver	—	10^{-2}	±10%
Gold	—	10^{-1}	±5%
Black	0	1	—
Brown	1	10	±1%
Red	2	10^2	±2%
Orange	3	10^3	—
Yellow	4	10^4	—
Green	5	10^5	±0.5%
Blue	6	10^6	±0.25%
Violet	7	10^7	±0.1%
Grey	8	10^8	—
White	9	10^9	—
None	—	—	±20%

Table 41.2

Resistance Value	Marked as:
0.47 Ω	R47
1 Ω	1R0
4.7 Ω	4R7
47 Ω	47R
100 Ω	100R
1 kΩ	1K0
10 kΩ	10 K
10 MΩ	10 M

yellow-violet-orange-red indicates 47 kΩ with a tolerance of ±2% (Note that the first band is the one nearest the end of the resistor)

(ii) For a **five-band fixed resistor** (i.e. resistance values with three significant figures):

red-yellow-white-orange-brown indicates 249 kΩ with a tolerance of ±1% (Note that the fifth band is 1.5 to 2 times wider than the other bands)

(b) Letter and digit code for resistors

Another way of indicating the value of resistors is the letter and digit code shown in Table 41.2

Tolerance is indicated as follows:

$$F = ±1\%, \quad G = ±2\%,$$
$$J = ±5\%, \quad K = ±10\% \text{ and}$$
$$M = ±20\%$$

For example, $R33M = 0.33Ω ± 20\%$

$4R7K = 4.7Ω ± 10\%$

$390RJ = 390Ω ± 5\%$

42 Chemical Effects of Electricity

Introduction

A material must contain **charged particles** to be able to conduct electric current. In **solids**, electrons carry the current. Copper, lead, aluminium, iron and carbon are some examples of solid conductors. In **liquids and gases**, the current is carried by the part of a molecule that has acquired an electric charge, called **ions**. These can possess a positive or negative charge, and examples include hydrogen ion H^+, copper ion Cu^{++} and hydroxyl ion OH^-. Distilled water contains no ions and is a poor conductor of electricity, whereas salt water contains ions and is a fairly good conductor of electricity.

Electrolysis

Electrolysis is the decomposition of a liquid compound by the passage of electric current through it. Practical applications of electrolysis include the electroplating of metals, the refining of copper and the extraction of aluminium from its ore.

An **electrolyte** is a compound that will undergo electrolysis. Examples include salt water, copper sulphate and sulphuric acid.

The **electrodes** are the two conductors carrying current to the electrolyte. The positive-connected electrode is called the **anode** and the negative-connected electrode the **cathode**.

When two copper wires connected to a battery are placed in a beaker containing a salt-water solution, current will flow through the solution. Air bubbles appear around the wires as the water is changed into hydrogen and oxygen by electrolysis.

Electroplating

Electroplating uses the principle of electrolysis to apply a thin coat of one metal to another metal. Some practical applications include the tin-plating of steel, silver-plating of nickel alloys and chromium plating of steel. If two copper electrodes connected to a battery are placed in a beaker containing copper sulphate as the electrolyte it is found that the cathode (i.e. the electrode connected to the negative terminal of the battery) gains copper whilst the anode loses copper.

The Simple Cell

The purpose of an **electric cell** is to convert chemical energy into electrical energy.

A **simple cell** comprises two dissimilar conductors (electrodes) in an electrolyte. Such a cell is shown in Figure 42.1, comprising copper and zinc electrodes. An electric current is found to flow between the electrodes. Other possible electrode pairs exist, including zinc-lead and zinc-iron. The electrode potential (i.e. the p.d. measured between the electrodes) varies for each pair of metals. By knowing the e.m.f. of each metal with respect to some standard electrode, the e.m.f. of any pair of metals may be determined. The standard used is the hydrogen electrode. The **electrochemical series** is a way of listing elements in order of electrical potential, and Table 42.1 shows a number of elements in such a series.

Table 42.1 Part of the electrochemical series

Potassium
sodium
aluminium
zinc
iron
lead
hydrogen
copper
silver
carbon

In a simple cell two faults exist — those due to **polarisation** and **local action**.

Polarisation

If the simple cell shown in Figure 42.1 is left connected for some time, the current I decreases fairly rapidly. This is because of the formation of a film of hydrogen bubbles on the copper anode. This effect is known as the **polarisation** of the cell. The hydrogen prevents full contact between the copper electrode and the electrolyte and this increases the internal resistance of the

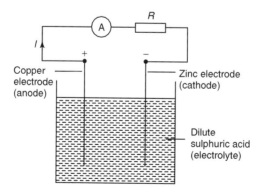

Figure 42.1

cell. The effect can be overcome by using a chemical depolarising agent or depolariser, such as potassium dichromate that removes the hydrogen bubbles as they form. This allows the cell to deliver a steady current. For more on polarisation, see chapter 78.

Local action

When commercial zinc is placed in dilute sulphuric acid, hydrogen gas is liberated from it and the zinc dissolves. The reason for this is that impurities, such as traces of iron, are present in the zinc that set up small primary cells with the zinc. These small cells are short-circuited by the electrolyte, with the result that localised currents flow causing corrosion. This action is known as **local action** of the cell. This may be prevented by rubbing a small amount of mercury on the zinc surface, which forms a protective layer on the surface of the electrode.

When two metals are used in a simple cell the electrochemical series may be used to predict the behaviour of the cell:

(i) The metal that is higher in the series acts as the negative electrode, and vice-versa. For example, the zinc electrode in the cell shown in Figure 42.1 is negative and the copper electrode is positive.

(ii) The greater the separation in the series between the two metals the greater is the e.m.f. produced by the cell.

The electrochemical series is representative of the order of reactivity of the metals and their compounds:

(i) The higher metals in the series react more readily with oxygen and vice-versa.

(ii) When two metal electrodes are used in a simple cell the one that is higher in the series tends to dissolve in the electrolyte.

Corrosion

Corrosion is the gradual destruction of a metal in a damp atmosphere by means of simple cell action. In addition to the presence of moisture and air required for rusting, an electrolyte, an anode and a cathode are required for corrosion. Thus, if metals widely spaced in the electrochemical series, are used in contact with each other in the presence of an electrolyte, corrosion will occur. For example, if a brass valve is fitted to a heating system made of steel, corrosion will occur.

The **effects of corrosion** include the weakening of structures, the reduction of the life of components and materials, the wastage of materials and the expense of replacement.

Corrosion may be **prevented** by coating with paint, grease, plastic coatings and enamels, or by plating with tin or chromium. Also, iron may be galvanised, i.e. plated with zinc, the layer of zinc helping to prevent the iron from corroding.

E.m.f. and Internal Resistance of a Cell

The **electromotive force (e.m.f.)**, E, of a cell is the p.d. between its terminals when it is not connected to a load (i.e. the cell is on 'no load').

The e.m.f. of a cell is measured by using a **high resistance voltmeter** connected in parallel with the cell. The voltmeter must have a high resistance otherwise it will pass current and the cell will not be on 'no-load'. For example, if the resistance of a cell is 1 Ω and that of a voltmeter 1 MΩ then the equivalent resistance of the circuit is 1 MΩ + 1 Ω, i.e. approximately 1 MΩ, hence no current flows and the cell is not loaded.

The voltage available at the terminals of a cell falls when a load is connected. This is caused by the **internal resistance** of the cell, that is, the opposition of the material of the cell to the flow of current. The internal resistance acts in series with other resistances in the circuit. Figure 42.2 shows a cell of e.m.f. E volts and internal resistance, r, and XY represents the terminals of the cell.

When a load (shown as resistance R) is not connected, no current flows and the terminal p.d., $V = E$. When R is connected a current I flows which causes a voltage drop in the cell, given by Ir. The p.d. available at the cell terminals is less than the e.m.f. of the cell and is given by:

$$\boxed{V = E - Ir}$$

Thus, if a battery of e.m.f. 12 volts and internal resistance 0.01 Ω delivers a current of 100 A, the terminal p.d.,

$$V = 12 - (100)(0.01) = 12 - 1 = 11 \text{ V}$$

When different values of potential difference V across a cell or power supply are measured for different values of current I, a graph may be plotted as shown in Figure 42.3. Since the e.m.f. E of the cell or power supply is the p.d. across its terminals on no load (i.e. when $I = 0$), then E is as shown by the broken line.

Since $V = E - Ir$ then the internal resistance may be calculated from

$$\boxed{r = \frac{E - V}{I}}$$

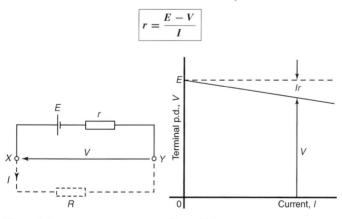

Figure 42.2   **Figure 42.3**

When a current is flowing in the direction shown in Figure 42.2 the cell is said to be **discharging** ($E > V$).

When a current flows in the opposite direction to that shown in Figure 42.2 the cell is said to be **charging** ($V > E$).

A **battery** is a combination of more than one cell. The cells in a battery may be connected in series or in parallel.

(i) **For cells connected in series:**
Total e.m.f. = sum of cell's e.m.f.'s
Total internal resistance = sum of cell's internal resistance's
(ii) **For cells connected in parallel:**
If each cell has the same e.m.f. and internal resistance:
Total e.m.f. = e.m.f. of one cell
Total internal resistance of n cells

$$= \frac{1}{n} \times \text{internal resistance of one cell}$$

For example, eight cells, each with an internal resistance of 0.2 Ω and an e.m.f. of 2.2 V are connected (a) in series, (b) in parallel.

(a) When connected in series, total e.m.f. = sum of cell's e.m.f.
$$= 2.2 \times 8 = \textbf{17.6 V}$$
Total internal resistance = sum of cell's internal resistance
$$= 0.2 \times 8 = \textbf{1.6 } \boldsymbol{\Omega}$$
(b) When connected in parallel, total e.m.f. = e.m.f. of one cell
$$= \textbf{2.2 V}$$

Total internal resistance of 8 cells

$$= \tfrac{1}{8} \times \text{internal resistance of one cell}$$

$$= \tfrac{1}{8} \times 0.2 = \textbf{0.025 } \boldsymbol{\Omega}$$

Primary Cells

Primary cells cannot be recharged, that is, the conversion of chemical energy to electrical energy is irreversible and the cell cannot be used once the chemicals are exhausted. Examples of primary cells include the Leclanché cell and the mercury cell.

Lechlanché cell

A typical dry Lechlanché cell is shown in Figure 42.4. Such a cell has an e.m.f. of about 1.5 V when new, but this falls rapidly if in continuous use due to polarisation. The hydrogen film on the carbon electrode forms faster than can be dissipated by the depolarizer. The Lechlanché cell is suitable only for intermittent use, applications including torches, transistor radios, bells, indicator circuits, gas lighters, controlling switch-gear, and so on. The cell is the most commonly used of primary cells, is cheap, requires little maintenance and has a shelf life of about 2 years.

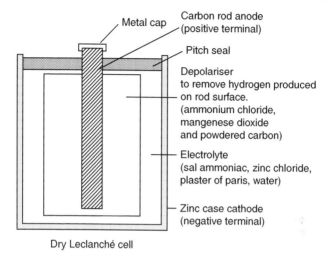

Metal cap

Carbon rod anode
(positive terminal)

Pitch seal

Depolariser
to remove hydrogen produced
on rod surface.
(ammonium chloride,
mangenese dioxide
and powdered carbon)

Electrolyte
(sal ammoniac, zinc chloride,
plaster of paris, water)

Zinc case cathode
(negative terminal)

Dry Leclanché cell

Figure 42.4

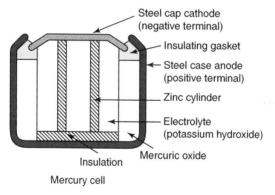

Steel cap cathode
(negative terminal)

Insulating gasket

Steel case anode
(positive terminal)

Zinc cylinder

Electrolyte
(potassium hydroxide)

Insulation   Mercuric oxide

Mercury cell

Figure 42.5

Mercury cell

A typical mercury cell is shown in Figure 42.5. Such a cell has an e.m.f. of about 1.3 V which remains constant for a relatively long time. Their main advantage over the Lechlanché cell is its smaller size and its long shelf life. Typical practical applications include hearing aids, medical electronics, cameras and for guided missiles.

Secondary Cells

Secondary cells can be recharged after use, that is, the conversion of chemical energy to electrical energy is reversible and the cell may be used many times. Examples of secondary cells include the lead-acid cell and the alkaline cell. Practical applications of such cells include car batteries, telephone circuits and for traction purposes — such as milk delivery vans and fork lift trucks.

Lead-acid cell

A typical lead-acid cell is constructed of:

(i) A container made of glass, ebonite or plastic.
(ii) **Lead plates**
    (a) the negative plate (cathode) consists of spongy lead
    (b) the positive plate (anode) is formed by pressing lead peroxide into the lead grid.
    The plates are interleaved as shown in the plan view of Figure 42.6 to increase their effective cross-sectional area and to minimise internal resistance.
(iii) **Separators** made of glass, celluloid or wood.
(iv) An **electrolyte** which is a mixture of sulphuric acid and distilled water.

The relative density (or specific gravity) of a lead-acid cell, which may be measured using a hydrometer, varies between about 1.26 when the cell is fully charged to about 1.19 when discharged. The terminal p.d. of a lead-acid cell is about 2 V.

When a cell supplies current to a load it is said to be **discharging**. During discharge:

(i) the lead peroxide (positive plate) and the spongy lead (negative plate) are converted into lead sulphate, and
(ii) the oxygen in the lead peroxide combines with hydrogen in the electrolyte to form water. The electrolyte is therefore weakened and the relative density falls.

The terminal p.d. of a lead-acid cell when fully discharged is about 1.8 V. A cell is **charged** by connecting a d.c. supply to its terminals, the positive

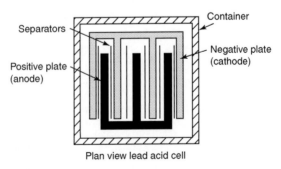

Plan view lead acid cell

Figure 42.6

terminal of the cell being connected to the positive terminal of the supply. The charging current flows in the reverse direction to the discharge current, and the chemical action is reversed. During charging:

(i) the lead sulphate on the positive and negative plates is converted back to lead peroxide and lead respectively, and
(ii) the water content of the electrolyte decreases as the oxygen released from the electrolyte combines with the lead of the positive plate. The relative density of the electrolyte thus increases.

The colour of the positive plate when fully charged is dark brown and when discharged is light brown. The colour of the negative plate when fully charged is grey and when discharged is light grey.

Alkaline cell

There are two main types of alkaline cell — the nickel-iron cell and the nickel-cadmium cell. In both types the positive plate is made of nickel hydroxide enclosed in finely perforated steel tubes, the resistance being reduced by the addition of pure nickel or graphite. The tubes are assembled into nickel-steel plates.

In the nickel-iron cell, (sometimes called the **Edison cell** or **nife cell**), the negative plate is made of iron oxide, with the resistance being reduced by a little mercuric oxide, the whole being enclosed in perforated steel tubes and assembled in steel plates. In the nickel-cadmium cell the negative plate is made of cadmium. The electrolyte in each type of cell is a solution of potassium hydroxide that does not undergo any chemical change and thus the quantity can be reduced to a minimum. The plates are separated by insulating rods and assembled in steel containers that are then enclosed in a non-metallic crate to insulate the cells from one another. The average discharge p.d. of an alkaline cell is about 1.2 V.

Advantages of an alkaline cell (for example, a nickel-cadmium cell or a nickel-iron cell) over a lead-acid cell include:

(i) More robust construction
(ii) Capable of withstanding heavy charging and discharging currents without damage
(iii) Has a longer life
(iv) For a given capacity is lighter in weight
(v) Can be left indefinitely in any state of charge or discharge without damage
(vi) Is not self-discharging

Disadvantages of an alkaline cell over a lead-acid cell include:

(i) Is relatively more expensive
(ii) Requires more cells for a given e.m.f.
(iii) Has a higher internal resistance
(iv) Must be kept sealed
(v) Has a lower efficiency

Alkaline cells may be used in extremes of temperature, in conditions where vibration is experienced or where duties require long idle periods or heavy discharge currents. Practical examples include traction and marine work, lighting in railway carriages, military portable radios and for starting diesel and petrol engines.

However, the lead-acid cell is the most common one in practical use.

Cell Capacity

The **capacity** of a cell is measured in ampere-hours (Ah). A fully charged 50 battery rated for 10 h discharge can be discharged at a steady current of 5 A for 10 h, but if the load current is increased to 10 A then the battery is discharged in 3–4 h, since the higher the discharge current, the lower is the effective capacity of the battery. Typical discharge characteristics for a lead-acid cell are shown in Figure 42.7.

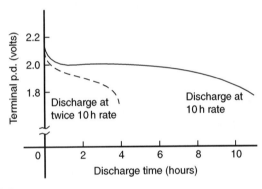

Figure 42.7

43 Series and Parallel Networks

Series Circuits

Figure 43.1 shows three resistors R_1, R_2 and R_3 connected end to end, i.e. in series, with a battery source of V volts. Since the circuit is closed a current I will flow and the p.d. across each resistor may be determined from the voltmeter readings V_1, V_2 and V_3

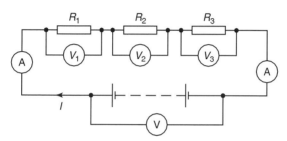

Figure 43.1

In a series circuit

(a) the current I is the same in all parts of the circuit and hence the same reading is found on each of the ammeters shown, and
(b) the sum of the voltages V_1, V_2 and V_3 is equal to the total applied voltage, V, i.e.

$$\boxed{V = V_1 + V_2 + V_3}$$

From Ohm's law: $V_1 = IR_1$, $V_2 = IR_2$, $V_3 = IR_3$ and $V = IR$ where R is the total circuit resistance.

Since $V = V_1 + V_2 + V_3$ then $IR = IR_1 + IR_2 + IR_3$

Dividing throughout by I gives: $\boxed{R = R_1 + R_2 + R_3}$

Thus for a series circuit, the total resistance is obtained by adding together the values of the separate resistance's.

Potential Divider

The voltage distribution for the circuit shown in Figure 43.2(a) is given by:

$$V_1 = \left(\frac{R_1}{R_1 + R_2} \right) V \text{ and } V_2 = \left(\frac{R_2}{R_1 + R_2} \right) V$$

228

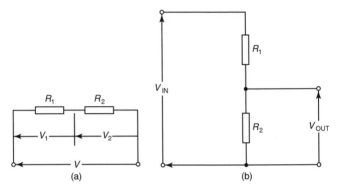

Figure 43.2

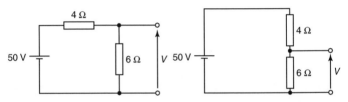

Figure 43.3   **Figure 43.4**

The circuit shown in Figure 43.2(b) is often referred to as a **potential divider** circuit. Such a circuit can consist of a number of similar elements in series connected across a voltage source, voltages being taken from connections between the elements. Frequently the divider consists of two resistors as shown in Figure 43.2(b), where

$$V_{OUT} = \left(\frac{R_2}{R_1 + R_2} \right) V_{IN}$$

For example, to determined the value of voltage V shown in Figure 43.3:
    Redrawing the circuit as shown in Figure 43.4

gives: voltage $V = \left(\dfrac{6}{6+4} \right) (50) = \mathbf{30\ V}$

Parallel Networks

Figure 43.5 shows three resistors, R_1, R_2 and R_3 connected across each other, i.e. in parallel, across a battery source of V volts.

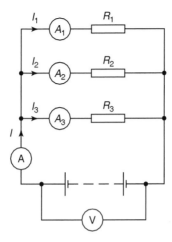

Figure 43.5

In a parallel circuit:

(a) the sum of the currents I_1, I_2 and I_3 is equal to the total circuit current, I, i.e. $\boxed{I = I_1 + I_2 + I_3}$ and

(b) the source p.d., V volts, is the same across each of the resistors.

From Ohm's law: $I_1 = \dfrac{V}{R_1}$, $I_2 = \dfrac{V}{R_2}$, $I_3 = \dfrac{V}{R_3}$ and $I = \dfrac{V}{R}$ where R is the total circuit resistance.

Since $I = I_1 + I_2 + I_3$ then $\dfrac{V}{R} = \dfrac{V}{R_1} + \dfrac{V}{R_2} + \dfrac{V}{R_3}$

Dividing throughout by V gives: $\boxed{\dfrac{1}{R} = \dfrac{1}{R_1} + \dfrac{1}{R_2} + \dfrac{1}{R_3}}$

This equation must be used when finding the total resistance R of a parallel circuit.

For the special case of **two resistors in parallel**

$$\dfrac{1}{R} = \dfrac{1}{R_1} + \dfrac{1}{R_2} = \dfrac{R_2 + R_1}{R_1 R_2}$$

Hence $\boxed{R = \dfrac{R_1 R_2}{R_1 + R_2}}$ $\left(\text{i.e. } \dfrac{\text{product}}{\text{sum}} \right)$

Current Division

For the circuit shown in Figure 43.6,

$$I_1 = \left(\frac{R_2}{R_1 + R_2}\right)(I) \text{ and } I_2 = \left(\frac{R_1}{R_1 + R_2}\right)(I)$$

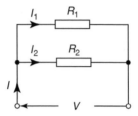

Figure 43.6

For example, for the series-parallel arrangement shown in Figure 43.7:

The equivalent resistance R_x of R_2 and R_3 in parallel is: $R_x = \dfrac{6 \times 2}{6 + 2} = 1.5 \ \Omega$

The equivalent resistance R_T of R_1, R_x and R_4 in series is:

$$R_T = 2.5 + 1.5 + 4 = 8 \ \Omega$$

and the supply current $I = \dfrac{V}{R_T} = \dfrac{200}{8} = \textbf{25 A}$

The current flowing through R_1 and R_4 is 25 A

The current flowing through $R_2 = \left(\dfrac{R_3}{R_2 + R_3}\right)I = \left(\dfrac{2}{6 + 2}\right)25 = \textbf{6.25 A}$

The current flowing through $R_3 = \left(\dfrac{R_2}{R_2 + R_3}\right)I = \left(\dfrac{6}{6 + 2}\right)25 = \textbf{18.75 A}$

(Note that the currents flowing through R_2 and R_3 must add up to the total current flowing into the parallel arrangement, i.e. 25 A).

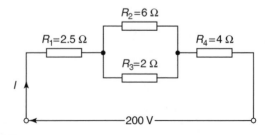

Figure 43.7

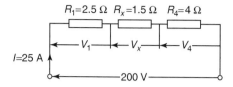

Figure 43.8

The equivalent circuit of Figure 43.7 is shown in Figure 43.8.

p.d. across R_1, i.e. $V_1 = IR_1 = (25)(2.5) = \mathbf{62.5\ V}$

p.d. across R_x, i.e. $V_x = IR_x = (25)(1.5) = \mathbf{37.5\ V}$

p.d. across R_4, i.e. $V_4 = IR_4 = (25)(4) = \mathbf{100\ V}$

Hence the p.d. across R_2 = p.d. across $R_3 = \mathbf{37.5\ V}$

Wiring Lamps in Series and in Parallel

Series connection

Figure 43.9 shows three lamps, each rated at 240 V, connected in series across a 240 V supply.

(i) Each lamp has only $\dfrac{240}{3}$ V, i.e. 80 V across it and thus each lamp glows dimly.

(ii) If another lamp of similar rating is added in series with the other three lamps then each lamp now has $\dfrac{240}{4}$ V, i.e. 60 V across it and each now glows even more dimly.

(iii) If a lamp is removed from the circuit or if a lamp develops a fault (i.e. an open circuit) or if the switch is opened, then the circuit is broken, no current flows, and the remaining lamps will not light up.

(iv) Less cable is required for a series connection than for a parallel one.

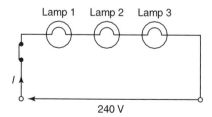

Figure 43.9

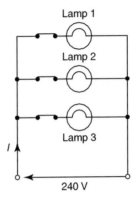

Figure 43.10

The series connection of lamps is usually limited to decorative lighting such as for Christmas tree lights.

Parallel connection

Figure 43.10 shows three similar lamps, each rated at 240 V, connected in parallel across a 240 V supply.

(i) Each lamp has 240 V across it and thus each will glow brilliantly at their rated voltage.

(ii) If any lamp is removed from the circuit or develops a fault (open circuit) or a switch is opened, the remaining lamps are unaffected.

(iii) The addition of further similar lamps in parallel does not affect the brightness of the other lamps.

(iv) More cable is required for parallel connection than for a series one.

The parallel connection of lamps is the most widely used in electrical installations.

44 Capacitors and Capacitance

Electrostatic Field

Figure 44.1 represents two parallel metal plates, A and B, charged to different potentials. If an electron that has a negative charge is placed between the plates, a force will act on the electron tending to push it away from the negative plate B towards the positive plate, A. Similarly, a positive charge would be acted on by a force tending to move it toward the negative plate. Any region such as that shown between the plates in Figure 44.1, in which an electric charge experiences a force, is called an **electrostatic field**. The direction of the field is defined as that of the force acting on a positive charge placed in the field. In Figure 44.1, the direction of the force is from the positive plate to the negative plate. Such a field may be represented in magnitude and direction by **lines of electric force** drawn between the charged surfaces. The closeness of the lines is an indication of the field strength. Whenever a p.d. is established between two points, an electric field will always exist.

Figure 44.2(a) shows a typical field pattern for an isolated point charge, and Figure 44.2(b) shows the field pattern for adjacent charges of opposite polarity. Electric lines of force (often called electric flux lines) are continuous and start and finish on point charges; also, the lines cannot cross each other. When a charged body is placed close to an uncharged body, an induced charge of opposite sign appears on the surface of the uncharged body. This is because lines of force from the charged body terminate on its surface.

The concept of field lines or lines of force is used to illustrate the properties of an electric field. However, it should be remembered that they are only aids to the imagination.

The **force of attraction or repulsion** between two electrically charged bodies is proportional to the magnitude of their charges and inversely proportional to the square of the distance separating them,

i.e. force $\propto \dfrac{q_1 q_2}{d^2}$ or $\boxed{\text{force} = k\,\dfrac{q_1 q_2}{d^2}}$ where constant $k \approx 9 \times 10^9$

This is known as **Coulomb's law**.

Hence the force between two charged spheres in air with their centres 16 mm apart and each carrying a charge of $+1.6\ \mu C$ is given by:

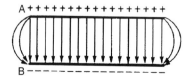

Figure 44.1

234

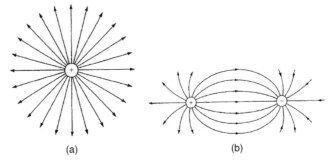

Figure 44.2

$$\text{force} = k\frac{q_1 q_2}{d^2} \approx (9 \times 10^9)\frac{(1.6 \times 10^{-6})^2}{(16 \times 10^{-3})^2} = \mathbf{90\ newtons}$$

Electric Field Strength

Figure 44.3 shows two parallel conducting plates separated from each other by air. They are connected to opposite terminals of a battery of voltage V volts. There is therefore an electric field in the space between the plates. If the plates are close together, the electric lines of force will be straight and parallel and equally spaced, except near the edge where fringing will occur (see Figure 44.1). Over the area in which there is negligible fringing,

$$\boxed{\textbf{Electric field strength, } E = \frac{V}{d} \textbf{ volts/metre}}$$

where d is the distance between the plates. Electric field strength is also called **potential gradient**.

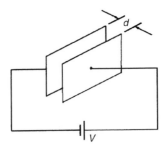

Figure 44.3

Capacitance

Static electric fields arise from electric charges, electric field lines beginning and ending on electric charges. Thus the presence of the field indicates the presence of equal positive and negative electric charges on the two plates of Figure 44.3. Let the charge be $+Q$ coulombs on one plate and $-Q$ coulombs on the other. The property of this pair of plates which determines how much charge corresponds to a given p.d. between the plates is called their capacitance:

$$\text{capacitance } C = \frac{Q}{V}$$

The **unit of capacitance** is the **farad F** (or more usually $\mu F = 10^{-6}$ F or $pF = 10^{-12}$ F), which is defined as the capacitance when a p.d. of one volt appears across the plates when charged with one coulomb.

For example, the p.d. across a 4 μF capacitor when charged with 5 mC is determined as follows:

$$\text{Since } C = \frac{Q}{V} \text{ then } V = \frac{Q}{C} = \frac{5 \times 10^{-3}}{4 \times 10^{-6}} = \frac{5 \times 10^6}{4 \times 10^3} = \frac{5000}{4}$$

$$= \textbf{1250 V or 1.25 kV}$$

Capacitors

Every system of electrical conductors possesses capacitance. For example, there is capacitance between the conductors of overhead transmission lines and also between the wires of a telephone cable. In these examples the capacitance is undesirable but has to be accepted, minimised or compensated for. There are other situations where capacitance is a desirable property.

Devices specially constructed to possess capacitance are called **capacitors** (or condensers, as they used to be called). In its simplest form a capacitor consists of two plates that are separated by an insulating material known as a **dielectric**. A capacitor has the ability to store a quantity of static electricity.

The symbols for a fixed capacitor and a variable capacitor used in electrical circuit diagrams are shown in Figure 44.4

The **charge Q** stored in a capacitor is given by:

$$Q = I \times t \text{ coulombs}$$

where I is the current in amperes and t the time in seconds.

Fixed capacitor   Variable capacitor

Figure 44.4

Electric Flux Density

Unit flux is defined as emanating from a positive charge of 1 coulomb. Thus electric flux ψ is measured in coulombs, and for a charge of Q coulombs, the flux $\psi = Q$ coulombs.

Electric flux density D is the amount of flux passing through a defined area A that is perpendicular to the direction of the flux:

$$\text{electric flux density, } D = \frac{Q}{A} \text{ coulombs/metre}^2$$

Electric flux density is also called **charge density, σ.**

For example, two parallel rectangular plates measuring 20 cm by 40 cm carry an electric charge of 0.2 μC, hence

$$\text{electric flux density } D = \frac{Q}{A} = \frac{0.2 \times 10^{-6}}{800 \times 10^{-4}} = \frac{0.2 \times 10^4}{800 \times 10^6}$$

$$= \textbf{2.5 μC/m}^2$$

Permittivity

At any point in an electric field, the electric field strength E maintains the electric flux and produces a particular value of electric flux density D at that point. For a field established in **vacuum** (or for practical purposes in air), the ratio D/E is a constant ε_o, i.e.

$$\frac{D}{E} = \varepsilon_o$$

where ε_o is called the **permittivity of free space** or the free space constant. The value of ε_o is $\textbf{8.85} \times \textbf{10}^{-12}$ F/m.

When an insulating medium, such as mica, paper, plastic or ceramic, is introduced into the region of an electric field the ratio of D/E is modified:

$$\frac{D}{E} = \varepsilon_o \, \varepsilon_r$$

where ε_r, the **relative permittivity** of the insulating material, indicates its insulating power compared with that of vacuum:

$$\text{relative permittivity } \varepsilon_r = \frac{\text{flux density in material}}{\text{flux density in vacuum}}$$

ε_r has no unit. Typical values of ε_r include air, 1.00; polythene, 2.3; mica, 3–7; glass, 5–10; water, 80; ceramics, 6–1000.

The product $\varepsilon_o \varepsilon_r$ is called the **absolute permittivity**, ε,

i.e. $$\varepsilon = \varepsilon_o \, \varepsilon_r$$

The insulating medium separating charged surfaces is called a **dielectric**. Compared with conductors, dielectric materials have very high resistivities. They are therefore used to separate conductors at different potentials, such as capacitor plates or electric power lines.

For example, if two parallel plates having a p.d. of 200 V between them are spaced 0.8 mm apart, then

$$\textbf{electric field strength } E = \frac{V}{d} = \frac{200}{0.8 \times 10^{-3}} = \textbf{250 kV/m}$$

If the dielectric between the plates is polythene of relative permittivity 2.3, then since $\dfrac{D}{E} = \varepsilon_o \varepsilon_r$, **electric flux density**

$$\textbf{\textit{D}} = E\varepsilon_o \varepsilon_r = (250 \times 10^3 \times 8.85 \times 10^{-12} \times 2.3) = \textbf{5.089 μC/m}^2$$

The Parallel Plate Capacitor

For a parallel-plate capacitor, as shown in Figure 44.5(a),

$$\textbf{capacitance, } C = \frac{\varepsilon_o \varepsilon_r A}{d} \textbf{ farads}$$

where $\varepsilon_o = 8.85 \times 10^{-12}$ F/m (constant), ε_r = relative permittivity, A = area of one of the plates, in m^2, and d = thickness of dielectric in m.

Another method used to increase the capacitance is to interleave several plates as shown in Figure 44.5(b). Ten plates are shown, forming nine capacitors with a capacitance nine times that of one pair of plates.

If such an arrangement has n plates then capacitance $C \propto (n-1)$.

Thus,
$$\textbf{capacitance, } C = \frac{\varepsilon_o \varepsilon_r A(n-1)}{d} \textbf{ farads}$$

For example, a parallel plate capacitor has nineteen interleaved plates each 75 mm by 75 mm separated by mica sheets 0.2 mm thick. Assuming the relative permittivity of the mica is 5, the capacitance of the capacitor is

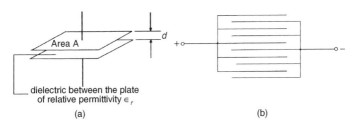

(a)　　　　　　　　　　　　　　(b)

Figure 44.5

given by:

$$C = \frac{\varepsilon_o \varepsilon_r A(n-1)}{d} = \frac{8.85 \times 10^{-12} \times 5 \times 75^2 \times 10^{-6} \times (19-1)}{0.2 \times 10^{-3}}$$

$$= \textbf{0.0224 } \boldsymbol{\mu}\textbf{F or 22.4 nF}$$

Capacitors Connected in Parallel and Series

(a) Capacitors connected in parallel

Figure 44.6 shows three capacitors, C_1, C_2 and C_3, connected in parallel with a supply voltage V applied across the arrangement.

When the charging current I reaches point A it divides, some flowing into C_1, some flowing into C_2 and some into C_3. Hence the total charge $Q_T (= I \times t)$ is divided between the three capacitors. The capacitors each store a charge and these are shown as Q_1, Q_2 and Q_3 respectively.

Hence $\quad Q_T = Q_1 + Q_2 + Q_3$

But $\quad Q_T = CV, \quad Q_1 = C_1 V, Q_2 = C_2 V$ and $Q_3 = C_3 V$

Therefore $CV = C_1 V + C_2 V + C_3 V$ where C is the total equivalent circuit capacitance, i.e. $C = C_1 + C_2 + C_3$

It follows that for n parallel-connected capacitors,

$$\boxed{C = C_1 + C_2 + C_3 \cdots + C_n}$$

(Note that this formula is similar to that used for **resistors** connected in **series**).

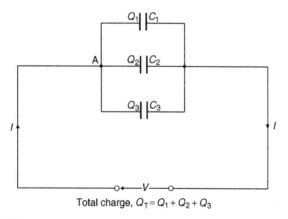

Total charge, $Q_T = Q_1 + Q_2 + Q_3$

Figure 44.6

For example, capacitance's of 1 μF, 3 μF, 5 μF and 6 μF are connected in parallel to a direct voltage supply of 100 V.

The equivalent capacitance $C = C_1 + C_2 + C_3 + C_4 = 1 + 3 + 5 + 6$

$$= 15 \text{ μF}$$

Total charge $Q_T = CV$ where C is the equivalent circuit capacitance

i.e. $$Q_T = 15 \times 10^{-6} \times 100 = 1.5 \times 10^{-3} C = \textbf{1.5 mC}$$

The charge on the 1 μF capacitor $Q_1 = C_1 V = 1 \times 10^{-6} \times 100$

$$= \textbf{0.1 mC}$$

The charge on the 3 μF capacitor $Q_2 = C_2 V = 3 \times 10^{-6} \times 100$

$$= \textbf{0.3 mC}$$

The charge on the 5 μF capacitor $Q_3 = C_3 V = 5 \times 10^{-6} \times 100$

$$= \textbf{0.5 mC}$$

The charge on the 6 μF capacitor $Q_4 = C_4 V = 6 \times 10^{-6} \times 100$

$$= \textbf{0.6 mC}$$

(b) Capacitors connected in series

Figure 44.7 shows three capacitors, C_1, C_2 and C_3, connected in series across a supply voltage V. Let the p.d. across the individual capacitors be V_1, V_2 and V_3 respectively as shown.

Let the charge on plate 'a' of capacitor C_1 be $+Q$ coulombs. This induces an equal but opposite charge of $-Q$ coulombs on plate 'b'. The conductor between plates 'b' and 'c' is electrically isolated from the rest of the circuit so that an equal but opposite charge of $+Q$ coulombs must appear on plate 'c', which, in turn, induces an equal and opposite charge of $-Q$ coulombs on plate 'd', and so on. Hence when capacitors are connected in series the charge on each is the same.

In a series circuit: $V = V_1 + V_2 + V_3$

Since $V = \dfrac{Q}{C}$ then $\dfrac{Q}{C} = \dfrac{Q}{C_1} + \dfrac{Q}{C_2} + \dfrac{Q}{C_3}$

where C is the total equivalent circuit capacitance,

i.e. $$\dfrac{1}{C} = \dfrac{1}{C_1} + \dfrac{1}{C_2} + \dfrac{1}{C_3}$$

It follows that for n series-connected capacitors:

$$\boxed{\dfrac{1}{C} = \dfrac{1}{C_1} + \dfrac{1}{C_2} + \dfrac{1}{C_3} + \cdots + \dfrac{1}{C_n}}$$

(Note that this formula is similar to that used for **resistors** connected in **parallel**).

For example, capacitance's of 3 μF, 6 μF and 12 μF are connected in series across a 350 V supply. The circuit diagram is shown in Figure 44.8.

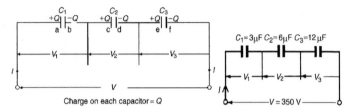

Figure 44.7   **Figure 44.8**

The equivalent circuit capacitance C for three capacitors in series is given by: $\dfrac{1}{C} = \dfrac{1}{3} + \dfrac{1}{6} + \dfrac{1}{12} = \dfrac{4+2+1}{12} = \dfrac{7}{12}$ from which,

the equivalent circuit capacitance $C = \dfrac{12}{7} = 1.714\ \mu F$

Total charge $Q_T = CV$, hence $Q_T = \dfrac{12}{7} \times 10^{-6} \times 350 = 600\ \mu C$ or 0.6 mC

Since the capacitors are connected in series 0.6 mC is the charge on each of them.

The voltage across the 3 μF capacitor, $V_1 = \dfrac{Q}{C_1} = \dfrac{0.6 \times 10^{-3}}{3 \times 10^{-6}}$

$$= \mathbf{200\ V}$$

The voltage across the 6 μF capacitor, $V_2 = \dfrac{Q}{C_2} = \dfrac{0.6 \times 10^{-3}}{6 \times 10^{-6}}$

$$= \mathbf{100\ V}$$

The voltage across the 12 μF capacitor, $V_3 = \dfrac{Q}{C_3} = \dfrac{0.6 \times 10^{-3}}{12 \times 10^{-6}}$

$$= \mathbf{50\ V}$$

In practice, capacitors are rarely connected in series unless they are of the same capacitance. The reason for this can be seen from above where the lowest valued capacitor (i.e. 3 μF) has the highest p.d. across it (i.e. 200 V) which means that if all the capacitors have an identical construction they must all be rated at the highest voltage.

For the special case of **two capacitors in series:**

$$\boxed{C = \frac{C_1 C_2}{C_1 + C_2}} \quad \left(\text{i.e. } \frac{\text{product}}{\text{sum}}\right)$$

For example, the equivalent capacitance of two capacitors of 6 μF and 4 μF connected in series is given by: $C = \dfrac{6 \times 4}{6 + 4} = \dfrac{24}{10} = \mathbf{2.4\ \mu F}$

Dielectric Strength

The maximum amount of field strength that a dielectric can withstand is called the dielectric strength of the material.

$$\text{Dielectric strength,} \quad \boxed{E_m = \frac{V_m}{d}}$$

Energy Stored in Capacitors

The energy, W, stored by a capacitor is given by:

$$\boxed{W = \tfrac{1}{2}CV^2 \text{ joules}}$$

For example, the energy stored in a 3 μF capacitor when charged to 400 V is given by: **energy stored $W = \tfrac{1}{2}CV^2 = \tfrac{1}{2} \times 3 \times 10^{-6} \times 400^2 = 0.24$ J**

Practical Types of Capacitor

Practical types of capacitor are characterised by the material used for their dielectric. The main types include: variable air, mica, paper, ceramic, plastic, titanium oxide and electrolytic.

1. **Variable air capacitors**. These usually consist of two sets of metal plates (such as aluminium), one fixed, the other variable. The set of moving plates rotate on a spindle as shown by the end view of Figure 44.9.
   As the moving plates are rotated through half a revolution, the meshing, and therefore the capacitance, varies from a minimum to a maximum value. Variable air capacitors are used in radio and electronic circuits where very low losses are required, or where a variable capacitance is needed. The maximum value of such capacitors is between 500 pF and 1000 pF.
2. **Mica capacitors**. A typical older type construction is shown in Figure 44.10. Usually the whole capacitor is impregnated with wax and placed in a bakelite case. Mica is easily obtained in thin sheets and is a good insulator.

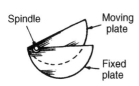

Figure 44.9

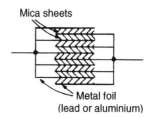

Mica sheets

Metal foil (lead or aluminium)

Figure 44.10

However, mica is expensive and is not used in capacitors above about 0.2 μF. A modified form of mica capacitor is the silvered mica type. The mica is coated on both sides with a thin layer of silver that forms the plates. Capacitance is stable and less likely to change with age. Such capacitors have a constant capacitance with change of temperature, a high working voltage rating and a long service life and are used in high frequency circuits with fixed values of capacitance up to about 1000 pF.

3. **Paper capacitors**. A typical paper capacitor is shown in Figure 44.11 where the length of the roll corresponds to the capacitance required. The whole is usually impregnated with oil or wax to exclude moisture, and then placed in a plastic or aluminium container for protection. Paper capacitors are made in various working voltages up to about 150 kV and are used where loss is not very important. The maximum value of this type of capacitor is between 500 pF and 10 μF. Disadvantages of paper capacitors include variation in capacitance with temperature change and a shorter service life than most other types of capacitor.

4. **Ceramic capacitors**. These are made in various forms, each type of construction depending on the value of capacitance required. For high values, a tube of ceramic material is used as shown in the cross section of Figure 44.12. For smaller values the cup construction is used as shown in Figure 44.13, and for still smaller values the disc construction shown in Figure 44.14 is used. Certain ceramic materials have a very high permittivity and this enables capacitors of high capacitance to be made which are of small physical size with a high working voltage rating. Ceramic capacitors are available in the range 1 pF to 0.1 μF and may be used in high frequency electronic circuits subject to a wide range of temperatures.

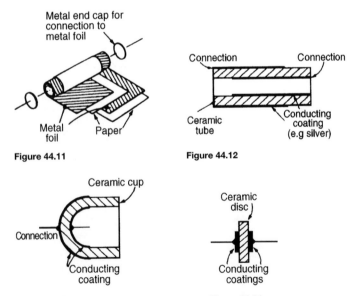

Figure 44.11

Figure 44.12

Figure 44.13

Figure 44.14

5. **Plastic capacitors**. Some plastic materials such as polystyrene and Teflon can be used as dielectrics. Construction is similar to the paper capacitor but using a plastic film instead of paper. Plastic capacitors operate well under conditions of high temperature, provide a precise value of capacitance, a very long service life and high reliability.

6. **Titanium oxide capacitors** have a very high capacitance with a small physical size when used at a low temperature.

7. **Electrolytic capacitors**. Construction is similar to the paper capacitor with aluminium foil used for the plates and with a thick absorbent material, such as paper, impregnated with an electrolyte (ammonium borate), separating the plates. The finished capacitor is usually assembled in an aluminium container and hermetically sealed. Its operation depends on the formation of a thin aluminium oxide layer on the positive plate by electrolytic action when a suitable direct potential is maintained between the plates. This oxide layer is very thin and forms the dielectric. (The absorbent paper between the plates is a conductor and does not act as a dielectric.) Such capacitors **must always be used on d.c.** and must be connected with the correct polarity; if this is not done the capacitor will be destroyed since the oxide layer will be destroyed. Electrolytic capacitors are manufactured with working voltage from 6 V to 600 V, although accuracy is generally not very high. These capacitors possess a much larger capacitance than other types of capacitors of similar dimensions due to the oxide film being only a few microns thick. The fact that they can be used only on d.c. supplies limit their usefulness.

Discharging Capacitors

When a capacitor has been disconnected from the supply it may still be charged and it may retain this charge for some considerable time. Thus precautions must be taken to ensure that the capacitor is automatically discharged after the supply is switched off. Connecting a high value resistor across the capacitor terminals does this.

45 Magnetic Circuits

Magnetic Fields

A **permanent magnet** is a piece of ferromagnetic material (such as iron, nickel or cobalt) that has properties of attracting other pieces of these materials. A permanent magnet will position itself in a north and south direction when freely suspended. The north-seeking end of the magnet is called the **north pole, N**, and the south-seeking end the **south pole, S**.

The area around a magnet is called the **magnetic field** and it is in this area that the effects of the **magnetic force** produced by the magnet can be detected. A magnetic field cannot be seen, felt, smelt or heard and therefore is difficult to represent. Michael Faraday suggested that the magnetic field could be represented pictorially, by imagining the field to consist of **lines of magnetic flux**, which enables investigation of the distribution and density of the field to be carried out.

The distribution of a magnetic field can be investigated by using some iron filings. A bar magnet is placed on a flat surface covered by, say, cardboard, upon which is sprinkled some iron filings. If the cardboard is gently tapped the filings will assume a pattern similar to that shown in Figure 45.1. If a number of magnets of different strength are used, it is found that the stronger the field the closer are the lines of magnetic flux and vice versa. Thus a magnetic field has the property of exerting a force, demonstrated in this case by causing the iron filings to move into the pattern shown. The strength of the magnetic field decreases as we move away from the magnet. It should be realised, of course, that the magnetic field is three dimensional in its effect, and not acting in one plane as appears to be the case in this experiment.

If a compass is placed in the magnetic field in various positions, the direction of the lines of flux may be determined by noting the direction of the compass pointer. The direction of a magnetic field at any point is taken as that in which the north-seeking pole of a compass needle points when suspended in the field. The direction of a line of flux is from the north pole to the south pole on the outside of the magnet and is then assumed to continue through the magnet back to the point at which it emerged at the north pole. Thus such lines of flux always form complete closed loops or paths, they never intersect and always have a definite direction.

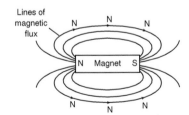

Figure 45.1

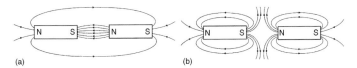

Figure 45.2

The laws of magnetic attraction and repulsion can be demonstrated by using two bar magnets. In Figure 45.2(a), **with unlike poles adjacent, attraction takes place**. Lines of flux are imagined to contract and the magnets try to pull together. The magnetic field is strongest in between the two magnets, shown by the lines of flux being close together. In Figure 45.2(b), **with similar poles adjacent (i.e. two north poles), repulsion occurs**, i.e. the two north poles try to push each other apart, since magnetic flux lines running side by side in the same direction repel.

Magnetic Flux and Flux Density

Magnetic flux is the amount of magnetic field (or the number of lines of force) produced by a magnetic source. The symbol for magnetic flux is Φ (Greek letter 'phi'). The unit of magnetic flux is the **weber, Wb**.

Magnetic flux density is the amount of flux passing through a defined area that is perpendicular to the direction of the flux:

$$\textbf{Magnetic flux density} = \frac{\textbf{magnetic flux}}{\textbf{area}}$$

The symbol for magnetic flux density is B. The unit of magnetic flux density is the tesla, T, where $1\ T = 1\ \text{Wb/m}^2$. Hence

$$\boxed{B = \frac{\Phi}{A}\ \textbf{tesla}} \quad \text{where A(m}^2\text{) is the area}$$

For example, a magnetic pole face has a rectangular section having dimensions 200 mm by 100 mm. If the total flux emerging from the pole is 150 μWb, the flux density, $B = \dfrac{\Phi}{A} = \dfrac{150 \times 10^{-6}}{200 \times 100 \times 10^{-6}} = \textbf{0.0075 T or 7.5 mT}$

Magnetomotive Force and Magnetic Field Strength

Magnetomotive force (m.m.f.) is the cause of the existence of a magnetic flux in a magnetic circuit,

$$\boxed{\text{m.m.f. } F_m = NI \text{ amperes}}$$

where N is the number of conductors (or turns) and I is the current in amperes. The unit of m.m.f. is sometimes expressed as 'ampere-turns'. However since 'turns' have no dimensions, the S.I. unit of m.m.f. is the ampere.

Magnetic field strength (or **magnetising force**),

$$H = \frac{NI}{l} \text{ ampere per metre}$$

where l is the mean length of the flux path in metres.

Thus

$$\text{m.m.f.} = NI = Hl \text{ amperes}$$

For example, a magnetising force of 8000 A/m is applied to a circular magnetic circuit of mean diameter 30 cm by passing a current through a coil wound on the circuit. The coil is uniformly wound around the circuit and has 750 turns. To determine the current in the coil:

$$\text{Since } H = \frac{NI}{l} \text{ then } I = \frac{Hl}{N} = \frac{8000 \times \pi \times 30 \times 10^{-2}}{750} = \mathbf{10.05 \text{ A}}$$

Permeability and B-H Curves

For air, or any non-magnetic medium, the ratio of magnetic flux density to magnetising force is a constant, i.e. B/H = a constant. This constant is μ_0, the **permeability of free space** (or the magnetic space constant) and is equal to

$4\pi \times 10^{-7}$ H/m, i.e. **for air, or any non-magnetic medium,** $\boxed{\dfrac{\mathbf{B}}{\mathbf{H}} = \mu_0}$

(Although all non-magnetic materials, including air, exhibit slight magnetic properties, these can effectively be neglected.)

For all media other than free space, $\boxed{\dfrac{\mathbf{B}}{\mathbf{H}} = \mu_0 \mu_r}$

where u_r is the relative permeability, and is defined as

$$\mu_r = \frac{\textbf{flux density in material}}{\textbf{flux density in a vacuum}}$$

μ_r varies with the type of magnetic material and, since it is a ratio of flux densities, it has no unit. From its definition, μ_r for a vacuum is 1.

$$\mu_0\mu_r = \mu, \text{ called the \textbf{absolute permeability}}$$

By plotting measured values of flux density B against magnetic field strength H, a **magnetisation curve** (or **B-H curve**) is produced. For non-magnetic materials this is a straight line. Typical curves for four magnetic materials are shown in Figure 45.3

For example, a uniform ring of cast iron has a cross-sectional area of 10 cm^2 and a mean circumference of 20 cm. The m.m.f. necessary to produce a flux of 0.3 mWb in the ring is determined as follows:

$$\text{Flux density B} = \frac{\Phi}{A} = \frac{0.3 \times 10^{-3}}{10 \times 10^{-4}} = 0.3 \text{ T}$$

From the magnetisation curve for cast iron, when B = 0.3 T, H = 1000 A/m, hence **m.m.f.** = Hl = 1000 × 0.2 = **200 A**

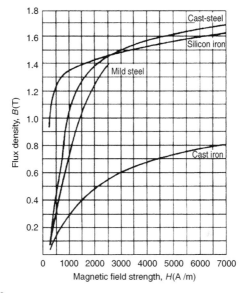

Figure 45.3

The **relative permeability** of a ferromagnetic material is proportional to the slope of the B-H curve and thus varies with the magnetic field strength. The approximate range of values of relative permeability μ_r for some common magnetic materials are:

Cast iron	$\mu_r = 100\text{--}250$	Mild steel	$\mu_r = 200\text{--}800$
Silicon iron	$\mu_r = 1000\text{--}5000$	Cast steel	$\mu_r = 300\text{--}900$
Mumetal	$\mu_r = 200\text{--}5000$	Stalloy	$\mu_r = 500\text{--}6000$

For example, a flux density of 1.2 T is produced in a piece of cast steel by a magnetising force of 1250 A/m. The relative permeability of the steel under these conditions is determined using $\mathbf{B} = \mu_0 \mu_r H$,

from which, $\mu_r = \dfrac{B}{\mu_0 H} = \dfrac{1.2}{(4\pi \times 10^{-7})(1250)} = \mathbf{764}$

Reluctance

Reluctance S (or R_M) is the 'magnetic resistance' of a magnetic circuit to the presence of magnetic flux.

$$\textbf{Reluctance } S = \frac{F_M}{\Phi} = \frac{NI}{\Phi} = \frac{Hl}{BA} = \frac{l}{\dfrac{B}{H}A} = \frac{\ell}{\mu_0 \mu_r A}$$

The unit of reluctance is 1/H (or H^{-1}) or A/Wb.

For example, the reluctance of a piece of mumetal of length 150 mm and cross-sectional area 1800 mm^2 when the relative permeability is 4000 is given by:

$$\text{reluctance } S = \frac{l}{\mu_0 \mu_r A} = \frac{150 \times 10^{-3}}{(4\pi \times 10^{-7})(4000)(1800 \times 10^{-6})}$$

$$= \mathbf{16580/H}$$

Ferromagnetic materials have a low reluctance and can be used as **magnetic screens** to prevent magnetic fields affecting materials within the screen.

Composite Series Magnetic Circuits

For a series magnetic circuit having n parts, the **total reluctance S** is given by:

$$S = S_1 + S_2 + \cdots + S_n$$

(This is similar to resistors connected in series in an electrical circuit)

For example, a closed magnetic circuit of cast steel contains a 6 cm long path of cross-sectional area 1 cm^2 and a 2 cm path of cross-sectional area 0.5 cm^2. A coil of 200 turns is wound around the 6 cm length of the circuit and a current of 0.4 A flows. If the relative permeability of the cast steel is 750 the flux density in the 2 cm path is determined as follows:

For the 6 cm long path:

$$\text{reluctance } S_1 = \frac{l_1}{\mu_0 \mu_r A_1} = \frac{6 \times 10^{-2}}{(4\pi \times 10^{-7})(750)(1 \times 10^{-4})}$$

$$= 6.366 \times 10^5 / H$$

For the 2 cm long path:

$$\text{reluctance } S_2 = \frac{l_2}{\mu_0 \mu_r A_2} = \frac{2 \times 10^{-2}}{(4\pi \times 10^{-7})(750)(0.5 \times 10^{-4})}$$

$$= 4.244 \times 10^5 / H$$

Total circuit reluctance $S = S_1 + S_2 = (6.366 + 4.244) \times 10^5$
$$= 10.61 \times 10^5 / H$$

$$S = \frac{\text{m.m.f.}}{\Phi} \text{ i.e. } \Phi = \frac{\text{m.m.f.}}{S} = \frac{NI}{S} = \frac{200 \times 0.4}{10.61 \times 10^5}$$

$$= 7.54 \times 10^{-5} \text{ Wb}$$

Hence the **flux density in the 2 cm path, B** $= \dfrac{\Phi}{A} = \dfrac{7.54 \times 10^{-5}}{0.5 \times 10^{-4}} = \mathbf{1.51 \ T}$

Comparison between Electrical and Magnetic Quantities

Electrical circuit	Magnetic circuit
e.m.f. E (V)	m.m.f. F_m (A)
current I (A)	flux Φ (Wb)
resistance R (Ω)	reluctance S (H^{-1})
$I = \dfrac{E}{R}$	$\Phi = \dfrac{m.m.f.}{S}$
$R = \dfrac{\rho l}{A}$	$S = \dfrac{l}{\mu_0 \mu_r A}$

Hysteresis and Hysteresis Loss

Hysteresis loop

Let a ferromagnetic material which is completely demagnetised, i.e. one in which B = H = 0, be subjected to increasing values of magnetic field strength H and the corresponding flux density B measured. The resulting relationship between B and H is shown by the curve Oab in Figure 45.4. At a particular value of H, shown as Oy, it becomes difficult to increase the flux density any further. The material is said to be saturated. Thus by is the **saturation flux density**.

If the value of H is now reduced it is found that the flux density follows curve bc. When H is reduced to zero, flux remains in the iron. This

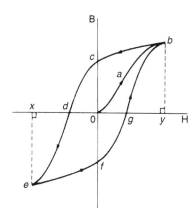

Figure 45.4

remanent flux density or **remanence** is shown as **O**c in Figure 45.4. When H is increased in the opposite direction, the flux density decreases until, at a value shown as O*d*, the flux density has been reduced to zero. The magnetic field strength **O**d required to remove the residual magnetism, i.e. reduce B to zero, is called the **coercive force**.

Further increase of H in the reverse direction causes the flux density to increase in the reverse direction until saturation is reached, as shown by curve *de*. If H is varied backwards from O*x* to O*y*, the flux density follows the curve *efgb*, similar to curve *bcde*.

It is seen from Figure 45.4 that the flux density changes lag behind the changes in the magnetic field strength. This effect is called **hysteresis**. The closed figure *bcdefgb* is called the **hysteresis loop** (or the B/H loop).

Hysteresis loss

A disturbance in the alignment of the domains (i.e. groups of atoms) of a ferromagnetic material causes energy to be expended in taking it through a cycle of magnetisation. This energy appears as heat in the specimen and is called the **hysteresis loss**

The energy loss associated with hysteresis is proportional to the area of the hysteresis loop.

If the hysteresis loop is plotted to a scale of 1 cm = α ampere/metre along the horizontal axis and 1 cm = β tesla along the vertical axis, and if A represents the area of the loop in square centimetres, then

$$\boxed{\text{hysteresis loss/cycle} = A\alpha\beta \text{ joules per metre}^3} \tag{1}$$

For example, the area of a hysteresis loop obtained from a ferromagnetic specimen is 12.5 cm^2. The scales used are: horizontal axis 1 cm = 500 A/m; vertical axis 1 cm = 0.2 T.

From equation (1), hysteresis loss per cycle = $A\alpha\beta$ = (12.5)(500)(0.2) = **1250J/m^3**

(Note that, since $\alpha = 500$ A/m per centimetre and $\beta = 0.2$ T per centimetre, then 1 cm^2 of the loop represents

$$500\frac{A}{m} \times 0.2 \text{ T} = 100 \frac{A}{m}\frac{Wb}{m^2} = 100 \frac{AVs}{m^3} = 100 \frac{Ws}{m^3} = 100 \text{ J/m}^3$$

Hence 12.5 cm^2 represents 12.5×100 = **1250 J/m^3**)

At, say, 50 Hz frequency, hysteresis loss = (1250 J/m^3) (50 1/s) = **62500 W/m^3**

If hysteresis loops for a given ferromagnetic material are determined for different maximum values of H, they are found to lie within one another as shown in Figure 45.5.

The maximum sized hysteresis loop for a particular material is obtained at saturation. If, for example, the maximum flux density is reduced to half its value at saturation, the area of the resulting loop is considerably less than the area of the loop at saturation. From the areas of a number of such hysteresis loops, as shown in Figure 45.5, the hysteresis loss per cycle was found by Steinmetz (an American electrical engineer) to be proportional to $(B_m)^n$, where n is called the **Steinmetz index** and can have a value between about 1.6 and

3.0, depending on the quality of the ferromagnetic material and the range of flux density over which the measurements are made.

It is found that the hysteresis loss is given by:

$$\boxed{\text{hysteresis loss, } P_h = k_h \, vf \, (B_m)^n \text{ watts}} \tag{2}$$

where v = volume in cubic metres, f = frequency in hertz, and k_h is a constant for a given specimen and given range of B.

For example, if a ferromagnetic specimen has a hysteresis loss of 62500 Wb/m^3 when the maximum flux density is 1.5 T at a frequency of 50 Hz, then the hysteresis loss per m^3 for a maximum flux density of 1.1 T and frequency of 25 Hz, assuming the Steinmetz index to be 1.6, is determined as follows:

From equation (2), hysteresis loss $P_h = k_h vf (B_m)^n$

Thus $62\,500 = k_h(1)(50)(1.5)^{1.6}$, from which,

$$\text{constant } k_h = \frac{62\,500}{(50)(1.5)^{1.6}} = 653.4$$

When $f = 25$ Hz and $B_m = 1.1$ T,

$$\text{hysteresis loss, } P_h = k_h vf (B_m)^n = (653.4)(1)(25)(1.1)^{1.6}$$

$$= \mathbf{19026 \ W/m^3}$$

The magnitude of the hysteresis loss depends on the composition of the specimen and on the heat treatment and mechanical handling to which the specimen has been subjected.

Figure 45.6 shows typical hysteresis loops for (a) hard steel, which has a high remanence Oc and a large coercivity Od, (b) soft steel, which has a large remanence and small coercivity and (c) ferrite, this being a ceramic-like magnetic substance made from oxides of iron, nickel, cobalt, magnesium, aluminium and manganese. The hysteresis of ferrite is very small.

For a.c.-excited devices the hysteresis loop is repeated every cycle of alternating current. Thus a hysteresis loop with a large area (as with hard steel) is often unsuitable since the energy loss would be considerable.

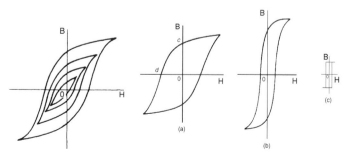

Figure 45.5   **Figure 45.6**

Silicon steel has a narrow hysteresis loop, and thus small hysteresis loss, and is suitable for transformer cores and rotating machine armatures.

Eddy Current Loss

If a coil is wound on a ferromagnetic core (such as in a transformer) and alternating current is passed through the coil, an alternating flux is set up in the core. The alternating flux induces an e.m.f. e in the coil given by: $e = N\dfrac{d\phi}{dt}$. However, in addition to the desirable effect of inducing an e.m.f. in the coil, the alternating flux induces undesirable voltages in the iron core. These induced e.m.f.'s set up circulating currents in the core, known as **eddy currents**. Since the core possesses resistance, the eddy currents heat the core, and this represents wasted energy.

Eddy currents can be reduced by laminating the core, i.e. splitting it into thin sheets with very thin layers of insulating material inserted between each pair of the laminations (this may be achieved by simply varnishing one side of the lamination or by placing paper between each lamination). The insulation presents a high resistance and this reduces any induced circulating currents.

Eddy current loss is given by: $\boxed{P_e = k_e(B_m)^2 f^2 t^3 \text{ watts}}$ (3)

where k_e is a constant and t is the lamination thickness.

Alternatively, the eddy current loss per m^3 is given by:

$$\boxed{P_e = k_e(B_m)^2 f^2 t^2 \text{ watts per m}^3}$$ (4)

From equation (4) it is seen that eddy current loss is proportional to the square of the thickness of the core strip. It is therefore desirable to make lamination strips as thin as possible. However, at high frequencies where it is not practicable to make very thin laminations, using ferrite or dust cores may reduce core losses. Dust cores consist of fine particles of carbonyl iron or permalloy (i.e. nickel and iron), each particle of which is insulated from its neighbour by a binding material. Such materials have a very high value of resistivity.

For example, the core of a transformer operating at 50 Hz has an eddy current loss of 100 W/m^3 and the core laminations have a thickness of 0.50 mm. The core is redesigned so as to operate with the same eddy current loss but at a different voltage and at a frequency of 250 Hz. Assuming that at the new voltage the maximum flux density is one-third of its original value and the resistivity of the core remains unaltered, the necessary new thickness of the laminations is determined as follows:

From equation (4), $P_e = k_e(B_m)^2 f^2 t^2$ watts per m^3.

Hence, at 50 Hz frequency, $100 = k_e(B_m)^2(50)^2(0.50 \times 10^{-3})^2$ from which, $k_e = \dfrac{100}{(B_m)^2(50)^2(0.50 \times 10^{-3})^2}$

At 250 Hz frequency,

$$100 = k_e \left(\frac{B_m}{3}\right)^2 (250)^2 (t)^2$$

i.e.
$$100 = \left(\frac{100}{(B_m)^2(50)^2(0.50 \times 10^{-3})^2}\right) \left(\frac{B_m}{3}\right)^2 (250)^2 (t)^2$$

$$= \frac{100(250)^2(t)^2}{(3)^2(50)^2(0.50 \times 10^{-3})^2}$$

from which,

$$t^2 = \frac{(100)(3)^2(50)^2(0.50 \times 10^{-3})}{(100)(250)^2}$$

i.e. **lamination thickness,**

$$t = \frac{(3)(50)(0.50 \times 10^{-3})}{250} = 0.3 \times 10^{-3} \text{ m or } \textbf{0.30 mm}$$

Separation of Hysteresis and Eddy Current Losses

From equation (2), hysteresis loss, $P_h = k_h v f (B_m^n)$
From equation (3), eddy current loss, $P_e = k_e (B_m)^2 f^2 t^3$
The total core loss P_c is given by $P_c = P_h + P_e$
If for a particular inductor or transformer, the core flux density is maintained constant, then $P_h = k_1 f$, where constant $k_1 = k_h v (B_m)^n$ and

$$P_e = k_2 f^2, \text{ where constant } k_2 = k_e (B_m)^2 t^3$$

Thus the total core loss,

$$P_c = k_1 f + k_2 f^2 \text{ and } \boxed{\frac{P_c}{f} = k_1 + k_2 f}$$

which is of the straight line form $y = mx + c$. Thus if P_c/f is plotted vertically against f horizontally, a straight line graph results having a gradient k_2 and a vertical-axis intercept k_1

If the total core loss P_c is measured over a range of frequencies, then k_1 and k_2 may be determined from the graph of P_c/f against f. Hence the hysteresis loss $P_h(= k_1 f)$ and the eddy current loss $P_e(= k_2 f^2)$ at a given frequency may be determined.

The above method of separation of losses is an approximate one since the Steinmetz index n is not a constant value but tends to increase with increase of frequency. However, a reasonable indication of the relative magnitudes of the hysteresis and eddy current losses in an iron core may be determined.

For example, the total core loss of a ferromagnetic cored transformer winding is measured at different frequencies and the results obtained are:

Total core loss, P_c (watts)	45	105	190	305
Frequency, f (hertz)	30	50	70	90

To obtain a straight line graph, values of P_c/f are plotted against f.

f (Hz)	30	50	70	90
P_c/f	1.5	2.1	2.7	3.4

A graph of P_c/f against f is shown in Figure 45.7. The graph is a straight line of the form $P_c/f = k_1 + k_2 f$

The vertical axis intercept at $f = 0$, $\mathbf{k_1 = 0.5}$

The gradient of the graph, $k_2 = \dfrac{a}{b} = \dfrac{3.7 - 0.5}{100} = \mathbf{0.032}$

Since $P_c/f = k_1 + k_2 f$, then $P = k_1 f + k_2 f^2$
i.e. total core losses = hysteresis loss + eddy current loss
At a frequency of, say, 50 Hz,

$$\text{hysteresis loss} = k_1 f = (0.5)(50) = \mathbf{25\ W}$$

$$\text{eddy current loss} = k_2 f^2 = (0.032)(50)^2 = \mathbf{80\ W}$$

and at a frequency of, say, 60 Hz,

$$\text{hysteresis loss} = k_1 f = (0.5)(60) = \mathbf{30\ W}$$

$$\text{eddy current loss} = k_2 f^2 = (0.032)(60)^2 = \mathbf{115.2\ W}$$

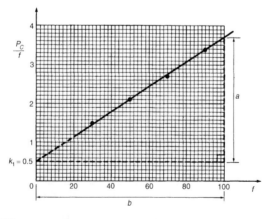

Figure 45.7

46 Magnetic Materials

Magnetic Properties of Materials

The full theory of magnetism is one of the most complex of subjects. However the phenomenon may be satisfactorily explained by the use of a simple model. Bohr and Rutherford, who discovered atomic structure, suggested that electrons move around the nucleus confined to a plane, like planets around the sun. An even better model is to consider each electron as having a surface, which may be spherical or elliptical or something more complicated.

Magnetic effects in materials are due to the electrons contained in them, the electrons giving rise to magnetism in the following two ways:

(i) by revolving around the nucleus
(ii) by their angular momentum about their own axis, called spin.

In each of these cases the charge of the electron can be thought of as moving round in a closed loop and therefore acting as a current loop.

The main measurable quantity of an atomic model is the **magnetic moment**. When applied to a loop of wire carrying a current,

$$\text{magnetic moment} = \text{current} \times \text{area of the loop}$$

Electrons associated with atoms possess magnetic moment which gives rise to their magnetic properties.

Diamagnetism is a phenomenon exhibited by materials having a relative permeability less than unity. When electrons move more or less in a spherical orbit around the nucleus, the magnetic moment due to this orbital is zero, all the current due to moving electrons being considered as averaging to zero. If the net magnetic moment of the electron spins were also zero then there would be no tendency for the electron motion to line up in the presence of a magnetic field. However, as a field is being turned on, the flux through the electron orbitals increases. Thus, considering the orbital as a circuit, there will be, by Faraday's laws, an e.m.f. induced in it which will change the current in the circuit. The flux change will accelerate the electrons in its orbit, causing an induced magnetic moment. By Lenz's law the flux due to the induced magnetic moment will be such as to oppose the applied flux. As a result, the net flux through the material becomes less than in a vacuum. Since relative permeability is defined as $\dfrac{\text{flux density in material}}{\text{flux density in vacuum}}$ **with diamagnetic materials the relative permeability is less than one**.

Paramagnetism is a phenomenon exhibited by materials where the relative permeability is greater than unity. Paramagnetism occurs in substances where atoms have a permanent magnetic moment. This may be caused by the orbitals not being spherical or by the spin of the electrons. Electron spins tend to pair up and cancel each other. However, there are many atoms with odd numbers of electrons, or in which pairing is incomplete. Such atoms have what is called a permanent dipole moment. When a field is applied to them

256

they tend to line up with the field, like compass needles, and so strengthen the flux in that region. (Diamagnetic materials do not tend to line up with the field in this way.) When this effect is stronger than the diamagnetic effect, the overall effect is to make the relative permeability greater than one. Such materials are called paramagnetic.

Ferromagnetic materials

Ferromagnetism is the phenomenon exhibited by materials having a relative permeability which is considerably greater than 1 and which varies with flux density. Iron, cobalt and nickel are the only elements that are ferromagnetic at ordinary working temperatures, but there are several alloys containing one or more of these metals as constituents, with widely varying ferromagnetic properties.

Consider the simple model of a single iron atom represented in Figure 46.1. It consists of a small heavy central nucleus surrounded by a total of 26 electrons. Each electron has an orbital motion about the nucleus in a limited region, or shell, such shells being represented by circles K, L, M and N. The numbers in Figure 46.1 represent the number of electrons in each shell. The outer shell N contains two loosely held electrons, these electrons becoming the carriers of electric current, making iron electrically conductive. There are 14 electrons in the M shell and it is this group that is responsible for magnetism. An electron carries a negative charge and a charge in motion constitutes an electric current with which is associated a magnetic field. Magnetism would therefore result from the orbital motion of each electron in the atom. However, experimental evidence indicates that the resultant magnetic effect due to all the orbital motions in the metal solid is zero; thus the orbital currents may be disregarded.

In addition to the orbital motion, each electron spins on its own axis. A rotating charge is equivalent to a circular current and gives rise to a magnetic field. In any atom, all the axes about which the electrons spin are parallel, but rotation may be in either direction. In the single atom shown in Figure 46.1, in each of the K, L and N shells equal numbers of electrons spin in the clockwise and anticlockwise directions respectively and therefore these shells are magnetically neutral. However, in shell M, nine of the electrons spin in one direction while five spin in the opposite direction. There is therefore a resultant effect due to four electrons.

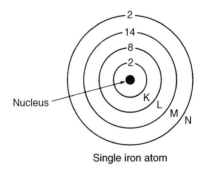

Single iron atom

Figure 46.1

The atom of cobalt has 15 electrons in the M shell, nine spinning in one direction and six in the other. Thus with cobalt there is a resultant effect due to 3 electrons. A nickel atom has a resultant effect due to 2 electrons. The atoms of the paramagnetic elements, such as manganese, chromium or aluminium, also have a resultant effect for the same reasons as that of iron, cobalt and nickel. However, in the diamagnetic materials there is an exact equality between the clockwise and anticlockwise spins.

The total magnetic field of the resultant effect due to the four electrons in the iron atom is large enough to influence other atoms. Thus the orientation of one atom tends to spread through the material, with atoms acting together in groups instead of behaving independently. These groups of atoms, called **domains** (which tend to remain permanently magnetised), act as units. Thus, when a field is applied to a piece of iron, these domains as a whole tend to line up and large flux densities can be produced. This means that the relative permeability of such materials is much greater than one. As the applied field is increased, more and more domains align and the induced flux increases.

The overall magnetic properties of iron alloys and materials containing iron, such as ferrite (ferrite is a mixture of iron oxide together with other oxides — lodestone is a ferrite), depend upon the structure and composition of the material. However, the presence of iron ensures marked magnetic properties of some kind in them. Ferromagnetic effects decrease with temperature, as do those due to paramagnetism. The loss of ferromagnetism with temperature is more sudden, however; the temperature at which it has all disappeared is called the **Curie temperature**. The ferromagnetic properties reappear on cooling, but any magnetism will have disappeared. Thus a permanent magnet will be demagnetised by heating above the Curie temperature (1040 K for iron) but can be remagnetised after cooling. Above the Curie temperature, ferromagnetics behave as paramagnetics.

Nonpermanent Magnetic Materials

General

Nonpermanent magnetic materials are those in which magnetism may be induced. With the magnetic circuits of electrical machines, transformers and heavy current apparatus a high value of flux density B is desirable so as to limit the cross-sectional area A ($\Phi = BA$) and therefore the weight and cost involved. At the same time the magnetic field strength $H \left(= \dfrac{NI}{l} \right)$ should be as small as possible so as to limit the I^2R losses in the exciting coils. The relative permeability $\left(\mu_r = \dfrac{B}{\mu_0 H} \right)$ and the saturation flux density should therefore be high. Also, when flux is continually varying, as in transformers, inductors and armature cores, low hysteresis and eddy current losses are essential.

Silicon-iron alloys

In the earliest electrical machines the magnetic circuit material used was iron with low content of carbon and other impurities. However, it was later discovered that the deliberate addition of silicon to the iron brought about a great improvement in magnetic properties. The laminations now used in electrical machines and in transformers at supply frequencies are made of silicon-steel in which the silicon in different grades of the material varies in amounts from about 0.5% to 4.5% by weight. The silicon added to iron increases the resistivity. This in turn increases the resistance $\left(R = \dfrac{\rho l}{A} \right)$ and thus helps to reduce eddy current loss. The hysteresis loss is also reduced; however, the silicon reduces the saturation flux density.

A limit to the amount of silicon which may be added in practice is set by the mechanical properties of the material, since the addition of silicon causes a material to become brittle. Also the brittleness of a silicon-iron alloy depends on temperature. About 4.5% silicon is found to be the upper practical limit for silicon-iron sheets. Lohys is a typical example of a silicon-iron alloy and is used for the armatures of d.c. machines and for the rotors and stators of a.c. machines. Stalloy, which has a higher proportion of silicon and lower losses, is used for transformer cores.

Silicon steel sheets are often produced by a hot-rolling process. In these finished materials the constituent crystals are not arranged in any particular manner with respect, for example, to the direction of rolling or the plane of the sheet. If silicon steel is reduced in thickness by rolling in the cold state and the material is then annealed it is possible to obtain a finished sheet in which the crystals are nearly all approximately parallel to one another. The material has strongly directional magnetic properties, the rolling direction being the direction of highest permeability. This direction is also the direction of lowest hysteresis loss. This type of material is particularly suitable for use in transformers, since the axis of the core can be made to correspond with the rolling direction of the sheet and thus full use is made of the high permeability, low loss direction of the sheet.

With silicon-iron alloys a maximum magnetic flux density of about 2 T is possible. With cold-rolled silicon steel, used for large machine construction, a maximum flux density of 2.5 T is possible, whereas the maximum obtainable with the hot-rolling process is about 1.8 T. (In fact, with any material, only under the most abnormal of conditions will the value of flux density exceed 3 T).

It should be noted that the term 'iron-core' implies that the core is made of iron; it is, in fact, almost certainly made from steel, pure iron being extremely hard to come by. Equally, an iron alloy is generally a steel and so it is preferred to describe a core as being a steel rather than an iron core.

Nickel-iron alloys

Nickel and iron are both ferromagnetic elements and when they are alloyed together in different proportions a series of useful magnetic alloys is obtained. With about 25%–30% nickel content added to iron, the alloy tends to be very hard and almost nonmagnetic at room temperature. However, when the nickel content is increased to, say, 75%–80% (together with small amounts of molybdenum and copper), very high values of initial and maximum permeabilities and very low values of hysteresis loss are obtainable if the alloys are given suitable heat treatment. For example, Permalloy, having a content of 78% nickel, 3% molybdenum and the remainder iron, has an initial permeability of

20 000 and a maximum permeability of 100 000 compared with values of 250 and 5000 respectively for iron. The maximum flux density for Permalloy is about 0.8 T. Mumetal (76% nickel, 5% copper and 2% chromium) has similar characteristics. Such materials are used for the cores of current and a.f. transformers, for magnetic amplifiers and also for magnetic screening. However, nickel-iron alloys are limited in that they have a low saturation value when compared with iron. Thus, in applications where it is necessary to work at a high flux density, nickel-iron alloys are inferior to both iron and silicon-iron. Also nickel-iron alloys tend to be more expensive than silicon-iron alloys.

Eddy current loss is proportional to the thickness of lamination squared, thus using laminations as thin as possible can reduce such losses. Nickel-iron alloy strip as thin as 0.004 mm, wound in a spiral, may be used.

Dust cores

In many circuits high permeability may be unnecessary or it may be more important to have a very high resistivity. Where this is so, metal powder or dust cores are widely used up to frequencies of 150 MHz. These consist of particles of nickel-iron-molybdenum for lower frequencies and iron for the higher frequencies. The particles, which are individually covered with an insulating film, are mixed with an insulating, resinous binder and pressed into shape.

Ferrites

Magnetite, or ferrous ferrite, is a compound of ferric oxide and ferrous oxide and possesses magnetic properties similar to those of iron. However, being a semiconductor, it has a very high resistivity. Manufactured ferrites are compounds of ferric oxide and an oxide of some other metal such as manganese, nickel or zinc. Ferrites are free from eddy current losses at all but the highest frequencies (i.e. >100 MHz) but have a much lower initial permeability compared with nickel-iron alloys or silicon-iron alloys. Ferrites have typically a maximum flux density of about 0.4 T. Ferrite cores are used in audio-frequency transformers and inductors.

Permanent Magnetic Materials

A permanent magnet is one in which the material used exhibits magnetism without the need for excitation by a current-carrying coil. The silicon-iron and nickel-iron alloys discussed earlier are 'soft' magnetic materials having high permeability and hence low hysteresis loss. The opposite characteristics are required in the 'hard' materials used to make permanent magnets. In permanent magnets, high remanent flux density and high coercive force, after magnetisation to saturation, are desirable in order to resist demagnetisation. The hysteresis loop should embrace the maximum possible area. Possibly the best criterion of the merit of a permanent magnet is its maximum energy product $(BH)_m$, i.e. the maximum value of the product of the flux density B and the magnetic field strength H along the demagnetisation curve (shown as cd in Figure 45.4, page 249). A rough criterion is the product of coercive force and remanent flux density, i.e. $(Od)(Oc)$ in Figure 45.4. The earliest materials used for permanent magnets were tungsten and chromium steel, followed by a series of cobalt steels, to give both a high remanent flux density and a high value of $(BH)_m$

Alni was the first of the aluminium-nickel-iron alloys to be discovered, and with the addition of cobalt, titanium and niobium, the Alnico series of magnets was developed, the properties of which vary according to composition. These materials are very hard and brittle. Many alloys with other compositions and trade names are commercially available.

A considerable advance was later made when it was found that directional magnetic properties could be induced in alloys of suitable composition if they were heated in a strong magnetic field. This discovery led to the powerful Alcomex and Hycomex series of magnets. By using special casting techniques to give a grain-oriented structure, even better properties are obtained if the field applied during heat treatment is parallel to the columnar crystals in the magnet. The values of coercivity, the remanent flux density and hence $(BH)_m$ are high for these alloys.

The most recent and most powerful permanent magnets discovered are made by powder metallurgy techniques and are based on an intermetallic compound of cobalt and samarium. These are very expensive and are only available in a limited range of small sizes.

47 Electromagnetism

Magnetic Field due to an Electric Current

Magnetic fields can be set up not only by permanent magnets, as shown in chapter 45, but also by electric currents.

Let a piece of wire be arranged to pass vertically through a horizontal sheet of cardboard on which is placed some iron filings, as shown in Figure 47.1(a). If a current is now passed through the wire, then the iron filings will form a definite circular field pattern with the wire at the centre, when the cardboard is gently tapped. By placing a compass in different positions the lines of flux are seen to have a definite direction as shown in Figure 47.1(b). If the current direction is reversed, the direction of the lines of flux is also reversed. The effect on both the iron filings and the compass needle disappears when the current is switched off. The electric current thus produces the magnetic field. The magnetic flux produced has the same properties as the flux produced by a permanent magnet. If the current is increased the strength of the field increases and, as for the permanent magnet, the field strength decreases as we move away from the current-carrying conductor.

In Figure 47.1, the effect of only a small part of the magnetic field is shown. If the whole length of the conductor is similarly investigated it is found that the magnetic field round a straight conductor is in the form of concentric cylinders as shown in Figure 47.2, the field direction depending on the direction of the current flow.

When dealing with magnetic fields formed by electric current it is usual to portray the effect as shown in Figure 47.3. The convention adopted is:

(i) Current flowing away from the viewer, i.e. into the paper, is indicated by ⊗. This may be thought of as the feathered end of the shaft of an arrow. See Figure 47.3(a)

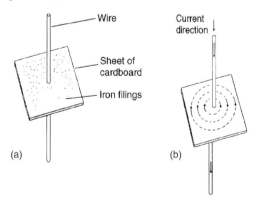

(a)   (b)

Figure 47.1

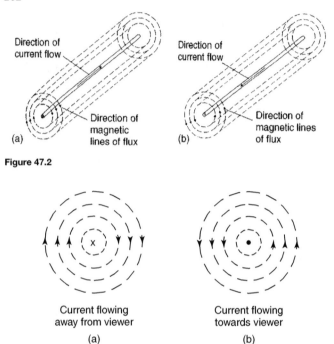

Figure 47.2

Current flowing
away from viewer

(a)

Current flowing
towards viewer

(b)

Figure 47.3

(ii) Current flowing towards the viewer, i.e. out of the paper, is indicated by
●. This may be thought of as the point of an arrow. See Figure 47.3(b).

The direction of the magnetic lines of flux is best remembered by the **screw
rule** which states that:

*If a normal right-hand thread screw is screwed along the conductor in
the direction of the current, the direction of rotation of the screw is in
the direction of the magnetic field.*

For example, with current flowing away from the viewer (Figure 47.3(a)) a
right-hand thread screw driven into the paper has to be rotated clockwise.
Hence the direction of the magnetic field is clockwise.

A magnetic field set up by a long coil, or **solenoid**, is shown in Figure
47.4(a) and is seen to be similar to that of a bar magnet. If the solenoid is
wound on an iron bar, as shown in Figure 47.4(b), an even stronger magnetic
field is produced, the iron becoming magnetised and behaving like a permanent
magnet. The direction of the magnetic field produced by the current I in the
solenoid may be found by either of two methods, i.e. the screw rule or the
grip rule.

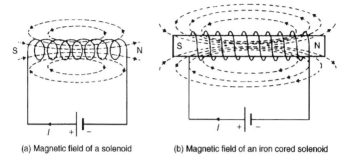

(a) Magnetic field of a solenoid   (b) Magnetic field of an iron cored solenoid

Figure 47.4

(a) **The screw rule** states that if a normal right-hand thread screw is placed along the axis of the solenoid and is screwed in the direction of the current it moves in the direction of the magnetic field **inside** the solenoid. The direction of the magnetic field **inside** the solenoid is from south to north. Thus in Figures 47.4(a) and (b) the north pole is to the right.

(b) **The grip rule** states that if the coil is gripped with the **right** hand, with the fingers pointing in the direction of the current, then the thumb, outstretched parallel to the axis of the solenoid, points in the direction of the magnetic field **inside** the solenoid.

Electromagnets

The solenoid is very important in electromagnetic theory since the magnetic field inside the solenoid is practically uniform for a particular current, and is also versatile, in as much that a variation of the current can alter the strength of the magnetic field. An electromagnet, based on the solenoid, provides the basis of many items of electrical equipment, examples of which include electric bells, relays, lifting magnets and telephone receivers.

(i) Electric bell

There are various types of electric bell, including the single-stroke bell, the trembler bell, the buzzer and a continuously ringing bell, but all depend on the attraction exerted by an electromagnet on a soft iron armature. A typical single stroke bell circuit is shown in Figure 47.5. When the push button is operated a current passes through the coil. Since the iron-cored coil is energised the soft iron armature is attracted to the electromagnet. The armature also carries a striker that hits the gong. When the circuit is broken the coil becomes demagnetised and the spring steel strip pulls the armature back to its original position. The striker will only operate when the push is operated.

(ii) Relay

A relay is similar to an electric bell except that contacts are opened or closed by operation instead of a gong being struck. A typical simple relay is shown

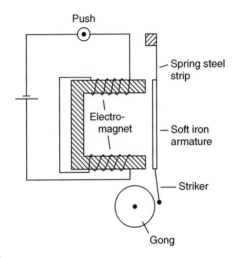

Figure 47.5

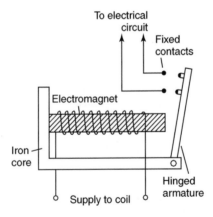

Figure 47.6

in Figure 47.6, which consists of a coil wound on a soft iron core. When the coil is energised the hinged soft iron armature is attracted to the electromagnet and pushes against two fixed contacts so that they are connected together, thus closing some other electrical circuit.

(iii) Lifting magnet

Lifting magnets, incorporating large electromagnets, are used in iron and steel works for lifting scrap metal. A typical robust lifting magnet, capable of

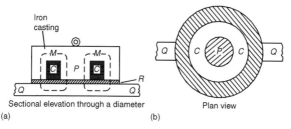

Sectional elevation through a diameter
(a)

Plan view
(b)

Figure 47.7

exerting large attractive forces, is shown in the elevation and plan view of Figure 47.7 where a coil, C, is wound round a central core, P, of the iron casting. Over the face of the electromagnet is placed a protective non-magnetic sheet of material, R. The load, Q, which must be of magnetic material is lifted when the coils are energised, the magnetic flux paths, M, being shown by the broken lines.

(iv) Telephone receiver

Whereas a transmitter or microphone changes sound waves into corresponding electrical signals, a telephone receiver converts the electrical waves back into sound waves. A typical telephone receiver is shown in Figure 47.8 and consists of a permanent magnet with coils wound on its poles. A thin, flexible diaphragm of magnetic material is held in position near to the magnetic poles but not touching them. Variation in current from the transmitter varies the magnetic field and the diaphragm consequently vibrates. The vibration produces sound variations corresponding to those transmitted.

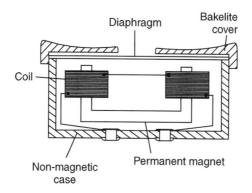

Figure 47.8

Force on a Current-carrying Conductor

If a current-carrying conductor is placed in a magnetic field produced by permanent magnets, then the fields due to the current-carrying conductor and the permanent magnets interact and cause a force to be exerted on the conductor. The force on the current-carrying conductor in a magnetic field depends upon:

(a) the flux density of the field, B teslas
(b) the strength of the current, I amperes,
(c) the length of the conductor perpendicular to the magnetic field, l metres, and
(d) the directions of the field and the current.

When the magnetic field, the current and the conductor are mutually at right angles then:

$$\boxed{\text{Force } F = BIl \text{ newtons}}$$

When the conductor and the field are at an angle $\theta°$ to each other then:

$$\boxed{\text{Force } F = Bil \sin \theta \text{ newtons}}$$

Since when the magnetic field, current and conductor are mutually at right angles, $F = BIl$, the magnetic flux density B may be defined by $B = \dfrac{F}{Il}$, i.e., the flux density is 1 T if the force exerted on 1 m of a conductor when the conductor carries a current of 1 A is 1 N.

For example, a conductor carries a current of 20 A and is at right angles to a magnetic field having a flux density of 0.9 T. If the length of the conductor in the field is 30 cm, the force acting on the conductor is given by:

$F = BIl = (0.9)(20)(0.30) = \textbf{5.4 N}$ when the conductor is at right angles to the field, as shown in Figure 47.9(a)

When the conductor is inclined at 30° to the field, as shown in Figure 47.9(b), then

$$F = BIl \sin \theta = (0.9)(20)(0.30) \sin 30° = \textbf{2.7 N}$$

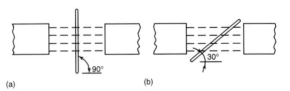

(a)   (b)

Figure 47.9

Loudspeaker

A simple application of the above force is the moving coil loudspeaker. The loudspeaker is used to convert electrical signals into sound waves.

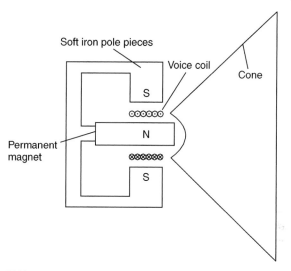

Figure 47.10

Figure 47.10 shows a typical loudspeaker having a magnetic circuit comprising a permanent magnet and soft iron pole pieces so that a strong magnetic field is available in the short cylindrical air gap. A moving coil, called the voice or speech coil, is suspended from the end of a paper or plastic cone so that it lies in the gap. When an electric current flows through the coil it produces a force that tends to move the cone backwards and forwards according to the direction of the current. The cone acts as a piston, transferring this force to the air, and producing the required sound waves.

If the current-carrying conductor shown in Figure 47.3(a) is placed in the magnetic field shown in Figure 47.11(a), then the two fields interact and cause a force to be exerted on the conductor as shown in Figure 47.11(b). The field is strengthened above the conductor and weakened below, thus tending to move the conductor downwards. This is the basic principle of operation of the electric motor and the moving-coil instrument.

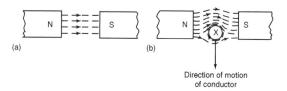

Figure 47.11

268

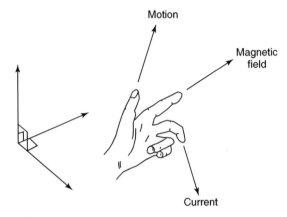

Motion

Magnetic field

Current

Figure 47.12

The direction of the force exerted on a conductor can be pre-determined by using **Fleming's left-hand rule** (often called the motor rule) which states:

Let the thumb, first finger and second finger of the left hand be extended such that they are all at right angles to each other, (as shown in Figure 47.12). If the first finger points in the direction of the magnetic field, the second finger points in the direction of the current, then the thumb will point in the direction of the motion of the conductor.

Summarising:

First finger — Field
SeCond finger — Current
ThuMb — Motion

Principle of Operation of a Simple d.c. Motor

A rectangular coil that is free to rotate about a fixed axis is shown placed inside a magnetic field produced by permanent magnets in Figure 47.13. A direct current is fed into the coil via carbon brushes bearing on a commutator, which consists of a metal ring split into two halves separated by insulation. When current flows in the coil a magnetic field is set up around the coil that interacts with the magnetic field produced by the magnets. This causes a force F to be exerted on the current-carrying conductor, which, by Fleming's left-hand rule, is downward between points A and B and upward between C and D for the current direction shown. This causes a torque and the coil rotates anticlockwise. When the coil has turned through 90° from the position shown in Figure 47.13 the brushes connected to the positive and negative terminals

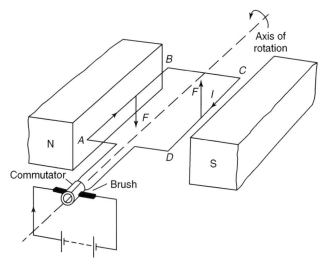

Figure 47.13

of the supply make contact with different halves of the commutator ring, thus reversing the direction of the current flow in the conductor. If the current is not reversed and the coil rotates past this position the forces acting on it change direction and it rotates in the opposite direction thus never making more than half a revolution. The current direction is reversed every time the coil swings through the vertical position and thus the coil rotates anti-clockwise for as long as the current flows. This is the principle of operation of a d.c. motor which is thus a device that takes in electrical energy and converts it into mechanical energy.

Principle of Operation of a Moving-coil Instrument

A moving-coil instrument operates on the motor principle. When a conductor carrying current is placed in a magnetic field, a force F is exerted on the conductor, given by $F = BIl$. If the flux density B is made constant (by using permanent magnets) and the conductor is a fixed length (say, a coil) then the force will depend only on the current flowing in the conductor.

In a moving-coil instrument a coil is placed centrally in the gap between shaped pole pieces as shown by the front elevation in Figure 47.14(a). (The air-gap is kept as small as possible, although for clarity it is shown exaggerated in Figure 47.14). Steel pivots, resting in jewel bearings, on a cylindrical iron core, support the coil. Current is led into and out of the coil by two phosphor bronze spiral hairsprings which are wound in opposite directions to minimise the effect of temperature change and to limit the coil swing (i.e. to **control** the

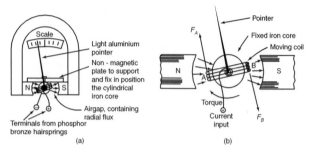

Figure 47.14

movement) and return the movement to zero position when no current flows. Current flowing in the coil produces forces as shown in Figure 47.14(b), the directions being obtained by Fleming's left-hand rule. The two forces, F_A and F_B, produce a torque that will move the coil in a clockwise direction, i.e. move the pointer from left to right. Since force is proportional to current the scale is linear.

When the aluminium frame, on which the coil is wound, is rotated between the poles of the magnet, small currents (called eddy currents) are induced into the frame, and this provides automatically the necessary **damping** of the system due to the reluctance of the former to move within the magnetic field.

The moving-coil instrument will measure only direct current or voltage and the terminals are marked positive and negative to ensure that the current passes through the coil in the correct direction to deflect the pointer 'up the scale'. The range of this sensitive instrument is extended by using shunts and multipliers (see Chapter 50)

Force on a Charge

When a charge of Q coulombs is moving at a velocity of v m/s in a magnetic field of flux density B teslas, the charge moving perpendicular to the field, then the magnitude of the force F exerted on the charge is given by:

$$\boxed{F = QvB \textbf{ newtons}}$$

For example, an electron in a television tube has a charge of 1.6×10^{-19} coulombs and travels at 3×10^7 m/s perpendicular to a field of flux density 18.5 µT. The force F exerted on the electron in the field is given by:

$$F = QvB = 1.6 \times 10^{-19} \times 3 \times 10^7 \times 18.5 \times 10^{-6}$$

$$= 1.6 \times 3 \times 18.5 \times 10^{-18} = 88.8 \times 10^{-18}$$

$$= \mathbf{8.88 \times 10^{-17} \ N}$$

48 Electromagnetic Induction and Inductance

Introduction to Electromagnetic Induction

When a conductor is moved across a magnetic field so as to cut through the lines of force (or flux), an electromotive force (e.m.f.) is produced in the conductor. If the conductor forms part of a closed circuit then the e.m.f. produced causes an electric current to flow round the circuit. Hence, an e.m.f. (and thus current) is 'induced' in the conductor as a result of its movement across the magnetic field. This effect is known as **'electromagnetic induction'**.

Figure 48.1(a) shows a coil of wire connected to a centre-zero galvanometer, which is a sensitive ammeter with the zero-current position in the centre of the scale.

(a) When the magnet is moved at constant speed towards the coil (Figure 48.1(a)), a deflection is noted on the galvanometer showing that a current has been produced in the coil.

(b) When the magnet is moved at the same speed as in (a) but away from the coil the same deflection is noted but is in the opposite direction (see Figure 48.1(b)).

(c) When the magnet is held stationary, even within the coil, no deflection is recorded.

(d) When the coil is moved at the same speed as in (a) and the magnet held stationary the same galvanometer deflection is noted.

(e) When the relative speed is, say, doubled, the galvanometer deflection is doubled.

(f) When a stronger magnet is used, a greater galvanometer deflection is noted.

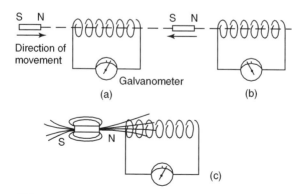

Figure 48.1

(g) When the number of turns of wire of the coil is increased, a greater galvanometer deflection is noted.

Figure 48.1(c) shows the magnetic field associated with the magnet. As the magnet is moved towards the coil, the magnetic flux of the magnet moves across, or cuts, the coil. **It is the relative movement of the magnetic flux and the coil that causes an e.m.f. and thus current, to be induced in the coil**. This effect is known as electromagnetic induction. The laws of electromagnetic induction evolved from experiments such as those described above.

Laws of Electromagnetic Induction

Faraday's laws of electromagnetic induction state:

 (i) *An induced e.m.f. is set up whenever the magnetic field linking that circuit changes.*
(ii) *The magnitude of the induced e.m.f. in any circuit is proportional to the rate of change of the magnetic flux linking the circuit.*

Lenz's law states:

The direction of an induced e.m.f. is always such that it tends to set up a current opposing the motion or the change of flux responsible for inducing that e.m.f.

An alternative method to Lenz's law of determining relative directions is given by **Fleming's Right-hand rule** (often called the gene<u>R</u>ator rule) which states:

Let the thumb, first finger and second finger of the right hand be extended such that they are all at right angles to each other (as shown in Figure 48.2). If the first finger points in the direction of the magnetic field and the thumb points in the direction of motion of the conductor relative to the magnetic field, then the second finger will point in the direction of the induced e.m.f.

Summarising:

> First finger — <u>F</u>ield
> Thu<u>M</u>b — <u>M</u>otion
> S<u>E</u>cond finger — <u>E</u>.m.f.

In a generator, conductors forming an electric circuit are made to move through a magnetic field. By Faraday's law, an e.m.f. is induced in the conductors, and thus a source of e.m.f. is created. A generator converts mechanical energy into electrical energy. (The action of a simple a.c. generator is described in Chapter 54).

The induced e.m.f. E set up between the ends of the conductor shown in Figure 48.3 is given by:

$$E = Blv \text{ volts}$$

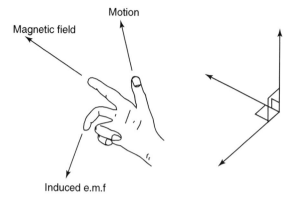

Motion

Magnetic field

Induced e.m.f

Figure 48.2

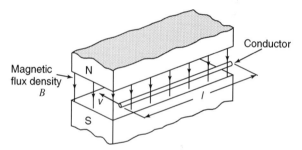

Conductor

Magnetic flux density B

N

v

l

S

Figure 48.3

where B, the flux density, is measured in teslas, l, the length of conductor in the magnetic field, is measured in metres, and v, the conductor velocity, is measured in metres per second.

If the conductor moves at an angle $\theta°$ to the magnetic field (instead of at $90°$ as assumed above) then:

$$\boxed{E = Blv \sin \theta \text{ volts}}$$

For example, a conductor moves with a velocity of 15 m/s at an angle of $90°$ to a magnetic field produced between two square-faced poles of side length 2 cm. If the flux leaving a pole face is 5 μWb, the magnitude of the induced e.m.f., is given by:

$$E_{90} = Blv \sin 90° = \left(\frac{\Phi}{A} \right) lv \sin 90° = \left(\frac{5 \times 10^{-6}}{4 \times 10^{-4}} \right) (0.02)(15)(1)$$

$$= \textbf{3.75 mV}$$

If the conductor moves at an angle of, say, 30° then:

$$E_{30} = Blv \sin 30° = E_{90} \sin 30° = 3.75 \sin 30° = \mathbf{1.875 \ mV}$$

Inductance

Inductance is the name given to the property of a circuit whereby there is an e.m.f. induced into the circuit by the change of flux linkages produced by a current change.

When the e.m.f. is induced in the same circuit as that in which the current is changing, the property is called **self inductance, L**

When the e.m.f. is induced in a circuit by a change of flux due to current changing in an adjacent circuit, the property is called **mutual inductance, M** (see chapter 49 following).The unit of inductance is the **henry, H.**

A circuit has an inductance of one henry when an e.m.f. of one volt is induced in it by a current changing at the rate of one ampere per second

Induced e.m.f. in a coil of N turns,

$$\boxed{E = -N \frac{d\Phi}{dt} \text{volts}}$$

where $d\Phi$ is the change in flux in Webers, and dt is the time taken for the flux to change in seconds (i.e. $\frac{d\Phi}{dt}$ is the rate of change of flux).

The minus sign in the equation remind us of its direction (given by Lenz's law).

For example, the e.m.f. E induced in a coil of 200 turns when there is a change of flux of 25 mWb linking with it in 50 ms, is given by:

$$E = -N \frac{d\Phi}{dt} = -(200) \left(\frac{25 \times 10^{-3}}{50 \times 10^{-3}} \right) = \mathbf{-100 \ volts}$$

Induced e.m.f. in a coil of inductance L henrys,

$$\boxed{E = -L \frac{dI}{dt} \text{ volts}}$$

where dI is the change in current in amperes and dt is the time taken for the current to change in seconds (i.e. $\frac{dI}{dt}$ is the rate of change of current).

For example, the e.m.f. E induced in a coil of inductance 12 H by a current changing at the rate of 4 A/s, is given by:

$$E = -L \frac{dI}{dt} = -(12)(4) = \mathbf{-48 \ volts}$$

Inductors

A component called an inductor is used when the property of inductance is required in a circuit. The basic form of an inductor is simply a coil of wire.
Factors that affect the inductance of an inductor include:

(i) the number of turns of wire–the more turns the higher the inductance
(ii) the cross-sectional area of the coil of wire — the greater the cross-sectional area the higher the inductance
(iii) the presence of a magnetic core–when the coil is wound on an iron core the same current sets up a more concentrated magnetic field and the inductance is increased
(iv) the way the turns are arranged–a short thick coil of wire has a higher inductance than a long thin one.

Two examples of practical inductors are shown in Figure 48.4, and the standard electrical circuit diagram symbols for air-cored and iron-cored inductors are shown in Figure 48.5.
An iron-cored inductor is often called a **choke** since, when used in a.c. circuits, it has a choking effect, limiting the current flowing through it.
Inductance is often undesirable in a circuit. To reduce inductance to a minimum the wire may be bent back on itself, as shown in Figure 48.6, so that the magnetising effect of one conductor is neutralised by that of the adjacent conductor. The wire may be coiled around an insulator, as shown,

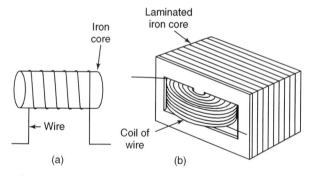

Iron core

Laminated iron core

Wire

Coil of wire

(a)   (b)

Figure 48.4

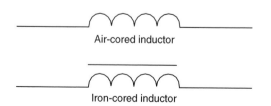

Air-cored inductor

Iron-cored inductor

Figure 48.5

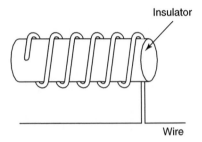

Insulator

Wire

Figure 48.6

without increasing the inductance. Standard resistors may be non-inductively wound in this manner.

Energy Stored

An inductor possesses an ability to store energy. The energy stored, W, in the magnetic field of an inductor is given by:

$$W = \tfrac{1}{2}LI^2 \text{ joules}$$

For example, the energy is stored in the magnetic field of an 8 H inductor which has a current of 3 A flowing through it, is given by:

$$\text{energy stored, } W = \tfrac{1}{2}LI^2 = \tfrac{1}{2}(8)(3)^2 = \mathbf{36 \text{ joules}}$$

Inductance of a Coil

The inductance of a coil, L, is given by:

$$L = \frac{N\Phi}{I} \text{ henrys}$$

For example, the coil inductance when a current of 4 A in a coil of 800 turns produces a flux of 5 mWb linking with the coil, is given by:

$$\text{inductance } L = \frac{N\Phi}{I} = \frac{(800)(5 \times 10^{-3})}{4} = \mathbf{1 \text{ H}}$$

49 Magnetically Coupled Circuits

Introduction

When the interaction between two loops of a circuit takes place through a magnetic field instead of through common elements, the loops are said to be inductively or **magnetically coupled**. The windings of a transformer, for example, are magnetically coupled (see Chapter 60).

Mutual Inductance

Mutual inductance is said to exist between two circuits when a changing current in one induces, by electromagnetic induction, an e.m.f. in the other. An ideal equivalent circuit of a mutual inductor is shown in Figure 49.1.

L_1 and L_2 are the self inductances of the two circuits and M the mutual inductance between them. The mutual inductance M is defined by the relationship:

$$E_2 = -M \frac{dI_1}{dt} \text{ or } E_1 = -M \frac{dI_2}{dt} \tag{1}$$

where E_2 is the e.m.f. in circuit 2 due to current I_1 in circuit 1 and E_1 is the e.m.f. in circuit 1 due to the current I_2 in circuit 2.

The unit of M is the **henry**.

For example, two coils have a mutual inductance of 0.2 H; if the current in one coil is changed from 10 A to 4 A in 10 ms, the average induced e.m.f. in the second coil,

$$E_2 = -M \frac{dI_1}{dt} = -(0.2)\left(\frac{10 - 4}{10 \times 10^{-3}}\right) = -120 \text{ V}$$

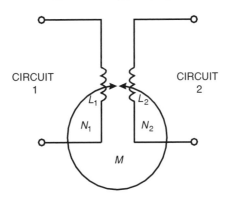

CIRCUIT 1

CIRCUIT 2

L_1 L_2

N_1 N_2

M

Figure 49.1

It may also be shown that:

$$M = N_2 \frac{d\Phi_2}{dI_1} \text{ and } M = N_1 \frac{d\Phi_1}{dI_2} \tag{2}$$

If the coils are linked with air as the medium, the flux and current are linearly related and equation (2) become:

$$M = \frac{N_2 \Phi_2}{I_1} \text{ and } M = \frac{N_1 \Phi_1}{I_2} \tag{3}$$

For example, A and B are two coils in close proximity. A has 1200 turns and B has 1000 turns. When a current of 0.8 A flows in coil A a flux of 100 μWb links with coil A and 75% of this flux links coil B. Then

self inductance of coil A, $L_A = \dfrac{N_A \Phi_A}{I_A}$ (from chapter 48)

$$= \frac{(1200)(100 \times 10^{-6})}{0.80} = \textbf{0.15 H}$$

and **mutual inductance,**

$$M = \frac{N_B \Phi_B}{I_A} = \frac{(1000)(0.75 \times 100 \times 10^{-6})}{0.80} = \textbf{93.75 mH}$$

Coupling Coefficient

The coupling coefficient k is the degree or fraction of magnetic coupling that occurs between circuits: $k = \dfrac{\text{flux linking two circuits}}{\text{total flux produced}}$

When there is no magnetic coupling, $k = 0$. If the magnetic coupling is perfect, i.e. all the flux produced in the primary links with the secondary then $k = 1$. Coupling coefficient is used in communications engineering to denote the degree of coupling between two coils. If the coils are close together, most of the flux produced by current in one coil passes through the other, and the coils are termed **tightly coupled**. If the coils are spaced apart, only a part of the flux links with the second, and the coils are termed **loosely coupled**.

It may be shown that:

$$\boxed{M = k \sqrt{L_1 L_2}} \tag{4}$$

or, $\boxed{\textbf{coefficient of coupling, } k = \dfrac{M}{\sqrt{L_1 L_2}}}$ $\qquad (5)$

For example, if two coils have self inductances of 250 mH and 400 mH respectively and if their mutual inductance is 80 mH, then

$$\text{coupling coefficient, } k = \frac{M}{\sqrt{L_1 L_2}} = \frac{80 \times 10^{-3}}{\sqrt{(250 \times 10^{-3})(400 \times 10^{-3})}}$$

$$= \frac{80 \times 10^{-3}}{\sqrt{0.1}} = \mathbf{0.253}$$

Coils Connected in Series

Figure 49.2 shows two coils 1 and 2 wound on an insulating core with terminals B and C joined. The fluxes in each coil produced by current i are in the same direction and the coils are termed **cumulatively coupled**.

Let the self inductance of coil 1 be L_1 and that of coil 2 be L_2 and let their mutual inductance be M.

If the winding between terminals A and D in Figure 49.2 are considered as a single circuit having a self inductance L_A henrys then it may be shown that:

$$L_A = L_1 + L_2 + 2M \tag{6}$$

If terminals B and D are joined as shown in Figure 49.3 the direction of the current in coil 2 is reversed and the coils are termed **differentially coupled**.

If L_B is the self inductance of the whole circuit between terminals A and C in Figure 49.3 then it may be shown that:

$$L_B = L_1 + L_2 - 2M \tag{7}$$

Thus the total inductance L of inductively coupled circuits is given by:

$$\boxed{L = L_1 + L_2 \pm 2M} \tag{8}$$

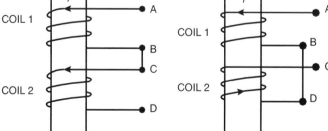

Figure 49.2   **Figure 49.3**

For example, two coils connected in series have self inductance of 40 mH and 10 mH respectively. The total inductance of the circuit is found to be 60 mH.

Then from equation (8),

total inductance, $L = L_1 + L_2 \pm 2M$

Hence $60 = 40 + 10 \pm 2M$

Since $(L_1 + L_2) < L$ then $60 = 40 + 10 + 2M$

from which $2M = 60 - 40 - 10 = 10$

and **mutual inductance,** $M = \dfrac{10}{2} = 5$ **mH**

Equation (6)–equation (7) gives:

$$L_A - L_B = (L_1 + L_2 + 2M) - (L_1 + L_2 - 2M)$$

i.e. $L_A - L_B = 2M - -2M = 4M$

from which,

$$\boxed{\text{mutual inductance, } M = \frac{L_A - L_B}{4}} \qquad (9)$$

An experimental method of determining the mutual inductance is indicated by equation (9), i.e. connect the coils both ways and determine the equivalent inductances L_A and L_B using an a.c. bridge. The mutual inductance is then given by a quarter of the difference between the two values of inductance.

Coupled Circuits

The magnitude of the secondary e.m.f. in Figure 49.4 is given by:

$$E_2 = M \frac{\mathrm{d}I_1}{\mathrm{d}t} \qquad \text{from chapter 48}$$

If the current I_1 is sinusoidal, i.e. $I_1 = I_{1m} \sin \omega t$ then

$E_2 = M \dfrac{\mathrm{d}}{\mathrm{d}t}(I_{1m} \sin \omega t) = M \omega I_{1m} \cos \omega\, t$

Since $\cos \omega t = \sin(\omega t + 90°)$ then $\cos \omega t = j \sin \omega t$ in complex form.

Hence $E_2 = M \omega I_{1m}(j \sin \omega t) = j\omega M (I_{1m} \sin \omega t)$

i.e. $\boxed{E_2 = j\omega M I_1} \qquad (10)$

If L_1 is the self inductance of the primary winding in Figure 49.4, there will be an e.m.f. generated equal to $j\omega L_1 I_1$ induced into the primary winding since the flux set up by the primary current also links with the primary winding.

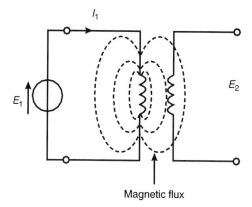

Figure 49.4

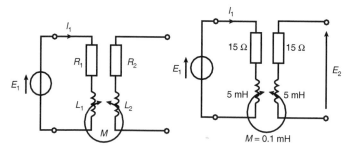

Figure 49.5

Figure 49.6

(a) Secondary open-circuited

For the circuit shown in Figure 49.5:

$$E_2 = \frac{j\omega M E_1}{R_1 + j\omega L_1} \qquad (11)$$

For example, for the circuit shown in Figure 49.6, the p.d. which appears across the open-circuited secondary winding, given that $E_1 = 8\sin 2500t$ volts, is determined as follows:

Impedance of primary,

$$Z_1 = R_1 + j\omega L_1 = 15 + j(2500)(5 \times 10^{-3})$$

$$= (15 + j12.5)\ \Omega \ \text{or} \ 19.53\angle 39.81°\ \Omega$$

Primary current $I_1 = \dfrac{E_1}{Z_1} = \dfrac{8\angle 0°}{19.53\angle 39.81°}\text{A}$

From equation (10),

$$E_2 = j\omega M I_1 = \frac{j\omega M E_1}{R_1 + j\omega L_1} = \frac{j(2500)(0.1 \times 10^{-3})(8\angle 0°)}{19.53\angle 39.81°}$$

$$= \frac{2\angle 90°}{19.53\angle 39.81°}$$

$$= \mathbf{0.102\angle 50.19°\ V}$$

(b) Secondary terminals having load impedance

In the circuit shown in Figure 49.7 a load resistor R_L is connected across the secondary terminals. Let $R_2' + R_L = R_2$

When an e.m.f. is induced into the secondary winding a current I_2 flows and this will induce an e.m.f. into the primary winding.

The effective primary impedance $Z_{1(eff)}$ of the circuit is given by:

$$Z_{1(eff)} = \frac{E_1}{I_1} = R_1 + \frac{\omega^2 M^2 R_2}{R_2^2 + \omega^2 L_2^2} + j\left(\omega L_1 - \frac{\omega^3 M^2 L_2}{R_2^2 + \omega^2 L_2^2}\right) \quad (12)$$

In equation (12), the primary impedance is $(R_1 + j\omega L_1)$. The remainder, i.e. $\left(\dfrac{\omega^2 M^2 R_2}{R_2^2 + \omega^2 L_2^2} - j\dfrac{\omega^3 M^2 L_2}{R_2^2 + \omega^2 L_2^2}\right)$ is known as the **reflected impedance** since it represents the impedance reflected back into the primary side by the presence of the secondary current.

Hence reflected impedance $= \dfrac{\omega^2 M^2 R_2}{R_2^2 + \omega^2 L_2^2} - j\dfrac{\omega^3 M^2 L_2}{R_2^2 + \omega^2 L_2^2}$

$$= \omega^2 M^2 \left(\frac{R_2 - j\omega L_2}{R_2^2 + \omega^2 L_2^2}\right)$$

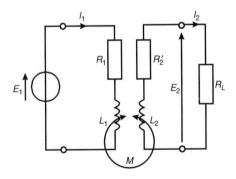

Figure 49.7

$$= \omega^2 M^2 \left(\frac{(R_2 - j\omega L_2)}{(R_2 + j\omega L_2)(R_2 - j\omega L_2)} \right)$$

$$= \frac{\omega^2 M^2}{(R_2 + j\omega L_2)}$$

i.e. reflected impedance, $Z_r = \dfrac{\omega^2 M^2}{Z_2}$ (13)

For example, for the coupled circuit shown in Figure 49.8, the self impedance of the primary circuit, $Z_1 = 300 + j(500)(0.2 + 0.5)$

i.e. $Z_1 = (300 + j350)\ \Omega$

Self impedance of secondary circuit,

$$Z_2 = 500 + j \left[(500)(0.3) - \frac{1}{(500)(5 \times 10^{-6})} \right]$$

$$= 500 + j(150 - 400)$$

i.e. $Z_2 = (500 - j250)\ \Omega$

From equation (13), **reflected impedance**,

$$Z_r = \frac{\omega^2 M^2}{Z_2} = \frac{(500)^2(0.2)^2}{500 - j250} = \frac{10^4(500 + j250)}{500^2 + 250^2} = (16 + j8)\ \Omega$$

$$\left. \begin{array}{l} \text{Effective primary } Z_{1(eff)} \\ \text{impedance,} \end{array} \right\} = Z_1 + Z_r \text{(note this is} \\ \text{equivalent to equation 12)}$$

$$= (300 + j350) + (16 + j8)$$

i.e. $Z_{1(eff)} = (316 + j358)\ \Omega$

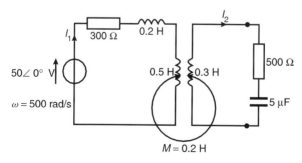

Figure 49.8

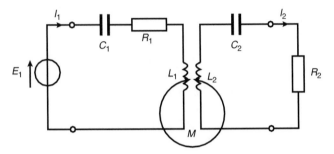

Figure 49.9

Primary current $I_1 = \dfrac{E_1}{Z_{1(eff)}} = \dfrac{50\angle 0°}{316 + j358} = \dfrac{50\angle 0°}{477.51\angle 48.57°}$

$$= \mathbf{0.105\angle - 48.57°\ A}$$

Secondary current, $I_2 = \dfrac{E_2}{Z_2}$ where $E_2 = j\omega M I_1$ from equation (10)

Hence $\quad \mathbf{I_2} = \dfrac{j\omega M I_1}{Z_2} = \dfrac{j(500)(0.2)(0.105\angle - 48.57°)}{500 - j250}$

$$= \dfrac{(100\angle 90°)(0.105\angle - 48.57°)}{559.02\angle - 26.57°} = \mathbf{0.0188\angle 68°\ A}$$

(c) Resonance by tuning capacitors

Tuning capacitors may be added to the primary and/or secondary circuits to cause it to resonate at particular frequencies. These may be connected either in series or in parallel with the windings. Figure 49.9 shows each winding tuned by series-connected capacitors C_1 and C_2. The expression for the effective primary impedance, $Z_{1(eff)}$ i.e. equation (12) applies except that ωL_1 becomes $\left(\omega L_1 - \dfrac{1}{\omega C_1}\right)$ and ωL_2 becomes $\left(\omega L_2 - \dfrac{1}{\omega C_2}\right)$

Dot Rule for Coupled Circuits

Applying Kirchhoff's voltage law to each mesh of the circuit shown in Figure 49.10 gives:

$$E_1 = I_1 (R_1 + j\omega L_1) \pm j\omega M I_2$$

and $\quad 0 = I_2 (R_2 + R_L + j\omega L_2) \pm j\omega M I_1$

In these equations the 'M' terms have been written as $\pm$ because it is not possible to state whether the magnetomotive forces due to currents I_1 and I_2

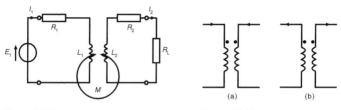

Figure 49.10   **Figure 49.11**

are added or subtracted. To make this clearer a **dot notation** is used whereby the polarity of the induced e.m.f. due to mutual inductance is identified by placing a dot on the diagram adjacent to that end of each equivalent winding which bears the same relationship to the magnetic flux.

The **dot rule** determines the sign of the voltage of mutual inductance in the Kirchhoff's law equations shown above, and states:

(i) *when both currents enter, or both currents leave, a pair of coupled coils at the dotted terminals, the signs of the 'M' terms will be the same as the signs of the 'L' terms, or*

(ii) *when one current enters at a dotted terminal and one leaves by a dotted terminal, the signs of the 'M' terms are opposite to the signs of the 'L' terms*

Thus Figure 49.11 shows two cases in which the signs of M and L are the same, and Figure 49.12 shows two cases where the signs of M and L are opposite. In Figure 49.10, therefore, if dots had been placed at the top end of coils L_1 and L_2 then the terms $j\omega MI_2$ and $j\omega MI_1$ in the Kirchhoff's equations would be negative (since current directions are similar to Figure 49.12(a)).

For example, for the coupled circuit shown in Figure 49.13, the values of currents I_1 and I_2 are determined as follows:

The position of the dots and the current directions correspond to Figure 49.12(a), and hence the signs of M and L terms are opposite. Applying

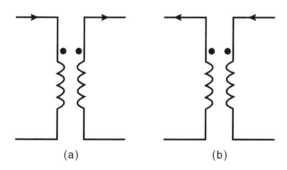

(a)   (b)

Figure 49.12

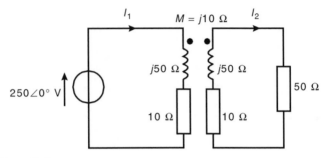

Figure 49.13

Kirchhoff's law to the primary circuit gives:

$$250\angle 0° = (10 + j50)I_1 - j10I_2 \tag{14}$$

and applying Kirchhoff's voltage law to the secondary circuit gives:

$$0 = (10 + 50 + j50)I_2 - j10I_1 \tag{15}$$

From equation (15),

$$j10I_1 = (60 + j50)I_2$$

and $$I_1 = \frac{(60 + j50)I_2}{j10} = \left(\frac{60}{j10} + \frac{j50}{j10}\right)I_2 = (-j6 + 5)I_2$$

i.e. $$I_1 = (5 - j6)I_2 \tag{16}$$

Substituting for I_1 in equation (14) gives:

$$250\angle 0° = (10 + j50)(5 - j6)I_2 - j10I_2$$

$$= (50 - j60 + j250 + 300 - j10)I_2 = (350 + j180)I_2$$

from which,

$$I_2 = \frac{250\angle 0°}{350 + j180} = \frac{250\angle 0°}{393.57\angle 27.22°} = \mathbf{0.635\angle - 27.22° \ A}$$

From equation (16),

$$I_1 = (5 - j6)I_2 = (5 - j6)(0.635\angle - 27.22°)$$

$$= (7.810\angle - 50.19°)(0.635\angle - 27.22°)$$

i.e. $$\mathbf{I_2 = 4.959\angle - 77.41° \ A}$$

50 Electrical Measuring Instruments and Measurements

Introduction

Tests and measurements are important in designing, evaluating, maintaining and servicing electrical circuits and equipment. In order to detect electrical quantities such as current, voltage, resistance or power, it is necessary to transform an electrical quantity or condition into a visible indication. This is done with the aid of instruments (or meters) that indicate the magnitude of quantities either by the position of a pointer moving over a graduated scale (called an analogue instrument) or in the form of a decimal number (called a digital instrument).

Analogue Instruments

All analogue electrical indicating instruments require three essential devices:

(a) **A deflecting or operating device**. A mechanical force is produced by the current or voltage which causes the pointer to deflect from its zero position.
(b) **A controlling device**. The controlling force acts in opposition to the deflecting force and ensures that the deflection shown on the meter is always the same for a given measured quantity. It also prevents the pointer always going to the maximum deflection. There are two main types of controlling device — spring control and gravity control.
(c) **A damping device**. The damping force ensures that the pointer comes to rest in its final position quickly and without undue oscillation. There are three main types of damping used — eddy-current damping, air-friction damping and fluid-friction damping.

There are basically **two types of scale** — linear and non-linear.

A **linear scale** is shown in Figure 50.1(a), where the divisions or graduations are evenly spaced. The voltmeter shown has a range 0–100 V, i.e. a full-scale deflection (f.s.d.) of 100 V. A **non-linear scale** is shown in Figure 50.1(b) where the scale is cramped at the beginning and the graduations are uneven throughout the range. The ammeter shown has a f.s.d. of 10 A.

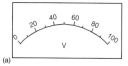
(a)

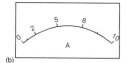
(b)

Figure 50.1

Moving-iron Instrument

(a) An **attraction type** of moving-iron instrument is shown diagrammatically in Figure 50.2(a). When current flows in the solenoid, a pivoted soft-iron disc is attracted towards the solenoid and the movement causes a pointer to move across a scale.

(b) In the **repulsion type** moving-iron instrument shown diagrammatically in Figure 50.2(b), two pieces of iron are placed inside the solenoid, one being fixed, and the other attached to the spindle carrying the pointer. When current passes through the solenoid, the two pieces of iron are magnetised in the same direction and therefore repel each other. The pointer thus moves across the scale. The force moving the pointer is, in each type, proportional to I^2 and because of this the direction of current does not matter. The moving-iron instrument can be used on d.c. or a.c.; the scale, however, is non-linear.

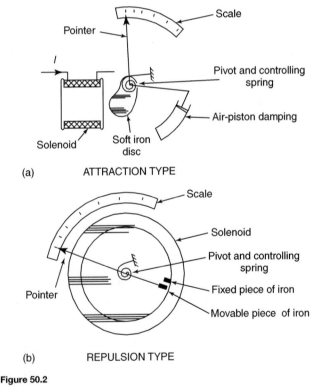

Figure 50.2

The Moving-coil Rectifier Instrument

A moving-coil instrument, which measures only d.c., may be used in conjunction with a bridge rectifier circuit as shown in Figure 50.3 to provide an indication of alternating currents and voltages (see chapter 54). The average value of the full wave rectified current is $0.637\ I_m$. However, a meter being used to measure a.c. is usually calibrated in r.m.s. values. For sinusoidal quantities the indication is $\dfrac{0.707\ I_m}{0.637\ I_m}$ i.e. 1.11 times the mean value. Rectifier instruments have scales calibrated in r.m.s. quantities and it is assumed by the manufacturer that the a.c. is sinusoidal.

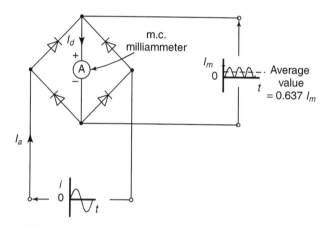

Figure 50.3

Comparison of moving-coil, moving-iron and moving-coil rectifier instruments

Type of instrument	Moving-coil	Moving-iron	Moving-coil rectifier
Suitable for measuring	Direct current and voltage	Direct and alternating currents and voltage (reading in r.m.s. value)	Alternating current and voltage (reads average value but scale is adjusted to give r.m.s. value for sinusoidal waveforms)
Scale	Linear	Non-linear	Linear

Type of instrument	Moving-coil	Moving-iron	Moving-coil rectifier
Method of control	Hairsprings	Hairsprings	Hairsprings
Method of damping	Eddy current	Air	Eddy current
Frequency limits	—	20–200 Hz	20–100 kHz
Advantages	1 Linear scale	1 Robust construction	1 Linear scale
	2 High sensitivity	2 Relatively cheap	2 High sensitivity
		3 Measures d.c. and a.c.	
	3 Well shielded from stray magnetic fields	4 In frequency range 20–100 Hz reads r.m.s. correctly regardless of supply wave-form	3 Well shielded from stray magnetic fields
	4 Low power consumption		4 Lower power consumption
			5 Good frequency range
Disadvantages	1 Only suitable for d.c.	1 Non-linear scale	1 More expensive than moving iron type
	2 More expensive than moving iron type	2 Affected by stray magnetic fields	2 Errors caused when supply is non-sinusoidal
	3 Easily damaged	3 Hysteresis errors in d.c. circuits	
		4 Liable to temperature errors	
		5 Due to the inductance of the solenoid, readings can be affected by variation of frequency	

(For the principle of operation of a moving-coil instrument, see chapter 47)

Shunts and Multipliers

An **ammeter**, which measures current, has a low resistance (ideally zero) and must be connected in series with the circuit.

A **voltmeter**, which measures p.d., has a high resistance (ideally infinite) and must be connected in parallel with the part of the circuit whose p.d. is required.

There is no difference between the basic instrument used to measure current and voltage since both use a milliammeter as their basic part. This is a sensitive instrument that gives f.s.d. for currents of only a few milliamperes. When an ammeter is required to measure currents of larger magnitude, a proportion of the current is diverted through a low-value resistance connected in parallel with the meter. Such a diverting resistor is called a **shunt**.

From Figure 50.4(a), $V_{PQ} = V_{RS}$. Hence $I_a r_a = I_S R_S$

Thus the value of the shunt, $$R_S = \frac{I_a r_a}{I_S} \text{ ohms}$$

The milliammeter is converted into a voltmeter by connecting a high value resistance (called a **multiplier**) in series with it as shown in Figure 50.4(b). From Figure 50.4(b), $V = V_a + V_M = I r_a + I R_M$

Thus the value of the multiplier, $$R_M = \frac{V - I r_a}{I} \text{ ohms}$$

For example, let a m.c. instrument have a f.s.d. of 20 mA and a resistance of 25 Ω. To enable the instrument to be used as a 0–10 A ammeter, a shunt resistance R_S needs to be connected in parallel with the instrument. From Figure 50.4(a), $I = 10$ A, $I_S = I - I_a = 10 - 0.020 = 9.98$ A.

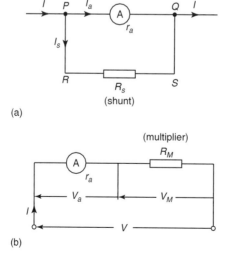

(a)

(b)

Figure 50.4

Hence the value of R_S is given by:

$$R_S = \frac{I_a r_a}{I_S} = \frac{(0.020)(25)}{9.98} = \textbf{50.10 m}\Omega$$

To enable the instrument to be used as a 0–100 V voltmeter, a multiplier R_M needs to be connected in series with the instrument, the value of R_M being given by:

$$R_M = \frac{V - I r_a}{I} = \frac{100 - (0.020)(25)}{0.020} = \textbf{4.975 k}\Omega$$

Electronic Instruments

Electronic measuring instruments have advantages over instruments such as the moving-iron or moving-coil meters, in that they have a much higher input resistance (some as high as 1000 MΩ) and can handle a much wider range of frequency (from d.c. up to MHz).

The **digital voltmeter (DVM)** is one that provides a digital display of the voltage being measured. Advantages of a DVM over analogue instruments include higher accuracy and resolution, no observational or parallex errors (see later) and a very high input resistance, constant on all ranges.

A **digital multimeter** is a DVM with additional circuitry that makes it capable of measuring a.c. voltage, d.c. and a.c. current and resistance.

Instruments for a.c. measurements are generally calibrated with a sinusoidal alternating waveform to indicate r.m.s. values when a sinusoidal signal is applied to the instrument. Some instruments, such as the moving-iron and electro-dynamic instruments, give a true r.m.s. indication. With other instruments the indication is either scaled up from the mean value (such as with the rectified moving-coil instrument) or scaled down from the peak value.

Sometimes quantities to be measured have complex waveforms (see chapter 76), and whenever a quantity is non-sinusoidal, errors in instrument readings can occur if the instrument has been calibrated for sine waves only. Using electronic instruments can largely eliminate such waveform errors.

The Ohmmeter

An **ohmmeter** is an instrument for measuring electrical resistance. A simple ohmmeter circuit is shown in Figure 50.5(a). Unlike the ammeter or voltmeter, the ohmmeter circuit does not receive the energy necessary for its operation from the circuit under test. In the ohmmeter this energy is supplied by a self-contained source of voltage, such as a battery. Initially, terminals XX are short-circuited and R adjusted to give f.s.d. on the milliammeter. If current I is at a maximum value and voltage E is constant, then resistance $R = E/I$ is at a minimum value. Thus f.s.d. on the milliammeter is made zero on the resistance scale. When terminals XX are open circuited no current flows and $R(= E/O)$ is infinity, ∞

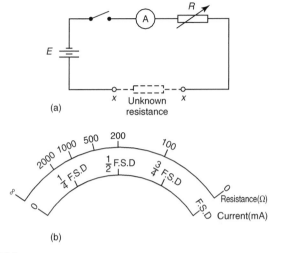

(a)

(b)

Figure 50.5

The milliammeter can thus be calibrated directly in ohms. A cramped (non-linear) scale results and is 'back to front', as shown in Figure 50.5(b). When calibrated, an unknown resistance is placed between terminals XX and its value determined from the position of the pointer on the scale. An ohmmeter designed for measuring low values of resistance is called a **continuity tester**. An ohmmeter designed for measuring high values of resistance (i.e. megohms) is called an **insulation resistance tester** (e.g. **'Megger'**)

Multimeters

Instruments are manufactured that combine a moving-coil meter with a number of shunts and series multipliers, to provide a range of readings on a single scale graduated to read current and voltage. If a battery is incorporated then resistance can also be measured. Such instruments are called **multimeters** or **universal instruments** or **multirange instruments**. An 'Avometer' is a typical example. A particular range may be selected either by the use of separate terminals or by a selector switch. Only one measurement can be performed at a time. Often such instruments can be used in a.c. as well as d.c. circuits when a rectifier is incorporated in the instrument.

Wattmeters

A **wattmeter** is an instrument for measuring electrical power in a circuit. Figure 50.6 shows typical connections of a wattmeter used for measuring power supplied to a load. The instrument has two coils:

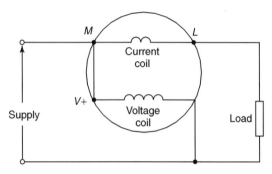

Figure 50.6

(i) a current coil, which is connected in series with the load, like an ammeter, and

(ii) a voltage coil, which is connected in parallel with the load, like a voltmeter.

Instrument 'Loading' Effect

Some measuring instruments depend for their operation on power taken from the circuit in which measurements are being made. Depending on the 'loading' effect of the instrument (i.e. the current taken to enable it to operate), the prevailing circuit conditions may change.

The resistance of voltmeters may be calculated since each have a stated sensitivity (or 'figure of merit'), often stated in 'kΩ per volt' of f.s.d. A voltmeter should have as high a resistance as possible (- ideally infinite). In a.c. circuits the impedance of the instrument varies with frequency and thus the loading effect of the instrument can change.

The Cathode Ray Oscilloscope

The **cathode ray oscilloscope** (c.r.o.) may be used in the observation of waveforms and for the measurement of voltage, current, frequency, phase and periodic time. For examining periodic waveforms the electron beam is deflected horizontally (i.e. in the X direction) by a sawtooth generator acting as a timebase. The signal to be examined is applied to the vertical deflection system (Y direction) usually after amplification.

Oscilloscopes normally have a transparent grid of 10 mm by 10 mm squares in front of the screen, called a graticule. Among the timebase controls is a 'variable' switch that gives the sweep speed as time per centimetre. This may be in s/cm, ms/cm or μs/cm, a large number of switch positions being available. Also on the front panel of a c.r.o. is a Y amplifier switch marked in volts per centimetre, with a large number of available switch positions.

(i) With **direct voltage measurements**, only the Y amplifier 'volts/cm' switch on the c.r.o. is used. With no voltage applied to the Y plates the position of the spot trace on the screen is noted. When a direct voltage is applied to the Y plates the new position of the spot trace is an indication of the magnitude of the voltage. For example, in Figure 50.7(a), with no voltage applied to the Y plates, the spot trace is in the centre of the screen (initial position) and then the spot trace moves 2.5 cm to the final position shown, on application of a d.c. voltage. With the 'volts/cm' switch on 10 volts/cm the magnitude of the direct voltage is 2.5 cm × 10 volts/cm, i.e. 25 volts.

(ii) With **alternating voltage measurements**, let a sinusoidal waveform be displayed on a c.r.o. screen as shown in Figure 50.7(b). If the time/cm switch is on, say, 5 ms/cm then the **periodic time** T of the sinewave is 5 ms/cm × 4 cm, i.e. **20 ms or 0.02 s**

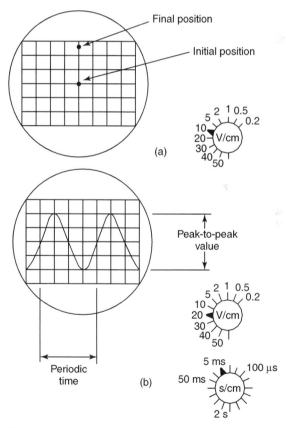

Figure 50.7

Since frequency $f = \dfrac{1}{T}$, **frequency** $= \dfrac{1}{0.02} = 50$ **Hz**

If the 'volts/cm' switch is on, say, 20 volts/cm then the **amplitude** or **peak value** of the sine wave shown is 20 volts/cm × 2 cm, i.e. 40 V.

Since r.m.s. voltage $= \dfrac{\text{peak voltage}}{\sqrt{2}}$, (see chapter 54),

r.m.s. voltage $= \dfrac{40}{\sqrt{2}} = 28.28$ **volts**

Double beam oscilloscopes are useful whenever two signals are to be compared simultaneously.

For example, for the double-beam oscilloscope displays shown in Figure 50.8, the 'time/cm' switch is on 100 μs/cm and the 'volts/cm' switch on 2 V/cm. The width of each complete cycle is 5 cm for both waveforms.

Hence the periodic time, T, of each waveform is 5 cm × 100 μs/cm = 0.5 ms

The frequency of each waveform, $f = \dfrac{1}{T} = \dfrac{1}{0.5 \times 10^{-3}} = 2$ **kHz**

The peak value of waveform A is 2 cm × 2 V/cm = **4 V**,

hence the r.m.s. value of waveform $A = \dfrac{4}{\sqrt{2}} = 2.83$ **V**

The peak value of waveform B is 2.5 cm × 2 V/cm = 5 V,

hence the r.m.s. value of waveform $B = \dfrac{5}{\sqrt{2}} = 3.54$ **V**

Since 5 cm represents 1 cycle, then 5 cm represents 360°, i.e. 1 cm represents $\dfrac{360}{5} = 72°$. The phase angle $\phi = 0.5$ cm = 0.5 cm × 72°/cm = 36°

Hence waveform A leads waveform B by 36°

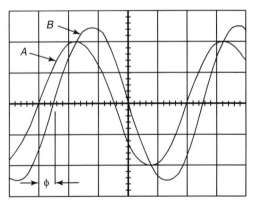

Figure 50.8

The c.r.o. demands **reasonable skill** in adjustment and use. However its greatest advantage is in observing the shape of a waveform —a feature not possessed by other measuring instruments.

Logarithmic Ratios

In electronic systems, the ratio of two similar quantities measured at different points in the system, are often expressed in logarithmic units. By definition, if the ratio of two powers P_1 and P_2 is to be expressed in **decibel (dB) units** then the number of decibels, X, is given by:

$$X = 10 \lg \left(\frac{P_2}{P_1} \right) \text{ dB} \tag{1}$$

Thus, when the power ratio, $\dfrac{P_2}{P_1} = 1$

then the decibel power ratio $= 10 \lg 1 = 0$

when the power ratio, $\dfrac{P_2}{P_1} = 100$

then the decibel power ratio $= 10 \lg 100$

$$= +20 \text{ (i.e. a power gain)}$$

and when the power ratio, $\dfrac{P_2}{P_1} = \dfrac{1}{100}$

then the decibel power ratio $= 10 \lg \dfrac{1}{100}$

$$= -20 \text{ (i.e. a power loss} \atop \text{or attenuation).}$$

For example, an amplifier has a gain of 14 dB and its input power is 8 mW. Its output power is determined as follows:

Decibel power ratio $= 10 \lg \left(\dfrac{P_2}{P_1} \right)$

Hence $\qquad 14 = 10 \lg \left(\dfrac{P_2}{8} \right)$ from which $1.4 = \lg \left(\dfrac{P_2}{8} \right)$

and $\qquad 10^{1.4} = \dfrac{P_2}{8}$ from the definition of a logarithm

i.e. $\qquad 25.12 = \dfrac{P_2}{8}$

Hence the output power, $P_2 = (25.12)(8) = $ **201 mW** or **0.201 W**

Logarithmic units may also be used for voltage and current ratios. The number of decibels, X, is given by:

$$X = 20 \lg \left(\frac{I_2}{I_1} \right) \text{ dB} \quad \text{or} \quad X = 20 \lg \left(\frac{V_2}{V_1} \right) \text{ dB}$$

From equation (1), X decibels is a logarithmic ratio of two similar quantities and is not an absolute unit of measurement. It is therefore necessary to state a **reference level** to measure a number of decibels above or below that reference. The most widely used reference level for power is 1 mW, and when power levels are expressed in decibels, above or below the 1 mW reference level, the unit given to the new power level is dBm.

A voltmeter can be re-scaled to indicate the power level directly in decibels. The scale is generally calibrated by taking a reference level of 0 dB when a power of 1 mW is dissipated in a 600 Ω resistor (this being the natural impedance of a simple transmission line). The reference voltage V is then obtained from $P = \dfrac{V^2}{R}$, i.e. $1 \times 10^{-3} = \dfrac{V^2}{600}$ from which, $V = 0.775$ volts.

In general, the number of dBm, $X = 20 \lg \left(\dfrac{V}{0.775} \right)$

Thus $V = 0.20$ V corresponds to $20 \lg \left(\dfrac{0.2}{0.775} \right) = -11.77$ dBm and

$V = 0.90$ V corresponds to $20 \lg \left(\dfrac{0.90}{0.775} \right) = +1.3$ dBm, and so on.

A typical **decibelmeter**, or **dB meter**, scale is shown in Figure 50.9. Errors are introduced with dB meters when the circuit impedance is not 600 Ω.

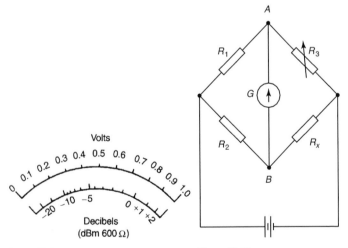

Volts

Decibels
(dBm 600 Ω)

Figure 50.9

Figure 50.10

Null Method of Measurement

A **null method of measurement** is a simple, accurate and widely used method which depends on an instrument reading being adjusted to read zero current only. The method assumes:

(i) if there is any deflection at all, then some current is flowing
(ii) if there is no deflection, then no current flows (i.e. a null condition).

Hence it is unnecessary for a meter sensing current flow to be calibrated when used in this way. A sensitive milliammeter or microammeter with centre zero position setting is called a **galvanometer**. Examples where the method is used are in the Wheatstone bridge, in the d.c. potentiometer and with a.c. bridges (see chapter 67)

Wheatstone Bridge

Figure 50.10 shows a **Wheatstone bridge** circuit that compares an unknown resistance R_x with others of known values, i.e. R_1 and R_2, which have fixed values, and R_3, which is variable. R_3 is varied until zero deflection is obtained on the galvanometer G. No current then flows through the meter, $V_A = V_B$, and the bridge is said to be 'balanced'

At balance, $R_1 R_x = R_2 R_3$ i.e. $\boxed{R_x = \dfrac{R_2 R_3}{R_1} \text{ ohms}}$

D.c. Potentiometer

The **d.c. potentiometer** is a null-balance instrument used for determining values of e.m.f.'s and p.d.s. by comparison with a known e.m.f. or p.d. In Figure 50.11(a), using a standard cell of known e.m.f. E_1, the slider S is moved along the slide wire until balance is obtained (i.e. the galvanometer deflection is zero), shown as length l_1

The standard cell is now replaced by a cell of unknown e.m.f. E_2 (see Figure 50.11(b)) and again balance is obtained (shown as l_2).

Since $E_1 \propto l_1$ and $E_2 \propto l_2$ then $\dfrac{E_1}{E_2} = \dfrac{l_1}{l_2}$ and

$$\boxed{E_2 = E_1 \left(\dfrac{l_2}{l_1} \right) \text{ volts}}$$

A potentiometer may be arranged as a resistive two-element potential divider in which the division ratio is adjustable to give a simple variable d.c. supply. Such devices may be constructed in the form of a resistive element carrying a sliding contact that is adjusted by a rotary or linear movement of the control knob.

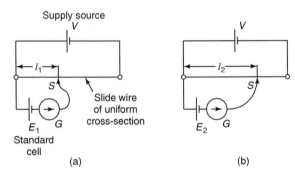

Figure 50.11

Q-meter

The **Q-factor** for a series L-C-R circuit is the voltage magnification at resonance, i.e. Q-factor $= \dfrac{\text{voltage across capacitor}}{\text{supply voltage}} = \dfrac{V_c}{V}$ (see chapter 68).

The simplified circuit of a **Q-meter**, used for measuring Q-factor, is shown in Figure 50.12. Current from a variable frequency oscillator flowing through a very low resistance r develops a variable frequency voltage, V_r, which is applied to a series L-R-C circuit. The frequency is then varied until resonance causes voltage V_c to reach a maximum value. At resonance V_r and V_c are noted.

Then   Q-factor $= \dfrac{V_c}{V_r} = \dfrac{V_c}{Ir}$

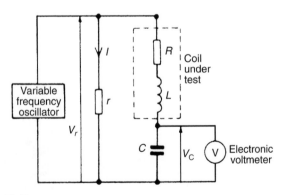

Figure 50.12

In a practical Q-meter, V_r is maintained constant and the electronic voltmeter can be calibrated to indicate the Q-factor directly. If a variable capacitor C is used and the oscillator is set to a given frequency, then C can be adjusted to give resonance. In this way inductance L may be calculated using $f_r = \dfrac{1}{2\pi\sqrt{LC}}$. Since $Q = \dfrac{2\pi f L}{R}$, then R may be calculated.

Q-meters operate at various frequencies and instruments exist with frequency ranges from 1 kHz to 50 MHz. Errors in measurement can exist with Q-meters since the coil has an effective parallel self-capacitance due to capacitance between turns. The accuracy of a Q-meter is approximately $\pm 5\%$.

Measurement Errors

Errors are always introduced when using instruments to measure electrical quantities. The errors most likely to occur in measurements are those due to:

(i) the limitations of the instrument
(ii) the operator
(iii) the instrument disturbing the circuit

(i) Errors in the limitations of the instrument

The **calibration accuracy** of an instrument depends on the precision with which it is constructed. Every instrument has a margin of error that is expressed as a percentage of the instruments full scale deflection. For example, industrial grade instruments have an accuracy of $\pm 2\%$ of f.s.d. Thus if a voltmeter has a f.s.d. of 100 V and it indicates 40 V say, then the actual voltage may be anywhere between $40 \pm (2\% \text{ of } 100)$, or 40 ± 2, i.e. between 38 V and 42 V.

When an instrument is calibrated, it is compared against a standard instrument and a graph is drawn of 'error' against 'meter deflection'. A typical graph is shown in Figure 50.13 where it is seen that the accuracy varies over the scale length. Thus, a meter with a $\pm 2\%$ f.s.d. accuracy would tend to have an accuracy which is much better than $\pm 2\%$ f.s.d. over much of the range.

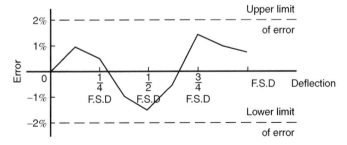

Figure 50.13

302

(ii) Errors by the operator

It is easy for an operator to misread an instrument. With linear scales the values of the sub-divisions are reasonably easy to determine; non-linear scale graduations are more difficult to estimate. Also, scales differ from instrument to instrument and some meters have more than one scale (as with multimeters) and mistakes in reading indications are easily made. When reading a meter scale it should be viewed from an angle perpendicular to the surface of the scale at the location of the pointer; a meter scale should not be viewed 'at an angle'.

(iii) Errors due to the instrument disturbing the circuit

Any instrument connected into a circuit will affect that circuit to some extent. Meters require some power to operate, but provided this power is small compared with the power in the measured circuit, then little error will result. Incorrect positioning of instruments in a circuit can be a source of errors. For example, let a resistance be measured by the voltmeter-ammeter method as shown in Figure 50.14. Assuming 'perfect' instruments, the resistance should be given by the voltmeter reading divided by the ammeter reading (i.e. $R = V/I$). However, in Figure 50.14(a), $V/I = R + r_a$ and in Figure 50.14(b) the current through the ammeter is that through the resistor plus that through the voltmeter. Hence the voltmeter reading divided by the ammeter reading will not give the true value of the resistance R for either method of connection.

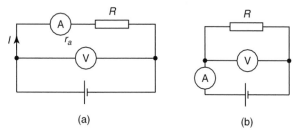

(a) (b)

Figure 50.14

51 Semiconductor Diodes

Types of Materials

Materials may be classified as **conductors, semiconductors** or **insulators**. The classification depends on the value of resistivity of the material. Good conductors are usually metals and have resistivities in the order of 10^{-7} to 10^{-8} Ωm. Semiconductors have resistivities in the order of 10^{-3} to 3×10^3 Ωm. The resistivities of insulators are in the order of 10^4 to 10^{14} Ωm. Some typical approximate values at normal room temperatures are:

Conductors:

Aluminium	2.7×10^{-8} Ωm
Brass (70 Cu/30 Zn)	8×10^{-8} Ωm
Copper (pure annealed)	1.7×10^{-8} Ωm
Steel (mild)	15×10^{-8} Ωm

Semiconductors:

Silicon $\quad\;\; 2.3 \times 10^3$ Ωm $\left.\right\}$ at $27°$C
Germanium $\;\; 0.45$ Ωm

Insulators:

Glass $\geq 10^{10}$ Ωm
Mica $\geq 10^{11}$ Ωm
PVC $\geq 10^{13}$ Ωm
Rubber (pure) 10^{12} to 10^{14} Ωm

In general, over a limited range of temperatures, the resistance of a conductor increases with temperature increase. The resistance of insulators remains approximately constant with variation of temperature. The resistance of semiconductor materials decreases as the temperature increases. For a specimen of each of these materials, having the same resistance (and thus completely different dimensions), at, say, $15°$C, the variation for a small increase in temperature to $t°$C is as shown in Figure 51.1.

Silicon and Germanium

The most important semiconductors used in the electronics industry are silicon and germanium. As the temperature of these materials is raised above room temperature, the resistivity is reduced and ultimately a point is reached where they effectively become conductors. For this reason, silicon should not operate at a working temperature in excess of $150°$C to $200°$C, depending on its purity, and germanium should not operate at a working temperature in excess of $75°$C to $90°$C, depending on its purity. As the temperature of a semiconductor is

304

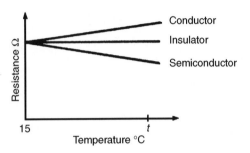

Figure 51.1

reduced below normal room temperature, the resistivity increases until at very low temperatures the semiconductor becomes an insulator.

---

n-type and *p*-type Materials

Adding extremely small amounts of impurities to pure semiconductors in a controlled manner is called **doping**. Antimony, arsenic and phosphorus are called n-type impurities and form an **n-type material** when any of these impurities are added to silicon or germanium. The amount of impurity added usually varies from 1 part impurity in 10^5 parts semiconductor material to 1 part impurity to 10^8 parts semiconductor material, depending on the resistivity required. Indium, aluminium and boron are called p-type impurities and form a **p-type material** when any of these impurities are added to a semiconductor.

In semiconductor materials, there are very few charge carriers per unit volume free to conduct. This is because the 'four electron structure' in the outer shell of the atoms (called valency electrons), form strong covalent bonds with neighbouring atoms, resulting in a tetrahedral structure with the electrons held fairly rigidly in place. A two-dimensional diagram depicting this is shown for germanium in Figure 51.2.

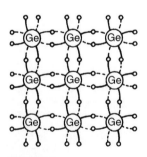

Figure 51.2

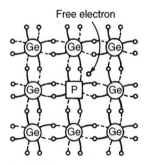

Figure 51.3

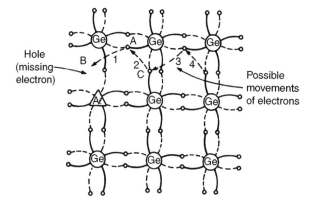

Figure 51.4

Arsenic, antimony and phosphorus have five valency electrons and when a semiconductor is doped with one of these substances, some impurity atoms are incorporated in the tetrahedral structure. The 'fifth' valency electron is not rigidly bonded and is free to conduct, the impurity atom donating a charge carrier. A two-dimensional diagram depicting this is shown in Figure 51.3, in which a phosphorus atom has replaced one of the germanium atoms. The resulting material is called n-type material, and contains free electrons.

Indium, aluminium and boron have three valency electrons and when a semiconductor is doped with one of these substances impurity atoms replace some of the semiconductor atoms. One of the four bonds associated with the semiconductor material is deficient by one electron and this deficiency is called a **hole**.

Holes give rise to conduction when a potential difference exists across the semiconductor material due to movement of electrons from one hole to another, as shown in Figure 51.4. In this figure, an electron moves from A to B, giving the appearance that the hole moves from B to A. Then electron C moves to A, giving the appearance that the hole moves to C, and so on. The resulting material is p-type material containing holes.

The p-n Junction

A **p-n junction** is a piece of semiconductor material in which part of the material is p-type and part is n-type. In order to examine the charge situation, assume that separate blocks of p-type and n-type materials are pushed together. Also assume that a hole is a positive charge carrier and that an electron is a negative charge carrier.

At the junction, the donated electrons in the n-type material, called **majority carriers**, diffuse into the p-type material (diffusion is from an area of high density to an area of lower density) and the acceptor holes in the p-type material diffuse into the n-type material as shown by the arrows in Figure 51.5.

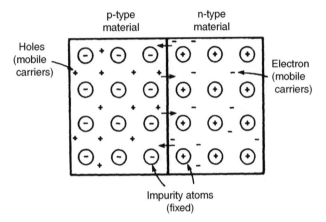

Figure 51.5

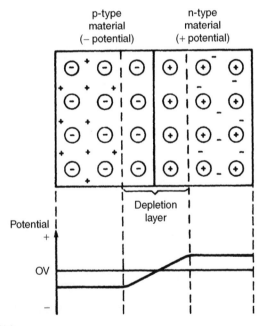

Figure 51.6

Because the n-type material has lost electrons, it acquires a positive potential with respect to the p-type material and thus tends to prevent further movement of electrons. The p-type material has gained electrons and becomes negatively charged with respect to the n-type material and hence tends to retain holes. Thus after a short while, the movement of electrons and holes stops due to the potential difference across the junction, called the **contact potential**. The area in the region of the junction becomes depleted of holes and electrons due to electron-hole recombinations, and is called a **depletion layer**, as shown in Figure 51.6.

Forward and Reverse Bias

When an external voltage is applied to a p-n junction making the p-type material positive with respect to the n-type material, as shown in Figure 51.7, the p-n junction is **forward biased**. The applied voltage opposes the contact potential, and, in effect, closes the depletion layer. Holes and electrons can now cross the junction and a current flows.

An increase in the applied voltage above that required to narrow the depletion layer (about 0.2 V for germanium and 0.6 V for silicon), results in a rapid rise in the current flow. Graphs depicting the current-voltage relationship for forward biased p-n junctions, for both germanium and silicon, called the **forward characteristics**, are shown in Figure 51.8.

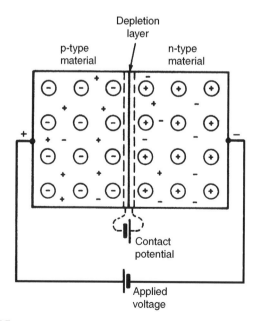

Figure 51.7

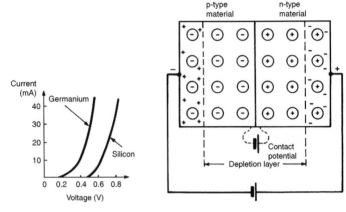

Figure 51.8   **Figure 51.9**

When an external voltage is applied to a p-n junction making the p-type material negative with respect to the n-type material as in shown in Figure 51.9, the p-n junction is **reverse biased**. The applied voltage is now in the same sense as the contact potential and opposes the movement of holes and electrons due to opening up the depletion layer. Thus, in theory, no current flows. However at normal room temperature certain electrons in the covalent bond lattice acquire sufficient energy from the heat available to leave the lattice, generating mobile electrons and holes. This process is called electron-hole generation by thermal excitation.

The electrons in the p-type material and holes in the n-type material caused by thermal excitation, are called **minority carriers** and these will be attracted by the applied voltage. Thus, in practice, a small current of a few microamperes for germanium and less than one microampere for silicon, at normal room temperature, flows under reverse bias conditions. Typical **reverse characteristics** are shown in Figure 51.10 for both germanium and silicon.

As the magnitude of the reverse voltage is increased a point will be reached where a large current suddenly starts to flow. The voltage at which this occurs is called the breakdown voltage. This current is due to two effects:

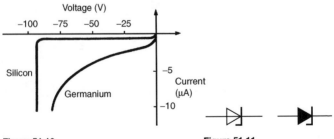

Figure 51.10   **Figure 51.11**

(i) the **zener effect**, resulting from the applied voltage being sufficient to break some of the covalent bonds, and

(ii) the **avalanche effect**, resulting from the charge carriers moving at sufficient speed to break covalent bonds by collision.

A **zener diode** is used for voltage reference purposes or for voltage stabilization. Two common circuit diagram symbols for a zener diode are shown in Figure 51.11.

Semiconductor Diodes

A semiconductor diode is a device having a p-n junction mounted in a container, suitable for conducting and dissipating the heat generated in operation and having connecting leads. Its operating characteristics are as shown in Figures 51.8 and 51.10. Two circuit diagram symbols for semiconductor diodes are in common use and are as shown in Figure 51.12. Sometimes the symbols are encircled as in Figure 51.13.

Rectification

The process of obtaining unidirectional currents and voltages from alternating currents and voltages is called **rectification**. Automatic switching in circuits is carried out by diodes.

Using a single diode, as shown in Figure 51.13, **half-wave rectification** is obtained. When P is sufficiently positive with respect to Q, diode D is switched on and current i flows. When P is negative with respect to Q, diode D is switched off. Transformer T isolates the equipment from direct connection with the mains supply and enables the mains voltage to be changed.

Two diodes may be used as shown in Figure 51.14 to obtain **full wave rectification**. A centre-tapped transformer T is used. When P is sufficiently positive with respect to Q, diode D_1 conducts and current flows (shown by the broken line in Figure 51.14). When S is positive with respect to Q, diode D_2 conducts and current flows (shown by the continuous line in Figure 51.14). The current flowing in R is in the same direction for both half cycles of the input. The output waveform is thus as shown in Figure 51.14.

Four diodes may be used in a **bridge rectifier** circuit, as shown in Figure 51.15 to obtain **full wave rectification**. As for the rectifier shown in

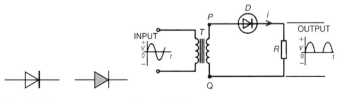

Figure 51.12   **Figure 51.13**

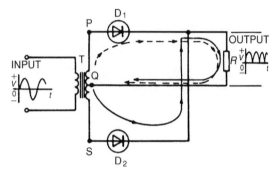

Figure 51.14

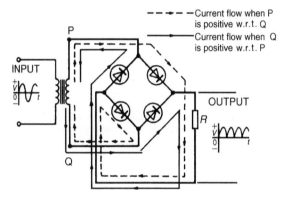

Figure 51.15

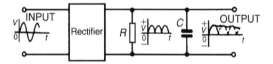

Figure 51.16

Figure 51.14, the current flowing in R is in the same direction for both half cycles of the input giving the output waveform shown.

To smooth the output of the rectifiers described above, capacitors having a large capacitance may be connected across the load resistor R. The effect of this is shown on the output in Figure 51.16.

52 Transistors

The Bipolar Junction Transistor

The bipolar junction transistor consists of three regions of semiconductor material. One type is called a p-n-p transistor, in which two regions of p-type material sandwich a very thin layer of n-type material. A second type is called an n-p-n transistor, in which two regions of n-type material sandwich a very thin layer of p-type material. Both of these types of transistors consist of two p-n junctions placed very close to one another in a back-to-back arrangement on a single piece of semiconductor material. Diagrams depicting these two types of transistors are shown in Figure 52.1.

The two p-type material regions of the p-n-p transistor are called the **emitter** and **collector** and the n-type material is called the **base**. Similarly, the two n-type material regions of the n-p-n transistor are called the emitter and collector and the p-type material region is called the base, as shown in Figure 52.1.

Transistors have three connecting leads and in operation an electrical input to one pair of connections, say the emitter and base connections can control the output from another pair, say the collector and emitter connections. This type of operation is achieved by appropriately biasing the two internal p-n junctions. When batteries and resistors are connected to a p-n-p transistor, as shown in Figure 52.2(a), the base-emitter junction is **forward biased** and the base-collector junction is **reverse biased**.

Similarly, an n-p-n transistor has its base-emitter junction forward biased and its base-collector junction reverse biased when the batteries are connected as shown in Figure 52.2(b).

For a silicon p-n-p transistor, biased as shown in Figure 52.2(a), if the base-emitter junction is considered on its own, it is forward biased and a

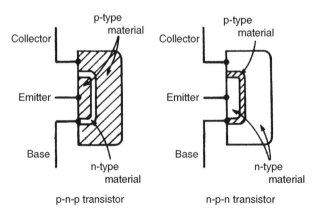

p-n-p transistor   n-p-n transistor

Figure 52.1

312

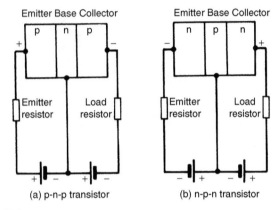

(a) p-n-p transistor   (b) n-p-n transistor

Figure 52.2

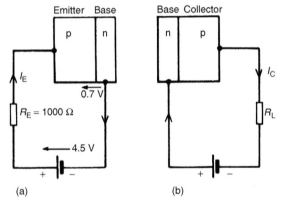

(a)   (b)

Figure 52.3

current flows. This is depicted in Figure 52.3(a). For example, if R_E is 1000 Ω, the battery is 4.5 V and the voltage drop across the junction is taken as 0.7 V, the current flowing is given by $\dfrac{4.5-0.7}{1000} = 3.8$ mA.

When the base-collector junction is considered on its own, as shown in Figure 52.3(b), it is reverse biased and the collector current is something less than 1 μA.

However, when both external circuits are connected to the transistor, most of the 3.8 mA of current flowing in the emitter, which previously flowed from the base connection, now flows out through the collector connection due to transistor action.

Transistor Action

In a **p-n-p transistor**, connected as shown in Figure 52.2(a), transistor action is accounted for as follows:

(a) The majority carriers in the emitter p-type material are holes
(b) The base-emitter junction is forward biased to the majority carriers and the holes cross the junction and appear in the base region
(c) The base region is very thin and is only lightly doped with electrons so although some electron-hole pairs are formed, many holes are left in the base region
(d) The base-collector junction is reverse biased to electrons in the base region and holes in the collector region, but forward biased to holes in the base region; these holes are attracted by the negative potential at the collector terminal
(e) A large proportion of the holes in the base region cross the base-collector junction into the collector region, creating a collector current; conventional current flow is in the direction of hole movement

The transistor action is shown diagrammatically in Figure 52.4. For transistors having very thin base regions, up to 99.5% of the holes leaving the emitter cross the base collector junction.

In an **n-p-n transistor**, connected as shown in Figure 52.2(b), transistor action is accounted for as follows:

(a) The majority carriers in the emitter p-type material are electrons
(b) The base-emitter junction is forward biased to these majority carriers and electrons cross the junction and appear in the base region
(c) The base region is very thin and only lightly doped with holes, so some recombination with holes occurs but many electrons are left in the base region
(d) The base-collector junction is reverse biased to holes in the base region and electrons in the collector region, but is forward biased to electrons in the base region; these electrons are attracted by the positive potential at the collector terminal
(e) A large proportion of the electrons in the base region cross the base-collector junction into the collector region, creating a collector current

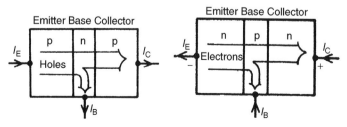

Figure 52.4   **Figure 52.5**

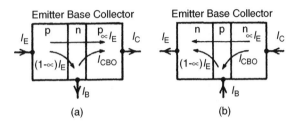

Figure 52.6

The transistor action is shown diagrammatically in Figure 52.5. As stated earlier, conventional current flow is taken to be in the direction of hole flow, that is, in the opposite direction to electron flow, hence the directions of the conventional current flow are as shown in Figure 52.5.

For a p-n-p transistor, the base-collector junction is reverse biased for majority carriers. However, a small leakage current, I_{CBO} flows from the base to the collector due to thermally generated minority carriers (electrons in the collector and holes in the base), being present.

The base-collector junction is forward biased to these minority carriers. If a proportion, α, (having a value of up to 0.995 in modern transistors), of the holes passing into the base from the emitter, pass through the base-collector junction, then the various currents flowing in a p-n-p transistor are as shown in Figure 52.6(a).

Similarly, for an n-p-n transistor, the base-collector junction is reversed biased for majority carriers, but a small leakage current, I_{CBO} flows from the collector to the base due to thermally generated minority carriers (holes in the collector and electrons in the base), being present. The base-collector junction is forward biased to these minority carriers. If a proportion, α, of the electrons passing through the base-emitter junction also pass through the base-collector junction then the currents flowing in an n-p-n transistor are as shown in Figure 52.6(b).

Transistor Symbols

Symbols are used to represent p-n-p and n-p-n transistors in circuit diagrams and are as shown in Figure 52.7. The arrowhead drawn on the emitter of the symbol is in the direction of conventional emitter current (hole flow). The potentials marked at the collector, base and emitter are typical values for a silicon transistor having a potential difference of 6 V between its collector and its emitter.

The voltage of 0.6 V across the base and emitter is that required to reduce the potential barrier and if it is raised slightly to, say, 0.62 V, it is likely that the collector current will double to about 2 mA. Thus a small change of voltage between the emitter and the base can give a relatively large change of current in the emitter circuit; because of this, transistors can be used as amplifiers.

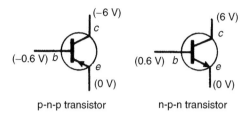

p-n-p transistor   n-p-n transistor

Figure 52.7

Transistor Connections

There are three ways of connecting a transistor, depending on the use to which it is being put. The ways are classified by the electrode that is common to both the input and the output. They are called:

(a) common-base configuration, shown in Figure 52.8(a)
(b) common-emitter configuration, shown in Figure 52.8(b)
(c) common-collector configuration, shown in Figure 52.8(c)

These configurations are for an n-p-n transistor. The current flows shown are all reversed for a p-n-p transistor.

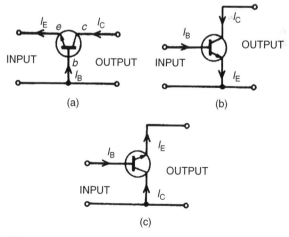

Figure 52.8

Transistor Characteristics

The effect of changing one or more of the various voltages and currents associated with a transistor circuit can be shown graphically and these graphs are called the **characteristics** of the transistor. As there are five variables (collector, base and emitter currents, and voltages across the collector and base and emitter and base) and also three configurations, many characteristics are possible. Some of the possible characteristics are given below.

(a) Common-base configuration

(i) **Input characteristic**. With reference to Figure 52.8(a), the input to a common-base transistor is the emitter current, I_E, and can be varied by altering the base emitter voltage V_{EB}. The base-emitter junction is essentially a forward biased junction diode, so as V_{EB} is varied, the current flowing is similar to that for a junction diode, as shown in Figure 52.9 for a silicon transistor. Figure 52.9 is called the input characteristic for an n-p-n transistor having common-base configuration. The variation of the collector-base voltage V_{CB} has little effect on the characteristic. A similar characteristic can be obtained for a p-n-p transistor, these having reversed polarities.

(ii) **Output characteristics**. The value of the collector current I_C is very largely determined by the emitter current, I_E. For a given value of I_E the collector-base voltage, V_{CB}, can be varied and has little effect on the value of I_C. If V_{CB} is made slightly negative, the collector no longer attracts the majority carriers leaving the emitter and I_C falls rapidly to zero. A family of curves for various values of I_E are possible and some of these are shown in Figure 52.10. Figure 52.10 is called the output characteristics for an n-p-n transistor having common-base configuration. Similar characteristics can be obtained for a p-n-p transistor, these having reversed polarities.

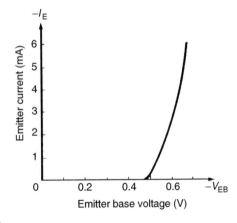

Figure 52.9

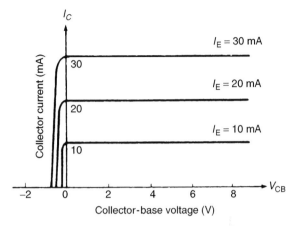

Figure 52.10

(b) Common-emitter configuration

(i) **Input characteristic**. In a common-emitter configuration (see Figure 52.8(b)), the base current is now the input current. As V_{EB} is varied, the characteristic obtained is similar in shape to the input characteristic for a common-base configuration shown in Figure 52.9, but the values of current are far less. With reference to Figure 52.6(a), as long as the junctions are biased as described, the three currents I_E, I_C and I_B keep the ratio $1 : \alpha : (1 - \alpha)$, whichever configuration is adopted. Thus the base current changes are much smaller than the corresponding emitter current changes and the input characteristic for an n-p-n transistor is as shown in Figure 52.11. A similar characteristic can be obtained for a p-n-p transistor, these having reversed polarities.

(ii) **Output characteristics**. A family of curves can be obtained, depending on the value of base current I_B and some of these for an n-p-n transistor are shown in Figure 52.12. A similar set of characteristics can be obtained for a p-n-p transistor, these having reversed polarities. These characteristics differ from the common base output characteristics in two ways:

    (a) the collector current reduces to zero without having to reverse the collector voltage, and

    (b) the characteristics slope upwards indicating a lower output resistance (usually kilohms for a common-emitter configuration compared with megohms for a common-base configuration).

A circuit diagram for obtaining the input and output characteristics for an n-p-n transistor connected in common-base configuration is shown in Figure 52.13. The input characteristic can be obtained by varying R_1, which varies V_{EB}, and noting the corresponding values of I_E. This is repeated for various values of V_{CB}. It will be found that the input characteristic is almost independent of V_{CB} and it is usual to give only one characteristic, as shown in Figure 52.9.

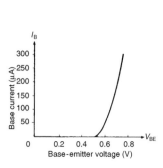

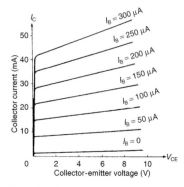

Figure 52.11   **Figure 52.12**

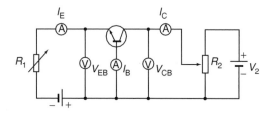

Figure 52.13

To obtain the output characteristics, as shown in Figure 52.10, I_E is set to a suitable value by adjusting R_1. For various values of V_{CB}, set by adjusting R_2, I_C is noted. This procedure is repeated for various values of I_E. To obtain the full characteristics, the polarity of battery V_2 has to be reversed to reduce I_C to zero. This must be done very carefully or else values of I_C will rapidly increase in the reverse direction and burn out the transistor.

The Transistor as an Amplifier

The amplifying properties of a transistor depend upon the fact that current flowing in a low-resistance circuit is transferred to a high-resistance circuit with negligible change in magnitude. If the current then flows through a load resistance, a voltage is developed. This voltage can be many times greater than the input voltage that caused the original current flow.

(a) Common-base amplifier

The basic circuit for a transistor is shown in Figure 52.14 where an n-p-n transistor is biased with batteries b_1 and b_2. A sinusoidal alternating input signal, v_e, is placed in series with the input bias voltage, and a load resistor, R_L,

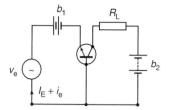

Figure 52.14

is placed in series with the collector bias voltage. The input signal is therefore the sinusoidal current i_e resulting from the application of the sinusoidal voltage v_e **superimposed** on the direct current I_E established by the base-emitter voltage V_{BE}.

Let the signal voltage v_e be 100 mV and the base-emitter circuit resistance be 50 Ω. Then the emitter signal current will be $\dfrac{100}{50} = 2$ mA. Let the load resistance $R_L = 2.5$ kΩ. About 0.99 of the emitter current will flow in R_L. Hence the collector signal current will be about $0.99 \times 2 = 1.98$ mA and the signal voltage across the load will be $2500 \times 1.98 \times 10^{-3} = 4.95$ V. Thus a signal voltage of 100 mV at the emitter has produced a voltage of 4950 mV across the load. The voltage amplification or gain is therefore $4950/100 = 49.5$ times. This example illustrates the action of a common-base amplifier where the input signal is applied between emitter and base and the output is taken from between collector and base.

(b) Common-emitter amplifier

The basic circuit arrangement of a common-emitter amplifier is shown in Figure 52.15. Although two batteries are shown, it is more usual to employ only one to supply all the necessary bias. The input signal is applied between base and emitter, and the load resistor R_L is connected between collector and emitter. Let the base bias battery provides a voltage which causes a base current I_B of 0.1 mA to flow. This value of base current determines the mean d.c. level upon which the a.c. input signal will be superimposed. This is the **d.c. base current operating point**.

Let the static current gain of the transistor, α_E, be 50. Since 0.1 mA is the steady base current, the collector current I_C will be $\alpha_E \times I_B = 50 \times 0.1 = 5$ mA. This current will flow through the load resistor $R_L (= 1$ kΩ), and there will be a steady voltage drop across R_L given by $I_C R_L = 5 \times 10^{-3} \times 1000 = 5$ V. The voltage at the collector, V_{CE}, will therefore be $V_{CC} - I_C R_L = 12 - 5 = 7$ V. This value of V_{CE} is the mean (or quiescent) level about which the output signal voltage will swing alternately positive and negative. This is the **collector voltage d.c. operating point**. Both of these d.c. operating points can be pin-pointed on the input and output characteristics of the transistor. Figure 52.16 shows the I_B/V_{BE} characteristic with the operating point X positioned at $I_B = 0.1$ mA, $V_{BE} = 0.75$ V, say.

Figure 52.17 shows the I_C/V_{CE} characteristics, with the operating point Y positioned at $I_C = 5$ mA, $V_{CE} = 7$ V. It is usual to choose the operating points Y somewhere near the centre of the graph.

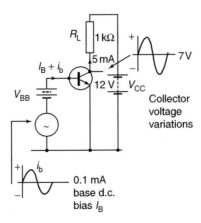

Figure 52.15

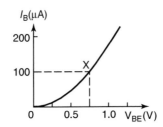

Figure 52.16

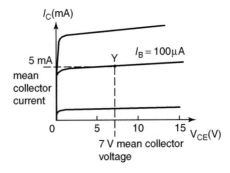

Figure 52.17

It is possible to remove the bias battery V_{BB} and obtain base bias from the collector supply battery V_{CC} instead. The simplest way to do this is to connect a bias resistor R_B between the positive terminal of the V_{CC} supply and the base as shown in Figure 52.18. The resistor must be of such a value that it allows 0.1 mA to flow in the base-emitter diode.

For a silicon transistor, the voltage drop across the junction for forward bias conditions is about 0.6 V. The voltage across R_B must then be $12 - 0.6 = 11.4$ V. Hence, the value of R_B must be such that $I_B \times R_B = 11.4$ V, i.e. $R_B = \dfrac{11.4}{I_B} = \dfrac{11.4}{0.1 \times 10^{-3}} = 114$ kΩ. With the inclusion of the 1 kΩ load resistor, R_L, a steady 5 mA collector current, and a collector-emitter voltage of 7 V, the d.c. conditions are established.

An alternating input signal (v_i) can now be applied. In order not to disturb the bias condition established at the base, the input must be fed to the base by way of a capacitor C_1. This will permit the alternating signal to pass to the base but will prevent the passage of direct current. The reactance of this capacitor must be such that it is very small compared with the input resistance of the transistor. The circuit of the amplifier is now as shown in Figure 52.19. The a.c. conditions can now be determined.

When an alternating signal voltage v_1 is applied to the base via capacitor C_1 the base current i_b varies. When the input signal swings positive, the base current increases; when the signal swings negative, the base current decreases. The base current consists of two components: I_B, the static base bias established by R_B, and i_b, the signal current. The current variation i_b will in turn vary the collector current, i_C. The relationship between i_C and i_b is given by

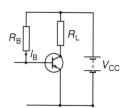

Figure 52.18

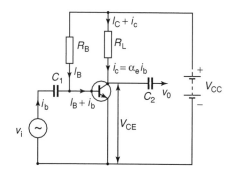

Figure 52.19

$i_C = \alpha_e i_b$, where α_e is the **dynamic current gain** of the transistor and is not quite the same as the static current gain α_E; the difference is usually small enough to be insignificant.

The current through the load resistor R_L also consists of two components: I_C, the static collector current, and i_C, the signal current. As i_b increases, so does i_C and so does the voltage drop across R_L. Hence, from the circuit:

$$V_{CE} = V_{CC} - (I_C + i_C)R_L$$

The d.c. components of this equation, though necessary for the amplifier to operate at all, need not be considered when the a.c. signal conditions are being examined. Hence, the signal voltage variation relationship is:

$$v_{ce} = -\alpha_e \times i_b \times R_L = i_C R_L$$

the negative sign being added because v_{ce} decreases when i_b increases and vice versa. The signal output and input voltages are of opposite polarity i.e. a phase shift of 180° has occurred. So that the collector d.c. potential is not passed on to the following stage, a second capacitor, C_2, is added as shown in Figure 52.19. This removes the direct component but permits the signal voltage $v_o = i_C R_L$ to pass to the output terminals.

The Load Line

The relationship between the collector-emitter voltage (V_{CE}) and collector current (I_C) is given by the equation: $V_{CE} = V_{CC} - I_C R_L$ in terms of the **d.c. conditions**. Since V_{CC} and R_L are constant in any given circuit, this represents the equation of a straight line which can be written in the $y = mx + c$ form. Transposing $V_{CE} = V_{CC} - I_C R_L$ for I_C gives:

$$I_C = \frac{V_{CC} - V_{CE}}{R_L} = \frac{V_{CC}}{R_L} - \frac{V_{CE}}{R_L} = -\left(\frac{1}{R_L}\right)V_{CE} + \frac{V_{CC}}{R_L}$$

i.e. $\quad I_C = -\left(\frac{1}{R_L}\right)V_{CE} + \frac{V_{CC}}{R_L}$

which is of the straight line form $y = mx + c$; hence if I_C is plotted vertically and V_{CE} horizontally, then the gradient is given by $-\left(\frac{1}{R_L}\right)$ and the vertical axis intercept is $\frac{V_{CC}}{R_L}$

A family of collector static characteristics drawn on such axes is shown in Figure 52.12 on page 318, and so the line may be superimposed on these as shown in Figure 52.20.

The reason why this line is necessary is because the static curves relate I_C to V_{CE} for a series of fixed values of I_B. When a signal is applied to the base of the transistor, the base current varies and can instantaneously take any of the values between the extremes shown. Only two points are necessary to draw the line and these can be found conveniently by considering extreme

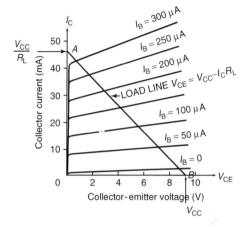

Figure 52.20

conditions. From the equation:

$$V_{CE} = V_{CC} - I_C R_L \quad \text{(i) when } I_C = 0, \quad V_{CE} = V_{CC}$$

$$\text{(ii) when } V_{CE} = 0, \quad I_C = \frac{V_{CC}}{R_L}$$

Thus the points A and B respectively are located on the axes of the I_C/V_{CE} characteristics. This line is called the **load line** and it is dependent for its position upon the value of V_{CC} and for its gradient upon R_L. As the gradient is given by $-\left(\dfrac{1}{R_L}\right)$, the slope of the line is negative.

For every value assigned to R_L in a particular circuit there will be a corresponding (and different) load line. If V_{CC} is maintained constant, all the possible lines will start at the same point (B) but will cut the I_C axis at different points A. Increasing R_L will reduce the gradient of the line and vice-versa. Quite clearly the collector voltage can never exceed V_{CC} (point B) and equally the collector current can never be greater than that value which would make V_{CE} zero (point A).

Using the circuit example of Figure 52.15, we have

$$V_{CE} = V_{CC} = 12 \text{ V}, \quad \text{when } I_C = 0$$

$$I_C = \frac{V_{CC}}{R_L} = \frac{12}{1000} = 12 \text{ mA}, \quad \text{when } V_{CE} = 0$$

The load line is drawn on the characteristics shown in Figure 52.21, which we assume are characteristics for the transistor used in the circuit of Figure 52.15 earlier. Notice that the load line passes through the operating point X, as it should, since every position on the line represents a relationship between V_{CE} and I_C for the particular values of V_{CC} and R_L given. Suppose

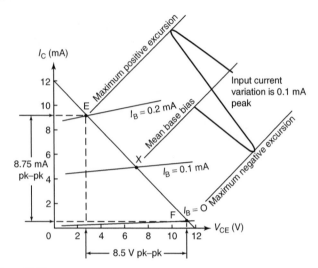

Figure 52.21

that the base current is caused to vary ±0.1 mA about the d.c. base bias of 0.1 mA. The result is I_B changes from 0 mA to 0.2 mA and back again to 0 mA during the course of each input cycle. Hence the operating point moves up and down the load line in phase with the input current and hence the input voltage. A sinusoidal input cycle is shown on Figure 52.21.

Current and Voltage Gains

The output signal voltage (v_{ce}) and current (i_C) can be obtained by projecting vertically from the load line on to V_{CE} and I_C axes respectively. When the input current i_b varies sinusoidally as shown in Figure 52.21, then v_{ce} varies sinusoidally if the points E and F at the extremities of the input variations are equally spaced on either side of X.

The peak to peak output voltage is seen to be 8.5 V, giving an r.m.s. value of 3 V $\left(\text{i.e. } 0.707 \times \dfrac{8.5}{2} \right)$. The peak-to-peak output current is 8.75 mA, giving an r.m.s. value of 3.1 mA. From these figures the voltage and current amplifications can be obtained.

The **dynamic current gain** A_i ($=\alpha_e$) as opposed to the static gain α_E, is given by:

$$A_i = \frac{\textbf{change in collector current}}{\textbf{change in base current}}$$

This always leads to a different figure from that obtained by the direct division of $\dfrac{I_C}{I_B}$ which assumes that the collector load resistor is zero. From Figure 52.21 the peak input current is 0.1 mA and the peak output current is 4.375 mA. Hence $A_i = \dfrac{4.375 \times 10^{-3}}{0.1 \times 10^{-3}} = 43.75$

The **voltage gain A_v** is given by:

$$A_v = \frac{\text{change in collector voltage}}{\text{change in base voltage}}$$

This cannot be calculated from the data available, but if we assume that the base current flows in the input resistance, then the base voltage can be determined. The input resistance can be determined from an input characteristic such as was shown earlier.

Then $\qquad R_i = \dfrac{\text{change in } V_{BC}}{\text{change in } I_B}$

and $\qquad v_i = i_b R_C$ and $v_o = i_C R_L$

and $\qquad A_v = \dfrac{i_c R_L}{I_b R_i} = \alpha_e \dfrac{R_L}{R_i}$

For a resistive load, **power gain, A_p**, is given by:

$$A_p = A_v \times A_i$$

Thermal Runaway

When a transistor is used as an amplifier it is necessary to ensure that it does not overheat. Overheating can arise from causes outside of the transistor itself, such as the proximity of radiators or hot resistors, or within the transistor as the result of dissipation by the passage of current through it. Power dissipated within the transistor, which is given approximately by the product $I_C \, V_{CE}$, is wasted power; it contributes nothing to the signal output power and merely raises the temperature of the transistor. Such overheating can lead to very undesirable results.

The increase in the temperature of a transistor will give rise to the production of hole electron pairs, hence an increase in leakage current represented by the additional minority carriers. In turn, this leakage current leads to an increase in collector current and this increases the product $I_C \, V_{CE}$. The whole effect thus becomes self-perpetuating and results in **thermal runaway**. This rapidly leads to the destruction of the transistor.

Two basic methods of preventing thermal runaway are available and either or both may be used in a particular application:

Method 1. The use of a single biasing resistor R_B as shown earlier in Figure 52.18 is not particularly good practice. If the temperature of the transistor increases, the leakage current also increases. The collector current, collector voltage and base current are thereby changed, the base current decreasing as

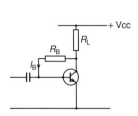

Figure 52.22

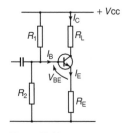

Figure 52.23

I_C increases. An alternative is shown in Figure 52.22. Here the resistor R_B is returned, not to the V_{CC} line, but to the collector itself.

If the collector current increases for any reason, the collector voltage V_{CE} will fall. Therefore, the d.c. base current I_B will fall, since $I_B = \dfrac{V_{CE}}{R_B}$. Hence the collector current $I_C = \alpha_E I_B$ will also fall and compensate for the original increase.

A commonly used bias arrangement is shown in Figure 52.23. If the total resistance value of resistors R_1 and R_2 is such that the current flowing through the divider is large compared with the d.c. bias current I_B, then the base voltage V_{BE} will remain substantially constant regardless of variations in collector current. The emitter resistor R_E in turn determines the value of emitter current which flows for a given base voltage at the junction of R_1 and R_2. Any increase in I_C produces an increase in I_E and a corresponding increase in the voltage drop across R_E. This reduces the forward bias voltage V_{BE} and leads to a compensating reduction in I_C

Method 2. This method concerns some means of keeping the transistor temperature down by external cooling. For this purpose, a heat sink is employed, as shown in Figure 52.24 If the transistor is clipped or bolted to a large conducting area of aluminium or copper plate (which may have cooling fins), cooling is achieved by convection and radiation.

Heat sinks are usually blackened to assist radiation and are normally used where large power dissipation's are involved. With small transistors, heat sinks are unnecessary. Silicon transistors particularly have such small leakage currents that thermal problems rarely arise.

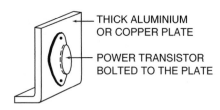

THICK ALUMINIUM
OR COPPER PLATE

POWER TRANSISTOR
BOLTED TO THE PLATE

Figure 52.24

53 D.c. Circuit Theory

Introduction

The laws that determine the currents and voltage drops in d.c. networks are:
(a) Ohm's law (see chapter 40), (b) the laws for resistors in series and in
parallel (see chapter 43), and (c) Kirchhoff's laws. In addition, there are a
number of circuit theorems that have been developed for solving problems in
electrical networks. These include:

(i) the superposition theorem
(ii) Thévenin's theorem
(iii) Norton's theorem
(iv) the maximum power transfer theorem

Kirchhoff's Laws

Kirchhoff's laws state:

(a) **Current Law**. *At any junction in an electric circuit the total current flowing
towards that junction is equal to the total current flowing away from the
junction, i.e.* $\Sigma I = 0$
Thus, referring to Figure 53.1:

$$I_1 + I_2 = I_3 + I_4 + I_5 \quad \text{or} \quad I_1 + I_2 - I_3 - I_4 - I_5 = 0$$

(b) **Voltage Law**. *In any closed loop in a network, the algebraic sum of the
voltage drops (i.e. products of current and resistance) taken around the
loop is equal to the resultant e.m.f. acting in that loop.*
Thus, referring to Figure 53.2:

$$E_1 - E_2 = IR_1 + IR_2 + IR_3$$

(Note that if current flows away from the positive terminal of a source, that
source is considered by convention to be positive. Thus moving anticlockwise around the loop of Figure 53.2, E_1 is positive and E_2 is negative)

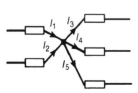

Figure 53.1

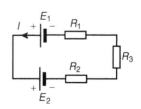

Figure 53.2

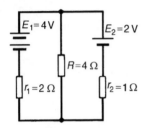

Figure 53.3

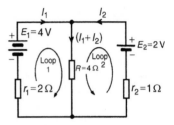

Figure 53.4

For example, using Kirchhoff's laws to determine the currents flowing in each branch of the network shown in Figure 53.3, the procedure is as follows:

1. Use Kirchhoff's current law and label current directions on the original circuit diagram. The directions chosen are arbitrary, but it is usual, as a starting point, to assume that current flows from the positive terminals of the batteries. This is shown in Figure 53.4 where the three branch currents are expressed in terms of I_1 and I_2 only, since the current through R is $(I_1 + I_2)$

2. Divide the circuit into two loops and apply Kirchhoff's voltage law to each. From loop 1 of Figure 53.4, and moving in a clockwise direction as indicated (the direction chosen does not matter), gives

$$E_1 = I_1 r_1 + (I_1 + I_2)R \quad \text{i.e.} \quad 4 = 2I_1 + 4(I_1 + I_2),$$

$$\text{i.e. } 6I_1 + 4I_2 = 4 \tag{1}$$

From loop 2 of Figure 53.4, and moving in an anticlockwise direction as indicated (once again, the choice of direction does not matter; it does not have to be in the same direction as that chosen for the first loop), gives:

$$E_2 = I_2 r_2 + (I_1 + I_2)R, \quad \text{i.e.} \quad 2 = I_2 + 4(I_1 + I_2),$$

$$\text{i.e. } 4I_1 + 5I_2 = 2 \tag{2}$$

3. Solve equations (1) and (2) for I_1 and I_2

$$2 \times (1) \text{ gives:} \quad 12I_1 + 8I_2 = 8 \tag{3}$$

$$3 \times (2) \text{ gives:} \quad 12I_1 + 15I_2 = 6 \tag{4}$$

$$(3) - (4) \text{ gives: } -7I_2 = 2$$

$$\text{hence } I_2 = -\tfrac{2}{7}$$

$$= -0.286 \text{ A}$$

(i.e. I_2 is flowing in the opposite direction to that shown in Figure 53.4)

From (1) $\qquad\qquad 6I_1 + 4(-0.286) = 4$

$$6I_1 = 4 + 1.144$$

Hence $\quad I_1 = \dfrac{5.144}{6}$

$$= 0.857 \text{ A}$$

Current flowing through resistance R is $(I_1 + I_2) = 0.857 + (-0.286)$

$$= \mathbf{0.571 \ A}$$

Note that a third loop is possible, as shown in Figure 53.5, giving a third equation which can be used as a check:

$$E_1 - E_2 = I_1 r_1 - I_2 r_2$$

$$4 - 2 = 2I_1 - I_2$$

$$2 = 2I_1 - I_2$$

[Check: $2I_1 - I_2 = 2(0.857) - (-0.286) = 2$]

See chapter 70 for the use of Kirchhoff's laws with a.c. networks.

The Superposition Theorem

The **superposition theorem** states:

In any network made up of linear resistances and containing more than one source of e.m.f., the resultant current flowing in any branch is the algebraic sum of the currents that would flow in that branch if each source was considered separately, all other sources being replaced at that time by their respective internal resistances.

For example, to determine the current in each branch of the network shown in Figure 53.6, using the superposition theorem the procedure is as follows:

1. Redraw the original circuit with source E_2 removed, being replaced by r_2 only, as shown in Figure 53.7(a)
2. Label the currents in each branch and their directions as shown in Figure 53.7(a) and determine their values. (Note that the choice of current

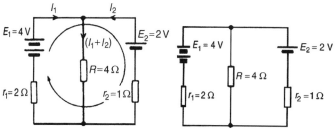

Figure 53.5   Figure 53.6

330

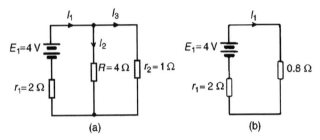

(a)　　　　　　　　　　(b)

Figure 53.7

directions depends on the battery polarity, which, by convention is taken as flowing from the positive battery terminal as shown)

R in parallel with r_2 gives an equivalent resistance of $\dfrac{4 \times 1}{4 + 1} = 0.8\ \Omega$

From the equivalent circuit of Figure 53.7(b),

$$I_1 = \frac{E_1}{r_1 + 0.8} = \frac{4}{2 + 0.8} = 1.429\ \text{A}$$

From Figure 53.7(a),　$I_2 = \left(\dfrac{1}{4 + 1}\right)I_1 = \dfrac{1}{5}(1.429) = 0.286\ \text{A}$

and　　　　　　$I_3 = \left(\dfrac{4}{4 + 1}\right)I_1 = \dfrac{4}{5}(1.429) = 1.143\ \text{A}$

by current division

3. Redraw the original circuit with source E_1 removed, being replaced by r_1 only, as shown in Figure 53.8(a)
4. Label the currents in each branch and their directions as shown in Figure 53.8(a) and determine their values. r_1 in parallel with R gives an equivalent resistance of $\dfrac{2 \times 4}{2 + 4} = \dfrac{8}{6} = 1.333\ \Omega$

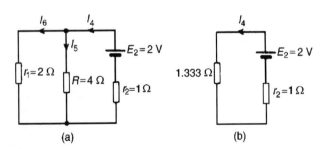

(a)　　　　　　　　　　(b)

Figure 53.8

From the equivalent circuit of Figure 53.8(b),

$$I_4 = \frac{E_2}{1.333 + r_2} = \frac{2}{1.333 + 1} = 0.857 \text{ A}$$

From Figure 53.8(a), $I_5 = \left(\frac{2}{2+4}\right) I_4 = \frac{2}{6}(0.857) = 0.286 \text{ A}$

$$I_6 = \left(\frac{4}{2+4}\right) I_4 = \frac{4}{6}(0.857) = 0.571 \text{ A}$$

5. Superimpose Figure 53.8(a) on to Figure 53.7(a) as shown in Figure 53.9.
6. Determine the algebraic sum of the currents flowing in each branch.
   Resultant current flowing through source 1, i.e.

$$I_1 - I_6 = 1.429 - 0.571 = \textbf{0.858 A (discharging)}$$

Resultant current flowing through source 2, i.e.

$$I_4 - I_3 = 0.857 - 1.143 = \textbf{−0.286 A (charging)}$$

Resultant current flowing through resistor R, i.e.

$$I_2 + I_5 = 0.286 + 0.286 = \textbf{0.572 A}$$

The resultant currents with their directions are shown in Figure 53.10. See chapter 72 for the use of the superposition theorem with a.c. networks.

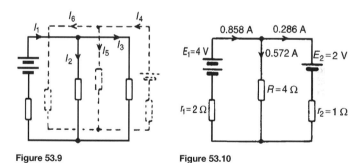

Figure 53.9   **Figure 53.10**

General d.c. Circuit Theory

The following points involving d.c. circuit analysis need to be appreciated before proceeding with problems using Thévenin's and Norton's theorems:

(i) The open-circuit voltage, E, across terminals AB in Figure 53.11 is equal to 10 V, since no current flows through the 2 Ω resistor and hence no voltage drop occurs.

332

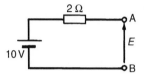

Figure 53.11

(ii) The open-circuit voltage, E, across terminals AB in Figure 53.12(a) is the same as the voltage across the 6 Ω resistor. The circuit may be redrawn as shown in Figure 53.12(b)

$E = \left(\dfrac{6}{6+4}\right)$ (50) by voltage division in a series circuit, i.e. $E = \mathbf{30\ V}$

(iii) For the circuit shown in Figure 53.13(a) representing a practical source supplying energy, $V = E - Ir$, where E is the battery e.m.f., V is the battery terminal voltage and r is the internal resistance of the battery (as shown in chapter 42). For the circuit shown in Figure 53.13(b),

$$V = E - (-I)r, \text{ i.e. } V = E + Ir$$

(iv) The resistance 'looking-in' at terminals AB in Figure 53.14(a) is obtained by reducing the circuit in stages as shown in Figures 53.14(b) to (d). Hence the equivalent resistance across AB is 7 Ω.

(v) For the circuit shown in Figure 53.15(a), the 3 Ω resistor carries no current and the p.d. across the 20 Ω resistor is 10 V. Redrawing the circuit gives Figure 53.14(b), from which $E = \left(\dfrac{4}{4+6}\right) \times 10 = \mathbf{4\ V}$

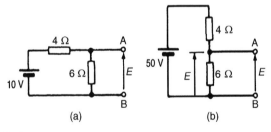

Figure 53.12

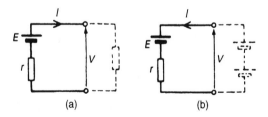

Figure 53.13

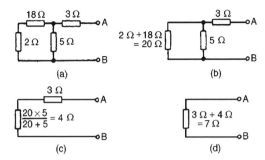

Figure 53.14

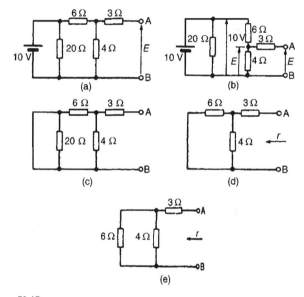

Figure 53.15

(vi) If the 10 V battery in Figure 53.15(a) is removed and replaced by a short-circuit, as shown in Figure 53.15(c), then the 20 Ω resistor may be removed. The reason for this is that a short-circuit has zero resistance, and 20 Ω in parallel with zero ohms gives an equivalent resistance of $\dfrac{20 \times 0}{20 + 0}$ i.e. 0 Ω. The circuit is then as shown in Figure 53.15(d), which is redrawn in Figure 53.15(e). From Figure 53.15(e), the equivalent resistance across AB, $r = \dfrac{6 \times 4}{6 + 4} + 3 = 2.4 + 3 = \mathbf{5.4\ \Omega}$.

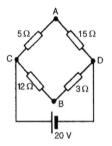

Figure 53.16

(vii) To find the voltage across AB in Figure 53.16:
Since the 20 V supply is across the 5 Ω and 15 Ω resistors in series then, by voltage division, the voltage drop across AC,

$$V_{AC} = \left(\frac{5}{5+15}\right)(20) = 5 \text{ V}$$

Similarly, $V_{CB} = \left(\frac{12}{12+3}\right)(20) = 16 \text{ V}$.

V_C is at a potential of $+20$ V.

$$V_A = V_C - V_{AC} = +20 - 5 = 15 \text{ V and}$$

$$V_B = V_C - V_{BC} = +20 - 16 = 4 \text{ V}.$$

Hence the voltage between AB is $V_A - V_B = 15 - 4 = 11$ V and current would flow from A to B since A has a higher potential than B.

(viii) In Figure 53.17(a), to find the equivalent resistance across AB the circuit may be redrawn as in Figures 53.17(b) and (c). From Figure 53.16(c), the equivalent resistance across

$$AB = \frac{5 \times 15}{5 + 15} + \frac{12 \times 3}{12 + 3} = 3.75 + 2.4 = \mathbf{6.15 \ \Omega}$$

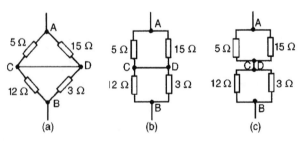

Figure 53.17

Thévenin's Theorem

Thévenin's theorem states:

> *The current in any branch of a network is that which would result if an e.m.f. equal to the p.d. across a break made in the branch, were introduced into the branch, all other e.m.f.'s being removed and represented by the internal resistances of the sources.*

The **procedure** adopted when using Thévenin's theorem is summarised below. To determine the current in any branch of an active network (i.e. one containing a source of e.m.f.):

 (i) remove the resistance R from that branch,
 (ii) determine the open-circuit voltage, E, across the break,
(iii) remove each source of e.m.f. and replace them by their internal resistances and then determine the resistance, r, 'looking-in' at the break,
(iv) determine the value of the current from the equivalent circuit shown in Figure 53.18, i.e. $I = \dfrac{E}{R + r}$

For example, using Thévenin's theorem to determine the current in the 4 Ω resistor shown in Figure 53.19, using the above procedure:

 (i) The 4 Ω resistor is removed from the circuit as shown in Figure 53.20(a)
 (ii) Current $I_1 = \dfrac{E_1 - E_2}{r_1 + r_2} = \dfrac{4 - 2}{2 + 1} = \dfrac{2}{3}$ A
 P.d. across AB, $E = E_1 - I_1 r_1 = 4 - \frac{2}{3}(2) = 2\frac{2}{3}$ V
 (Alternatively, p.d. across AB, $E = E_2 + I_1 r_2 = 2 + \frac{2}{3}(1) = 2\frac{2}{3}$ V)
(iii) Removing the sources of e.m.f. gives the circuit shown in Figure 53.20(b), from which, resistance $r = \dfrac{2 \times 1}{2 + 1} = \dfrac{2}{3}$ Ω

(iv) The equivalent Thévenin's circuit is shown in Figure 53.20(c), from

which, **current, $I = \dfrac{E}{r + R} = \dfrac{2\frac{2}{3}}{\dfrac{2}{3} + 4} = \dfrac{\frac{8}{3}}{\dfrac{14}{3}} = \dfrac{8}{14} = 0.571$ A**

See chapter 73 for the use of Thévenin's theorem in a.c. networks.

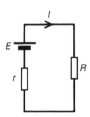

Figure 53.18

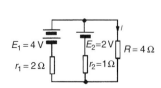

Figure 53.19

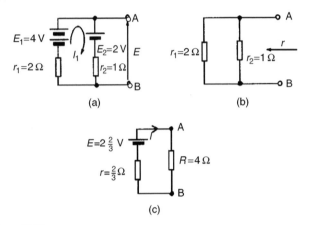

(a)   (b)

(c)

Figure 53.20

Constant Current Source

A source of electrical energy can be represented by a source of e.m.f. in series with a resistance. In the above section, the Thévenin constant-voltage source consisted of a constant e.m.f. E in series with an internal resistance r. However this is not the only form of representation. A source of electrical energy can also be represented by a constant-current source in parallel with a resistance. It may be shown that the two forms are equivalent. An **ideal constant-voltage generator** is one with zero internal resistance so that it supplies the same voltage to all loads. An **ideal constant-current generator** is one with infinite internal resistance so that it supplies the same current to all loads.

Note the symbol for an ideal current source (BS 3939,1985), shown in Figure 53.21.

Norton's Theorem

Norton's theorem states:

> *The current that flows in any branch of a network is the same as that which would flow in the branch if it were connected across a source of electrical energy, the short-circuit current of which is equal to the current that would flow in a short-circuit across the branch, and the internal resistance of which is equal to the resistance which appears across the open-circuited branch terminals.*

The **procedure** adopted when using Norton's theorem is summarised below.

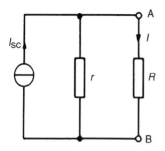

Figure 53.21

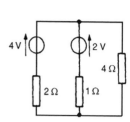

Figure 53.22

To determine the current flowing in a resistance R of a branch AB of an active network:

(i) short-circuit branch AB
(ii) determine the short-circuit current I_{SC} flowing in the branch
(iii) remove all sources of e.m.f. and replace them by their internal resistance (or, if a current source exists, replace with an open-circuit), then determine the resistance r, 'looking-in' at a break made between A and B
(iv) determine the current I flowing in resistance R from the Norton equivalent network shown in Figure 53.21, i.e.

$$I = \left(\frac{r}{r+R}\right)I_{SC}$$

For example, using Norton's theorem to determine the current in the 4 Ω resistance shown in Figure 53.22, using the above procedure:

(i) The 4 Ω branch is short-circuited as shown in Figure 53.23(a)
(ii) From Figure 53.23(a), $I_{SC} = I_1 + I_2 = \frac{4}{2} + \frac{2}{1} = 4$ A
(iii) If the sources of e.m.f. are removed the resistance 'looking-in' at a break made between A and B is given by: $r = \frac{2 \times 1}{2 + 1} = \frac{2}{3}$ Ω

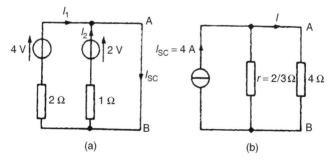

(a)   (b)

Figure 53.23

338

(iv) From the Norton equivalent network shown in Figure 53.23(b) the current in the 4 Ω resistance is given by:

$$I = \left(\frac{\frac{2}{3}}{\frac{2}{3} + 4} \right)(4) = \mathbf{0.571\ A}$$

See chapter 73 for the use of Norton's theorem in a.c. networks.

Thévenin and Norton Equivalent Networks

The Thévenin and Norton networks shown in Figure 53.24 are equivalent to each other. The resistance 'looking-in' at terminals AB is the same in each of the networks, i.e. r

If terminals AB in Figure 53.24(a) are short-circuited, the short-circuit current is given by $\frac{E}{r}$. If terminals AB in Figure 53.24(b) are short-circuited, the short-circuit current is I_{SC}. For the circuit shown in Figure 53.24(a) to be equivalent to the circuit in Figure 53.24(b) the same short-circuit current must flow. Thus $I_{SC} = \frac{E}{r}$

For example, the circuit of Figure 53.25(a) is equivalent to the circuit of Figure 53.25(b).

Similarly, the circuit of Figure 53.26(a) is equivalent to the circuit of Figure 53.26(b).

In another example, the circuit to the left of terminals AB in Figure 53.27 is converted to an equivalent Thévenin circuit as follows:

For the branch containing the 12 V source, converting to a Norton equivalent circuit gives $I_{SC} = \frac{12}{3} = 4$ A and $r_1 = 3$ Ω. For the branch containing the 24 V source, converting to a Norton equivalent circuit gives $I_{SC2} = \frac{24}{2} = 12$ A and $r_2 = 2$ Ω.

Thus Figure 53.28(a) shows a network equivalent to Figure 53.27.

From Figure 53.28(a) the total short-circuit current is $4 + 12 = 16$ A and the total resistance is given by: $\frac{3 \times 2}{3 + 2} = \mathbf{1.2\ \Omega}$

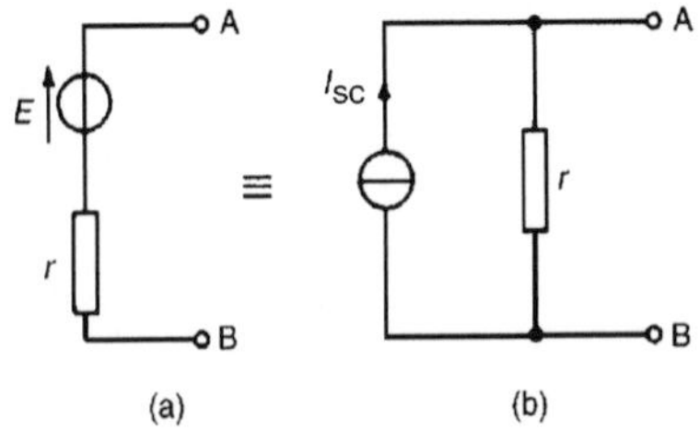

(a)　　　　　(b)

Figure 53.24

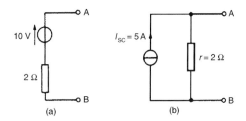

Figure 53.25

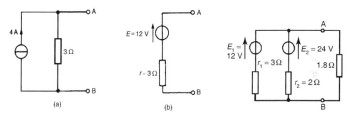

Figure 53.26   **Figure 53.27**

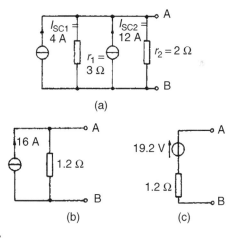

Figure 53.28

Thus Figure 53.28(a) simplifies to Figure 53.28(b).

The open-circuit voltage across AB of Figure 53.28(b), $E = (16)(1.2) =$ **19.2 V**, and the resistance 'looking-in' at AB is 1.2 Ω. Hence the Thévenin equivalent circuit is as shown in Figure 53.28(c).

When the 1.8 Ω resistance is connected between terminals A and B of Figure 53.28(c), the current I flowing is given by: $I = \left(\dfrac{19.2}{1.2 + 1.8}\right) = \mathbf{6.4\ A}$

Maximum Power Transfer Theorem

The **maximum power transfer theorem** states:

The power transferred from a supply source to a load is at its maximum when the resistance of the load is equal to the internal resistance of the source. Hence, in Figure 53.29, when $\boldsymbol{R = r}$ the power transferred from the source to the load is a maximum.

For example, a d.c. source has an open-circuit voltage of 30 V and an internal resistance of 1.5 Ω as shown in Figure 53.30.

From the maximum power transfer theorem, for maximum power dissipation,

$$R_L = r = \mathbf{1.5\ \Omega}$$

From Figure 53.30, current $I = \dfrac{E}{r + R_L} = \dfrac{30}{1.5 + 1.5} = 10$ A and the maximum power dissipated, $P = I^2 R_L = (10)^2(1.5) = \mathbf{150\ W}$

See chapter 75 for the use of the maximum power transfer theorem in a.c. circuits.

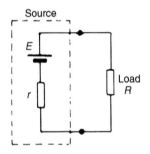

Figure 53.29

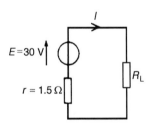

Figure 53.30

54 Alternating Voltages and Currents

Introduction

Electricity is produced by generators at power stations and then distributed by a vast network of transmission lines (called the National Grid system) to industry and for domestic use. It is easier and cheaper to generate alternating current (a.c.) than direct current (d.c.) and a.c. is more conveniently distributed than d.c. since its voltage can be readily altered using transformers. Whenever d.c. is needed in preference to a.c., devices called rectifiers are used for conversion (see chapter 51).

The a.c. Generator

Let a single turn coil be free to rotate at constant angular velocity symmetrically between the poles of a magnet system as shown in Figure 54.1.

An e.m.f. is generated in the coil (from Faraday's laws) which varies in magnitude and reverses its direction at regular intervals. The reason for this is shown in Figure 54.2. In positions (a), (e) and (i) the conductors of the loop are effectively moving along the magnetic field, no flux is cut and hence no e.m.f. is induced. In position (c) maximum flux is cut and hence maximum e.m.f. is induced. In position (g), maximum flux is cut and hence maximum e.m.f. is again induced. However, using Fleming's right-hand rule, the induced e.m.f. is in the opposite direction to that in position (c) and is thus shown as $-E$. In positions (b), (d), (f) and (h) some flux is cut and hence some e.m.f. is induced. If all such positions of the coil are considered, in one revolution of the coil, one cycle of alternating e.m.f. is produced as shown. This is the principle of operation of the a.c. generator (i.e. the alternator).

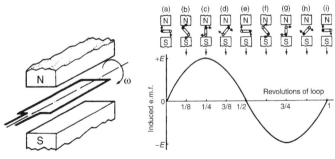

Figure 54.1   **Figure 54.2**

Waveforms

If values of quantities that vary with time t are plotted to a base of time, the resulting graph is called a **waveform**. Some typical waveforms are shown in Figure 54.3. Waveforms (a) and (b) are **unidirectional waveforms**, for, although they vary considerably with time, they flow in one direction only (i.e. they do not cross the time axis and become negative). Waveforms (c) to (g) are called **alternating waveforms** since their quantities are continually changing in direction (i.e. alternately positive and negative).

A waveform of the type shown in Figure 54.3(g) is called a **sine wave**. It is the shape of the waveform of e.m.f. produced by an alternator and thus the mains electricity supply is of 'sinusoidal' form.

One complete series of values is called a **cycle** (i.e. from O to P in Figure 54.3(g)).

The time taken for an alternating quantity to complete one cycle is called the **period** or the **periodic time, T,** of the waveform.

The number of cycles completed in one second is called the **frequency,** f, of the supply and is measured in **hertz, Hz**. The standard frequency of the electricity supply in Great Britain is 50 Hz

$$T = \frac{1}{f} \text{ or } f = \frac{1}{T}$$

For example, if, say, an alternating current completes 5 cycles in 8 ms, then the time for 1 cycle = $\frac{8}{5}$ ms = 1.6 ms = periodic time T and the

frequency, $f = \frac{1}{T} = \frac{1}{1.6 \times 10^{-3}} = \frac{1000}{1.6} = \frac{10\,000}{16} = \textbf{625 Hz}$

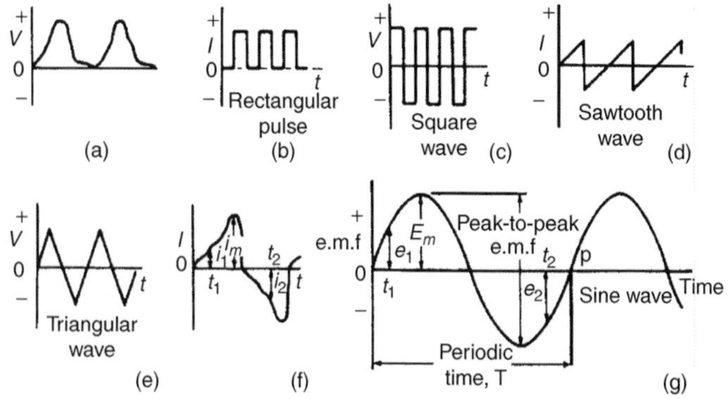

Figure 54.3

A.c. Values

Instantaneous values are the values of the alternating quantities at any instant of time. They are represented by small letters, i, v, e, etc., (see Figures 54.3(f) and (g)).

The largest value reached in a half cycle is called the **peak value** or the **maximum value** or the **crest value** or the **amplitude** of the waveform. Such values are represented by V_m, I_m, E_m, etc. (see Figures 54.3(f) and (g)). A **peak-to-peak** value of e.m.f. is shown in Figure 54.3(g) and is the difference between the maximum and minimum values in a cycle.

The **average** or **mean value** of a symmetrical alternating quantity, (such as a sine wave), is the average value measured over a half cycle, (since over a complete cycle the average value is zero).

$$\text{Average or mean value} = \frac{\textbf{area under the curve}}{\textbf{length of base}}$$

The area under the curve is found by approximate methods such as the trapezoidal rule, the mid-ordinate rule or Simpson's rule. Average values are represented by V_{AV}, I_{AV}, E_{AV}, etc.

For a sine wave:

average value = **0.637** × maximum value

$$\left(\text{i.e. } \frac{2}{\pi} \times \text{maximum value}\right)$$

For example, if the peak value of a sine wave is 200 V, the average or mean value is $0.637 \times 200 = \mathbf{127.4\ V}$

The **effective value** of an alternating current is that current which will produce the same heating effect as an equivalent direct current. The effective value is called the **root mean square (r.m.s.) value** and whenever an alternating quantity is given, it is assumed to be the r.m.s. value. For example, the domestic mains supply in Great Britain is 240 V and is assumed to mean '240 V r.m.s.'. The symbols used for r.m.s. values are I, V, E, etc. For a non-sinusoidal waveform as shown in Figure 54.4 the r.m.s. value is given by:

$$I = \sqrt{\left(\frac{i_1^2 + i_2^2 + \cdots + i_n^2}{n}\right)}$$

where n is the number of intervals used.

For a sine wave:

rms value = **0.707** × maximum value

$$\left(\text{i.e. } \frac{1}{\sqrt{2}} \times \text{maximum value}\right)$$

344

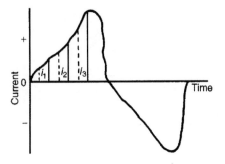

Figure 54.4

For example, if the peak value of a sine wave is 200 V, the r.m.s. value is $0.707 \times 200 = \mathbf{141.4\ V}$

$$\textbf{Form factor} = \frac{\textbf{r.m.s. value}}{\textbf{average value}}$$ For a sine wave, form factor $= 1.11$

$$\textbf{Peak factor} = \frac{\textbf{maximum value}}{\textbf{r.m.s. value}}$$ For a sine wave, peak factor $= 1.41$

The values of form and peak factors give an indication of the shape of waveforms.

For example, for the triangular waveform shown in Figure 54.5:
The time for 1 complete cycle $= 20$ ms $=$ periodic time, T
Hence the frequency

$$f = \frac{1}{T} = \frac{1}{20 \times 10^{-3}} = \frac{1000}{20} = \mathbf{50\ Hz}$$

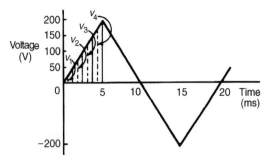

Figure 54.5

Area under the triangular waveform for a half cycle

$= \frac{1}{2} \times \text{base} \times \text{height} = \frac{1}{2} \times (10 \times 10^{-3} \times 200) = 1$ volt second

Average value of waveform $= \dfrac{\text{area under curve}}{\text{length of base}}$

$$= \frac{1 \text{ volt second}}{10 \times 10^{-3} \text{ second}} = \frac{1000}{10} = \textbf{100V}$$

In Figure 54.5, the first $\frac{1}{4}$ cycle is divided into 4 intervals.

Thus r.m.s. value $= \sqrt{\left(\dfrac{v_1^2 + v_2^2 + v_3^2 + v_4^2}{4} \right)}$

$$= \sqrt{\left(\frac{25^2 + 75^2 + 125^2 + 175^2}{4} \right)} = \textbf{114.6 V}$$

(Note that the greater the number of intervals chosen, the greater the accuracy of the result. For example, if twice the number of ordinates as that chosen above are used, the r.m.s. value is found to be 115.6 V)

The form factor $= \dfrac{\text{r.m.s. value}}{\text{average value}} = \dfrac{114.6}{100} = \textbf{1.15}$

The peak factor $= \dfrac{\text{maximum value}}{\text{r.m.s. value}} = \dfrac{200}{114.6} = \textbf{1.75}$

The Equation of a Sinusoidal Waveform

In Figure 54.6, OA represents a vector that is free to rotate anticlockwise about O at an angular velocity of ω rad/s. A rotating vector is known as a **phasor**.

After time t seconds the vector OA has turned through an angle ωt. If the line BC is constructed perpendicular to OA as shown, then

$$\sin \omega t = \frac{BC}{OB} \quad \text{i.e.} \quad BC = OB \sin \omega t$$

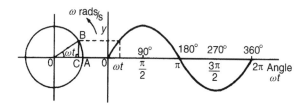

Figure 54.6

346

If all such vertical components are projected on to a graph of y against angle ωt (in radians), a sine curve results of maximum value OA. Any quantity that varies sinusoidally can thus be represented as a phasor.

A sine curve may not always start at $0°$. To show this a periodic function is represented by $y = \sin(\omega t \pm \phi)$, where ϕ is the phase (or angle) difference compared with $y = \sin \omega t$. In Figure 54.7(a), $y_2 = \sin(\omega t + \phi)$ starts ϕ radians earlier than $y_1 = \sin \omega t$ and is thus said to **lead** y_1 by ϕ radians. Phasors y_1 and y_2 are shown in Figure 54.7(b) at the time when $t = 0$.

In Figure 54.7(c), $y_4 = \sin(\omega t - \phi)$ starts ϕ radians later than $y_3 = \sin \omega t$ and is thus said to **lag** y_3 by ϕ radians. Phasors y_3 and y_4 are shown in Figure 54.7(d) at the time when $t = 0$.

Given the general sinusoidal voltage, $v = V_m \sin(\omega t \pm \phi)$, then

  (i) Amplitude or maximum value $= V_m$
 (ii) Peak to peak value $= 2V_m$
(iii) Angular velocity $= \omega$ rad/s
 (iv) Periodic time, $T = 2\pi/\omega$ seconds
  (v) Frequency, $f = \omega/2\pi$ Hz (since $\omega = 2\pi f$)
 (vi) $\phi =$ angle of lag or lead (compared with $v = V_m \sin \omega t$)

For example, an alternating voltage is given by:
$v = 75 \sin(200\pi t - 0.25)$ volts.

Comparing with the general expression $v = V_m \sin(\omega t \pm \phi)$ gives:

Amplitude, or peak value $=$ **75 V**

Peak-to-peak value $= 2 \times 75 =$ **150 V**

The r.m.s. value $= 0.707 \times$ maximum value $= 0.707 \times 75 =$ **53 V**

Angular velocity, $\omega = 200\pi$ rad/s

hence periodic time, $T = \dfrac{2\pi}{\omega} = \dfrac{2\pi}{200\pi} = \dfrac{1}{100} =$ **0.01 s or 10 ms**

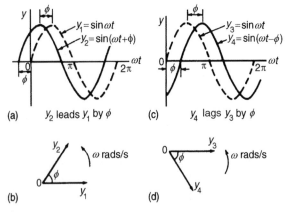

(a) y_2 leads y_1 by ϕ (c) y_4 lags y_3 by ϕ

(b)   (d)

Figure 54.7

Frequency, $f = \dfrac{1}{T} = \dfrac{1}{0.01} = \mathbf{100\ Hz}$

Phase angle, $\phi = 0.25$ radians lagging $75 \sin 200\pi t$ and

0.25 rads $= 0.25 \times \dfrac{180°}{\pi} = 14.32°$

Hence the phase angle is **14.32° lagging**

Combination of Waveforms

The resultant of the addition (or subtraction) of two sinusoidal quantities may be determined either:

(a) by plotting the periodic functions graphically, or
(b) by resolution of phasors by drawing or calculation

For example, currents $i_1 = 20 \sin \omega t$ and $i_2 = 10 \sin \left(\omega t + \dfrac{\pi}{3} \right)$ are shown plotted in Figure 54.8

To determine the resultant of $i_1 + i_2$, ordinates of i_1 and i_2 are added at, say, 15° intervals. For example,

$$\text{at } 30°, \quad i_1 + i_2 = 10 + 10 = 20 \text{ A}$$

$$\text{at } 60°, \quad i_1 + i_2 = 17.3 + 8.7 = 26 \text{ A}$$

$$\text{at } 150°, \quad i_1 + i_2 = 10 + (-5) = 5 \text{ A, and so on.}$$

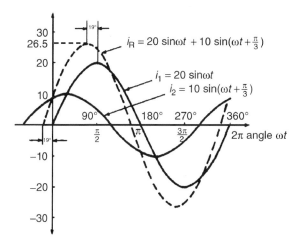

Figure 54.8

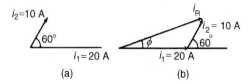

Figure 54.9

The resultant waveform for $i_1 + i_2$ is shown by the broken line in Figure 54.8. It has the same period, and hence frequency, as i_1 and i_2. The amplitude or peak value is 26.5 A. The resultant waveform leads the curve $i_1 = 20 \sin \omega t$ by 19° i.e. $\left(19 \times \dfrac{\pi}{180} \right)$ rads = 0.33 rads.

Hence the sinusoidal expression for the resultant $i_1 + i_2$ is given by:

$$i_R = i_1 + i_2 = 26.5 \sin(\omega t + 0.33) \ A$$

The relative positions of i_1 and i_2 at time $t = 0$ are shown as phasors in Figure 54.9(a). The phasor diagram in Figure 54.9(b) shows the resultant i_R, and i_R is measured as 26 A and angle ϕ as 19° or 0.33 rads leading i_1

Hence, **by drawing, $i_R = 26 \sin(\omega t + 0.33)$ A**

From Figure 54.9(b), by the cosine rule:

$$i_R^2 = 20^2 + 10^2 - 2(20)(10)(\cos 120°)$$

from which $i_R = \mathbf{26.46 \ A}$

By the sine rule:

$$\frac{10}{\sin \phi} = \frac{26.46}{\sin 120°} \quad \text{from which, } \phi = 19.10° = 0.333 \text{ rads}$$

Hence, **by calculation, $i_R = 26.46 \sin(\omega t + 0.333)$ A**

Rectification

The process of obtaining unidirectional currents and voltages from alternating currents and voltages is called **rectification**. Devices called diodes carry out automatic switching in circuits. Half and full-wave rectifiers are explained in chapter 51.

55 Single-phase Series a.c. Circuits

Purely Resistive a.c. Circuit

In a purely resistive a.c. circuit, the current I_R and applied voltage V_R are in phase. See Figure 55.1.

Purely Inductive a.c. Circuit

In a purely inductive a.c. circuit, the current I_L **lags** the applied voltage V_L by 90° $\left(\text{i.e. } \dfrac{\pi}{2} \text{ rads}\right)$. See Figure 55.2.

In a purely inductive circuit the opposition to the flow of alternating current is called the **inductive reactance, X_L**

$$X_L = \frac{V_L}{I_L} = 2\pi f L \ \Omega$$

where f is the supply frequency, in hertz, and L is the inductance, in henry's.

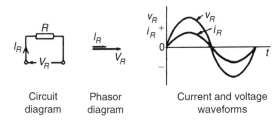

| Circuit diagram | Phasor diagram | Current and voltage waveforms |

Figure 55.1

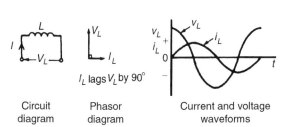

I_L lags V_L by 90°

| Circuit diagram | Phasor diagram | Current and voltage waveforms |

Figure 55.2

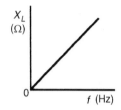

Figure 55.3

For example, if a coil has an inductance of 40 mH and negligible resistance and is connected to a 240 V, 50 Hz supply:

the inductive reactance, $X_L = 2\pi f L = 2\pi(50)(40 \times 10^{-3})$

$$= 12.57 \ \Omega$$

and the current, $I = \dfrac{V}{X_L} = \dfrac{240}{12.57} = 19.09 \ A$

X_L is proportional to f as shown in Figure 55.3.

Purely Capacitive a.c. Circuit

In a purely capacitive a.c. circuit, the current I_C **leads** the applied voltage V_C by 90° (i.e. $\pi/2$ rads). See Figure 55.4.

In a purely capacitive circuit the opposition to the flow of alternating current is called the **capacitive reactance**, X_C

$$\boxed{X_C = \frac{V_C}{I_C} = \frac{1}{2\pi f C} \ \Omega}$$

where C is the capacitance in farads.

For example, the capacitive reactance of a capacitor of 10μF when connected to a circuit of frequency 50 Hz is given by: capacitive reactance,

$$X_C = \frac{1}{2\pi f C} = \frac{1}{2\pi(50)(10 \times 10^{-6})} = \frac{10^6}{2\pi(50)(10)} = 318.3 \ \Omega$$

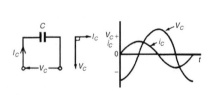

Figure 55.4

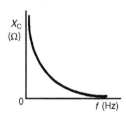

Figure 55.5

If the frequency is, say, 20 kHz, then

$$X_C = \frac{1}{2\pi f C} = \frac{1}{2\pi(20 \times 10^3)(10 \times 10^{-6})} = \frac{10^6}{2\pi(20 \times 10^3)(10)}$$

$$= \mathbf{0.796 \; \Omega}$$

X_C varies with frequency f as shown in Figure 55.5.

R-L Series a.c. Circuit

In an a.c. circuit containing inductance L and resistance R, the applied voltage V is the phasor sum of V_R and V_L (see Figure 55.6), and thus the current I lags the applied voltage V by an angle lying between 0° and 90° (depending on the values of V_R and V_L), shown as angle ϕ. In any a.c. series circuit the current is common to each component and is thus taken as the reference phasor.

From the phasor diagram of Figure 55.6, the **'voltage triangle'** is derived.

For the *R-L* circuit: $V = \sqrt{V_R^2 + V_L^2}$ (by Pythagoras' theorem)

and $\tan\phi = \dfrac{V_L}{V_R}$ (by trigonometric ratios)

For example, in a series *R-L* circuit the p.d. across the resistance R is 12 V and the p.d. across the inductance L is 5 V

From the voltage triangle of Figure 55.6,
supply voltage $\mathbf{V} = \sqrt{12^2 + 5^2} = \mathbf{13 \; V}$
(Note that in a.c. circuits, the supply voltage is **not** the arithmetic sum of the p.d's across components. It is, in fact, the **phasor sum**).

$$\tan\phi = \frac{V_L}{V_R} = \frac{5}{12},$$

from which, circuit phase angle $\phi = \tan^{-1}\left(\dfrac{5}{12}\right) = \mathbf{22.62° \; lagging}$

('Lagging' infers that the current is 'behind' the voltage, since phasors revolve anticlockwise).

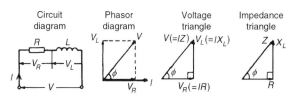

Figure 55.6

In an a.c. circuit, the ratio applied voltage V to current I is called the

$$\text{impedance}, Z, \text{i.e.} \quad \boxed{Z = \frac{V}{I} \ \Omega}$$

If each side of the voltage triangle in Figure 55.6 is divided by current I then the 'impedance triangle' is derived.

For the R-L circuit:

$$Z = \sqrt{R^2 + X_L^2}, \quad \tan\phi = \frac{X_L}{R}, \quad \sin\phi = \frac{X_L}{Z} \quad \text{and} \quad \cos\phi = \frac{R}{Z}$$

For example, a coil takes a current of 2 A from a 12 V d.c. supply. When connected to a 240 V, 50 Hz supply the current is 20 A.

Hence, resistance $\quad R = \dfrac{\text{d.c. voltage}}{\text{d.c. current}} = \dfrac{12}{2} = 6 \ \Omega$

and impedance $\quad Z = \dfrac{\text{a.c. voltage}}{\text{a.c. current}} = \dfrac{240}{20} = 12 \ \Omega$

Since $Z = \sqrt{R^2 + X_L^2}$, inductive reactance, $X_L = \sqrt{Z^2 - R^2}$

$$= \sqrt{12^2 - 6^2} = 10.39 \ \Omega$$

Since $X_L = 2\pi f L$, inductance $L = \dfrac{X_L}{2\pi f} = \dfrac{10.39}{2\pi(50)} = \textbf{33.1 mH}$

Phase angle, $\phi = \tan^{-1}\dfrac{X_L}{R} = \tan^{-1}\dfrac{10.39}{6} = 60°$ (lagging).

This example indicates a simple method for finding the inductance of a coil, i.e. firstly to measure the current when the coil is connected to a d.c. supply of known voltage, and then to repeat the process with an a.c. supply.

R-C Series a.c. Circuit

In an a.c. series circuit containing capacitance C and resistance R, the applied voltage V is the phasor sum of V_R and V_C (see Figure 55.7) and thus the current I leads the applied voltage V by an angle lying between 0° and 90° (depending on the values of V_R and V_C), shown as angle α.

From the phasor diagram of Figure 55.7, the 'voltage triangle' is derived.

For the R-C circuit: $\quad V = \sqrt{V_R^2 + V_C^2} \quad$ (by Pythagoras' theorem)

and $\quad\quad\quad\quad \tan\alpha = \dfrac{V_C}{V_R} \quad$ (by trigonometric ratios)

As stated above, in an a.c. circuit, the ratio applied voltage V to current I is called the **impedance** Z, i.e. $Z = \dfrac{V}{I} \ \Omega$

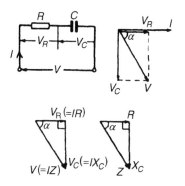

Figure 55.7

If each side of the voltage triangle in Figure 55.7 is divided by current I, then the **'impedance triangle'** is derived.

For the R-C circuit:

$$Z = \sqrt{R^2 + X_C^2}, \quad \tan\alpha = \frac{X_C}{R}, \quad \sin\alpha = \frac{X_C}{Z} \text{ and } \cos\alpha = \frac{R}{Z}$$

For example, if a resistor of 25 Ω is connected in series with a capacitor of 45 μF across a 240 V, 50 Hz supply, then

the capacitive reactance, $X_C = \dfrac{1}{2\pi f C} = \dfrac{1}{2\pi(50)(45 \times 10^{-6})}$

$$= 70.74 \ \Omega$$

Hence, the impedance $Z = \sqrt{R^2 + X_C^2} = \sqrt{25^2 + 70.74^2}$

$$= \mathbf{75.03 \ \Omega}$$

and current $I = \dfrac{V}{Z} = \dfrac{240}{75.03} = \mathbf{3.20 \ A}$

The phase angle between the supply voltage and current, $\alpha = \tan^{-1}\left(\dfrac{X_C}{R}\right)$

hence $\alpha = \tan^{-1}\left(\dfrac{70.74}{25}\right) = \mathbf{70.54° \ leading}$

('Leading' infers that the current is 'ahead' of the voltage, since phasors revolve anticlockwise)

R-L-C series a.c. circuit

In an a.c. series circuit containing resistance R, inductance L and capacitance C, the applied voltage V is the phasor sum of V_R, V_L and V_C (see Figure 55.8).

V_L and V_C are anti-phase, i.e. displaced by 180°, and there are three phasor diagrams possible — each depending on the relative values of V_L and V_C.

When $X_L > X_C$ (Figure 55.8(b)): $Z = \sqrt{R^2 + (X_L - X_C)^2}$ and

$$\tan \phi = \frac{X_L - X_C}{R}$$

When $X_C > X_L$ (Figure 55.8(c)): $Z = \sqrt{R^2 + (X_C - X_L)^2}$ and

$$\tan \alpha = \frac{X_C - X_L}{R}$$

When $X_L = X_C$ (Figure 55.8(d)), the applied voltage V and the current I are in phase. This effect is called **series resonance** (see later).

Series connected impedances

For series-connected impedances the total circuit impedance can be represented as a single L-C-R circuit by combining all values of resistance together, all values of inductance together and all values of capacitance together, (remembering that for series connected capacitors $\dfrac{1}{C} = \dfrac{1}{C_1} + \dfrac{1}{C_2} + \ldots .$).

For example, the following three impedances are connected in series across a 40 V, 20 kHz supply: (i) a resistance of 8 Ω, (ii) a coil of inductance 130 μH and 5 Ω resistance, and (iii) a 10 Ω resistor in series with a 0.25 μF capacitor. The circuit diagram is shown in Figure 55.9(a). Since the total circuit resistance is $8 + 5 + 10$, i.e. 23 Ω, an equivalent circuit diagram may be drawn as shown in Figure 55.9(b)

Inductive reactance, $X_L = 2\pi f L = 2\pi(20 \times 10^3)(130 \times 10^{-6}) = 16.34 \ \Omega$

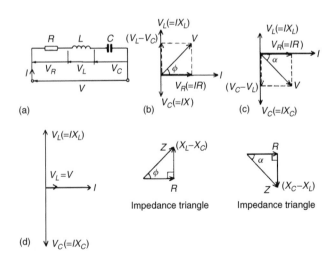

(a)   (b)   (c)

Impedance triangle   Impedance triangle

(d)

Figure 55.8

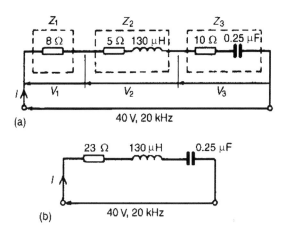

Figure 55.9

Capacitive reactance,

$$X_C = \frac{1}{2\pi f C} = \frac{1}{2\pi(20 \times 10^3)(0.25 \times 10^{-6})} = 31.83 \ \Omega$$

Since $X_C > X_L$, the circuit is capacitive
(see phasor diagram in Figure 55.8(c)).

$$X_C - X_L = 31.83 - 16.34 = 15.49 \ \Omega$$

Circuit impedance,

$$Z = \sqrt{R^2 + (X_C - X_L)^2} = \sqrt{23^2 + 15.49^2} = 27.73 \ \Omega$$

Circuit current, $I = \dfrac{V}{Z} = \dfrac{40}{27.73} = \textbf{1.442 A}$

From Figure 55.8(c), circuit phase angle, $\alpha = \tan^{-1}\left(\dfrac{X_C - X_L}{R}\right)$

i.e. $\alpha = \tan^{-1}\left(\dfrac{15.49}{23}\right) = \textbf{33.96}^\circ$ **leading**

From Figure 55.9(a), $V_1 = IR_1 = (1.442)(8) = \textbf{11.54 V}$

$$V_2 = IZ_2 = I\sqrt{5^2 + 16.34^2} = (1.442)(17.09) = \textbf{24.64 V}$$

$$V_3 = IZ_3 = I\sqrt{10^2 + 31.83^2} = (1.442)(33.36) = \textbf{48.11 V}$$

The 40 V supply voltage is the phasor sum of V_1, V_2 and V_3.

Series resonance

As stated earlier, for an R-L-C series circuit, when $X_L = X_C$ (Figure 55.8(d)), the applied voltage V and the current I are in phase.

This effect is called **series resonance**. At resonance:

(i) $V_L = V_C$

(ii) $Z = R$ (i.e. the minimum circuit impedance possible in an L-C-R circuit)

(iii) $I = \dfrac{V}{R}$ (i.e. the maximum current possible in an L-C-R circuit)

(iv) Since $X_L = X_C$, then $2\pi f_r L = \dfrac{1}{2\pi f_r C}$

from which, $f_r^2 = \dfrac{1}{(2\pi)^2 LC}$

and
$$\boxed{f_r = \dfrac{1}{2\pi\sqrt{LC}} \ \textbf{Hz}}$$

where f_r is the resonant frequency.

(v) The series resonant circuit is often described as an **acceptor circuit** since it has its minimum impedance, and thus maximum current, at the resonant frequency.

(vi) Typical graphs of current I and impedance Z against frequency are shown in Figure 55.10.

For example, a coil having a resistance of 10 Ω and an inductance of 125 mH is connected in series with a 60 µF capacitor across a 120 V supply. Thus, the resonant frequency,

$$f_r = \dfrac{1}{2\pi\sqrt{LC}} \ \text{Hz} = \dfrac{1}{2\pi\sqrt{\left[\left(\dfrac{125}{10^3}\right)\left(\dfrac{60}{10^6}\right)\right]}} \ \text{Hz}$$

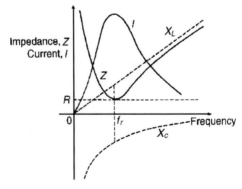

Figure 55.10

$$= \frac{1}{2\pi\sqrt{\left(\dfrac{125 \times 6}{10^8}\right)}}$$

$$= \frac{1}{2\pi\left(\dfrac{\sqrt{(125)(6)}}{10^4}\right)} = \frac{10^4}{2\pi\sqrt{(125)(6)}} = \textbf{58.12 Hz}$$

At resonance, $X_L = X_C$ and impedance $Z = R$, hence current,

$$I = \frac{V}{R} = \frac{120}{10} = \textbf{12 A}$$

Q-factor

At resonance, if R is small compared with X_L and X_C, it is possible for V_L and V_C to have voltages many times greater than the supply voltage (see Figure 55.8(d), page 354).

> **Voltage magnification at resonance** $= \dfrac{\text{voltage across } L \text{ (or } C)}{\text{supply voltage } V}$

This ratio is a measure of the quality of a circuit (as a resonator or tuning device) and is called the **Q-factor**. It may be shown that:

$$Q\text{-factor} = \frac{2\pi f_r}{R} = \frac{1}{2\pi f_r CR} = \frac{1}{R}\sqrt{\frac{L}{C}}$$

For example, a coil of inductance 80 mH and negligible resistance is connected in series with a capacitance of 0.25 μF and a resistor of resistance 12.5 Ω across a 100 V, variable frequency supply. Thus, the resonant frequency

$$f_r = \frac{1}{2\pi\sqrt{\left[\left(\dfrac{80}{10^3}\right)\left(\dfrac{0.25}{10^6}\right)\right]}} = \frac{1}{2\pi\sqrt{\left[\dfrac{(8)(0.25)}{10^8}\right]}} = \textbf{1.1254 kHz}$$

and current at resonance $I = \dfrac{V}{R} = \dfrac{100}{12.5} = \textbf{8 A}$

Voltage across inductance, at resonance,

$$V_L = IX_L = (I)(2\pi fL) = (8)(2\pi(1125.4))(80 \times 10^{-3}) = 4525.5 \text{ V}$$

(Also, voltage across capacitor,

$$V_C = IX_C = \frac{I}{2\pi fC} = \frac{8}{2\pi(1125.4)(0.25 \times 10^{-6})} = 4525.5 \text{ V})$$

Voltage magnification at resonance $= \dfrac{V_L}{V}$ or $\dfrac{V_C}{V} = \dfrac{4525.5}{100} = \mathbf{45.255}$ **V** i.e. at resonance, the voltage across the reactance's are 45.255 times greater than the supply voltage. Hence **the Q-factor of the circuit is 45.255**

Alternatively, Q-factor $= \dfrac{1}{R}\sqrt{\dfrac{L}{C}} = \dfrac{1}{12.5}\sqrt{\left(\dfrac{80 \times 10^{-3}}{0.25 \times 10^{-6}}\right)} = \mathbf{45.255}$

Bandwidth

Figure 55.11 shows how current I varies with frequency in an R-L-C series circuit. At the resonant frequency f_r, current is a maximum value, shown as I_r. Also shown are the points A and B where the current is 0.707 of the maximum value at frequencies f_1 and f_2. The power delivered to the circuit is I^2R. At $I = 0.707\, I_r$, the power is $(0.707\, I_r)^2 R = 0.5 I_r^2 R$, i.e. half the power that occurs at frequency f_r. The points corresponding to f_1 and f_2 are called the **half-power points**. The distance between these points, i.e. $(f_2 - f_1)$, is called the **bandwidth**.

It may be shown that $\boxed{Q = \dfrac{f_r}{(f_2 - f_1)}}$ or $\boxed{(f_2 - f_1) = \dfrac{f_r}{Q}}$

For example, a filter in the form of a series L-R-C circuit is designed to operate at a resonant frequency of 5 kHz. Included within the filter is a 20 mH inductance and 10 Ω resistance. Thus, Q-factor at resonance is given by:

$$Q_r = \frac{\omega_r L}{R} = \frac{(2\pi \times 5000)(20 \times 10^{-3})}{10} = 62.83$$

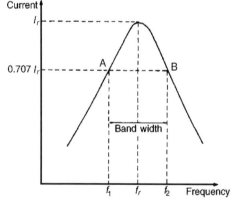

Bandwidth and half-power points f_1, f_2

Figure 55.11

and since $Q_r = \dfrac{f_r}{(f_2 - f_1)}$, **bandwidth**, $(f_2 - f_1) = \dfrac{f_r}{Q} = \dfrac{5000}{62.83} =$ **79.6 Hz**

Selectivity

Selectivity is the ability of a circuit to respond more readily to signals of a particular frequency to which it is tuned than to signals of other frequencies. The response becomes progressively weaker as the frequency departs from the resonant frequency. The higher the Q-factor, the narrower the bandwidth and the more selective is the circuit. Circuits having high Q-factors (say, in the order of 100 to 300) are therefore useful in communications engineering. A high Q-factor in a series power circuit has disadvantages in that it can lead to dangerously high voltages across the insulation and may result in electrical breakdown.

For more on Q-factor, bandwidth and selectivity, see Chapter 68.

Power in a.c. circuits

In Figures 55.12(a)–(c), the value of power at any instant is given by the product of the voltage and current at that instant, i.e. the instantaneous power, $p = vi$, as shown by the broken lines.

For a purely resistive a.c. circuit, the average power dissipated, P, is given by: $P = VI = I^2 R = \dfrac{V^2}{R}$ **watts** (V and I being r.m.s. values). See Figure 55.13(a)

For a purely inductive a.c. circuit, the average power is zero. See Figure 55.12(b)

For a purely capacitive a.c. circuit, the average power is zero. See Figure 55.12(c)

Figure 55.13 shows current and voltage waveforms for an R-L circuit where the current lags the voltage by angle ϕ. The waveform for power (where

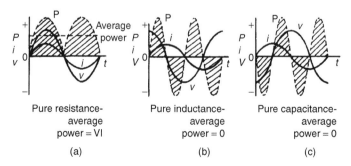

Figure 55.12

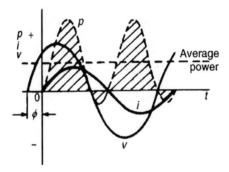

Figure 55.13

$p = vi$) is shown by the broken line, and its shape, and hence average power, depends on the value of angle ϕ.

For an *R-L*, *R-C* or *R-L-C* series a.c. circuit, the average power P is given by:

$$\boxed{P = VI \cos\phi \text{ watts}}$$

or $\boxed{P = I^2 R \text{ watts}}$ (V and I being r.m.s. values)

For example, a series circuit of resistance 60 Ω and inductance 75 mH is connected to a 110 V, 60 Hz supply. Thus, the inductive reactance,

$$X_L = 2\pi f L = 2\pi(60)(75 \times 10^{-3}) = 28.27 \ \Omega$$

and impedance, $Z = \sqrt{R^2 + X_L^2} = \sqrt{60^2 + 28.27^2} = 66.33 \ \Omega$

Current, $I = \dfrac{V}{Z} = \dfrac{110}{66.33} = 1.658$ A

Hence, power dissipated is given by:

either $P = I^2 R = (1.658)^2 (60) = \mathbf{165 \ W}$

or $P = VI \cos\phi$ where

$$\cos\phi = \frac{R}{Z} = \frac{60}{66.33} = 0.9046$$

i.e. $P = (110)(1.658)(0.9046) = \mathbf{165 \ W}$

Power triangle and power factor

Figure 55.14(a) shows a phasor diagram in which the current I lags the applied voltage V by angle ϕ. The horizontal component of V is $V \cos\phi$ and the

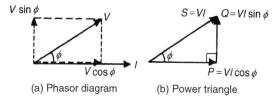

(a) Phasor diagram (b) Power triangle

Figure 55.14

vertical component of V is $V \sin\phi$. If each of the voltage phasors is multiplied by I, Figure 55.14(b) is obtained and is known as the '**power triangle**'.

Apparent power,	$S = VI$ **voltamperes (VA)**
True or active power,	$P = VI \cos\phi$ **watts (W)**
Reactive power,	$Q = VI \sin\phi$ **reactive voltamperes (var)**
Power factor	$= \dfrac{\textbf{True power P}}{\textbf{Apparent power S}}$

For more on the power triangle, see Chapter 66, page.

For sinusoidal voltages and currents,

$$\text{power factor} = \frac{P}{S} = \frac{VI \cos\phi}{VI}$$

i.e. $\boxed{p.f. = \cos\phi = \dfrac{R}{Z}}$ (from Figure 55.6)

The relationships stated above are also true when current I leads voltage V.

For example, if a transformer has a rated output of 200 kVA at a power factor of 0.8, then:

$$VI = 200 \text{ kVA} = 200 \times 10^3$$

and $\qquad\qquad$ p.f. $= 0.8 = \cos\phi$

Power output, $\qquad P = VI \cos\phi = (200 \times 10^3)(0.8) = \textbf{160 kW}$

Reactive power, $\qquad Q = VI \sin\phi$

If $\cos\phi = 0.8$, then $\phi = \cos^{-1} 0.8 = 36.87°$, hence $\sin\phi = \sin 36.87° = 0.6$

Thus, **reactive power,** $\qquad Q = (200 \times 10^3)(0.6) = \textbf{120 kvar}$

56 Single-phase Parallel a.c. Circuits

Introduction

In parallel circuits, such as those shown in Figures 56.1 and 56.2, the voltage is common to each branch of the network and is thus taken as the reference phasor when drawing phasor diagrams. For any parallel a.c. circuit:

True or active power, $P = VI \cos\phi$ watts (W)

or $P = I_R^2 R$ watts

Apparent power, $S = VI$ voltamperes (VA)

Reactive power, $Q = VI \sin\phi$ reactive voltamperes (var)

$$\text{Power factor} = \frac{\text{true power}}{\text{apparent power}} = \frac{P}{S} = \cos\phi$$

These formulae are the same as for series a.c. circuits as used in chapter 55.

R-L Parallel a.c. Circuit

In the two branch parallel circuit containing resistance R and inductance L shown in Figure 56.1, the current flowing in the resistance, I_R, is in-phase with the supply voltage V and the current flowing in the inductance, I_L, lags

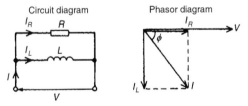

Figure 56.1

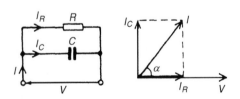

Figure 56.2

the supply voltage by 90°. The supply current I is the phasor sum of I_R and I_L and thus the current I lags the applied voltage V by an angle lying between 0° and 90° (depending on the values of I_R and I_L), shown as angle ϕ in the phasor diagram. From the phasor diagram:

$$I = \sqrt{I_R^2 + I_L^2} \quad \text{(by Pythagoras' theorem)}$$

where

$$I_R = \frac{V}{R} \text{ and } I_L = \frac{V}{X_L}, \tan\phi = \frac{I_L}{I_R}, \sin\phi = \frac{I_L}{I} \text{ and } \cos\phi = \frac{I_R}{I}$$

(by trigonometric ratios)

Circuit impedance, $Z = \dfrac{V}{I}$

For example, a 20 Ω resistor is connected in parallel with an inductance of 2.387 mH across a 60 V, 1 kHz supply. Thus, the current flowing in the resistor,

$$I_R = \frac{V}{R} = \frac{60}{20} = \textbf{3 A},$$

and the current flowing in the inductance,

$$I_L = \frac{V}{X_L} = \frac{V}{2\pi f L} = \frac{60}{2\pi(1000)(2.387 \times 10^{-3})} = \textbf{4 A}$$

From the phasor diagram, supply current,

$$I = \sqrt{I_R^2 + I_L^2} = \sqrt{3^2 + 4^2} = \textbf{5 A}$$

The circuit phase angle, $\phi = \tan^{-1} \dfrac{I_L}{I_R} = \tan^{-1} \dfrac{4}{3} = \textbf{53.13° lagging}$

Circuit impedance, $Z = \dfrac{V}{I} = \dfrac{60}{5} = \textbf{12 Ω}$

Power consumed $P = VI\cos\phi = (60)(5)(\cos 53.13°) = \textbf{180 W}$
(Alternatively, power consumed $P = I_R^2 R = (3)^2(20) = \textbf{180 W}$)

R-C Parallel a.c. Circuit

In the two branch parallel circuit containing resistance R and capacitance C shown in Figure 56.2, I_R is in-phase with the supply voltage V and the current flowing in the capacitor, I_C, leads V by 90°. The supply current I is the phasor sum of I_R and I_C and thus the current I leads the applied voltage V by an angle lying between 0° and 90° (depending on the values of I_R and I_C), shown as angle α in the phasor diagram. From the phasor diagram:

$$I = \sqrt{I_R^2 + I_C^2}, \quad \text{(by Pythagoras' theorem)}$$

where

$$I_R = \frac{V}{R} \text{ and } I_C = \frac{V}{X_C}, \tan \alpha = \frac{I_C}{I_R}, \sin \alpha = \frac{I_C}{I} \text{ and } \cos \alpha = \frac{I_R}{I}$$

(by trigonometric ratios)

Circuit impedance $Z = \dfrac{V}{I}$

For example, a 30 μF capacitor is connected in parallel with an 80 Ω resistor across a 240 V, 50 Hz supply. Thus, the current in resistor,

$$I_R = \frac{V}{R} = \frac{240}{80} = \mathbf{3\ A},$$

and the current in capacitor,

$$I_C = \frac{V}{X_C} = \frac{V}{\left(\dfrac{1}{2\pi f C}\right)} = 2\pi f C V$$

$$= 2\pi(50)(30 \times 10^{-6})(240) = \mathbf{2.262\ A}$$

The supply current, $I = \sqrt{I_R^2 + I_C^2} = \sqrt{3^2 + 2.262^2} = \mathbf{3.757\ A}$
The circuit phase angle,

$$\alpha = \tan^{-1} \frac{I_C}{I_R} = \tan^{-1} \frac{2.262}{3} = \mathbf{37.02°\ leading}$$

Circuit impedance, $Z = \dfrac{V}{I} = \dfrac{240}{3.757} = \mathbf{63.88\ \Omega}$

True or active power dissipated,

$$P = VI \cos \alpha = (240)(3.757) \cos 37.02° = \mathbf{720\ W}$$

(Alternatively, true power $P = I_R^2 R = (3)^2(80) = 720$ W)
Apparent power, $S = VI = (240)(3.757) = \mathbf{901.7\ VA}$

L-C Parallel Circuit

In the two branch parallel circuit containing inductance L and capacitance C shown in Figure 56.3, I_L lags V by 90° and I_C leads V by 90°.

Theoretically there are three phasor diagrams possible-each depending on the relative values of I_L and I_C:

(i) $I_L > I_C$ (giving a supply current, $I = I_L - I_C$ lagging V by 90°)
(ii) $I_C > I_L$ (giving a supply current, $I = I_C - I_L$ leading V by 90°)
(iii) $I_L = I_C$ (giving a supply current, $I = 0$)

The latter condition is not possible in practice due to circuit resistance inevitably being present.

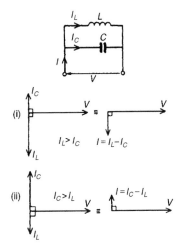

Figure 56.3

For the L-C parallel circuit,

$$I_L = \frac{V}{X_L}, \; I_C = \frac{V}{X_C},$$

I = phasor difference between I_L and I_C, and $Z = \dfrac{V}{I}$

When $X_L < X_C$ then $I_L > I_C$ and I lags V by $90°$
When $X_L > X_C$ then $I_L < I_C$ and I leads V by $90°$
  In a parallel circuit containing no resistance the power consumed is zero.

LR-C Parallel a.c. Circuit

In the two branch circuit containing capacitance C in parallel with inductance L and resistance R in series (such as a coil) shown in Figure 56.4(a), the phasor diagram for the LR branch alone is shown in Figure 56.4(b) and the phasor diagram for the C branch is shown alone in Figure 56.4(c). Rotating each and superimposing on one another gives the complete phasor diagram shown in Figure 56.5(d).

  The current I_{LR} of Figure 56.4(d) may be resolved into horizontal and vertical components. The horizontal component, shown as op is $I_{LR} \cos \phi_1$ and the vertical component, shown as pq is $I_{LR} \sin \phi_1$. There are three possible conditions for this circuit:

(i) $I_C > I_{LR} \sin \phi_1$ (giving a supply current I leading V by angle ϕ—as shown in Figure 56.4(e))

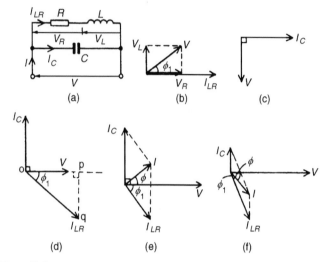

Figure 56.4

(ii) $I_{LR} \sin \phi_1 > I_C$ (giving I lagging V by angle ϕ—as shown in Figure 56.4(f))

(iii) $I_C = I_{LR} \sin \phi_1$ (this is called **parallel resonance**, see later).

There are two methods of finding the phasor sum of currents I_{LR} and I_C in Figure 56.4(e) and (f). These are: (i) by a scaled phasor diagram, or (ii) by resolving each current into their 'in-phase' (i.e. horizontal) and 'quadrature' (i.e. vertical) components.

With reference to the phasor diagrams of Figure 56.4:

Impedance of LR branch, $Z_{LR} = \sqrt{R^2 + X_L^2}$ and current, $I_{LR} = \dfrac{V}{Z_{LR}}$ and $I_C = \dfrac{V}{X_C}$

Supply current I = phasor sum of I_{LR} and I_C (by drawing)

$$= \sqrt{(I_{LR} \cos \phi_1)^2 + (I_{LR} \sin \phi_1 \sim I_C)^2}$$

(by calculation)

where $\sim$ means 'the difference between'

Circuit impedance $\quad Z = \dfrac{V}{I}$,

$$\tan \phi_1 = \frac{V_L}{V_R} = \frac{X_L}{R}, \sin \phi_1 = \frac{X_L}{Z_{LR}} \text{ and } \cos \phi_1 = \frac{R}{Z_{LR}}$$

$$\tan \phi = \frac{I_{LR} \sin \phi_1 \sim I_C}{I_{LR} \cos \phi_1} \text{ and } \cos \phi = \frac{I_{LR} \cos \phi_1}{I}$$

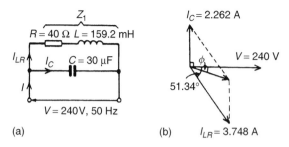

Figure 56.5

For example, a coil of inductance 159.2 mH and resistance 40 Ω is connected in parallel with a 30 μF capacitor across a 240 V, 50 Hz supply. The circuit diagram is shown in Figure 56.5(a). For the coil, inductive reactance

$$X_L = 2\pi f L = 2\pi(50)(159.2 \times 10^{-3}) = 50 \ \Omega$$

Impedance $\quad Z_1 = \sqrt{R^2 + X_L^2} = \sqrt{40^2 + 50^2} = 64.03 \ \Omega$

Current in coil, $\qquad I_{LR} = \dfrac{V}{Z_1} = \dfrac{240}{64.03} = \textbf{3.748 A}$

Branch phase angle $\quad \phi_1 = \tan^{-1}\dfrac{X_L}{R} = \tan^{-1}\dfrac{50}{40} = \tan^{-1} 1.25$

$$= \textbf{51.34° lagging}$$

(see phasor diagram in Figure 56.5(b))

Capacitive reactance, $\quad X_C = \dfrac{1}{2\pi f C} = \dfrac{1}{2\pi(50)(30 \times 10^{-6})}$

$$= 106.1 \ \Omega$$

Current in capacitor, $\qquad I_C = \dfrac{V}{X_C} = \dfrac{240}{106.1}$

$$= \textbf{2.262 A leading the supply voltage by 90°}$$

(see phasor diagram of Figure 56.5(b))

The supply current I is the phasor sum of I_{LR} and I_C. This may be obtained by drawing the phasor diagram to scale and measuring the current I and its phase angle relative to V. (Current I will always be the diagonal of the parallelogram formed as in Figure 56.5(b)).

Alternatively, the horizontal component of I_{LR} is:

$$I_{LR} \cos 51.34° = 3.748 \cos 51.34° = 2.341 \ \text{A}$$

The horizontal component of I_C is $I_C \cos 90° = 0$
Thus, the total horizontal component, $I_H = \textbf{2.341 A}$

The vertical component of I_{LR}

$$= -I_{LR}\sin 51.34° = -3.748\sin 51.34° = -2.927 \text{ A}$$

The vertical component of I_C

$$= I_C\sin 90° = 2.262\sin 90° = 2.262 \text{ A}$$

Thus, the total vertical component,

$$I_V = -2.927 + 2.262 = -0.665 \text{ A}$$

I_H and I_V are shown in Figure 56.6, from which,

$$I = \sqrt{2.341^2 + (-0.665)^2} = 2.434 \text{ A}$$

$$\text{Angle } \phi = \tan^{-1}\frac{0.665}{2.341} = 15.86° \text{ lagging}$$

Hence **the supply current $I = 2.434$ A lagging V by $15.86°$**

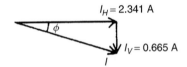

Figure 56.6

Parallel Resonance and Q-factor

Parallel resonance

Resonance occurs in the two-branch network containing capacitance C in parallel with inductance L and resistance R in series (see Figure 56.4(a)) when the quadrature (i.e. vertical) component of current I_{LR} is equal to I_C. At this condition the supply current I is in-phase with the supply voltage V.

Resonant frequency

When the quadrature component of I_{LR} is equal to I_C then: $I_C = I_{LR}\sin\phi_1$ (see Figure 56.7)
It may be shown that parallel resonant frequency,

$$\boxed{f_r = \frac{1}{2\pi}\sqrt{\left(\frac{1}{LC} - \frac{R^2}{L^2}\right)}}$$

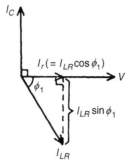

Figure 56.7

When R is negligible, then $f_r = \dfrac{1}{2\pi\sqrt{LC}}$, which is the same as for series resonance

$$\textbf{Current at resonance, } I_r = \frac{VRC}{L}$$

The current is at a **minimum** at resonance.

Dynamic resistance

Since the current at resonance is in-phase with the voltage the impedance of the circuit acts as a resistance. This resistance is known as the **dynamic resistance,** R_D (or sometimes, the dynamic impedance). Dynamic resistance,

$$\boxed{R_D = \frac{L}{RC} \textbf{ ohms}}$$

Rejector circuit

The parallel resonant circuit is often described as a **rejector** circuit since it presents its maximum impedance at the resonant frequency and the resultant current is a minimum.

Q-factor

Currents higher than the supply current can circulate within the parallel branches of a parallel resonant circuit, the current leaving the capacitor and establishing the magnetic field of the inductor, this then collapsing and recharging the capacitor, and so on. The *Q*-**factor** of a parallel resonant circuit is the ratio of the current circulating in the parallel branches of the circuit to the supply current, i.e. the current magnification.

Q-factor at resonance = current magnification

$$= \frac{\text{circulating current}}{\text{supply current}}$$

370

or $\boxed{\textbf{\textit{Q}-factor at resonance} = \dfrac{2\pi f_r L}{R}}$ (which is the same as for a

series circuit)

Note that in a **parallel** circuit the Q-factor is a measure of **current magnification**, whereas in a **series** circuit it is a measure of **voltage magnification**.

At mains frequencies the Q-factor of a parallel circuit is usually low, typically less than 10, but in radio-frequency circuits the Q-factor can be very high.

For example, a coil of inductance 0.20 H and resistance 60 Ω is connected in parallel with a 20 μF capacitor across a 20 V, variable frequency supply. Thus, the parallel resonant frequency,

$$f_r = \frac{1}{2\pi}\sqrt{\left(\frac{1}{LC} - \frac{R^2}{L^2}\right)}$$

$$= \frac{1}{2\pi}\sqrt{\left(\frac{1}{(0.20)(20\times 10^{-6})} - \frac{(60)^2}{(0.20)^2}\right)}$$

$$= \frac{1}{2\pi}\sqrt{250\,000 - 90\,000} = \textbf{63.66 Hz}$$

The dynamic resistance, $R_D = \dfrac{L}{RC} = \dfrac{0.20}{(60)(20\times 10^{-6})}$

$$= \textbf{166.7 } \Omega$$

The current at resonance, $I_r = \dfrac{V}{R_D} = \dfrac{20}{166.7} = \textbf{0.12 A}$

The circuit $\boldsymbol{Q}$-**factor** at resonance $= \dfrac{2\pi f_r L}{R} = \dfrac{2\pi(63.66)(0.20)}{60}$

$$= \textbf{1.33}$$

Alternatively, Q-factor at resonance = current magnification (for a parallel circuit) $= \dfrac{I_C}{I_r}$

$$I_c = \frac{V}{X_C} = \frac{V}{\left(\dfrac{1}{2\pi f_r C}\right)} = 2\pi f_r CV$$

$$= 2\pi(63.66)(20\times 10^{-6})(20) = 0.16\ A$$

Hence $\boldsymbol{Q}$-**factor** $= \dfrac{I_C}{I_r} = \dfrac{0.16}{0.12} = \textbf{1.33}$, as obtained above.

For more on parallel resonance and Q-factor, see Chapter 69.

57 D.c. Transients

Introduction

When a d.c. voltage is applied to a capacitor C and resistor R connected in series, there is a short period of time immediately after the voltage is connected, during which the current flowing in the circuit and voltages across C and R are changing.

Similarly, when a d.c. voltage is connected to a circuit having inductance L connected in series with resistance R, there is a short period of time immediately after the voltage is connected, during which the current flowing in the circuit and the voltages across L and R are changing.

These changing values are called **transients**.

Charging a Capacitor

(a) The circuit diagram for a series connected C-R circuit is shown in Figure 57.1 When switch S is closed then by Kirchhoff's voltage law:

$$V = v_C + v_R \qquad (1)$$

(b) The battery voltage V is constant. The capacitor voltage v_C is given by $\dfrac{q}{C}$, where q is the charge on the capacitor. The voltage drop across R is given by iR, where i is the current flowing in the circuit. Hence at all times:

$$V = \frac{q}{C} + iR \qquad (2)$$

At the instant of closing S, (initial circuit condition), assuming there is no initial charge on the capacitor, q_0 is zero, hence v_{C_o} is zero. Thus from equation (1), $V = 0 + v_{Ro}$, i.e. $v_{Ro} = V$
This shows that the resistance to current is solely due to R, and the initial current flowing, $i_0 = I = \dfrac{V}{R}$

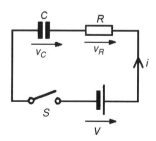

Figure 57.1

(c) A short time later at time t_1 seconds after closing S, the capacitor is partly charged to, say, q_1 coulombs because current has been flowing. The voltage v_{C1} is now $\dfrac{q_1}{C}$ volts. If the current flowing is i_1 amperes, then the voltage drop across R has fallen to $i_1 R$ volts.

Thus, equation (2) is now $V = \dfrac{q_1}{C} + i_1 R$

(d) A short time later still, say at time t_2 seconds after closing the switch, the charge has increased to q_2 coulombs and v_C has increased to $\dfrac{q_2}{C}$ volts. Since $V = v_C + v_R$ and V is a constant, then v_R decreases to $i_2 R$, Thus v_C is increasing and i and v_R are decreasing as time increases.

(e) Ultimately, a few seconds after closing S, (i.e. at the final or **steady state** condition), the capacitor is fully charged to, say, Q coulombs, current no longer flows, i.e. $i = 0$, and hence $v_R = iR = 0$. It follows from equation (1) that $v_C = V$

(f) Curves showing the changes in v_C, v_R and i with time are shown in Figure 57.2 The curve showing the variation of v_C with time is called an **exponential growth curve** and the graph is called the 'capacitor voltage/time' characteristic. The curves showing the variation of v_R and i with time are called **exponential decay curves**, and the graphs are called 'resistor voltage/time' and 'current/time' characteristics respectively. (The name 'exponential' shows that the shape can be expressed mathematically by an exponential mathematical equation, as shown later)

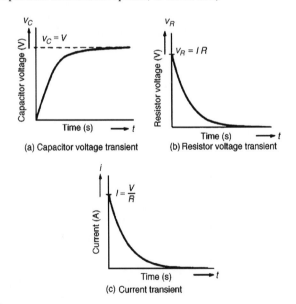

(a) Capacitor voltage transient

(b) Resistor voltage transient

(c) Current transient

Figure 57.2

Time Constant for a *C-R* Circuit

Time constant is defined as:

the time taken for a transient to reach its final state if the initial rate of change is maintained

For a series connected *C-R* circuit,

> **time constant $\tau = CR$ seconds**

Transient Curves for a *C-R* Circuit

There are two main methods of drawing transient curves graphically, these being:

(a) the **tangent method** — this method is shown in an example below
(b) the **initial slope and three point method**, which is based on the following properties of a transient exponential curve:
  (i) for a growth curve, the value of a transient at a time equal to one time constant is 0.632 of its steady state value (usually taken as 63% of the steady state value), at a time equal to two and a half time constants is 0.918 of its steady state value (usually taken as 92% of its steady state value) and at a time equal to five time constants is equal to its steady state value,
  (ii) for a decay curve, the value of a transient at a time equal to one time constant is 0.368 of its initial value (usually taken as 37% of its initial value), at a time equal to two and a half time constants is 0.082 of its initial value (usually taken as 8% of its initial value) and at a time equal to five time constants is equal to zero.

The transient curves shown in Figure 57.2 have mathematical equations, obtained by solving the differential equations representing the circuit. The equations of the curves are:

> **growth of capacitor voltage, $v_C = V(1 - e^{-t/CR})$**
> $$= V(1 - e^{-t/\tau})$$
> **decay of resistor voltage, $v_R = Ve^{-t/CR} = Ve^{-t/\tau}$ and**
> **decay of current flowing, $i = Ie^{-t/CR} = Ie^{-t/\tau}$**

For example, a 15 μF uncharged capacitor is connected in series with a 47 kΩ resistor across a 120 V, d.c. supply. Thus,

374

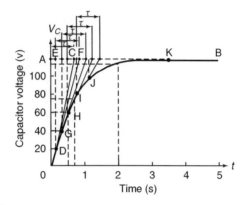

Figure 57.3

the time constant $= CR = 15\ \mu F \times 47\ k\Omega$

$$= 15 \times 10^{-6} \times 47 \times 10^3 = 0.705\ s$$

Steady state value of $v_C = V$, i.e. $v_C = 120$ V

With reference to Figure 57.3, the scale of the horizontal axis is drawn so that it spans at least five time constants, i.e. 5×0.705 or about 3.5 seconds. The scale of the vertical axis spans the change in the capacitor voltage, that is, from 0 to 120 V. A broken line AB is drawn corresponding to the final value of v_C Point C is measured along AB so that AC is equal to 1τ, i.e. AC = 0.705 s.

Straight line OC is drawn. Assuming that about five intermediate points are needed to draw the curve accurately, a point D is selected on OC corresponding to a v_C value of about 20 V. DE is drawn vertically. EF is made to correspond to 1τ, i.e. EF = 0.705 s. A straight line is drawn joining DF. This procedure of

(a) drawing a vertical line through point selected,
(b) at the steady-state value, drawing a horizontal line corresponding to 1τ, and
(c) joining the first and last points, is repeated for v_C values of 40, 60, 80 and 100 V, giving points G, H, I and J

The capacitor voltage effectively reaches its steady-state value of 120 V after a time equal to five time constants, shown as point K. Drawing a smooth curve through points O, D, G, H, I, J and K gives the exponential growth curve of capacitor voltage.

From the graph, the value of capacitor voltage at a time equal to the time constant is about **75 V**. It is a characteristic of all exponential growth curves, that after a time equal to one time constant, the value of the transient is 0.632 of its steady-state value. In this example, $0.632 \times 120 = 75.84$ V. Also from the graph, when t is two seconds, v_C is about **115 Volts**. This value may be checked using the equation $v_C = V(1 - e^{-t/\tau})$, where $V = 120$ V, $\tau = 0.705$ s and $t = 2$ s. This calculation gives $v_C = 112.97$ V.

The time for v_C to rise to one half of its final value, i.e. 60 V, can be determined from the graph and is about **0.5 s**. This value may be checked using $v_C = V(1 - e^{-t/\tau})$ where $V = 120$ V, $v_C = 60$ V and $\tau = 0.705$ s, giving $t = 0.489$ s.

Discharging a Capacitor

When a capacitor is charged (i.e. with the switch in position A in Figure 57.4), and the switch is then moved to position B, the electrons stored in the capacitor keep the current flowing for a short time. Initially, at the instant of moving from A to B, the current flow is such that the capacitor voltage v_C is balanced by an equal and opposite voltage $v_R = iR$. Since initially $v_C = v_R = V$, then $i = I = V/R$. During the transient decay, by applying Kirchhoff's voltage law to Figure 52.4, $v_C = v_R$. Finally the transients decay exponentially to zero, i.e. $v_C = v_R = 0$. The transient curves representing the voltages and current are as shown in Figure 57.5.

The equations representing the transient curves during the discharge period of a series connected C-R circuit are:

> **decay of voltage,** $v_C = v_R = Ve^{(-t/CR)} = Ve^{(-t/\tau)}$
>
> **decay of current,** $i = Ie^{(-t/CR)} = Ie^{(-t/\tau)}$

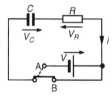

Figure 57.4

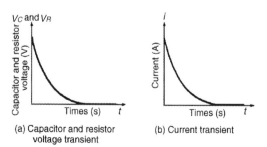

(a) Capacitor and resistor voltage transient

(b) Current transient

Figure 57.5

For example, a capacitor is charged to 100 V and then discharged through a 50 kΩ resistor. If the time constant of the circuit is 0.8 s, then, since time constant, $\tau = CR$, capacitance,

$$C = \frac{\tau}{R} = \frac{0.8}{50 \times 10^3} = \textbf{16 μF}$$

The time for the capacitor voltage to fall to, say, 20 V is determined using $v_C = Ve^{-t/\tau}$ then $20 = 100e^{-t/0.8}$ from which $\frac{1}{5} = e^{-t/0.8}$

Thus $e^{t/0.8} = 5$ and taking natural logarithms of each side, gives

$$\frac{t}{0.8} = \ln 5 \quad \text{and} \quad \text{time}, t = 0.8 \ln 5 = \textbf{1.29 s}$$

When a capacitor has been disconnected from the supply it may still be charged and it may retain this charge for some considerable time. Thus precautions must be taken to ensure that the capacitor is automatically discharged after the supply is switched off. Connecting a high value resistor across the capacitor terminals does this.

In a d.c. circuit, a capacitor blocks the current except during the times that there are changes in the supply voltage.

Current growth in an L-R circuit

(a) The circuit diagram for a series connected *L-R* circuit is shown in Figure 57.6. When switch *S* is closed, then by Kirchhoff's voltage law:

$$V = v_L + v_R \tag{3}$$

(b) The battery voltage V is constant. The voltage across the inductance is the induced voltage, i.e.

$$v_L = L \times \frac{\text{change of current}}{\text{change of time}} = L\frac{di}{dt}$$

The voltage drop across R, v_R is given by iR. Hence, at all times:

$$V = L\frac{di}{dt} + iR \tag{4}$$

(c) At the instant of closing the switch, the rate of change of current is such that it induces an e.m.f. in the inductance which is equal and opposite

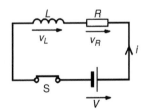

Figure 57.6

to V, hence $V = v_L + 0$, i.e. $v_L = V$. From equation (3), because $v_L = V$, then $v_R = 0$ and $i = 0$

(d) A short time later at time t_1 seconds after closing S, current i_1 is flowing, since there is a rate of change of current initially, resulting in a voltage drop of $i_1 R$ across the resistor. Since V (which is constant) $= v_L + v_R$ the induced e.m.f. is reduced, and equation (4) becomes:

$$V = L\frac{di_1}{dt_1} + i_1 R$$

(e) A short time later still, say at time t_2 seconds after closing the switch, the current flowing is i_2, and the voltage drop across the resistor increases to $i_2 R$. Since v_R increases, v_L decreases.

(f) Ultimately, a few seconds after closing S, the current flow is entirely limited by R, the rate of change of current is zero and hence v_L is zero. Thus $V = iR$. Under these conditions, steady state current flows, usually signified by I. Thus, $I = \dfrac{V}{R}$, $v_R = IR$ and $v_L = 0$ at steady state conditions.

(g) Curves showing the changes in v_L, v_R and i with time are shown in Figure 57.7 and indicate that v_L is a maximum value initially (i.e. equal to V), decaying exponentially to zero, whereas v_R and i grow exponentially from zero to their steady state values of V and $I = \dfrac{V}{R}$ respectively.

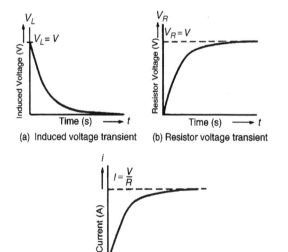

(a) Induced voltage transient

(b) Resistor voltage transient

(c) Current transient

Figure 57.7

Time Constant for an *L-R* Circuit

The time constant of a series connected *L-R* circuit is defined in the same way as the time constant for a series connected *C-R* circuit. Its value is given by:

$$\text{time constant, } \tau = \frac{L}{R} \text{ seconds}$$

Transient Curves for an *L-R* Circuit

Transient curves representing the induced voltage/time, resistor voltage/time and current/time characteristics may be drawn graphically, as outlined earlier. Each of the transient curves shown in Figure 57.7 have mathematical equations, and these are:

$$\text{decay of induced voltage, } v_L = Ve^{(-Rt/L)} = Ve^{(-t/\tau)}$$

$$\text{growth of resistor voltage, } v_R = V(1 - e^{-Rt/L}) = V(1 - e^{-t/\tau})$$

$$\text{growth of current flow, } i = I(1 - e^{-Rt/L}) = I(1 - e^{-t/\tau})$$

For example, a relay has an inductance of 100 mH and a resistance of 20 Ω. It is connected to a 60 V, d.c. supply.

$$\text{Time constant, } \tau = \frac{L}{R} = \frac{100 \times 10^{-3}}{20} = 5 \text{ ms}$$

$$\text{Final value of current, } I = \frac{V}{R} = \frac{60}{20} = 3 \text{ A}$$

(a) The scales should span at least five time constants (horizontally), i.e. 25 ms, and 3 A (vertically).
(b) With reference to Figure 57.8, the initial slope is obtained by making AB equal to 1 time constant, (i.e. 5 ms), and joining OB.
(c) At a time of 1 time constant, CD is $0.632 \times I = 0.632 \times 3 = 1.896$ A.
At a time of 2.5 time constants, EF is $0.918 \times I = 0.918 \times 3 = 2.754$ A.
At a time of 5 time constants, GH is $I = 3$ A.
(d) A smooth curve is drawn through points 0, D, F and H and this curve is the current/time characteristic.

From the characteristic, when $t = 2\tau$, $i \approx$ **2.6 A**. This may be checked by calculation using $i = I(1 - e^{-t/\tau})$, where $I = 3$ and $t = 2\tau$, giving $i = 2.59$ A.
Also, when the current is 1.5 A, the corresponding time is about **3.6 ms**. Again, this may be checked by calculation, using $i = I(1 - e^{-t/\tau})$ where $i = 1.5$, $I = 3$ and $\tau = 5$ ms, giving $t = 3.466$ ms.

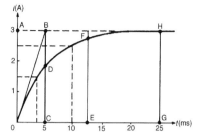

Figure 57.8

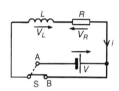

Figure 57.9

Current Decay in an *L-R* Circuit

When a series connected *L-R* circuit is connected to a d.c. supply as shown with *S* in position A of Figure 57.9, a current $I = \dfrac{V}{R}$ flows after a short time, creating a magnetic field ($\Phi \propto I$) associated with the inductor. When *S* is moved to position B, the current value decreases, causing a decrease in the strength of the magnetic field. Flux linkages occur, generating a voltage v_L, equal to $L\dfrac{di}{dt}$. By Lenz's law, this voltage keeps current *i* flowing in the circuit, its value being limited by *R*. Thus $v_L = v_R$. The current decays exponentially to zero and since v_R is proportional to the current flowing, v_R decays exponentially to zero. Since $v_L = v_R$, v_L also decays exponentially to zero. The curves representing these transients are similar to those shown in Figure 57.5. The equations representing the decay transient curves are:

---

decay of voltages, $v_L = v_R = Ve^{(-Rt/L)} = Ve^{(-t/\tau)}$

decay of current, $i = Ie^{(-Rt/L)} = Ie^{(-t/\tau)}$

For example, a coil having an inductance of 6 H and a resistance of R Ω is connected in series with a resistor of 10 Ω to a 120 V, d.c. supply. The time constant of the circuit is 300 ms. When steady-state conditions have been reached, the supply is replaced instantaneously by a short-circuit. Thus,

$$\text{the time constant,} \quad \tau = \frac{\text{circuit inductance}}{\text{total circuit resistance}} = \frac{L}{R + 10}$$

$$\text{from which,} \quad R = \frac{L}{\tau} - 10 = \frac{6}{0.3} - 10 = \mathbf{10\ \Omega}$$

The steady-state current, $I = \dfrac{V}{R} = \dfrac{120}{10 + 10} = 6$ A

The transient current after 1 s,

$$i = Ie^{-t/\tau} = 6e^{-1/0.3} = 6 \times 0.03567 = \mathbf{0.214\ A}$$

The time taken for the current to fall to, say, 10% of its initial value is determined as follows:

10% of the initial value of the current is $\dfrac{10}{100} \times 6$, i.e. 0.6 A

Using the equation $i = Ie^{-t/\tau}$ gives: $0.6 = 6e^{-t/0.3}$

$$\text{and } e^{t/0.3} = \frac{6}{0.6} = 10$$

Taking natural logarithms of each side of this equation gives: $\dfrac{t}{0.3} = \ln 10$ from which, time, $t = 0.3 \ln 10 = \mathbf{0.691}$ s

---

Switching Inductive Circuits

Energy stored in the magnetic field of an inductor exists because a current provides the magnetic field. When the d.c. supply is switched off the current falls rapidly, the magnetic field collapses causing a large induced e.m.f. which will either cause an arc across the switch contacts or will break down the insulation between adjacent turns of the coil. The high induced e.m.f. acts in a direction which tends to keep the current flowing, i.e. in the same direction as the applied voltage. The energy from the magnetic field will thus be aided by the supply voltage in maintaining an arc, which could cause severe damage to the switch. To reduce the induced e.m.f. when the supply switch is opened, a discharge resistor R_D is connected in parallel with the inductor as shown in Figure 57.10. The magnetic field energy is dissipated as heat in R_D and R and arcing at the switch contacts is avoided.

The Effects of Time Constant on a Rectangular Waveform

Integrator circuit

By varying the value of either C or R in a series connected C-R circuit, the time constant ($\tau = CR$), of a circuit can be varied. If a rectangular waveform varying from $+E$ to $-E$ is applied to a C-R circuit as shown in Figure 57.11, output waveforms of the capacitor voltage have various shapes, depending on the value of R.

When R is small, $\tau = CR$ is small and an output waveform such as that shown in Figure 57.12(a) is obtained. As the value of R is increased, the waveform changes to that shown in Figure 57.12(b). When R is large, the waveform is as shown in Figure 57.12(c), the circuit then being described as an **integrator circuit**.

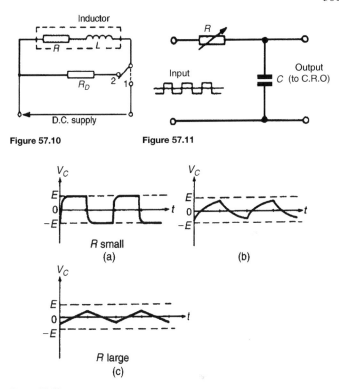

Figure 57.10

Figure 57.11

Figure 57.12

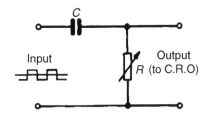

Figure 57.13

Differentiator circuit

If a rectangular waveform varying from $+E$ to $-E$ is applied to a series connected C-R circuit and the waveform of the voltage drop across the resistor is observed, as shown in Figure 57.13, the output waveform alters as R is varied due to the time constant, ($\tau = CR$), altering.

382

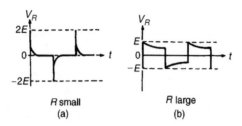

Figure 57.14

When R is small, the waveform is as shown in Figure 57.14(a), the voltage being generated across R by the capacitor discharging fairly quickly. Since the change in capacitor voltage is from $+E$ to $-E$, the change in discharge current is $2E/R$, resulting in a change in voltage across the resistor of $2E$. This circuit is called a **differentiator circuit**. When R is large, the waveform is as shown in Figure 57.14(b).

58 Operational Amplifiers

Introduction to Operational Amplifiers

Operational Amplifiers (usually called **'op amps'**) were originally made from discrete components, being designed to solve mathematical equations electronically, by performing operations such as addition and division in analogue computers. Now produced in integrated-circuit (IC) form, op amps have many uses, with one of the most important being as a high-gain d.c. and a.c. voltage amplifier.

The **main properties** of an op amp include:

(i) a very high open-loop voltage gain A_o of around 10^5 for d.c. and low frequency a.c., which decreases with frequency increase

(ii) a very high input impedance, typically 10^6 Ω to 10^{12} Ω, such that current drawn from the device, or the circuit supplying it, is very small and the input voltage is passed on to the op amp with little loss

(iii) a very low output impedance, around 100 Ω, such that its output voltage is transferred efficiently to any load greater than a few kilohms.

The **circuit diagram symbol** for an op amp is shown in Figure 58.1. It has one output, V_o, and two inputs; the **inverting input**, V_1, is marked $-$, and the non-**inverting input**, V_2, is marked $+$.

The operation of an op amp is most convenient from a dual balanced d.c. power supply $\pm V_S$ (i.e. $+V_S$, 0, $-V_S$); the centre point of the supply, i.e. 0 V, is common to the input and output circuits and is taken as their voltage reference level. The power supply connections are not usually shown in a circuit diagram.

An op amp is basically a **differential** voltage amplifier, i.e. it amplifies the difference between input voltages V_1 and V_2. Three situations are possible: (i) if $V_2 > V_1$, V_o is positive (ii) if $V_2 < V_1$, V_o is negative (iii) if $V_2 = V_1$, V_o is zero

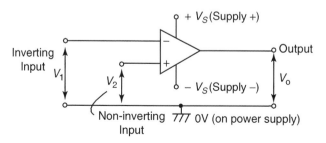

Figure 58.1

In general,

$$\boxed{V_o = A_o(V_2 - V_1)} \quad \text{or} \quad \boxed{A_o = \frac{V_o}{V_2 - V_1}} \tag{1}$$

where A_o is the open-loop voltage gain.

For example, if a differential amplifier has an open-loop voltage gain of 120 and the input signals are 2.45 V and 2.35 V, then the

output voltage, $V_o = A_o(V_2 - V_1)$

$$= 120(2.45 - 2.35) = (120)(0.1) = \mathbf{12\ V}$$

Transfer characteristic

A typical **voltage characteristic** showing how the output V_o varies with the input $(V_2 - V_1)$ is shown in Figure 58.2.

It is seen from Figure 58.2 that only within the very small input range P0Q is the output directly proportional to the input; it is in this range that the op amp behaves linearly and there is minimum distortion of the amplifier output. Inputs outside the linear range cause saturation and the output is then close to the maximum value, i.e. $+V_S$ or $-V_S$. The limited linear behaviour is due to the very high open-loop gain A_o, and the higher it is the greater is the limitation.

Negative feedback

Operational amplifiers nearly always use **negative feedback**, obtained by feeding back some, or all, of the output to the inverting (−) input (as shown in Figure 58.5 in the next section). The feedback produces an output voltage that opposes the one from which it is taken. This reduces the new output of the amplifier and the resulting closed-loop gain A is then less than the open-loop gain A_o.

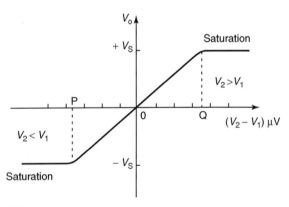

Figure 58.2

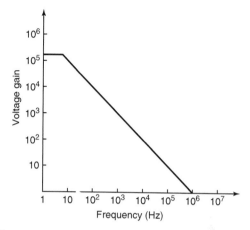

Figure 58.3

However, as a result, a wider range of voltages can be applied to the input for amplification. As long as $A_o \gg A$, negative feedback gives: (i) a constant and predictable voltage gain A, (ii) reduced distortion of the output, and (iii) better frequency response.

The advantages of using negative feedback outweigh the accompanying loss of gain that is easily increased by using two or more op amp stages.

Bandwidth

The open-loop voltage gain of an op amp is not constant at all frequencies; because of capacitive effects it falls at high frequencies. Figure 58.3 shows the gain/bandwidth characteristic of a 741 op amp. At frequencies below 10 Hz the gain is constant, but at higher frequencies the gain falls at a constant rate of 6 dB/octave (equivalent to a rate of 20 dB per decade) to 0 dB.

The gain-bandwidth product for any amplifier is the linear voltage gain multiplied by the bandwidth at that gain. The value of frequency at which the open-loop gain has fallen to unity is called the transition frequency f_T.

$$f_T = \text{closed-loop voltage gain} \times \text{bandwidth} \qquad (2)$$

In Figure 53.3, $f_T = 10^6$ Hz or 1 MHz; a gain of 20 dB (i.e. $20 \log_{10} 10$) gives a 100 kHz bandwidth, whilst a gain of 80 dB (i.e. $20 \log_{10} 10^4$) restricts the bandwidth to 100 Hz.

Some Op Amp Parameters

Input bias current

The input bias current, I_B, is the average of the currents into the two input terminals with the output at zero volts, which is typically around 80 nA (i.e.

80×10^{-9} A) for a 741 op amp. The input bias current causes a volt drop across the equivalent source impedance seen by the op amp input.

Input offset current

The input offset current, I_{os}, of an op amp is the difference between the two input currents with the output at zero volts. In a 741 op amp, I_{os} is typically 20 nA.

Input offset voltage

In the ideal op amp, with both inputs at zero there should be zero output. Due to imbalances within the amplifier this is not always the case and a small output voltage results. The effect can be nullified by applying a small offset voltage, V_{os}, to the amplifier. In a 741 op amp, V_{os} is typically 1 mV.

Common-mode rejection ratio

The output voltage of an op amp is proportional to the difference between the voltages applied to its two input terminals. Ideally, when the two voltages are equal, the output voltages should be zero. A signal applied to both input terminals is called a common-mode signal and it is usually an unwanted noise voltage. The ability of an op amp to suppress common-mode signals is expressed in terms of its common-mode rejection ratio (CMRR), which is defined by:

$$\text{CMRR} = 20\log_{10}\left(\frac{\text{differential voltage gain}}{\text{common mode gain}}\right) \text{ dB} \qquad (3)$$

In a 741 op amp, the CMRR is typically 90 dB.

The common-mode gain, A_{com}, is defined as:

$$A_{com} = \frac{V_o}{V_{com}} \qquad (4)$$

where V_{com} is the common input signal.

For example, if a differential amplifier has an open-loop voltage gain of 120 and a common input signal of 3.0 V to both terminals, and an output signal of 24 mV results, then the common-mode gain,

$$A_{com} = \frac{V_o}{V_{com}} = \frac{24 \times 10^{-3}}{3.0} = 8 \times 10^{-3} = \textbf{0.008}$$

and the $\text{CMRR} = 20\log_{10}\left(\frac{120}{0.008}\right) = 20\log_{10} 15\,000 = \textbf{83.52 dB}$

Slew Rate

The slew rate of an op amp is the maximum rate of change of output voltage following a step input voltage. Figure 58.4 shows the effects of slewing; it

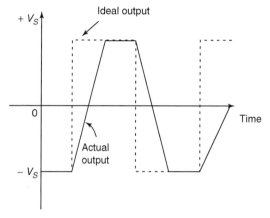

Figure 58.4

causes the output voltage to change at a slower rate that the input, such that the output waveform is a distortion of the input waveform. 0.5 V/µs is a typical value for the slew rate.

Op Amp Inverting Amplifier

The basic circuit for an inverting amplifier is shown in Figure 58.5 where the input voltage V_i (a.c. or d.c.) to be amplified is applied via resistor R_i to the inverting ($-$) terminal; the output voltage V_o is therefore in anti-phase with the input. The non-inverting ($+$) terminal is held at 0 V. Negative feedback is provided by the feedback resistor, R_f, feeding back a certain fraction of the output voltage to the inverting terminal.

Amplifier Gain

In an **ideal op amp** two assumptions are made, these being that:

(i) each input draws zero current from the signal source, i.e. their input impedance's are infinite, and
(ii) the inputs are both at the same potential if the op amp is not saturated, i.e. $V_A = V_B$ in Figure 58.5.

In Figure 58.5, $V_B = 0$, hence $V_A = 0$ and point X is called a **virtual earth**. Thus, $I_1 = \dfrac{V_i - 0}{R_i}$ and $I_2 = \dfrac{0 - V_o}{R_f}$

However, $I_1 = I_2$ from assumption (i) above.

Hence $\dfrac{V_i}{R_i} = -\dfrac{V_o}{R_f}$, the negative sign showing that V_o is negative when V_i is positive, and vice versa.

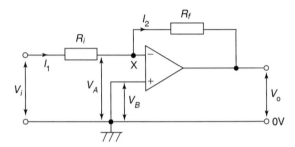

Figure 58.5

The **closed-loop gain** A is given by:

$$A = \frac{V_o}{V_i} = -\frac{R_f}{R_i} \qquad (5)$$

This shows that the gain of the amplifier depends only on the two resistors, which can be made with precise values, and not on the characteristics of the op amp, which may vary from sample to sample.

For example, if $R_i = 10$ kΩ and $R_f = 100$ kΩ, then the closed-loop gain,

$$A = \frac{-R_f}{R_i} = \frac{-100 \times 10^3}{10 \times 10^3} = -10$$

Thus an input of 100 mV will cause an output change of 1 V.

Input impedance

Since point X is a virtual earth (i.e. at 0 V), R_i may be considered to be connected between the inverting ($-$) input terminal and 0 V. The input impedance of the circuit is therefore R_i in parallel with the much greater input impedance of the op amp, i.e. effectively R_i. Simply changing the value of R_i can thus control the circuit input impedance.

For example, the op amp shown in Figure 58.6 has an input bias current of 100 nA at 20°C. Thus, from equation (5),

$$\textbf{voltage gain, } A = \frac{-R_f}{R_i} = \frac{-1 \times 10^6}{10 \times 10^3} = -100$$

The input bias current, I_B, causes a volt drop across the equivalent source impedance seen by the op amp input, in this case, R_i and R_f in parallel. Hence, the offset voltage, V_{os}, at the input due to the 100 nA input bias current, I_B, is given by:

$$V_{os} = I_B \left(\frac{R_i R_f}{R_i + R_f} \right) = (100 \times 10^{-9}) \left(\frac{10 \times 10^3 \times 1 \times 10^6}{(10 \times 10^3) + (1 \times 10^6)} \right)$$

$$= (10^{-7})(9.9 \times 10^3) = 9.9 \times 10^{-4} = \textbf{0.99 mV}$$

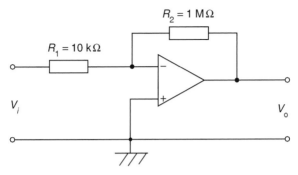

Figure 58.6

The effect of input bias current can be minimised by ensuring that both inputs 'see' the same driving resistance. This means that **a resistance of value of 9.9 kΩ should be placed between the non-inverting (+) terminal and earth** in Figure 58.6.

Op Amp Non-inverting Amplifier

The basic circuit for a non-inverting amplifier is shown in Figure 58.7 where the input voltage V_i (a.c. or d.c.) is applied to the non-inverting (+) terminal of the op amp. This produces an output V_o that is in phase with the input. Negative feedback is obtained by feeding back to the inverting (−) terminal, the fraction of V_o developed across R_i in the voltage divider formed by R_f and R_i across V_o.

Amplifier gain

In Figure 58.7, let the feedback factor, $\beta = \dfrac{R_i}{R_i + R_f}$

It may be shown that for an amplifier with open-loop gain A_o, the closed-loop voltage gain A is given by: $A = \dfrac{A_o}{1 + \beta A_o}$

For a typical op amp, $A_o = 10^5$, thus βA_o is large compared with 1, and the above expression approximates to:

$$A = \frac{A_o}{\beta A_o} = \frac{1}{\beta} \qquad (6)$$

Hence

$$A = \frac{V_o}{V_i} = \frac{R_i + R_f}{R_i} = 1 + \frac{R_f}{R_i} \qquad (7)$$

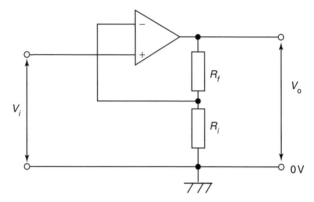

Figure 58.7

For example, if $R_i = 10$ kΩ and $R_f = 100$ kΩ, then

$$A = 1 + \frac{100 \times 10^3}{10 \times 10^3} = 1 + 10 = \mathbf{11}$$

Again, the gain depends only on the values of R_i and R_f and is independent of the open-loop gain A_o

Input impedance

Since there is no virtual earth at the non-inverting (+) terminal, the input impedance is much higher ($-$typically 50 MΩ) than that of the inverting amplifier. Also, it is unaffected if the gain is altered by changing R_f and/or R_i. This non-inverting amplifier circuit gives good matching when the input is supplied by a high impedance source.

Op Amp Voltage-follower

The **voltage-follower** is a special case of the non-inverting amplifier in which 100% negative feedback is obtained by connecting the output directly to the inverting ($-$) terminal, as shown in Figure 58.8. Thus R_f in Figure 58.7 is zero and R_i is infinite.

From equation (6), $A = \dfrac{1}{\beta}$ (when A_o is very large). Since all of the output is fed back, $\beta = 1$ and $A \approx 1$. Thus the voltage gain is nearly 1 and $V_o = V_i$ to within a few millivolts.

The circuit of Figure 58.8 is called a voltage-follower since, as with its transistor emitter-follower equivalent, V_o follows V_i. It has extremely high input impedance and a low output impedance. Its main use is as a **buffer amplifier**, giving current amplification, to match a high impedance source to a low impedance load. For example, it is used as the input stage of an analogue

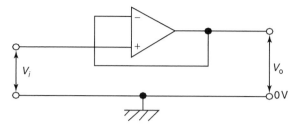

Figure 58.8

voltmeter where the highest possible input impedance is required so as not to disturb the circuit under test; the output voltage is measured by a relatively low impedance moving-coil meter.

Op Amp Summing Amplifier

Because of the existence of the virtual earth point, an op amp can be used to add a number of voltages (d.c. or a.c.) when connected as a multi-input inverting amplifier. This, in turn, is a consequence of the high value of the open-loop voltage gain A_o. Such circuits may be used as 'mixers' in audio systems to combine the outputs of microphones, electric guitars, pick-ups, etc. They are also used to perform the mathematical process of addition in analogue computing.

The circuit of an op amp summing amplifier having three input voltages V_1, V_2 and V_3 applied via input resistors R_1, R_2 and R_3 is shown in Figure 58.9. If it is assumed that the inverting $(-)$ terminal of the op amp draws no input current, all of it passing through R_f, then:

$$I = I_1 + I_2 + I_3$$

Since X is a virtual earth (i.e. at 0 V), it follows that:

$$\frac{-V_o}{R_f} = \frac{V_1}{R_1} + \frac{V_2}{R_2} + \frac{V_3}{R_3}$$

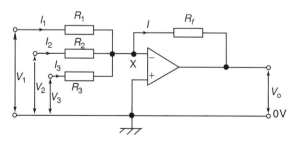

Figure 58.9

Hence

$$V_o = -\left(\frac{R_f}{R_1}V_1 + \frac{R_f}{R_2}V_2 + \frac{R_f}{R_3}V_3\right) = -R_f\left(\frac{V_1}{R_1} + \frac{V_2}{R_2} + \frac{V_3}{R_3}\right) \quad (8)$$

The three input voltages are thus added and amplified if R_f is greater than each of the input resistors; 'weighted' summation is said to have occurred. Alternatively, the input voltages are added and attenuated if R_f is less than each input resistor.

For example, if $\dfrac{R_f}{R_1} = 4$, $\dfrac{R_f}{R_2} = 3$ and $\dfrac{R_f}{R_3} = 1$ and $V_1 = V_2 = V_3 = +1$ V, then

$$\mathbf{V_o} = -\left(\frac{R_f}{R_1}V_1 + \frac{R_f}{R_2}V_2 + \frac{R_f}{R_3}V_3\right) = -(4+3+1) = \mathbf{-8\ V}$$

If $R_1 = R_2 = R_3 = R_i$, the input voltages are amplified or attenuated equally, and

$$V_o = -\frac{R_f}{R_i}(V_1 + V_2 + V_3)$$

If, also, $R_i = R_f$ then $V_o = -(V_1 + V_2 + V_3)$

The virtual earth is also called the **summing point** of the amplifier. It isolates the inputs from one another so that each behaves as if none of the others existed and none feeds any of the other inputs even though all the resistors are connected at the inverting $(-)$ input.

For example, for the summing op amp shown in Figure 58.10, the output voltage,

$$V_o = -R_f\left(\frac{V_1}{R_1} + \frac{V_2}{R_2} + \frac{V_3}{R_3}\right)$$

$$= -(50 \times 10^3)\left(\frac{0.5}{10 \times 10^3} + \frac{0.8}{20 \times 10^3} + \frac{1.2}{30 \times 10^3}\right) = \mathbf{-6.5\ V}$$

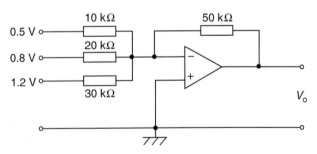

Figure 58.10

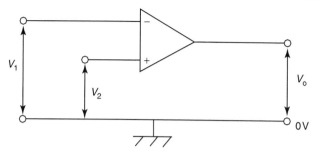

Figure 58.11

Op Amp Voltage Comparator

If both inputs of the op amp shown in Figure 58.11 are used simultaneously, then from equation (1), the output voltage is given by:

$$V_o = A_o(V_2 - V_1)$$

When $V_2 > V_1$ then V_o is positive, its maximum value being the positive supply voltage $+V_S$, which it has when $(V_2 - V_1) \geq \dfrac{V_S}{A_o}$. The op amp is then saturated. **For example**, if $V_S = +9$ V and $A_o = 10^5$, then saturation occurs when $(V_2 - V_1) \geq \dfrac{9}{10^5}$ i.e. when V_2 exceeds V_1 by 90 μV and $V_o \approx 9$ V.

When $V_1 > V_2$, then V_o is negative and saturation occurs if V_1 exceeds V_2 by $\dfrac{V_S}{A_o}$ i.e. around 90 μV in the above example; in this case, $V_o \approx -V_S = -9$ V.

A small change in $(V_2 - V_1)$ therefore causes V_o to switch between near $+V_S$ and near to $-V_S$ and enables the op amp to indicate when V_2 is greater or less than V_1, i.e. to act as a **differential amplifier** and compare two voltages. It does this in an electronic digital voltmeter.

Light-operated Alarm Circuit

A typical light-operated alarm circuit is shown in Figure 58.12. Resistor R and the light dependent resistor (LDR) form a voltage divider across the $+15/0/-15$ V supply. The op amp compares the voltage V_1 at the voltage divider junction, i.e. at the inverting $(-)$ input, with that at the non-inverting $(+)$ input, i.e. with V_2, which is 0 V. In the dark the resistance of the LDR is much greater than that of R, so more of the 30 V across the voltage divider is dropped across the LDR, causing V_1 to fall below 0 V. Now $V_2 > V_1$ and the output voltage V_o switches from near -15 V to near $+15$ V and the light emitting diode (LED) lights.

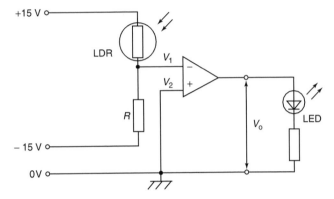

Figure 58.12

Op Amp Integrator

The circuit for the op amp integrator shown in Figure 58.13 is the same as for the op amp inverting amplifier shown in Figure 58.5, but feedback occurs via a capacitor C, rather than via a resistor.

The output voltage is given by:

$$V_o = -\frac{1}{CR} \int V_i \, \mathbf{d}t \qquad (9)$$

Since the inverting $(-)$ input is used in Figure 58.13, V_o is negative if V_i is positive, and vice versa, hence the negative sign in equation (9).

Since X is a virtual earth in Figure 58.13, i.e. at 0 V, the voltage across R is V_i and that across C is V_o. Assuming again that none of the input current I enters the op amp inverting $(-)$ input, then all of current I flows through C and charges it up. If V_i is constant, I will be a constant value given by $I = \dfrac{V_i}{R}$. Capacitor C therefore charges at a constant rate and the potential of

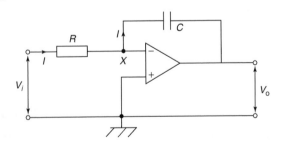

Figure 58.13

the output side of $C(=V_o$, since its input side is zero) charges so that the feedback path absorbs I. If Q is the charge on C at time t and the p.d. across it (i.e. the output voltage) changes from 0 to V_o in that time then:

$$Q = -V_o C = It \text{ (from chapter 44)}$$

i.e. $\quad -V_o C = \dfrac{V_i}{R} t$

i.e. $\quad V_o = -\dfrac{1}{CR} V_i t$

This result is the same as would be obtained from $V_o = -\dfrac{1}{CR}\displaystyle\int V_i \, dt$ if V_i is a constant value.

For example, if the input voltage $V_i = -2$ V and, say, $CR = 1$ s, then

$$V_o = -(-2)t = 2t$$

A graph of V_o/t will be ramp function as shown in Figure 58.14 ($V_o = 2t$ is of the straight line form $y = mx + c$; in this case $y = V_o$ and $x = t$, gradient, $m = 2$ and vertical axis intercept $c = 0$). V_o rises steadily by $+2$ V/s in Figure 58.14, and if the power supply is, say, ± 9 V, then V_o reaches $+9$ V after 4.5 s when the op amp saturates.

In another example, if a steady voltage of -0.75 V is applied to an op amp integrator having component values of $R = 200$ kΩ and $C = 2.5$ μF, and assuming that the initial capacitor charge is zero, the value of the output voltage, V_o, 100 ms after application of the input is given by:

$$V_o = -\frac{1}{CR}\int V_i \, dt = -\frac{1}{(2.5 \times 10^{-6})(200 \times 10^3)}\int (-0.75) dt$$

$$= -\frac{1}{0.5}\int (-0.75)\, dt = -2[-0.75t] = +1.5t$$

When time $t = 100$ ms,

output voltage, $V_o = (1.5)(100 \times 10^{-3}) = 0.15$ V

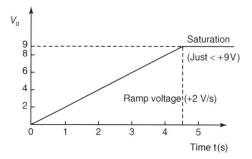

Figure 58.14

396

Op Amp Differential Amplifier

The circuit for an op amp differential amplifier is shown in Figure 58.15 where voltages V_1 and V_2 are applied to its two input terminals and the difference between these voltages is amplified.

(i) Let V_1 volts be applied to terminal 1 and 0 V be applied to terminal 2. The difference in the potentials at the inverting $(-)$ and non-inverting $(+)$ op amp inputs is practically zero and hence the inverting terminal must be at zero potential. Then $I_1 = \dfrac{V_1}{R_1}$. Since the op amp input resistance is high, this current flows through the feedback resistor R_f. The volt drop across R_f, which is the output voltage $V_o = \dfrac{V_1}{R_1} R_f$; hence, the closed loop voltage gain A is given by:

$$\boxed{A = \frac{V_o}{V_1} = -\frac{R_f}{R_1}} \qquad (10)$$

(ii) By similar reasoning, if V_2 is applied to terminal 2 and 0 V to terminal 1, then the voltage appearing at the non-inverting terminal will be $\left(\dfrac{R_3}{R_2 + R_3}\right) V_2$ volts. This voltage will also appear at the inverting $(-)$ terminal and thus the voltage across R_1 is equal to $-\left(\dfrac{R_3}{R_2 + R_3}\right) V_2$ volts.

Now the output voltage,

$$V_o = \left(\frac{R_3}{R_2 + R_3}\right) V_2 + \left[-\left(\frac{R_3}{R_2 + R_3}\right) V_2\right]\left(-\frac{R_f}{R_1}\right)$$

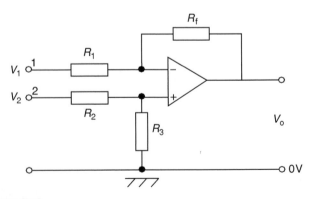

Figure 58.15

and the voltage gain,

$$A = \frac{V_o}{V_2} = \left(\frac{R_3}{R_2 + R_3}\right) + \left[-\left(\frac{R_3}{R_2 + R_3}\right)\right]\left(-\frac{R_f}{R_1}\right)$$

i.e. $\boxed{A = \frac{V_o}{V_2} = \left(\frac{R_3}{R_2 + R_3}\right)\left(1 + \frac{R_f}{R_1}\right)}$ (11)

(iii) Finally, if the voltages applied to terminals 1 and 2 are V_1 and V_2 respectively, then the difference between the two voltages will be amplified.

If $V_1 > V_2$, then: $V_o = (V_1 - V_2)\left(-\frac{R_f}{R_1}\right)$ (12)

If $V_2 > V_1$, then: $V_o = (V_2 - V_1)\left(\frac{R_3}{R_2 + R_3}\right)\left(1 + \frac{R_f}{R_1}\right)$

(13)

For example, for the differential amplifier shown in Figure 58.15, if $R_1 = 10$ kΩ, $R_2 = 10$ kΩ, $R_3 = 100$ kΩ and $R_f = 100$ kΩ, then the output voltage V_o if $V_1 = 5$ mV and $V_2 = 0$ is:

$$V_o = -\frac{R_f}{R_1}V_1 = -\left(\frac{100 \times 10^3}{10 \times 10^3}\right)(5)\text{ mV} = \mathbf{-50\ mV}$$

and if $V_1 = 25$ mV and $V_2 = 50$ mV then:

$$V_o = (V_2 - V_1)\left(\frac{R_3}{R_2 + R_3}\right)\left(1 + \frac{R_f}{R_1}\right)$$

$$= (50 - 25)\left(\frac{100}{100 + 10}\right)\left(1 + \frac{100}{10}\right)\text{mV} = \mathbf{+250\ mV}$$

Digital to Analogue (D/A) Conversion

There are a number of situations when digital signals have to be converted to analogue ones. For example, a digital computer often needs to produce a graphical display on the screen; this involves using a D/A converter to change the two-level digital output voltage from the computer, into a continuously varying analogue voltage for the input to the cathode ray tube, so that it can deflect the electron beam to produce screen graphics.

A binary weighted resistor D/A converter is shown in Figure 58.16 for a four-bit input. The values of the resistors, R, $2R$, $4R$, $8R$ increase according to the binary scale — hence the name of the converter. The circuit uses an op amp as a **summing amplifier** with a feedback resistor R_f. Digitally controlled electronic switches are shown as S_1 to S_4. Each switch connects the resistor in series with it to a fixed reference voltage V_{REF} when the input bit controlling it is a 1 and to ground (0 V) when it is a 0. The input voltages V_1 to V_4 applied to the op amp by the four-bit input via the resistors therefore have one of two values, i.e. either V_{REF} or 0 V.

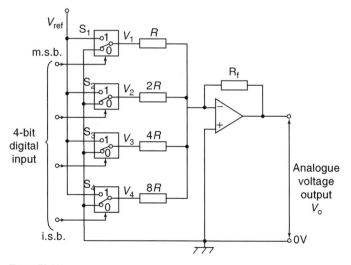

Figure 58.16

From equation (8), the analogue output voltage V_o is given by:

$$V_o = -\left(\frac{R_f}{R}V_1 + \frac{R_f}{2R}V_2 + \frac{R_f}{4R}V_3 + \frac{R_f}{8R}V_4\right)$$

Let $R_f = R = 1\text{ k}\Omega$, then: $V_o = -\left(V_1 + \frac{1}{2}V_2 + \frac{1}{4}V_3 + \frac{1}{8}V_4\right)$

With a four-bit input of 0001 (i.e. decimal 1), S_4 connects $8R$ to V_{REF}, i.e. $V_4 = V_{REF}$, and S_1, S_2 and S_3 connect R, $2R$ and $4R$ to 0 V, making $V_1 = V_2 = V_3 = 0$. Let $V_{REF} = -8$ V, then

output voltage, $\boldsymbol{V_o} = -(0 + 0 + 0 + \frac{1}{8}(-8)) = \boldsymbol{+1}$ **V**

With a four-bit input of 0101 (i.e. decimal 5), S_2 and S_4 connects $2R$ and $4R$ to V_{REF}, i.e. $V_2 = V_4 = V_{REF}$, and S_1 and S_3 connect R and $4R$ to 0 V, making $V_1 = V_3 = 0$. Again, if $V_{REF} = -8$ V, then

output voltage, $\boldsymbol{V_o} = -(0 + \frac{1}{2}(-8) + 0 + \frac{1}{8}(-8)) = \boldsymbol{+5}$ **V**

If the input is 0111 (i.e. decimal 7), the output voltage will be 7 V, and so on. From these examples, it is seen that the analogue output voltage, V_o, is directly proportional to the digital input.

V_o has a 'stepped' waveform, the waveform shape depending on the binary input. A typical waveform is shown in Figure 58.17.

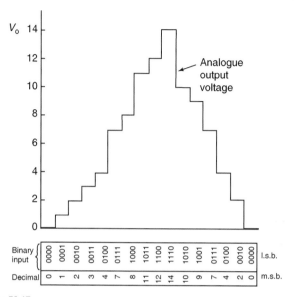

Figure 58.17

Analogue to Digital (A/D) Conversion

In a digital voltmeter, its input is in analogue form and the reading is displayed digitally. This is an example where an analogue to digital converter is needed.

A block diagram for a four-bit counter type A/D conversion circuit is shown in Figure 58.18. An op amp is again used, in this case as a **voltage comparator**. The analogue input voltage V_2, shown in Figure 58.19(a) as a steady d.c. voltage, is applied to the non-inverting $(+)$ input, whilst a sawtooth voltage V_1 supplies the inverting $(-)$ input.

The output from the comparator is applied to one input of an AND gate and is a 1 (i.e. 'high') until V_1 equals or exceeds V_2, when it then goes

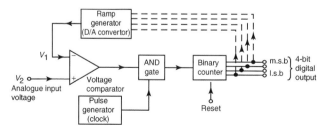

Figure 58.18

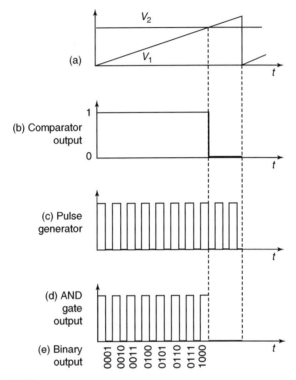

Figure 58.19

to 0 (i.e. 'low') as shown in Figure 58.19(b). The other input of the AND gate is fed by a steady train of pulses from a pulse generator, as shown in Figure 58.19(c). When both inputs to the AND gate are 'high', the gate 'opens' and gives a 'high' output, i.e. a pulse, as shown in Figure 58.19(d). The time taken by V_1 to reach V_2 is proportional to the analogue voltage if the ramp is linear. The output pulses from the AND gate are recorded by a binary counter and, as shown in Figure 58.19(e), are the digital equivalent of the analogue input voltage V_2. In practise, the ramp generator is a D/A converter that takes its digital input from the binary counter, shown by the broken lines in Figure 53.18. As the counter advances through its normal binary sequence, a staircase waveform with equal steps (i.e. a ramp) is built up at the output of the D/A converter (as shown by the first few steps in Figure 58.17.)

59 Three-phase Systems

Introduction

Generation, transmission and distribution of electricity via the National Grid system are accomplished by three-phase alternating currents.

The voltage induced by a single coil when rotated in a uniform magnetic field is shown in Figure 59.1 and is known as a **single-phase voltage**. Most consumers are fed by means of a single-phase a.c. supply. Two wires are used, one called the live conductor (usually coloured red) and the other is called the neutral conductor (usually coloured black). The neutral is usually connected via protective gear to earth, the earth wire being coloured green. The standard voltage for a single-phase a.c. supply is 240 V. The majority of single-phase supplies are obtained by connection to a three-phase supply (see Figure 59.6).

Three-phase Supply

A three-phase supply is generated when three coils are placed 120° apart and the whole rotated in a uniform magnetic field as shown in Figure 59.2(a). The result is three independent supplies of equal voltages which are each displaced by 120° from each other as shown in Figure 59.2(b).

(i) The convention adopted to identify each of the phase voltages is: R-red, Y-yellow, and B-blue, as shown in Figure 59.2.
(ii) The **phase-sequence** is given by the sequence in which the conductors pass the point initially taken by the red conductor. The national standard phase sequence is R, Y, B.

A three-phase a.c. supply is carried by three conductors, called **'lines'**, which are coloured red, yellow and blue. The currents in these conductors are

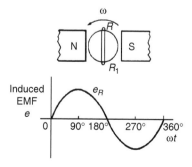

Figure 59.1

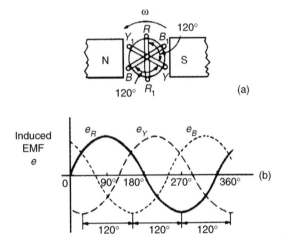

Figure 59.2

known as line currents (I_L) and the p.d.'s between them are known as line voltages (V_L). A fourth conductor, called the **neutral** (coloured black, and connected through protective devices to earth) is often used with a three-phase supply.

If the three-phase windings shown in Figure 59.2 are kept independent then six wires are needed to connect a supply source (such as a generator) to a load (such as motor). To reduce the number of wires it is usual to interconnect the three phases. There are two ways in which this can be done, these being: (a) a **star connection,** and (b) a **delta,** or **mesh, connection.** Sources of three-phase supplies, i.e. alternators, are usually connected in star, whereas three-phase transformer windings, motors and other loads may be connected either in star or delta.

Star Connection

  (i) A **star-connected load** is shown in Figure 59.3 where the three line conductors are each connected to a load and the outlets from the loads are joined together at N to form what is termed the **neutral point** or the **star point**.

 (ii) The voltages, V_R, V_Y and V_B are called **phase voltages** or line to neutral voltages. Phase voltages are generally denoted by V_p.

(iii) The voltages, V_{RY}, V_{YB} and V_{BR} are called **line voltages**.

(iv) From Figure 59.3 it can be seen that the phase currents (generally denoted by I_p) are equal to their respective line currents I_R, I_Y and I_B, i.e. for a star connection: $\boxed{I_L = I_p}$

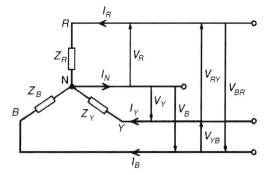

Figure 59.3

(v) For a balanced system:

$$I_R = I_Y = I_B, \qquad V_R = V_Y = V_B$$
$$V_{RY} = V_{YB} = V_{BR}, \quad Z_R = Z_Y = Z_B$$

and the current in the neutral conductor, $I_N = 0$

When a star-connected system is balanced, then the neutral conductor is unnecessary and is often omitted.

(vi) The line voltage, V_{RY}, shown in Figure 59.4(a) is given by $V_{RY} = V_R - V_Y$ (V_Y is negative since it is in the opposite direction to V_{RY}. In the phasor diagram of Figure 59.4(b), phasor V_Y is reversed (shown by the broken line) and then added phasorially to V_R (i.e. $V_{RY} = V_R + (-V_Y)$). By trigonometry, or by measurement, $V_{RY} = \sqrt{3}\,V_R$, i.e. for a balanced star connection:

$$\boxed{V_L = \sqrt{3}\,V_p}$$

A phasor diagram for a balanced, three-wire, star-connected, 3-phase load having a phase voltage of 240 V, a line current of 5 A and a lagging power factor of 0.966 is shown in Figure 59.5.

To construct the phasor diagram:

(a) Draw $V_R = V_Y = V_B = 240$ V and spaced 120° apart. (Note that V_R is shown vertically upwards — this however is immaterial for it may be drawn in any direction).

(b) Power factor $= \cos\phi = 0.966$ lagging. Hence the load phase angle is given by $\cos^{-1} 0.966$, i.e. 15° lagging. Hence $I_R = I_Y = I_B = 5$ A, lagging V_R, V_Y and V_B respectively by 15°.

(c) $V_{RY} = V_R - V_Y$ (phasorially). Hence V_Y is reversed and added phasorially to V_R. By measurement, $V_{RY} = 415$ V (i.e. $\sqrt{3} \times 240$) and leads V_R by 30°. Similarly, $V_{YB} = V_Y - V_B$ and $V_{BR} = V_B - V_R$.

(vii) The star connection of the three phases of a supply, together with a neutral conductor, allows the use of two voltages — the phase voltage and the line voltage. A 4-wire system is also used when the load is not balanced. The standard electricity supply to consumers in Great Britain is **415/240 V, 50 Hz, 3-phase, 4-wire alternating voltage**, and a diagram of connections is shown in Figure 59.6.

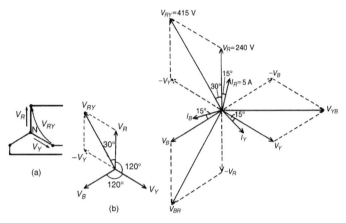

Figure 59.4

Figure 59.5

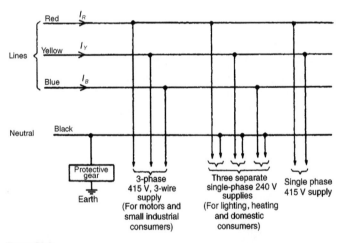

Figure 59.6

Delta connection

(i) A **delta (or mesh) connected load** is shown in Figure 59.7 where the end of one load is connected to the start of the next load.

(ii) From Figure 59.7, it can be seen that the line voltages V_{RY}, V_{YB} and V_{BR} are the respective phase voltages, i.e. for a delta connection:

$$\boxed{V_L = V_p}$$

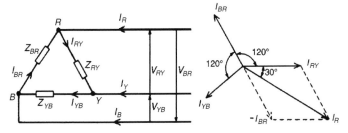

Figure 59.7   **Figure 59.8**

(iii) Using Kirchhoff's current law in Figure 59.7, $I_R = I_{RY} - I_{BR} = I_{RY} + (-I_{BR})$.

From the phasor diagram shown in Figure 59.8, by trigonometry or by measurement, $I_R = \sqrt{3}\,I_{RY}$, i.e. for a delta connection:

$$I_L = \sqrt{3}\,I_p$$

For example, three coils each having resistance 3 Ω and inductive reactance 4 Ω are connected (i) in star and (ii) in delta to a 415 V, 3-phase supply.

For a star connection: $I_L = I_p$ and $V_L = \sqrt{3}\,V_p$

A 415 V, 3-phase supply means that the line voltage, $V_L = \mathbf{415\ V}$

Phase voltage, $V_p = \dfrac{V_L}{\sqrt{3}} = \dfrac{415}{\sqrt{3}} = \mathbf{240\ V}$

Impedance per phase, $Z_p = \sqrt{R^2 + X_L^2} = \sqrt{3^2 + 4^2} = 5\ \Omega$

Phase current, $I_p = \dfrac{V_p}{Z_p} = \dfrac{240}{5} = \mathbf{48\ A}$

and line current, $I_L = I_p = \mathbf{48\ A}$

For a delta connection: $V_L = V_p$ and $I_L = \sqrt{3}\,I_p$

Line voltage, $V_L = 415\ V$ and phase voltage,

                                   $V_p = V_L = \mathbf{415\ V}$

Phase current, $I_p = \dfrac{V_p}{Z_p} = \dfrac{415}{5} = \mathbf{83\ A}$

Line current, $I_L = \sqrt{3}\,I_p = \sqrt{3}(83) = \mathbf{144\ A}$

Thus when the load is connected in delta, three times the line current is taken from the supply than is taken if connected in star.

Power in Three-phase Systems

For either a star or a delta balanced connection, the total power P is given by:

$$\boxed{P = \sqrt{3}\, V_L I_L \cos\phi \text{ watts} \quad \text{or} \quad P = 3I_p^2 R_p \text{ watts}}$$

Total volt-amperes, $\boxed{S = \sqrt{3}\, V_L I_L \text{ volt-amperes}}$

For example, three identical coils, each of resistance 10 Ω and inductance 42 mH are connected (a) in star and (b) in delta to a 415 V, 50 Hz, 3-phase supply.

Star connection

Inductive reactance, $\quad X_L = 2\pi f L = 2\pi(50)(42 \times 10^{-3})$

$$= 13.19 \ \Omega$$

Phase impedance, $\quad Z_p = \sqrt{R^2 + X_L^2} = \sqrt{10^2 + 13.19^2}$

$$= 16.55 \ \Omega$$

Line voltage, $\quad V_L = 415$ V

and phase voltage, $\quad V_P = \dfrac{V_L}{\sqrt{3}} = \dfrac{415}{\sqrt{3}} = 240$ V

Phase current, $\quad I_p = \dfrac{V_p}{Z_p} = \dfrac{240}{16.55} = 14.50$ A

Line current, $\quad I_L = I_p = 14.50$ A

Power factor $= \cos\phi = \dfrac{R_p}{Z_p} = \dfrac{10}{16.55} = 0.6042$ lagging

Power dissipated, $\quad \mathbf{P} = \sqrt{3}\, V_L I_L \cos\phi$

$$= \sqrt{3}(415)(14.50)(0.6042) = \textbf{6.3 kW}$$

(Alternatively, $\quad \mathbf{P} = 3I_p^2 R_p = 3(14.50)^2(10) = \textbf{6.3 kW}$)

Delta connection

$$V_L = V_p = 415 \text{ V}, \quad Z_p = 16.55 \ \Omega,$$

$$\cos\phi = 0.6042 \text{ lagging (from above)}$$

Phase current, $\quad I_p = \dfrac{V_p}{Z_p} = \dfrac{415}{16.55} = 25.08$ A

Line current, $\quad I_L = \sqrt{3}\, I_p = \sqrt{3}(25.08) = 43.44$ A

Power dissipated, $\quad P = \sqrt{3}\, V_L I_L \cos\phi = \sqrt{3}(415)(43.44)(0.6042)$

$$= \textbf{18.87 kW}$$

(Alternatively, $\qquad P = 3\, I_p^2 R_p = 3(25.08)^2 (10) = \textbf{18.87 kW}$)

Hence loads connected in delta dissipate three times the power than when connected in star, and also take a line current three times greater.

Measurement of Power in Three-phase Systems

Power in three-phase loads may be measured by the following methods:

(i) **One-wattmeter method for a balanced load**
Wattmeter connections for both star and delta are shown in Figure 59.9.

> **Total power = 3 × wattmeter reading**

(ii) **Two-wattmeter method for balanced or unbalanced loads**
A connection diagram for this method is shown in Figure 59.10 for a star-connected load. Similar connections are made for a delta-connected load.

> **Total power = sum of wattmeter readings = $P_1 + P_2$**

The power factor may be determined from:

$$\tan\phi = \sqrt{3}\left(\frac{P_1 - P_2}{P_1 + P_2}\right)$$

It is possible, depending on the load power factor, for one wattmeter to have to be 'reversed' to obtain a reading. In this case it is taken as a negative reading.
For example, two wattmeters indicate 10 kW and 3 kW respectively when connected to measure the input power to a 3-phase balanced

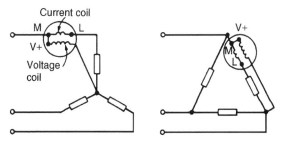

Figure 59.9

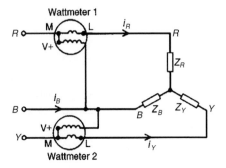

Figure 59.10

load, the reverse switch being operated on the meter indicating the 3 kW reading.

The 3 kW reading is thus taken as −3 kW.

Total input power, $P = P_1 + P_2 = 10 + (-3) = \textbf{7 kW}$

$$\tan\phi = \sqrt{3}\left(\frac{P_1 - P_2}{P_1 + P_2}\right) = \sqrt{3}\left(\frac{10 - (-3)}{10 + (-3)}\right)$$

$$= \sqrt{3}\left(\frac{13}{7}\right) = 3.2167$$

Angle $\phi = \tan^{-1} 3.2167 = 72.73°$

hence, power factor $= \cos\phi = \cos 72.73° = \textbf{0.297}$

(iii) **Three-wattmeter method for a three-phase, 4-wire system for balanced and unbalanced loads** (see Figure 59.11).

$$\boxed{\textbf{Total power} = P_1 + P_2 + P_3}$$

Comparison of Star and Delta Connections

(i) Loads connected in delta dissipate three times more power than when connected in star to the same supply.

(ii) For the same power, the phase currents must be the same for both delta and star connections (since power $= 3I_p^2 R_p$), hence the line current in the delta-connected system is greater than the line current in the corresponding star-connected system. To achieve the same phase current in a star-connected system as in a delta-connected system, the line voltage in the star system is $\sqrt{3}$ times the line voltage in the delta system.

Thus for a given power transfer, a delta system is associated with larger line currents (and thus larger conductor cross-sectional area) and a star system is associated with a larger line voltage (and thus greater insulation).

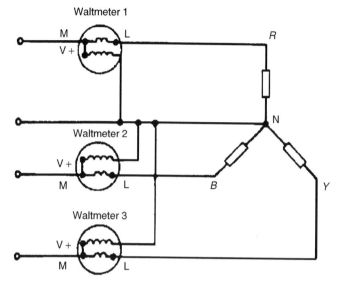

Figure 59.11

Advantages of Three-phase Systems

Advantages of three-phase systems over single-phase supplies include:

(i) For a given amount of power transmitted through a system, the three-phase system requires conductors with a smaller cross-sectional area. This means a saving of copper (or aluminium) and thus the original installation costs are less.
(ii) Two voltages are available.
(iii) Three-phase motors are very robust, relatively cheap, generally smaller, have self-starting properties, provide a steadier output and require little maintenance compared with single-phase motors.

60 Transformers

Introduction

A transformer is a device that uses the phenomenon of mutual induction (see chapters 48 and 49) to change the values of alternating voltages and currents. In fact, one of the main advantages of a.c. transmission and distribution is the ease with which an alternating voltage can be increased or decreased by transformers.

Losses in transformers are generally low and thus efficiency is high. Being static they have a long life and are very stable.

Transformers range in size from the miniature units used in electronic applications to the large power transformers used in power stations; the principle of operation is the same for each.

A transformer is represented in Figure 60.1(a) as consisting of two electrical circuits linked by a common ferromagnetic core. One coil is termed the **primary winding** which is connected to the supply of electricity, and the other the **secondary winding**, which may be connected to a load. A circuit diagram symbol for a transformer is shown in Figure 60.1(b).

Transformer Principle of Operation

When the secondary is an open-circuit and an alternating voltage V_1 is applied to the primary winding, a small current — called the no-load current I_0 — flows, which sets up a magnetic flux in the core. This alternating flux links with both primary and secondary coils and induces in them e.m.f.'s of E_1 and E_2 respectively by mutual induction.

The induced e.m.f. E in a coil of N turns is given by $E = -N\dfrac{d\Phi}{dt}$ volts, where $\dfrac{d\Phi}{dt}$ is the rate of change of flux. In an ideal transformer, the rate of change of flux is the same for both primary and secondary and thus $\dfrac{E_1}{N_1} = \dfrac{E_2}{N_2}$ i.e. **the induced e.m.f. per turn is constant**.

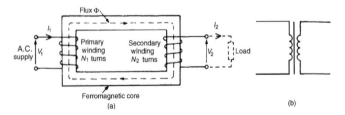

Figure 60.1

Assuming no losses, $E_1 = V_1$ and $E_2 = V_2$

Hence

$$\frac{V_1}{N_1} = \frac{V_2}{N_2} \quad \text{or} \quad \frac{V_1}{V_2} = \frac{N_1}{N_2} \tag{1}$$

$\dfrac{V_1}{V_2}$ is called the voltage ratio and $\dfrac{N_1}{N_2}$ the turns ratio, or the '**transformation ratio**' of the transformer. If N_2 is less than N_1 then V_2 is less than V_1 and the device is termed a **step-down transformer**. If N_2 is greater then N_1 then V_2 is greater than V_1 and the device is termed a **step-up transformer**.

When a load is connected across the secondary winding, a current I_2 flows. In an ideal transformer losses are neglected and a transformer is considered to be 100% efficient.

Hence input power = output power, or $V_1 I_1 = V_2 I_2$ i.e. in an ideal transformer, the **primary and secondary ampere-turns are equal**

Thus

$$\frac{V_1}{V_2} = \frac{I_2}{I_1} \tag{2}$$

Combining equations (1) and (2) gives:

$$\boxed{\frac{V_1}{V_2} = \frac{N_1}{N_2} = \frac{I_2}{I_1}} \tag{3}$$

For example, an ideal transformer has a turns ratio of $8:1$ and the primary current is 3 A when it is supplied at 240 V.

A turns ratio of $8:1$ means $\dfrac{N_1}{N_2} = \dfrac{8}{1}$ i.e. a step-down transformer.

$$\frac{N_1}{N_2} = \frac{V_1}{V_2}$$

and secondary voltage $\quad V_2 = V_1 \left(\dfrac{N_2}{N_1}\right) = 240 \left(\dfrac{1}{8}\right) = \mathbf{30\ volts}$

Also, $\quad \dfrac{N_1}{N_2} = \dfrac{I_2}{I_1}$

hence secondary current $\quad I_2 = I_1 \left(\dfrac{N_1}{N_2}\right) = 3 \left(\dfrac{8}{1}\right) = \mathbf{24\ A}$

The **rating** of a transformer is stated in terms of the volt-amperes that it can transform without overheating. With reference to Figure 60.1(a), the transformer rating is either $V_1 I_1$ or $V_2 I_2$, where I_2 is the full-load secondary current. In the above example, the rating is $(240)(3) = 720$ VA.

Transformer No-load Phasor Diagram

The core flux is common to both primary and secondary windings in a transformer and is thus taken as the reference phasor in a phasor diagram. On no-load the primary winding takes a small no-load current I_0 and since, with

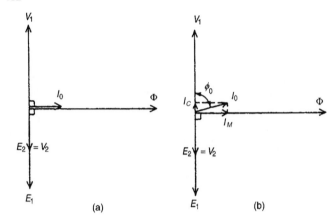

Figure 60.2

losses neglected, the primary winding is a pure inductor, this current lags the applied voltage V_1 by 90°. In the phasor diagram assuming no losses, shown in Figure 60.2(a), current I_0 produces the flux and is drawn in phase with the flux. The primary induced e.m.f. E_1 is in phase opposition to V_1 (by Lenz's law) and is shown 180° out of phase with V_1 and equal in magnitude. The secondary induced e.m.f. is shown for a 2 : 1 turns ratio transformer.

A no-load phasor diagram for a practical transformer is shown in Figure 60.2(b). If current flows then losses will occur. When losses are considered then the no load current I_0 is the phasor sum of two components — (i) I_M, **the magnetising component**, in phase with the flux, and (ii) I_c, **the core loss component** (supplying the hysteresis and eddy current losses). From Figure 60.2(b):

No-load current, $I_0 = \sqrt{I_M^2 + I_C^2}$ where $I_M = I_0\sin\phi_0$ and $I_C = I_0\cos\phi_0$

Power factor on no-load $= \cos\phi_0 = \dfrac{I_C}{I_0}$

The total core losses (i.e. iron losses) $= V_1 I_0 \cos\phi_0$

E.m.f. Equation of a Transformer

The r.m.s. value of e.m.f. induced in the primary,

$$\boxed{E_1 = 4.44 \, f \, \Phi_m N_1 \text{ volts}} \tag{4}$$

and the r.m.s. value of e.m.f. induced in the secondary,

$$\boxed{E_2 = 4.44 \, f \, \Phi_m N_2 \text{ volts}} \tag{5}$$

Dividing equation (4) by equation (5) gives:

$$\frac{E_1}{E_2} = \frac{N_1}{N_2} \quad \text{as previously obtained.}$$

For example, if a 100 kVA, 4000 V/200 V, 50 Hz single-phase transformer has 100 secondary turns, then from equation (5), $E_2 = 4.44 f \, \Phi_m N_2$ from which, maximum flux,

$$\Phi_m = \frac{E_2}{4.44 \, fN_2} = \frac{200}{(4.44)(50)(100)} \quad \text{(assuming } E_2 = V_2\text{)}$$

$$= \mathbf{9.01 \times 10^{-3} \ Wb \ or \ 9.01 \ mWb}$$

Transformer on-load Phasor Diagram

If the voltage drop in the windings of a transformer are assumed negligible, then the terminal voltage V_2 is the same as the induced e.m.f. E_2 in the secondary. Similarly, $V_1 = E_1$. Assuming an equal number of turns on primary and secondary windings, then $E_1 = E_2$, and let the load have a lagging phase angle ϕ_2.

In the phasor diagram of Figure 60.3, current I_2 lags V_2 by angle ϕ_2. When a load is connected across the secondary winding a current I_2 flows in the secondary winding. The resulting secondary e.m.f. acts so as to tend to reduce the core flux. However this does not happen since reduction of the core flux reduces E_1, hence a reflected increase in primary current I_1' occurs which provides a restoring m.m.f. Hence at all loads, primary and secondary m.m.f.'s are equal, but in opposition, and the core flux remains constant. I_1' is sometimes called the 'balancing' current and is equal, but in the opposite direction, to current I_2 as shown in Figure 60.3. I_0, shown at a phase angle

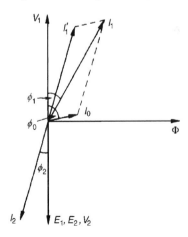

Figure 60.3

ϕ_0 to V_1, is the no-load current of the transformer. The phasor sum of I_1' and I_0 gives the supply current I_1 and the phase angle between V_1 and I_1 is shown as ϕ_1.

Transformer Construction

There are broadly two types of single-phase double-wound transformer constructions—the **core type** and the **shell type**, as shown in Figure 60.4. The low and high voltage windings are wound as shown to reduce leakage flux.

For **power transformers**, rated possibly at several MVA and operating at a frequency of 50 Hz in Great Britain, the core material used is usually laminated silicon steel or stalloy, the laminations reducing eddy currents and the silicon steel keeping hysteresis loss to a minimum.

Large power transformers are used in the main distribution system and in industrial supply circuits. Small power transformers have many applications, examples including welding and rectifier supplies, domestic bell circuits, imported washing machines, and so on.

For **audio frequency (a.f.) transformers**, rated from a few mVA to no more than 20 VA, and operating at frequencies up to about 15 kHz, the small core is also made of laminated silicon steel. A typical application of a.f. transformers is in an audio amplifier system.

Radio frequency (r.f.) transformers, operating in the MHz frequency region have either an air core, a ferrite core or a dust core. Ferrite is a ceramic material having magnetic properties similar to silicon steel, but having a high resistivity. Dust cores consist of fine particles of carbonyl iron or permalloy (i.e. nickel and iron), each particle of which is insulated from its neighbour. Applications of r.f. transformers are found in radio and television receivers.

Transformer **windings** are usually of enamel-insulated copper or aluminium. **Cooling** is achieved by air in small transformers and oil in large transformers.

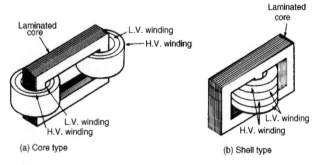

(a) Core type
(b) Shell type

Figure 60.4

Equivalent Circuit of a Transformer

Figure 60.5 shows an equivalent circuit of a transformer. R_1 and R_2 represent the resistances of the primary and secondary windings and X_1 and X_2 represent the reactances of the primary and secondary windings, due to leakage flux.

The core losses due to hysteresis and eddy currents are allowed for by resistance R which takes a current I_C, the core loss component of the primary current. Reactance X takes the magnetising component I_M.

In a simplified equivalent circuit shown in Figure 60.6, R and X are omitted since the no-load current I_0 is normally only about 3–5% of the full load primary current.

It is often convenient to assume that all of the resistance and reactance as being on one side of the transformer.

Resistance R_2 in Figure 60.6 can be replaced by inserting an additional resistance R_2' in the primary circuit such that the power absorbed in R_2' when carrying the primary current is equal to that in R_2 due to the secondary current, i.e. $I_1^2 R_2' = I_2^2 R_2$ from which,

$$R_2' = R_2 \left(\frac{I_2}{I_1} \right)^2 = R_2 \left(\frac{V_1}{V_2} \right)^2$$

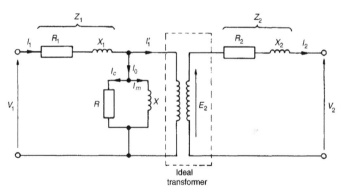

Figure 60.5

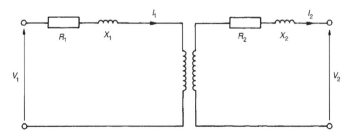

Figure 60.6

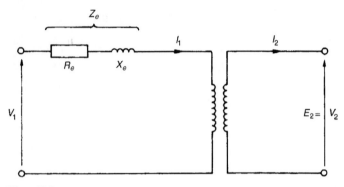

Figure 60.7

Then the total equivalent resistance in the primary circuit R_e is equal to the primary and secondary resistances of the actual transformer.

Hence $R_e = R_1 + R_2'$ i.e.

$$R_e = R_1 + R_2 \left(\frac{V_1}{V_2}\right)^2$$

By similar reasoning, the equivalent reactance in the primary circuit is given by $X_e = X_1 + X_2'$

i.e.

$$X_e = X_1 + X_2 \left(\frac{V_1}{V_2}\right)^2$$

The equivalent impedance Z_e of the primary and secondary windings referred to the primary is given by

$$Z_e = \sqrt{R_e^2 + X_e^2}$$

If ϕ_e is the phase angle between I_1 and the volt drop $I_1 Z_e$ then

$$\cos \phi_e = \frac{R_e}{Z_e}$$

The simplified equivalent circuit of a transformer is shown in Figure 60.7.

Regulation of a Transformer

When the secondary of a transformer is loaded, the secondary terminal voltage, V_2, falls. As the power factor decreases, this voltage drop increases. This is called the **regulation of the transformer** and it is usually expressed as a percentage of the secondary no-load voltage, E_2. For full-load conditions:

$$\boxed{\text{Regulation} = \left(\frac{E_2 - V_2}{E_2} \right) \times 100\%}$$

The fall in voltage, $(E_2 - V_2)$, is caused by the resistance and reactance of the windings.

Typical values of voltage regulation are about 3% in small transformers and about 1% in large transformers.

For example, if a 5 kVA, 200 V/400 V, single-phase transformer has a secondary terminal voltage of 387.6 volts when loaded, then

$$\text{regulation} = \frac{\begin{array}{c}\text{Noload secondary voltage}\\ - \text{ terminal voltage on load}\end{array}}{\text{no load secondary voltage}} \times 100\%$$

$$= \left(\frac{400 - 387.6}{400} \right) \times 100\% = \left(\frac{12.4}{400} \right) \times 100\% = \mathbf{3.1\%}$$

Transformer Losses and Efficiency

There are broadly two sources of **losses in transformers** on load, these being copper losses and iron losses.

(a) **Copper losses** are variable and result in a heating of the conductors, due to the fact that they possess resistance. If R_1 and R_2 are the primary and secondary winding resistances then the total copper loss is $I_1^2 R_1 + I_2^2 R_2$.

(b) **Iron losses** are constant for a given value of frequency and flux density and are of two types — hysteresis loss and eddy current loss.

   (i) **Hysteresis loss** is the heating of the core as a result of the internal molecular structure reversals which occur as the magnetic flux alternates. The loss is proportional to the area of the hysteresis loop and thus low loss nickel iron alloys are used for the core since their hysteresis loops have small areas. (See Chapters 45).

   (ii) **Eddy current loss** is the heating of the core due to e.m.f.'s being induced not only in the transformer windings but also in the core. These induced e.m.f.'s set up circulating currents, called eddy currents. Owing to the low resistance of the core, eddy currents can be quite considerable and can cause a large power loss and excessive heating of the core. Eddy current losses can be reduced by increasing the resistivity of the core material or, more usually, by laminating the core (i.e. splitting it into layers or leaves) when very thin layers of insulating material can be inserted between each pair of laminations. This increases the resistance of the eddy current path, and reduces the value of the eddy current.

Transformer efficiency, $\quad \eta = \dfrac{\text{output power}}{\text{input power}}$

$$= \frac{\text{input power} - \text{losses}}{\text{input power}}$$

$$\eta = 1 - \frac{\text{losses}}{\text{input power}}$$

i.e.

and is usually expressed as a percentage. It is not uncommon for power transformers to have efficiencies of between 95% and 98%.

Output power $= V_2 I_2 \cos \phi_2$

Total losses = copper loss + iron losses,

and   input power = output power + losses

For example, a 200 kVA rated transformer has a full-load copper loss of 1.5 kW and an iron loss of 1 kW. The efficiency at full load and 0.85 power factor is determined as follows:

Full-load output power $= VI \cos \phi = (200)(0.85) = 170$ kW

Total losses $= 1.5 + 1.0 = 2.5$ kW

Input power = output power + losses $= 170 + 2.5 = 172.5$ kW

Hence

$$\text{efficiency} = \left(1 - \frac{2.5}{172.5}\right) = 1 - 0.01449 = 0.9855 \text{ or } \mathbf{98.55\%}$$

The efficiency at half full load and 0.85 power factor is determined as follows:

Half full-load power output $= \frac{1}{2}(200)(0.85) = 85$ kW

Copper loss (or I^2R loss) is proportional to current squared. Hence the copper loss at half full-load is: $(\frac{1}{2})^2(1500) = 375$ W

Iron loss $= 1000$ W (constant)

Total losses $= 375 + 1000 = 1375$ W or 1.375 kW

$$\left.\begin{array}{l}\text{Input power at} \\ \text{half full-load}\end{array}\right\} = \text{output power at half full-load} + \text{losses}$$

$$= 85 + 1.375 = 86.375 \text{ kW}$$

Hence efficiency $= 1 - \dfrac{\text{losses}}{\text{input power}} = \left(1 - \dfrac{1.375}{86.375}\right) = 0.9841$ or $\mathbf{98.41\%}$

Maximum Efficiency

It may be shown that the efficiency of a transformer is a maximum when the variable copper loss (i.e. $I_1^2 R_1 + I_2^2 R_2$) is equal to the constant iron losses.

Resistance Matching

Varying a load resistance to be equal, or almost equal, to the source internal resistance is called **matching**. Examples where resistance matching is important include coupling an aerial to a transmitter or receiver, or in coupling a

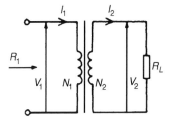

Figure 60.8

loudspeaker to an amplifier, where coupling transformers may be used to give maximum power transfer.

With d.c. generators or secondary cells, the internal resistance is usually very small. In such cases, if an attempt is made to make the load resistance as small as the source internal resistance, overloading of the source results.

A method of achieving maximum power transfer between a source and a load (see chapter 53), is to adjust the value of the load resistance to 'match' the source internal resistance. A transformer may be used as a **resistance matching device** by connecting it between the load and the source. With reference to Figure 60.8:

$$R_1 = \left(\frac{N_1}{N_2}\right)^2 R_L$$

Hence by varying the value of the turns ratio, the equivalent input resistance of a transformer can be 'matched' to the internal resistance of a load to achieve maximum power transfer.

For example, if a transformer having a turns ratio of 4 : 1 supplies a load of resistance 100 Ω, then the equivalent input resistance,

$$R_1 = \left(\frac{N_1}{N_2}\right)^2 R_L = \left(\frac{4}{1}\right)^2 (100) = \mathbf{1600}\ \Omega$$

Auto Transformers

An auto transformer is a transformer which has part of its winding common to the primary and secondary circuits. Figure 60.9(a) shows the circuit for a double-wound transformer and Figure 60.9(b) that for an auto transformer. The latter shows that the secondary is actually part of the primary, the current in the secondary being $(I_2 - I_1)$. Since the current is less in this section, the cross-sectional area of the winding can be reduced, which reduces the amount of material necessary.

Figure 60.10 shows the circuit diagram symbol for an auto transformer.

For example, a single-phase auto transformer has a voltage ratio 320 V : 250 V and supplies a load of 20 kVA at 250 V. Assuming an ideal transformer, the current in each section of the winding is determined as

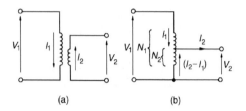

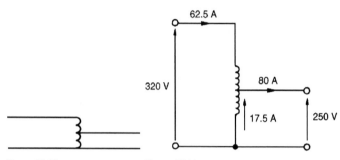

Figure 60.9

Figure 60.10   **Figure 60.11**

follows:

$$\text{Rating} = 20 \text{ kVA} = V_1 I_1 = V_2 I_2$$

Hence primary current, $I_1 = \dfrac{20 \times 10^3}{V_1} = \dfrac{20 \times 10^3}{320} = \textbf{62.5 A}$

and secondary current, $I_2 = \dfrac{20 \times 10^3}{V_2} = \dfrac{20 \times 10^3}{250} = \textbf{80 A}$

Hence current in common part of the winding $= 80 - 62.5 = \textbf{17.5 A}$

The current flowing in each section of the transformer is shown in Figure 60.11.

Saving of copper in an auto transformer

For the same output and voltage ratio, the auto transformer requires less copper than an ordinary double-wound transformer.

If $\dfrac{N_2}{N_1} = x$ then

> **(volume of copper in an auto transformer)**
> $= \textbf{(1} - \textbf{x)(volume of copper in a double-wound transformer)}$

If, say, $x = \frac{4}{5}$ then

    (volume of copper in auto transformer)

$$= \left(1 - \frac{4}{5}\right) \text{(volume of copper in a double-wound transformer)}$$

$$= \frac{1}{5}\text{(volume in double-wound transformer)}$$

i.e. a saving of 80%

    Similarly, if $x = \frac{1}{4}$, the saving is 25%, and so on.

    The closer N_2 is to N_1, the greater the saving in copper.

Advantages of auto transformers

The advantages of auto transformers over double-wound transformers include:

1. a saving in cost since less copper is needed (see above)
2. less volume, hence less weight
3. a higher efficiency, resulting from lower I^2R losses
4. a continuously variable output voltage is achievable if a sliding contact is used
5. a smaller percentage voltage regulation.

Disadvantages of auto transformers

The primary and secondary windings are not electrically separate, hence if an open-circuit occurs in the secondary winding the full primary voltage appears across the secondary.

Uses of auto transformers

Auto transformers are used for reducing the voltage when starting induction motors (see chapter 62) and for interconnecting systems that are operating at approximately the same voltage.

Isolating Transformers

Transformers not only enable current or voltage to be transformed to some different magnitude but provide a means of isolating electrically one part of a circuit from another when there is no electrical connection between primary and secondary windings. An **isolating transformer** is a 1 : 1 ratio transformer with several important applications, including bathroom shaver-sockets, portable electric tools, model railways, and so on.

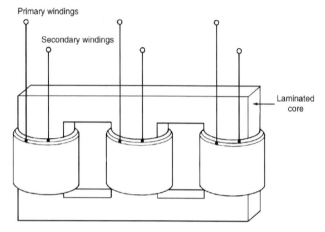

Figure 60.12

Three-phase Transformers

Three-phase double-wound transformers are mainly used in power transmission and are usually of the core type. They basically consist of three pairs of single-phase windings mounted on one core, as shown in Figure 60.12, which gives a considerable saving in the amount of iron used. The primary and secondary windings in Figure 60.12 are wound on top of each other in the form of concentric cylinders, similar to that shown in Figure 60.4(a). The windings may be with the primary delta-connected and the secondary star-connected, or star-delta, star-star or delta-delta, depending on its use.

A delta-connection is shown in Figure 60.13(a) and a star-connection in Figure 60.13(b).

Current Transformers

For measuring currents in excess of about 100 A, a current transformer is normally used. With a d.c. moving-coil ammeter the current required to give full scale deflection is very small — typically a few milliamperes. When larger currents are to be measured a shunt resistor is added to the circuit (see chapter 50). However, even with shunt resistors added it is not possible to measure very large currents. When a.c. is being measured a shunt cannot be used since the proportion of the current that flows in the meter will depend on its impedance, which varies with frequency.

In a double-wound transformer: $\dfrac{I_1}{I_2} = \dfrac{N_2}{N_1}$

from which, **secondary current $I_2 = I_1 \left(\dfrac{N_1}{N_2} \right)$**

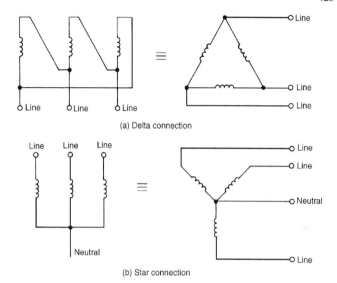

(a) Delta connection

(b) Star connection

Figure 60.13

In current transformers the primary usually consists of one or two turns whilst the secondary can have several hundred turns. A typical arrangement is shown in Figure 60.14.

For example, if the primary has 2 turns and the secondary 200 turns, then if the primary current is 500 A,

$$\text{secondary current}, I_2 = I_1 \left(\frac{N_1}{N_2} \right) = (500) \left(\frac{2}{200} \right) = 5 \text{ A}$$

Current transformers isolate the ammeter from the main circuit and allow the use of a standard range of ammeters giving full-scale deflections of 1 A, 2 A or 5 A. For very large currents the transformer core can be mounted around the conductor or bus-bar. Thus the primary then has just one turn.

It is very important to short-circuit the secondary winding before removing the ammeter. This is because if current is flowing in the primary, dangerously high voltages could be induced in the secondary should it be open-circuited.

Current transformer circuit diagram symbols are shown in Figure 60.15.

Voltage Transformers

For measuring voltages in excess of about 500 V it is often safer to use a voltage transformer. These are normal double-wound transformers with a large number of turns on the primary, which is connected to a high voltage

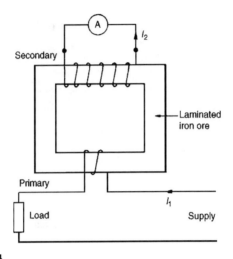

Figure 60.14

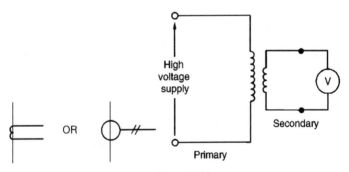

Figure 60.15   **Figure 60.16**

supply, and a small number of turns on the secondary. A typical arrangement is shown in Figure 60.16.

Since $\dfrac{V_1}{V_2} = \dfrac{N_1}{N_2}$ the **secondary voltage,** $V_2 = V_1 \left(\dfrac{N_2}{N_1} \right)$

Thus if the arrangement in Figure 60.16 has 4000 primary turns and 20 secondary turns then for a voltage of 22 kV on the primary, the voltage on the secondary,

$$V_2 = V_1 \left(\frac{N_2}{N_1} \right) = (22\,000) \left(\frac{20}{4000} \right) = \textbf{110 volts}$$

61 D.c. Machines

Introduction

When the input to an electrical machine is electrical energy, (seen as applying a voltage to the electrical terminals of the machine), and the output is mechanical energy, (seen as a rotating shaft), the machine is called an electric **motor**. Thus an electric motor converts electrical energy into mechanical energy. The principle of operation of a motor is explained in chapter 47, page 268. When the input to an electrical machine is mechanical energy, (seen as, say, a diesel motor, coupled to the machine by a shaft), and the output is electrical energy, (seen as a voltage appearing at the electrical terminals of the machine), the machine is called a **generator**. Thus, a generator converts mechanical energy to electrical energy.

The principle of operation of a generator is explained in chapter 48, page 272.

The Action of a Commutator

In an electric motor, conductors rotate in a uniform magnetic field. A single-loop conductor mounted between permanent magnets is shown in Figure 61.1. A voltage is applied at points A and B in Figure 61.1(a).

A force, F, acts on the loop due to the interaction of the magnetic field of the permanent magnets and the magnetic field created by the current flowing in the loop. This force is proportional to the flux density, B, the current flowing, I, and the effective length of the conductor, l, i.e. $F = BIl$. The force is made up of two parts, one acting vertically downwards due to the current flowing from C to D and the other acting vertically upwards due to the current flowing from E to F (from Fleming's left hand rule). If the loop is free to rotate, then when it has rotated through 180°, the conductors are as shown in Figure 61.1(b). For rotation to continue in the same direction, it is necessary for the current flow to be as shown in Figure 61.1(b), i.e. from D to C and from F to E. A process called **commutation** achieves this apparent reversal in the direction of current flow. With reference to Figure 61.2(a), when a direct voltage is applied at A and B, then as the single-loop conductor rotates, current flow will always be away from the commutator for the part of the conductor adjacent to the N-pole and towards the commutator for the part of the conductor adjacent to the S-pole. Thus the forces act to give continuous rotation in an anti-clockwise direction. The arrangement shown in Figure 61.2(a) is called a 'two-segment' commutator and the voltage is applied to the rotating segments by stationary **brushes**, (usually carbon blocks), which slide on the commutator material, (usually copper), when rotation takes place.

In practice, there are many conductors on the rotating part of a d.c. machine and these are attached to many commutator segments. A schematic diagram of a multi-segment commutator is shown in Figure 61.2(b).

Poor commutation results in sparking at the trailing edge of the brushes. This can be improved by using **interpoles** (situated between each pair of main

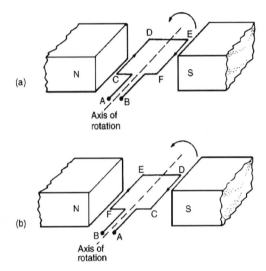

(a)

Axis of rotation

(b)

Axis of rotation

Figure 61.1

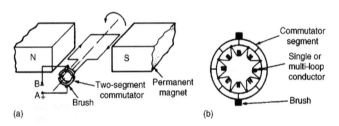

(a)

(b)

Commutator segment

Single or multi-loop conductor

Brush

Two-segment commutator

Permanent magnet

Brush

Figure 61.2

poles), high resistance brushes, or using brushes spanning several commutator segments.

D.c. Machine Construction

The basic parts of any d.c. machine are shown in Figure 61.3, and comprise:

(a) a stationary part called the **stator** having,
   (i) a steel ring called the **yoke**, to which are attached
   (ii) the magnetic **poles**, around which are the
   (iii) **field windings**, i.e. many turns of a conductor wound round the pole core; current passing through this conductor creates an electromagnet, (rather than the permanent magnets shown in Figures 61.1 and 61.2),

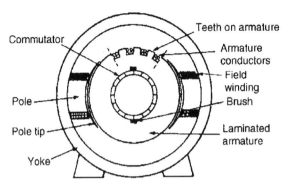

Figure 61.3

(b) a rotating part called the **armature** mounted in bearings housed in the
    stator and having,
   (iv) a laminated cylinder of iron or steel called the **core**, on which teeth
        are cut to house the
    (v) **armature winding**, i.e. a single or multi-loop conductor system, and
   (vi) the **commutator**.

Armature windings can be divided into two groups, depending on how the
wires are joined to the commutator. These are called **wave windings** and **lap
windings**.

(a) In **wave windings** there are two paths in parallel irrespective of the number
    of poles, each path supplying half the total current output.
    Wave wound generators produce high voltage, low current outputs.
(b) In **lap windings** there are as many paths in parallel as the machine has
    poles. The total current output divides equally between them. Lap wound
    generators produce high current, low voltage output.

Shunt, Series and Compound Windings

When the field winding of a d.c. machine is connected in parallel with the
armature, as shown in Figure 61.4(a), the machine is said to be **shunt** wound.
If the field winding is connected in series with the armature, as shown in
Figure 61.4(b), then the machine is said to be **series** wound. A **compound**
wound machine has a combination of series and shunt windings.

Depending on whether the electrical machine is series wound, shunt wound
or compound wound, it behaves differently when a load is applied. The
behaviour of a d.c. machine under various conditions is shown by means
of graphs, called characteristic curves or just **characteristics**. The characteristics shown in the following sections are theoretical, since they neglect the
effects of armature reaction.

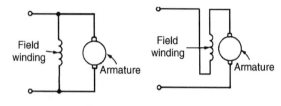

(a) Shunt-wound machine   (b) Series-wound machine

Figure 61.4

Armature reaction is the effect that the magnetic field produced by the armature current has on the magnetic field produced by the field system. In a generator, armature reaction results in a reduced output voltage, and in a motor, armature reaction results in increased speed.

A way of overcoming the effect of armature reaction is to fit compensating windings, located in slots in the pole face.

E.m.f. Generated in an Armature Winding

If Z = number of armature conductors, Φ = useful flux per pole, in webers,

     p = number of **pairs** of poles and n = armature speed in rev/s

     c = number of parallel paths through the winding between positive and negative brushes

then $\boxed{\textbf{generated e.m.f. } E = \dfrac{2p\,\Phi nZ}{c} \textbf{ volts}}$ (1)

where $c = 2$ **for a wave winding**

        $c = 2p$ **for a lap winding**

For example, an 8-pole, lap-wound armature has 1200 conductors and a flux per pole of 0.03 Wb. The e.m.f. generated when running at 500 rev/min is given by generated e.m.f.,

$$E = \frac{2p\Phi nZ}{c} = \frac{2p\Phi nZ}{2p} \text{ for a lap-wound machine,}$$

i.e. $E = \Phi nZ = (0.03)\left(\dfrac{500}{60}\right)(1200) = \textbf{300 volts}$

If the armature is wave-wound, generated e.m.f.

$$E = \frac{2p\Phi nZ}{c} = \frac{2p\Phi nZ}{2} \text{ (since } c = 2 \text{ for wave-wound)}$$

$$= p\Phi nZ = (4)(\Phi nZ) = (4)(300) = \textbf{1200 volts}$$

Since in equation (1), Z, p and c are constant for a given machine, then $E \propto \Phi n$.

However $2\pi n$ is the angular velocity ω in radians per second, hence the generated e.m.f. is proportional to Φ and ω,

i.e. $\boxed{\text{generated e.m.f. } E \propto \Phi \omega}$ (2)

D.c. Generators

D.c. generators are classified according to the method of their field excitation. These groupings are:

(i) **Separately-excited generators**, where the field winding is connected to a source of supply other than the armature of its own machine.
(ii) **Self-excited generators**, where the field winding receives its supply from the armature of its own machine, and which are sub-divided into (a) shunt, (b) series, and (c) compound wound generators.

Types of d.c. Generator and their Characteristics

(a) Separately-excited generator

A typical separately-excited generator circuit is shown in Figure 61.5.

When a load is connected across the armature terminals, a load current I_a will flow. The terminal voltage V will fall from its open-circuit e.m.f. E due to a volt drop caused by current flowing through the armature resistance, shown as R_a

i.e. $\boxed{\text{terminal voltage, } V = E - I_a R_a}$

or $\boxed{\text{generated e.m.f., } E = V + I_a R_a}$ (3)

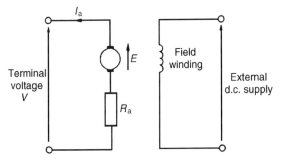

Figure 61.5

430

Characteristics

The two principal generator characteristics are the generated voltage/field current characteristics, called the **open-circuit characteristic** and the terminal voltage/load current characteristic, called the **load characteristic**.

A typical separately-excited generator **open-circuit characteristic** is shown in Figure 61.6(a) and a typical **load characteristic** is shown in Figure 61.6(b).

A separately-excited generator is used only in special cases, such as when a wide variation in terminal p.d. is required, or when exact control of the field current is necessary. Its disadvantage lies in requiring a separate source of direct current.

(b) Shunt wound generator

In a shunt wound generator the field winding is connected in parallel with the armature as shown in Figure 61.7. The field winding has a relatively high resistance and therefore the current carried is only a fraction of the armature current. For the circuit shown in Figure 61.7,

terminal voltage, $V = E - I_aR_a$

or   generated e.m.f., $E = V + I_aR_a$

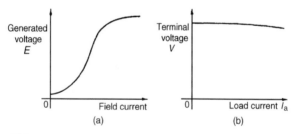

Figure 61.6

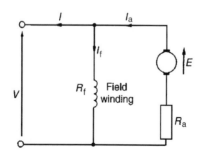

Figure 61.7

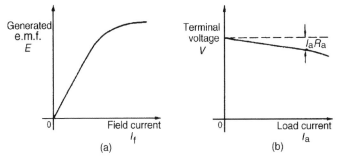

Figure 61.8

$I_a = I_f + I$ from Kirchhoff's current law, where I_a = armature current, I_f = field current $\left(= \dfrac{V}{R_f} \right)$ and I = load current.

Characteristics

The generated e.m.f., E, is proportional to $\Phi\omega$, hence at constant speed, since $\omega = 2\pi n$, $E \propto \Phi$. Also the flux Φ is proportional to field current I_f until magnetic saturation of the iron circuit of the generator occurs. Hence the open circuit characteristic is as shown in Figure 61.8(a)

As the load current on a generator having constant field current and running at constant speed increases, the value of armature current increases, hence the armature volt drop, $I_a R_a$ increases. The generated voltage E is larger than the terminal voltage V and the voltage equation for the armature circuit is $V = E - I_a R_a$. Since E is constant, V decreases with increasing load. The load characteristic is as shown in Figure 61.8(b). In practice, the fall in voltage is about 10% between no-load and full-load for many d.c. shunt-wound generators. The shunt-wound generator is the type most used in practice, but the load current must be limited to a value that is well below the maximum value. This then avoids excessive variation of the terminal voltage. Typical applications are with battery charging and motor car generators.

(c) Series-wound generator

In the series-wound generator the field winding is connected in series with the armature as shown in Figure 61.9.

Characteristic

The load characteristic is the terminal voltage/current characteristic. The generated e.m.f. E, is proportional to $\Phi\omega$ and at constant speed $\omega(=2\pi n)$ is a constant. Thus E is proportional to Φ. For values of current below magnetic saturation of the yoke, poles, air gaps and armature core, the flux Φ is proportional to the current, hence $E \propto I$. For values of current above those required

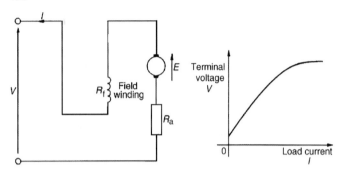

Figure 61.9   **Figure 61.10**

for magnetic saturation, the generated e.m.f. is approximately constant. The values of field resistance and armature resistance in a series wound machine are small, hence the terminal voltage V is very nearly equal to E. A typical load characteristic for a series generator is shown in Figure 61.10.

In a series-wound generator, the field winding is in series with the armature and it is not possible to have a value of field current when the terminals are open circuited, thus it is not possible to obtain an open-circuit characteristic. Series-wound generators are rarely used in practise, but can be used as a 'booster' on d.c. transmission lines.

(d) Compound-wound generator

In the compound-wound generator two methods of connection are used, both having a mixture of shunt and series windings, designed to combine the advantages of each. Figure 61.11(a) shows what is termed a **long-shunt** compound generator, and Figure 61.11(b) shows a **short-shunt** compound generator. The latter is the most generally used form of d.c. generator.

For example, a short-shunt compound generator supplies 80 A at 200 V. If the field resistance, $R_f = 40$ Ω, the series resistance, $R_{Se} = 0.02$ Ω and the armature resistance, $R_a = 0.04$ Ω, the e.m.f. generated is determined as follows:

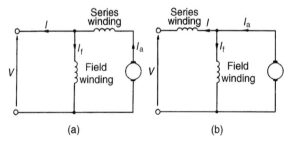

(a)   (b)

Figure 61.11

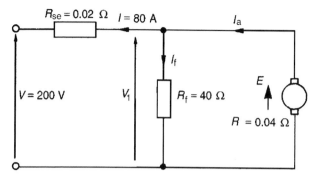

Figure 61.12

The circuit is shown in Figure 61.12.

Volt drop in series winding $= IR_{Se} = (80)(0.02) = 1.6$ V

P.d. across the field winding $=$ p.d. across armature

$$= V_1 = 200 + 1.6 = 201.6 \text{ V}$$

Field current $\qquad I_f = \dfrac{V_1}{R_f} = \dfrac{201.6}{40} = 5.04$ A

Armature current, $I_a = I + I_f = 80 + 5.04 = 85.04$ A

Hence, generated e.m.f., $E = V_1 + I_a R_a = 201.6 + (85.04)(0.04)$
$$= \textbf{205 volts}$$

Characteristics

In cumulative-compound machines the magnetic flux produced by the series and shunt fields are additive. Included in this group are **over-compounded**, **level-compounded** and **under-compounded machines** — the degree of compounding obtained depending on the number of turns of wire on the series winding.

A large number of series winding turns results in an over-compounded characteristic, as shown in Figure 61.13, in which the full-load terminal voltage exceeds the no-load voltage. A level-compound machine gives a full-load terminal voltage which is equal to the no-load voltage, as shown in Figure 61.13.

An under-compounded machine gives a full-load terminal voltage that is less than the no-load voltage, as shown in Figure 61.13. However even this latter characteristic is a little better than that for a shunt generator alone.

Compound-wound generators are used in electric arc welding, with lighting sets and with marine equipment.

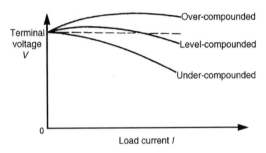

Figure 61.13

D.c. Machine Losses

As stated earlier, a generator is a machine for converting mechanical energy into electrical energy and a motor is a machine for converting electrical energy into mechanical energy. When such conversions take place, certain losses occur which are dissipated in the form of heat.

The principal **losses of machines** are:

(i) **Copper loss**, due to I^2R heat losses in the armature and field windings.
(ii) **Iron (or core) loss**, due to hysteresis and eddy-current losses in the armature. This loss can be reduced by constructing the armature of silicon steel laminations having a high resistivity and low hysteresis loss. At constant speed, the iron loss is assumed constant.
(iii) **Friction and windage losses**, due to bearing and brush contact friction and losses due to air resistance against moving parts (called windage). At constant speed, these losses are assumed to be constant.
(iv) **Brush contact loss** between the brushes and commutator. This loss is approximately proportional to the load current.

The total losses of a machine can be quite significant and operating efficiencies of between 80% and 90% are common.

Efficiency of a d.c. Generator

The efficiency of an electrical machine is the ratio of the output power to the input power and is usually expressed as a percentage. The Greek letter, 'η' (eta) is used to signify efficiency and since the units are $\dfrac{\text{power}}{\text{power}}$, then efficiency has no units. Thus

$$\text{efficiency, } \eta = \left(\frac{\text{output power}}{\text{input power}} \right) \times 100\%$$

If the total resistance of the armature circuit (including brush contact resistance) is R_a, then **the total loss in the armature circuit is $I_a^2 R_a$.**

If the terminal voltage is V and the current in the shunt circuit is I_f, then **the loss in the shunt circuit is $I_f V$.**

If the sum of the iron, friction and windage losses is C then **the total loss is given by: $I_a^2 R_a + I_f V + C$** ($I_a^2 R_a + I_f V$ is, in fact, the 'copper loss')

If the output current is I, then **the output power is VI**

Total input power $= VI + I_a^2 R_a + I_f V + C$

Hence,

$$\text{efficiency}, \eta = \frac{\textbf{output}}{\textbf{input}} = \left(\frac{VI}{VI + I_a^2 R + I_f V + C} \right) \times 100\% \qquad (4)$$

For example, a 10 kW shunt generator having an armature circuit resistance of 0.75 Ω and a field resistance of 125 Ω, generates a terminal voltage of 250 V at full load. The efficiency of the generator at full load, assuming the iron, friction and windage losses amount to 600 W, is determined as follows:

The circuit is shown in Figure 61.14.

Output power $= 10\,000\ W = VI$ from which,

load current $\qquad I = \dfrac{10\,000}{V} = \dfrac{10\,000}{250} = 40$ A

Field current, $\qquad I_f = \dfrac{V}{R_f} = \dfrac{250}{125} = 2$ A

Armature current, $\quad I_a = I_f + I = 2 + 40 = 42$ A

Efficiency, $\quad \eta = \left(\dfrac{VI}{VI + I_a^2 R + I_f V + C} \right) \times 100\%$

$\qquad\qquad = \left(\dfrac{10\,000}{10\,000 + (42)^2 (0.75) + (2)(250) + 600} \right) \times 100\%$

$\qquad\qquad = \textbf{80.50}\%$

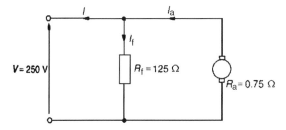

Figure 61.14

The **efficiency of a generator is a maximum** when the load is such that:

$$I_a^2 R_a = VI_f + C$$

i.e. when the variable loss = the constant loss

D.c. Motors

The construction of a d.c. motor is the same as a d.c. generator. The only difference is that in a generator the generated e.m.f. is greater than the terminal voltage, whereas in a motor the generated e.m.f. is less than the terminal voltage.

D.c. motors are often used in power stations to drive emergency stand-by pump systems which come into operation to protect essential equipment and plant should the normal a.c. supplies or pumps fail.

Back e.m.f.

When a d.c. motor rotates, an e.m.f. is induced in the armature conductors. By Lenz's law this induced e.m.f. E opposes the supply voltage V and is called a **back e.m.f.**, and the supply voltage, V is given by:

$$\boxed{V = E + I_a R_a} \quad \text{or} \quad \boxed{E = V - I_a R_a} \tag{5}$$

Torque of a d.c. Machine

From equation (5), for a d.c. motor, the supply voltage V is given by

$$V = E + I_a R_a$$

Multiplying each term by current I_a gives:

$$VI_a = EI_a + I_a^2 R_a$$

The term VI_a is the **total electrical power supplied to the armature**, the term $I_a^2 R_a$ is the **loss due to armature resistance** and the term EI_a is the **mechanical power developed by the armature**.

If T is the torque, in newton metres, then the mechanical power developed is given by $T\omega$ watts (see chapter 28)

Hence $$T\omega = 2\pi n T = EI_a$$

from which, $$\boxed{\text{torque } T = \frac{EI_a}{2\pi n} \text{ newton metres}} \tag{6}$$

From equation (1), the e.m.f. E generated is given by: $E = \dfrac{2p\Phi nZ}{c}$

Hence $$2\pi n T = EI_a = \left(\frac{2p\Phi nZ}{c}\right)I_a$$

Hence torque $\qquad T = \dfrac{\left(\dfrac{2p\Phi nZ}{c}\right)}{2\pi n} I_a$

i.e. $\qquad \boxed{T = \dfrac{p\Phi ZI_a}{\pi c} \text{ newton metres}}$ \hfill (7)

For a given machine, Z, c and p are fixed values

Hence torque, $\qquad \boxed{T \propto \Phi I_a}$ \hfill (8)

For example, an 8-pole d.c. motor has a wave-wound armature with 900 conductors. The useful flux per pole is 25 mWb. The torque exerted when a current of 30 A flows in each armature conductor is given by:

$$\text{torque, } T = \frac{p\Phi ZI_a}{\pi c} = \frac{(4)\left(25 \times 10^{-3}\right)(900)(30)}{\pi(2)} = \textbf{429.7 Nm}$$

$$\text{(since } c = 2 \text{ for a wave winding)}$$

Types of d.c. Motor and their Characteristics

(a) Shunt wound motor

In the shunt wound motor the field winding is in parallel with the armature across the supply as shown in Figure 61.15.

For the circuit shown in Figure 61.15,

Supply voltage, $\qquad V = E + I_a R_a$

or generated e.m.f., $\qquad E = V - I_a R_a$

Supply current, $\qquad I = I_a + I_f$ from Kirchhoff's current law

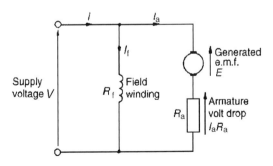

Figure 61.15

Characteristics

The two principal characteristics are the torque/armature current and speed/armature current relationships. From these, the torque/speed relationship can be derived.

(i) The theoretical torque/armature current characteristic can be derived from the expression $T \propto \Phi I_a$. For a shunt-wound motor, the field winding is connected in parallel with the armature circuit and thus the applied voltage gives a constant field current, i.e. a shunt-wound motor is a constant flux machine. Since Φ is constant, it follows that $T \propto I_a$, and the characteristic is as shown in Figure 61.16.

(ii) The armature circuit of a d.c. motor has resistance due to the armature winding and brushes, R_a ohms, and when armature current I_a is flowing through it, there is a voltage drop of $I_a R_a$ volts. In Figure 61.15 the armature resistance is shown as a separate resistor in the armature circuit to help understanding. Also, even though the machine is a motor, because conductors are rotating in a magnetic field, a voltage, $E \propto \Phi \omega$, is generated by the armature conductors. From equation (5), $V = E + I_a R_a$ or $E = V - I_a R_a$ However, $E \propto \Phi n$, hence $n \propto \dfrac{E}{\Phi}$

i.e. speed of rotation, $n \propto \dfrac{E}{\Phi} \propto \dfrac{V - I_a R_a}{\Phi}$ (9)

For a shunt motor, V, Φ and R_a are constants, hence as armature current I_a increases, $I_a R_a$ increases and $V - I_a R_a$ decreases, and the speed is proportional to a quantity which is decreasing and is as shown in Figure 61.17. As the load on the shaft of the motor increases, I_a increases and the speed drops slightly. In practice, the speed falls by about 10% between no-load and full-load on many d.c. shunt-wound motors. Due to this relatively small drop in speed, the d.c. shunt-wound motor is taken as basically being a constant-speed machine and may be used for driving lathes, lines of shafts, fans, conveyor belts, pumps, compressors, drilling machines and so on.

(iii) Since torque is proportional to armature current, (see (i) above), the theoretical speed/torque characteristic is as shown in Figure 61.18.

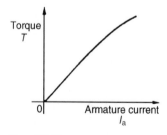

Figure 61.16

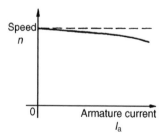

Figure 61.17

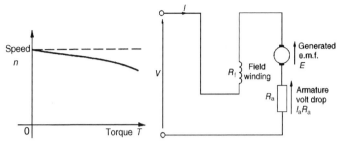

Figure 61.18   **Figure 61.19**

(b) Series-wound motor

In the series-wound motor the field winding is in series with the armature across the supply as shown in Figure 61.19.

Supply voltage $V = E + I(R_a + R_f)$

or generated e.m.f. $E = V - I(R_a + R_f)$

Characteristics

In a series motor, the armature current flows in the field winding and is equal to the supply current, I.

(i) **The torque/current characteristic**

It is shown earlier that torque $T \propto \Phi I_a$. Since the armature and field currents are the same current, I, in a series machine, then $T \propto \Phi I$ over a limited range, before magnetic saturation of the magnetic circuit of the motor is reached, (i.e. the linear portion of the B-H curve for the yoke, poles, air gap, brushes and armature in series). Thus $\Phi \propto I$ and $T \propto I^2$. After magnetic saturation, Φ almost becomes a constant and $T \propto I$. Thus the theoretical torque/current characteristic is as shown in Figure 61.20.

(ii) **The speed/current characteristic**

It is shown in equation (9) that $n \propto \dfrac{V - I_a R_a}{\Phi}$. In a series motor, $I_a = I$

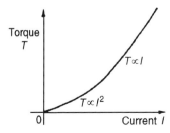

Figure 61.20   **Figure 61.21**

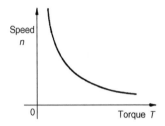

Figure 61.22

and below the magnetic saturation level, $\Phi \propto I$. Thus $n \propto \dfrac{V - IR}{I}$ where R is the combined resistance of the series field and armature circuit. Since IR is small compared with V, then an approximate relationship for the speed is $n \propto \dfrac{V}{I} \propto \dfrac{1}{I}$ since V is constant. Hence the theoretical speed/current characteristic is as shown in Figure 61.21. The high speed at small values of current indicate that this type of motor must not be run on very light loads and invariably, such motors are permanently coupled to their loads.

(iii) The theoretical speed/torque characteristic may be derived from (i) and (ii) above by obtaining the torque and speed for various values of current and plotting the co-ordinates on the speed/torque characteristics. A typical speed/torque characteristic is shown in Figure 61.22.

A d.c. series motor takes a large current on starting and the characteristic shown in Figure 61.20 shows that the series-wound motor has a large torque when the current is large. Hence these motors are used for traction (such as trains, milk delivery vehicles, etc.), driving fans and for cranes and hoists, where a large initial torque is required.

(c) Compound wound motor

There are two types of compound wound motor:

(i) **Cumulative compound**, in which the series winding is so connected that the field due to it assists that due to the shunt winding.
(ii) **Differential compound**, in which the series winding is so connected that the field due to it opposes that due to the shunt winding.

Figure 61.23(a) shows a **long-shunt** compound motor and Figure 61.23(b) a **short-shunt** compound motor.

Characteristics

A compound-wound motor has both a series and a shunt field winding, (i.e. one winding in series and one in parallel with the armature), and is usually

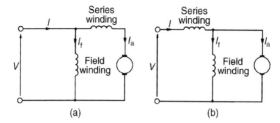

Figure 61.23

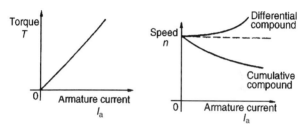

Figure 61.24

wound to have a characteristic similar in shape to a series wound motor (see Figures 61.20–61.22). A limited amount of shunt winding is present to restrict the no-load speed to a safe value. However, by varying the number of turns on the series and shunt windings and the directions of the magnetic fields produced by these windings (assisting or opposing), families of characteristics may be obtained to suit almost all applications. Generally, compound-wound motors are used for heavy duties, particularly in applications where sudden heavy load may occur such as for driving plunger pumps, presses, geared lifts, conveyors, hoists and so on. Typical compound motor torque and speed characteristics are shown in Figure 61.24.

The Efficiency of a d.c. Motor

It was stated earlier, that the efficiency of a d.c. machine is given by:

$$\text{efficiency, } \eta = \frac{\text{output power}}{\text{input power}} \times 100\%$$

Also, the total losses $= I_a^2 R_a + I_f V + C$ (for a shunt motor) where C is the sum of the iron, friction and windage losses.

For a motor, the input power $= VI$

and the output power $= VI - \text{losses}$

$$= VI - I_a^2 R_a - I_f V - C$$

Hence **efficiency**,

$$\boxed{\eta = \left(\frac{VI - I_a^2 R_a - I_f V - C}{VI} \right) \times 100\%}$$ (10)

For example, a d.c. series motor drives a load at 30 rev/s and takes a current of 10 A when the supply voltage is 400 V. If the total resistance of the motor is 2 Ω and the iron, friction and windage losses amount to 300 W, the efficiency of the motor is given by:

$$\text{efficiency,} \quad \eta = \left(\frac{VI - I^2 R - C}{VI} \right) \times 100\%$$

$$= \left(\frac{(400)(10) - (10)^2 (2) - 300}{(400)(10)} \right) \times 100\%$$

$$= \mathbf{87.5\%}$$

The **efficiency of a motor is a maximum** when the load is such that:

$$I_a^2 R_a = I_f V + C$$

D.c. Motor Starter

If a d.c. motor whose armature is stationary is switched directly to its supply voltage, it is likely that the fuses protecting the motor will burn out. This is because the armature resistance is small, frequently being less than one ohm. Thus, additional resistance must be added to the armature circuit at the instant of closing the switch to start the motor.

As the speed of the motor increases, the armature conductors are cutting flux and a generated voltage, acting in opposition to the applied voltage, is produced, which limits the flow of armature current. Thus the value of the additional armature resistance can then be reduced.

When at normal running speed, the generated e.m.f. is such that no additional resistance is required in the armature circuit. To achieve this varying resistance in the armature circuit on starting, a d.c. motor starter is used, as shown in Figure 61.25.

The starting handle is moved **slowly** in a clockwise direction to start the motor. For a shunt-wound motor, the field winding is connected to stud 1 or to L via a sliding contact on the starting handle, to give maximum field current, hence maximum flux, hence maximum torque on starting, since $T \propto \Phi I_a$. A similar arrangement without the field connection is used for series motors.

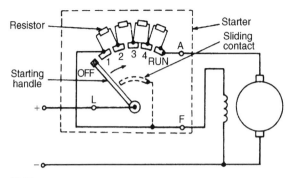

Figure 61.25

Speed Control of d.c. Motors

Shunt-wound motors

The speed of a shunt-wound d.c. motor, n, is proportional to $\dfrac{V - I_a R_a}{\Phi}$ (see equation (9)). The speed is varied either by varying the value of flux, Φ, or by varying the value of R_a. The former is achieved by using a variable resistor in series with the field winding, as shown in Figure 61.26(a) and such a resistor is called the **shunt field regulator**.

As the value of resistance of the shunt field regulator is increased, the value of the field current, I_f, is decreased. This results in a decrease in the value of flux, Φ, and hence an increase in the speed, since $n \propto 1/\Phi$. Thus only speeds **above** that given without a shunt field regulator can be obtained by this method. Speeds **below** those given by $\dfrac{V - I_a R_a}{\Phi}$ are obtained by increasing the resistance in the armature circuit, as shown in Figure 61.26(b), where $n \propto \dfrac{V - I_a (R_a + R)}{\Phi}$.

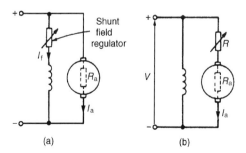

(a)   (b)

Figure 61.26

Since resistor R is in series with the armature, it carries the full armature current and results in a large power loss in large motors where a considerable speed reduction is required for long periods.

Series-wound motors

The speed control of series-wound motors is achieved using either (a) field resistance, or (b) armature resistance techniques.

(a) The speed of a d.c. series-wound motor is given by: $n = k\left(\dfrac{V - IR}{\Phi}\right)$

where k is a constant, V is the terminal voltage, R is the combined resistance of the armature and series field and Φ is the flux. Thus, a reduction in flux results in an increase in speed. This is achieved by putting a variable resistance in parallel with the field winding and reducing the field current, and hence flux, for a given value of supply current. A circuit diagram of this arrangement is shown in Figure 61.27(a). A variable resistor connected in parallel with the series-wound field to control speed is called a **diverter**. Speeds above those given with no diverter are obtained by this method.

(b) Speeds below normal are obtained by connecting a variable resistor in series with the field winding and armature circuit, as shown in Figure 61.27(b). This effectively increases the value of R in the equation $n = k\left(\dfrac{V - IR}{\Phi}\right)$ and thus reduces the speed. Since the additional resistor carries the full supply current, a large power loss is associated with large motors in which a considerable speed reduction is required for long periods.

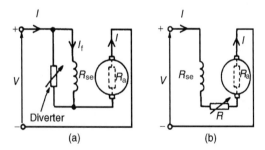

Figure 61.27

Motor Cooling

Motors are often classified according to the type of enclosure used, the type depending on the conditions under which the motor is used and the degree of ventilation required.

The most common type of protection is the **screen-protected type**, where ventilation is achieved by fitting a fan internally, with the openings at the end of the motor fitted with wire mesh.

A **drip-proof type** is similar to the screen-protected type but has a cover over the screen to prevent drips of water entering the machine.

A **flame-proof type** is usually cooled by the conduction of heat through the motor casing.

With a **pipe-ventilated type**, air is piped into the motor from a dust-free area, and an internally fitted fan ensures the circulation of this cool air.

62 A.c. Motors

Introduction

Two principal types of a.c. motors are in widespread use, these being **induction motors** and **synchronous motors**. Each of these types may be further sub-divided into those being run from a three-phase industrial supply, called three-phase motors and those largely in domestic use, called single-phase motors.

In d.c. motors, introduced in chapter 61, conductors on a rotating armature pass through a stationary magnetic field. In a **three-phase induction motor**, the magnetic field rotates and this has the advantage that no external electrical connections to the rotor need be made. Its name is derived from the fact that the current in the rotor is **induced** by the magnetic field instead of being supplied through electrical connections to the supply.

The result is a motor which: (i) is cheap and robust, (ii) is explosion proof, due to the absence of a commutator or slip-rings and brushes with their associated sparking, (iii) requires little or no skilled maintenance, and (iv) has self-starting properties when switched to a supply with no additional expenditure on auxiliary equipment. The principal disadvantage of a three-phase induction motor is that its speed cannot be readily adjusted.

Production of a Rotating Magnetic Field

When a three-phase supply is connected to symmetrical three-phase windings, the currents flowing in the windings produce a magnetic field. This magnetic field is constant in magnitude and rotates at constant speed as shown below, and is called the **synchronous speed**.

With reference to Figure 62.1, the windings are represented by three single-loop conductors, one for each phase, marked $R_S R_F$, $Y_S Y_F$ and $B_S B_F$, the S and F signifying start and finish. In practice, each phase winding comprises many turns and is distributed around the stator; the single-loop approach is for clarity only. When the stator windings are connected to a three-phase supply, the current flowing in each winding varies with time and is as shown in Figure 62.1(a). If the value of current in a winding is positive, the assumption is made that it flows from start to finish of the winding, i.e. if it is the red phase, current flows from R_S to R_F, i.e. away from the viewer in R_S and towards the viewer in R_F. When the value of current is negative, the assumption is made that it flows from finish to start, i.e. towards the viewer in an 'S' winding and away from the viewer in an 'F' winding. At time, say t_1, shown in Figure 62.1(a), the current flowing in the red phase is a maximum positive value. At the same time t_1, the currents flowing in the yellow and blue phases are both 0.5 times the maximum value and are negative.

The current distribution in the stator windings is therefore as shown in Figure 62.1(b), in which current flows away from the viewer, (shown as $\otimes$) in R_S since it is positive, but towards the viewer (shown as $\odot$) in Y_S and B_S,

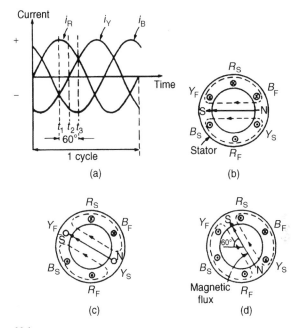

Figure 62.1

since these are negative. The resulting magnetic field is as shown, due to the 'solenoid' action and application of the corkscrew rule.

A short time later at time t_2, the current flowing in the red phase has fallen to about 0.87 times its maximum value and is positive, the current in the yellow phase is zero and the current in the blue phase is about 0.87 times its maximum value and is negative. Hence the currents and resultant magnetic field are as shown in Figure 62.1(c). At time t_3, the currents in the red and yellow phases are 0.5 of their maximum values and the current in the blue phase is a maximum negative value. The currents and resultant magnetic field are as shown in Figure 62.1(d).

Similar diagrams to Figure 62.1(b), (c) and (d) can be produced for all time values and these would show that the magnetic field travels through one revolution for each cycle of the supply voltage applied to the stator windings. By considering the flux values rather than the current values, it is shown below that the rotating magnetic field has a constant value of flux.

The three coils shown in Figure 62.2(a), are connected in star to a three-phase supply. Let the positive directions of the fluxes produced by currents flowing in the coils, be ϕ_A, ϕ_B and ϕ_C respectively. The directions of ϕ_A, ϕ_B and ϕ_C do not alter, but their magnitudes are proportional to the currents flowing in the coils at any particular time. At time t_1, shown in Figure 62.2(b), the currents flowing in the coils are: i_B, a maximum positive value, i.e. the flux is towards point P; i_A and i_C, half the maximum value and negative, i.e. the flux is away from point P.

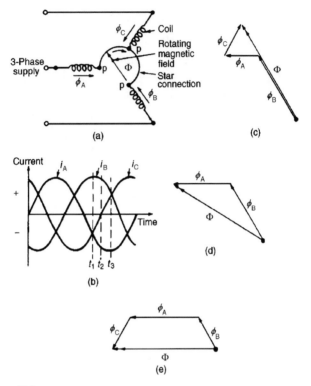

Figure 62.2

These currents give rise to the magnetic fluxes ϕ_A, ϕ_B and ϕ_C, whose magnitudes and directions are as shown in Figure 62.2(c). The resultant flux is the phasor sum of ϕ_A, ϕ_B and ϕ_C, shown as Φ in Figure 62.2(c). At time t_2, the currents flowing are:

i_B, 0.866 × maximum positive value, i_C, zero, and
i_A, 0.866 × maximum negative value.

The magnetic fluxes and the resultant magnetic flux are as shown in Figure 62.2(d).

At time t_3, i_B is 0.5 × maximum value and is positive
i_A is a maximum negative value, and
i_C is 0.5 × maximum value and is positive.

The magnetic fluxes and the resultant magnetic flux are as shown in Figure 62.2(e).

Inspection of Figures 62.2(c),(d) and (e) shows that the magnitude of the resultant magnetic flux, Φ, in each case is constant and is $1\frac{1}{2}$ × the maximum value of ϕ_A, ϕ_B or ϕ_C, but that its direction is changing. The process of determining the resultant flux may be repeated for all values of time and shows that the magnitude of the resultant flux is constant for all values of time and also that it rotates at constant speed, making one revolution for each cycle of the supply voltage.

Synchronous Speed

The rotating magnetic field produced by three-phase windings could have been produced by rotating a permanent magnet's north and south pole at synchronous speed, (shown as N and S at the ends of the flux phasors in Figures 62.1(b), (c) and (d)). For this reason, it is called a 2-pole system and an induction motor using three phase windings only is called a 2-pole induction motor.

If six windings displaced from one another by 60° are used, as shown in Figure 62.3(a), by drawing the current and resultant magnetic field diagrams at

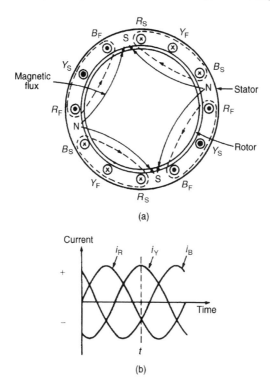

(a)

(b)

Figure 62.3

various time values, it may be shown that one cycle of the supply current to the stator windings causes the magnetic field to move through half a revolution. The current distribution in the stator windings are shown in Figure 62.3(a), for the time t shown in Figure 62.3(b).

It can be seen that for six windings on the stator, the magnetic flux produced is the same as that produced by rotating two permanent magnet north poles and two permanent magnet south poles at synchronous speed. This is called a 4-pole system and an induction motor using six phase windings is called a 4-pole induction motor. By increasing the number of phase windings the number of poles can be increased to any even number.

In general, if f is the frequency of the currents in the stator windings and the stator is wound to be equivalent to p **pairs** of poles, the speed of revolution of the rotating magnetic field, i.e. the synchronous speed, n_s is given by:

$$\boxed{n_s = \frac{f}{p} \text{ rev/s}}$$

For example, a three-phase four-pole induction motor is connected to a 50 Hz supply. The synchronous speed of the motor in rev/min is given by:

$$\text{synchronous speed, } n_s = \frac{50}{2} = 25 \text{ rev/s} = 25 \times 60 \text{ rev/min}$$

$$= \textbf{1500 rev/min}$$

Construction of a Three-phase Induction Motor

The stator of a three-phase induction motor is the stationary part corresponding to the yoke of a d.c. machine. It is wound to give a 2-pole, 4-pole, 6-pole, ... rotating magnetic field, depending on the rotor speed required. The rotor, corresponding to the armature of a d.c. machine, is built up of laminated iron, to reduce eddy currents.

In the type most widely used, known as a **squirrel-cage rotor**, copper or aluminium bars are placed in slots cut in the laminated iron, the ends of the bars being welded or brazed into a heavy conducting ring, (see Figure 62.4(a)). A cross-sectional view of a three-phase induction motor is shown in Figure 62.4(b).

The conductors are placed in slots in the laminated iron rotor core. If the slots are skewed, better starting and quieter running is achieved. This type of rotor has no external connections, which means that slip rings and brushes are not needed. The squirrel-cage motor is cheap, reliable and efficient.

Another type of rotor is the **wound rotor**. With this type there are phase windings in slots, similar to those in the stator. The windings may be connected in star or delta and the connections made to three slip rings. The slip rings are used to add external resistance to the rotor circuit, particularly for starting, but for normal running the slip rings are short-circuited.

The principle of operation is the same for both the squirrel cage and the wound rotor machines.

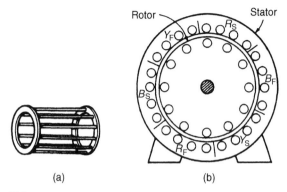

Figure 62.4

Principle of Operation of a Three-phase Induction Motor

When a three-phase supply is connected to the stator windings, a rotating magnetic field is produced. As the magnetic flux cuts a bar on the rotor, an e.m.f. is induced in it and since it is joined, via the end conducting rings, to another bar one pole pitch away, a current flows in the bars. The magnetic field associated with this current flowing in the bars interacts with the rotating magnetic field and a force is produced, tending to turn the rotor in the same direction as the rotating magnetic field, (see Figure 62.5). Similar forces are applied to all the conductors on the rotor, so that a torque is produced causing the rotor to rotate.

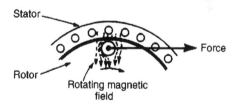

Figure 62.5

Slip

The force exerted by the rotor bars causes the rotor to turn in the direction of the rotating magnetic field. As the rotor speed increases, the rate at which the rotating magnetic field cuts the rotor bars is less and the frequency of the induced e.m.f.'s in the rotor bars is less. If the rotor runs at the same speed as the rotating magnetic field, no e.m.f.'s are induced in the rotor, hence there is no force on them and no torque on the rotor. Thus the rotor slows down. For this reason the rotor can never run at synchronous speed.

When there is no load on the rotor, the resistive forces due to windage and bearing friction are small and the rotor runs very nearly at synchronous speed. As the rotor is loaded, the speed falls and this causes an increase in the frequency of the induced e.m.f.'s in the rotor bars and hence the rotor current, force and torque increase. The difference between the rotor speed, n_r, and the synchronous speed, n_s, is called the **slip speed**, i.e.

$$\text{slip speed } = n_s - n_r \text{ rev/s}$$

The ratio $\dfrac{n_s - n_r}{n_s}$ is called the **fractional slip** or just the **slip**, s, and is usually expressed as a percentage. Thus

$$\boxed{\text{slip, } s = \left(\frac{n_s - n_r}{n_s} \right) \times 100\%}$$

Typical values of slip between no load and full load are about 4 to 5% for small motors and 1.5 to 2% for large motors.

For example, the stator of a 3-phase, 4-pole induction motor is connected to a 50 Hz supply and the rotor runs at 1455 rev/min at full load.

The **synchronous speed,** $\quad n_s = \dfrac{f}{p} = \dfrac{50}{2}$

$$= 25 \text{ rev/s}$$

The rotor speed, $n_r = \dfrac{1455}{60} = 24.25$ rev/s, hence

$$\textbf{slip, s} = \left(\frac{n_s - n_r}{n_s} \right) \times 100\% = \left(\frac{25 - 24.25}{25} \right) \times 100\% = \textbf{3}\%$$

Rotor e.m.f. and Frequency

Rotor e.m.f.

When an induction motor is stationary, the stator and rotor windings form the equivalent of a transformer as shown in Figure 62.6.

The rotor e.m.f. at standstill is given by

$$E_2 = \left(\frac{N_2}{N_1} \right) E_1 \tag{1}$$

where E_1 is the supply voltage per phase to the stator.

When an induction motor is running, the induced e.m.f. in the rotor is less since the relative movement between conductors and the rotating field is less.

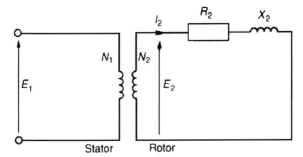

Figure 62.6

The induced e.m.f. is proportional to this movement, hence it must be proportional to the slip, s.

Hence **when running**, rotor e.m.f. per phase $= E_r = sE_2$

i.e. rotor e.m.f. per phase $= s\left(\dfrac{N_2}{N_1}\right)E_1$ (2)

Rotor frequency

The rotor e.m.f. is induced by an alternating flux and the rate at which the flux passes the conductors is the slip speed. Thus the frequency of the rotor e.m.f. is given by:

$$f_r = (n_s - n_r)p = \left(\frac{n_s - n_r}{n_s}\right)(n_s p)$$

However $\left(\dfrac{n_s - n_r}{n_s}\right)$ is the slip s and $(n_s p)$ is the supply frequency f, hence

$$\boxed{f_r = sf} \tag{3}$$

Rotor Impedance and Current

Rotor resistance

The rotor resistance R_2 is unaffected by frequency or slip, and hence remains constant.

Rotor reactance

Rotor reactance varies with the frequency of the rotor current.

At standstill, reactance per phase, $X_2 = 2\pi f L$

When running, reactance per phase, $X_r = 2\pi f_r L$

$$= 2\pi (sf)L \text{ from equation (3)}$$

$$= s(2\pi f L)$$

i.e. $\quad X_r = sX_2 \quad$ (4)

Figure 62.7 represents the rotor circuit when running.

Rotor Impedance

Rotor impedance per phase,

$$Z_r = \sqrt{R_2^2 + (sX_2)^2} \quad (5)$$

At standstill, slip $s = 1$, then

$$Z_2 = \sqrt{R_2^2 + X_2^2} \quad (6)$$

Rotor current

From Figures 62.6 and 62.7, at standstill, starting current,

$$\boxed{I_2 = \frac{E_2}{Z_2} = \frac{\left(\dfrac{N_2}{N_1}\right) E_1}{\sqrt{R_2^2 + X_2^2}}} \quad (7)$$

and when running, current,

$$\boxed{I_r = \frac{E_r}{Z_r} = \frac{s\left(\dfrac{N_2}{N_1}\right) E_1}{\sqrt{R_2^2 + (sX_2)^2}}} \quad (8)$$

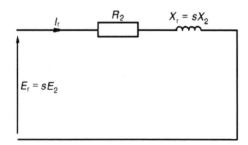

Figure 62.7

Rotor Copper Loss

Power $P = 2\pi nT$, where T is the torque in newton metres, hence torque $T = \dfrac{P}{2\pi n}$.

If P_2 is the power input to the rotor from the rotating field, and P_m is the mechanical power output (including friction losses)

$$\text{then } T = \frac{P_2}{2\pi n_s} = \frac{P_m}{2\pi n_r}$$

$$\text{from which, } \frac{P_2}{n_s} = \frac{P_m}{n_r}$$

$$\text{or } \frac{P_m}{P_2} = \frac{n_r}{n_s}$$

$$\text{Hence } 1 - \frac{P_m}{P_2} = 1 - \frac{n_r}{n_s}$$

$$\frac{P_2 - P_m}{P_2} = \frac{n_s - n_r}{n_s} = s$$

$P_2 - P_m$ is the electrical or copper loss in the rotor, i.e. $P_2 - P_m = I_r^2 R_2$
Hence

$$\boxed{\text{slip, } s = \frac{\text{rotor copper loss}}{\text{rotor input}} = \frac{I_r^2 R_2}{P_2}} \qquad (9)$$

or power input to the rotor,

$$\boxed{P_2 = \frac{I_r^2 R_2}{s}} \qquad (10)$$

Induction Motor Losses and Efficiency

Figure 62.8 summarises losses in induction motors.

Motor efficiency, $\boxed{\eta = \dfrac{\text{output power}}{\text{input power}} = \dfrac{P_m}{P_1} \times 100\%}$

For example, the power supplied to a three-phase induction motor is 32 kW and the stator losses are 1200 W. If the slip is 5% and friction and windage losses are 750 W, then

the input power to rotor = stator input power − stator losses

$$= 32 \text{ kW} - 1.2 \text{ kW} = 30.8 \text{ kW}$$

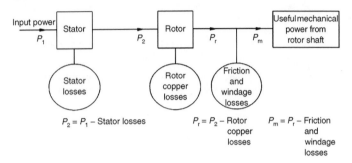

Figure 62.8

From equation (9),

$$\text{slip} = \frac{\text{rotor copper loss}}{\text{rotor input}}$$

i.e. $\dfrac{5}{100} = \dfrac{\text{rotor copper loss}}{30.8}$

from which, **rotor copper loss** $= (0.05)(30.8) = \textbf{1.54 kW}$

Total mechanical power developed by the rotor

$$= \text{rotor input power} - \text{rotor losses}$$

$$= 30.8 - 1.54 = \textbf{29.26 kW}$$

Output power of motor

$$= \text{power developed by the rotor} - \text{friction and windage losses}$$

$$= 29.26 - 0.75 = \textbf{28.51 kW}$$

Efficiency of induction motor,

$$\eta = \left(\frac{\text{output power}}{\text{input power}}\right) \times 100\%$$

$$= \left(\frac{28.51}{32}\right) \times 100\% = \textbf{89.10\%}$$

Torque Equation for an Induction Motor

Torque $T = \dfrac{P_2}{2\pi n_s} = \left(\dfrac{1}{2\pi n_s}\right)\left(\dfrac{I_r^2 R_2}{s}\right)$ (from equation (10))

From equation (8), $I_r = \dfrac{s\left(\dfrac{N_2}{N_1}\right)E_1}{\sqrt{R_2^2 + (sX_2)^2}}$

Hence torque per phase,

$$T = \left(\frac{1}{2\pi n_s}\right)\left(\frac{s^2\left(\dfrac{N_2}{N_1}\right)^2 E_1^2}{R_2^2 + (sX_2)^2}\right)\left(\frac{R_2}{s}\right)$$

$$= \left(\frac{1}{2\pi n_s}\right)\left(\frac{s\left(\dfrac{N_2}{N_1}\right)^2 E_1^2 R_2}{R_2^2 + (sX_2)^2}\right)$$

If there are m phases then

$$\boxed{T = \left(\frac{m\left(\dfrac{N_2}{N_1}\right)^2}{2\pi n_s}\right)\left(\frac{sE_1^2 R_2}{R_2^2 + (sX_2)^2}\right)} \qquad (11)$$

$$= k\left(\frac{sE_1^2 R_2}{R_2^2 + (sX_2)^2}\right)$$

where k is a constant for a particular machine, i.e.

$$\boxed{\text{torque, } T \propto \left(\frac{sE_1^2 R_2}{R_2^2 + (sX_2)^2}\right)} \qquad (12)$$

Under normal conditions, the supply voltage is usually constant, hence equation (12) becomes: $T \propto \dfrac{sR_2}{R_2^2 + (sX_2)^2} \propto \dfrac{R_2}{\dfrac{R_2^2}{s} + sX_2^2}$

The torque will be a maximum when the denominator is a minimum and this occurs when $\dfrac{R_2^2}{s} = sX_2^2$ i.e. when $s = \dfrac{R_2}{X_2}$ or $R_2 = sX_2 = X_r$ from equation (4). Thus **maximum torque** occurs when rotor resistance and rotor reactance are equal, i.e. when $R_2 = X_r$

Induction Motor Torque-speed Characteristics

The normal starting torque of an induction motor may be less than the full load torque. Also, the speed at which maximum torque occurs is determined by the

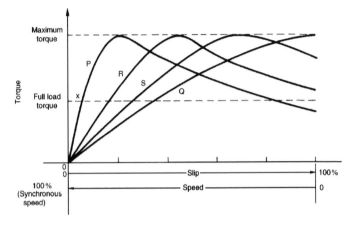

Figure 62.9

value of the rotor resistance. At synchronous speed, slip $s = 0$ and torque is zero. From these observations, the torque-speed and torque-slip characteristics of an induction motor are as shown in Figure 62.9.

The rotor resistance of an induction motor is usually small compared with its reactance (for example, $R_2 = 0.35 \, \Omega$ and $X_2 = 3.5 \, \Omega$ is not untypical), so that maximum torque occurs at a high speed, typically about 80% of synchronous speed.

Curve P in Figure 62.9 is a typical characteristic for an induction motor. The curve P cuts the full-load torque line at point X, showing that at full load the slip is about 4–5%. The normal operating conditions are between 0 and X, thus it can be seen that for normal operation the speed variation with load is quite small — the induction motor is an almost constant-speed machine. Redrawing the speed-torque characteristic between 0 and X gives the characteristic shown in Figure 62.10, which is similar to a d.c. shunt motor as shown in chapter 61.

If maximum torque is required at starting then a high resistance rotor is necessary, which gives characteristic Q in Figure 62.9. However, as can be seen, the motor has a full load slip of over 30%, which results in a drop in efficiency. Also such a motor has a large speed variation with variations of load. Curves R and S of Figure 62.9 are characteristics for values of rotor resistance's between those of P and Q. Better starting torque than for curve P is obtained, but with lower efficiency and with speed variations under operating conditions.

A **squirrel-cage induction motor** would normally follow characteristic P. This type of machine is highly efficient and about constant-speed under normal running conditions. However it has a poor starting torque and must be started off-load or very lightly loaded. Also, on starting, the current can be four or five times the normal full load current, due to the motor acting like a transformer with secondary short-circuited.

A **wound-rotor induction motor** would follow characteristic P when the slip-rings are short-circuited, which is the normal running condition. However,

459

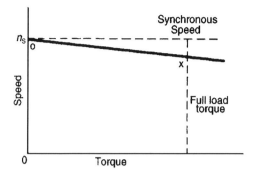

Figure 62.10

the slip-rings allow for the addition of resistance to the rotor circuit externally and, as a result, for starting, the motor can have a characteristic similar to curve Q in Figure 62.9 and the high starting current experienced by the cage induction motor can be overcome.

In general, for three-phase induction motors, the power factor is usually between about 0.8 and 0.9 lagging, and the full load efficiency is usually about 80–90%.

From equation (12), it is seen that torque is proportional to the square of the supply voltage. Any voltage variations therefore would seriously affect the induction motor performance.

Starting Methods for Induction Motors

Squirrel-cage rotor

(i) **Direct-on-line starting**
With this method, starting current is high and may cause interference with supplies to other consumers.

(ii) **Auto transformer starting**
With this method, an auto transformer is used to reduce the stator voltage, E_1, and thus the starting current (see equation (7)). However, the starting torque is seriously reduced (see equation (12)), so the voltage is reduced only sufficiently to give the required reduction of the starting current. A typical arrangement is shown in Figure 62.11. A double-throw switch connects the auto transformer in circuit for starting, and when the motor is up to speed the switch is moved to the run position which connects the supply directly to the motor.

(iii) **Star-delta starting**
With this method, for starting, the connections to the stator phase winding are star-connected, so that the voltage across each phase winding is $\dfrac{1}{\sqrt{3}}$ (i.e. 0.577) of the line voltage. For running, the windings are switched to

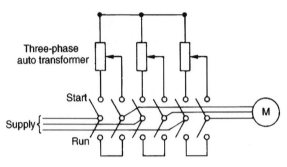

Figure 62.11

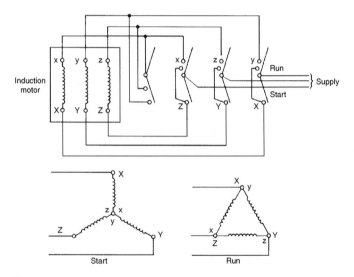

Figure 62.12

delta-connection. A typical arrangement is shown in Figure 62.12. This method of starting is less expensive than by auto transformer.

Wound rotor

When starting on load is necessary, a wound rotor induction motor must be used. This is because maximum torque at starting can be obtained by adding external resistance to the rotor circuit via slip rings. A face-plate type starter is used, and as the resistance is gradually reduced, the machine characteristics at each stage will be similar to Q, S, R and P of Figure 62.13. At each resistance step, the motor operation will transfer from one characteristic to the

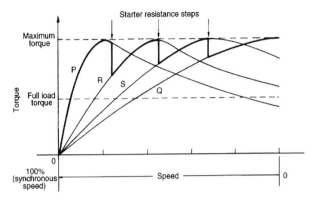

Figure 62.13

next so that the overall starting characteristic will be as shown by the bold line in Figure 62.13. For very large induction motors, a liquid type resistance achieves very gradual and smooth starting.

Advantages of Squirrel-Cage Induction Motors

The advantages of squirrel-cage motors compared with the wound rotor type are that they:

(i) are cheaper and more robust
(ii) have slightly higher efficiency and power factor
(iii) are explosion-proof, since the risk of sparking is eliminated by the absence of slip rings and brushes.

Advantages of Wound Rotor Induction Motors

The advantages of the wound rotor motor compared with the cage type are that they:

(i) have a much higher starting torque
(ii) have a much lower starting current
(iii) have a means of varying speed by use of external rotor resistance.

Double Cage Induction Motor

The advantages of squirrel-cage and wound rotor induction motors are combined in the double cage induction motor. This type of induction motor is specially constructed with the rotor having two cages, one inside the other.

462

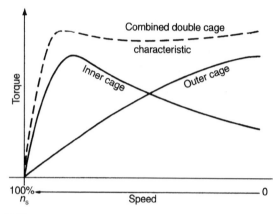

Figure 62.14

The outer cage has high resistance conductors so that maximum torque is achieved at or near starting. The inner cage has normal low resistance copper conductors but high reactance since it is embedded deep in the iron core. The torque-speed characteristic of the inner cage is that of a normal induction motor, as shown in Figure 62.14. At starting, the outer cage produces the torque, but when running the inner cage produces the torque. The combined characteristic of inner and outer cages is shown in Figure 62.14. The double cage induction motor is highly efficient when running.

Uses of Three-phase Induction Motors

Three-phase induction motors are widely used in industry and constitute almost all industrial drives where a nearly constant speed is required, from small workshops to the largest industrial enterprises.

Typical applications are with machine tools, pumps and mill motors. The squirrel cage rotor type is the most widely used of all a.c. motors.

Single-phase Induction Motor

The majority of 'fractional horse-power' motors in domestic use for driving refrigerators, hot water pumps, fans, hair dryers, and so on, are single-phase induction motors. A single-phase supply connected to the stator winding only produces a pulsating magnetic field rather than the rotating field of a three-phase supply and facilities such as additional windings have to be built-in to the stators of these motors to give some torque at zero speed. There are several devices used and motors are named after the particular device used. Some of these include: split-phase start,

capacitor start,
capacitor start and run,
shaded pole,
permanent-split capacitor induction motors, and so on.

Three-phase Synchronous Motor

Three-phase synchronous motors normally have two windings, a three-phase stator winding to produce a magnetic field and a rotor winding supplied by a direct current to magnetise the rotor. The rotor is locked magnetically to the rotating magnetic field, and for a constant frequency supply the rotor runs at constant speed, which is directly proportional to the supply frequency.

These motors may be used on many of the loads driven by induction motors but (i) are normally more expensive, (ii) require both a.c. and d.c. supplies, (iii) require additional equipment to run them up to near their normal running speed, so that the rotor magnetic field can be locked to the rotating magnetic field, and (iv) have higher maintenance costs than induction motors. However, they can run at higher efficiencies, and when used adjacent to several induction motors can lead to a higher efficiency of the whole system (power factor correction). Due to the disadvantages listed above, three-phase synchronous motors are normally used for applications requiring a large power input, where savings due to higher efficiency outweigh the disadvantages. These uses include driving large water pumps (power stations, water supply), driving rolling lines in steel mills and driving mine ventilating fans.

Single-phase Synchronous Motors

As for single-phase induction motors, single-phase synchronous motors require additional starting devices. They also require a direct current supply to the rotor or a permanent magnet built into the rotor. The former is rarely used and the principal uses of single-phase synchronous motors are largely limited to electric clocks, timing devices, record players and so on, these motors being called 'shaded-pole motors'. As for three-phase synchronous motors, they run at constant speed when connected to a constant frequency supply.

63 Revision of Complex Numbers

Introduction to Complex Numbers

A **complex number** is of the form $(a + jb)$ where a is a **real number** and jb is an **imaginary number**. Hence $(1 + j2)$ and $(5 - j7)$ are examples of complex numbers.

By definition, $\boxed{j = \sqrt{-1}}$ or $\boxed{j^2 = -1}$

Complex numbers are widely used in the analysis of series, parallel and series-parallel electrical networks supplied by alternating voltages (see chapters 64 to 66), in deriving balance equations with a.c. bridges (see chapter 67), in analysing a.c. circuits using Kirchhoff's laws (chapter 70), mesh and nodal analysis (chapter 71), the superposition theorem (chapter 72), with Thevenin's and Norton's theorems (chapter 73) and with delta-star and star-delta transforms (chapter 74) and in many other aspects of higher electrical engineering. The advantage of the use of complex numbers is that the manipulative processes become simply algebraic processes.

A complex number can be represented pictorially on an **Argand diagram**. In Figure 63.1 the line OA represents the complex number $(2 + j3)$, OB represents $(3 - j)$, OC represents $(-2 + j2)$ and OD represents $(-4 - j3)$.

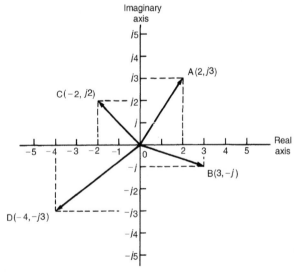

The Argand diagram

Figure 63.1

A complex number of the form $a + jb$ is called a **Cartesian or rectangular complex number**.

The significance of the j operator is shown in Figure 63.2. In Figure 63.2(a) the number 4, i.e. $(4 + j0)$, is shown drawn as a phasor horizontally to the right of the origin on the real axis. (Such a phasor could represent, for example, an alternating current, $i = 4 \sin \omega t$ amperes, when time t is zero).

The number $j4$, i.e. $(0 + j4)$ is shown in Figure 63.2(b) drawn vertically upwards from the origin on the imaginary axis. Hence multiplying the number 4 by the operator j results in an anticlockwise phase-shift of 90° without altering its magnitude.

Multiplying $j4$ by j gives $j^2 4$, i.e. -4, and is shown in Figure 63.2(c) as a phasor four units long on the horizontal real axis to the left of the origin — an anticlockwise phase-shift of 90° compared with the position shown in Figure 63.2(b). Thus multiplying by j^2 reverses the original direction of a phasor.

Multiplying $j^2 4$ by j gives $j^3 4$, i.e. $j(j)^2(4) = -j4$, and is shown in Figure 63.2(d) as a phasor four units long on the vertical, imaginary axis

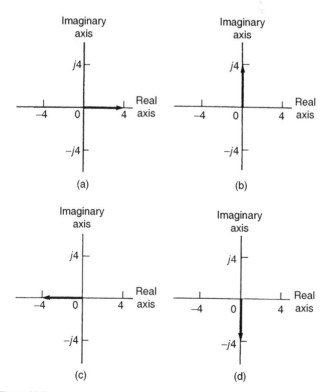

(a)

(b)

(c)

(d)

Figure 63.2

downward from the origin—an anticlockwise phase-shift of 90° compared with the position shown in Figure 63.2(c).

Multiplying $j^3 4$ by j gives $j^4 4$, i.e. $(j^2)^2(4) = 4$, which is the original position of the phasor shown in Figure 63.2(a).

Summarising, application of the operator j to any number rotates it 90° anticlockwise on the Argand diagram, multiplying a number by j^2 rotates it 180° anticlockwise, multiplying a number by j^3 rotates it 270° anticlockwise and multiplication by j^4 rotates it 360° anticlockwise, i.e. back to its original position. In each case the phasor is unchanged in its magnitude.

By similar reasoning, if a phasor is operated on by $-j$ then a phase shift of $-90°$ (i.e. clockwise direction) occurs, again without change of magnitude.

In electrical circuits, 90° phase shifts occur between voltage and current with pure capacitors and inductors; this is the key as to why j notation is used so much in the analysis of electrical networks. This is explained in chapter 64 following.

Operations Involving Cartesian Complex Numbers

(a) Addition and subtraction

$$(a + jb) + (c + jd) = (a + c) + j(b + d)$$
$$\text{and } (a + jb) - (c + jd) = (a - c) + j(b - d)$$

For example, $(3 + j2) + (2 - j4) = 3 + j2 + 2 - j4 = \mathbf{5 - j2}$ and
$$(3 + j2) - (2 - j4) = 3 + j2 - 2 + j4 = \mathbf{1 + j6}$$

(b) Multiplication

$$(a + jb)(c + jd) = ac + a(jd) + (jb)c + (jb)(jd)$$

$$= ac + jad + jbc + j^2 bd$$

But $j^2 = -1$, thus $(a + jb)(c + jd) = (ac - bd) + j(ad + bc)$

For example, $(3 + j2)(2 - j4) = 6 - j12 + j4 - j^2 8$

$$= 6 - j12 + j4 + 8 = 14 + j(-8)$$

$$= \mathbf{14 - j8}$$

(c) Complex conjugate

The complex conjugate of $(a + jb)$ is $(a - jb)$. For example, the conjugate of $(3 - j2)$ is $(3 + j2)$.

The product of a complex number and its complex conjugate is always a real number, and this is an important property used when dividing complex numbers.

Thus $(a + jb)(a - jb) = a^2 - jab + jab - j^2 b^2 = a^2 - (-b^2)$

$$= a^2 + b^2 \text{ (i.e. a real number)}$$

For example, $(1 + j2)(1 - j2) = 1^2 + 2^2 = \mathbf{5}$

and $(3 - j4)(3 + j4) = 3^2 + 4^2 = \mathbf{25}$

(d) Division

The expression of one complex number divided by another, in the form $a + jb$, is accomplished by multiplying the numerator and denominator by the complex conjugate of the denominator. This has the effect of making the denominator a real number.

For example,

$$\frac{2 + j4}{3 - j4} = \frac{2 + j4}{3 - j4} \times \frac{3 + j4}{3 + j4} = \frac{6 + j8 + j12 + j^2 16}{3^2 + 4^2}$$

$$= \frac{6 + j8 + j12 - 16}{25} = \frac{-10 + j20}{25}$$

$$= \frac{-10}{25} + j\frac{20}{25} \text{ or } -\mathbf{0.4} + j\mathbf{0.8}$$

The elimination of the imaginary part of the denominator by multiplying both the numerator and denominator by the conjugate of the denominator is often termed **'rationalising'**.

Complex Equations

If two complex numbers are equal, then their real parts are equal and their imaginary parts are equal. Hence, if $a + jb = c + jd$ then $a = c$ and $b = d$. This is a useful property, since equations having two unknown quantities can be solved from one equation. Complex equations are used when deriving balance equations with a.c. bridges (see chapter 67).

For example, to solve the complex equation:

$$(2 + j)(-2 + j) = x + jy$$

$$-4 + j2 - j2 + j^2 = x + jy$$

i.e. $-5 + j0 = x + jy$

Equating real and imaginary parts gives: $x = -\mathbf{5}, y = \mathbf{0}$

The Polar Form of a Complex Number

In Figure 63.3, $Z = x + jy = r\cos\theta + jr\sin\theta$ from trigonometry

$$= r(\cos\theta + j\sin\theta)$$

This latter form is usually abbreviated to $Z = r\angle\theta$, and is called the **polar form** of a complex number.

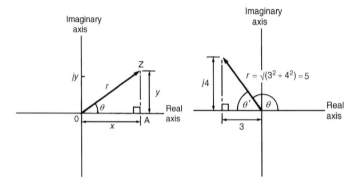

Figure 63.3   **Figure 63.4**

r is called the **modulus** (or magnitude of Z) and is written as mod Z or $|Z|$. r is determined from Pythagoras's theorem on triangle OAZ, i.e.

$$|Z| = r = \sqrt{x^2 + y^2}$$

The modulus is represented on the Argand diagram by the distance OZ.
θ is called the **argument** (or amplitude) of Z and is written as arg Z.

θ is also deduced from triangle OAZ: $\arg Z = \theta = \tan^{-1}\dfrac{y}{x}$

For example, the Cartesian complex number $(3 + j4)$ is equal to $r\angle\theta$ in polar form, where $r = \sqrt{3^2 + 4^2} = 5$ and $\theta = \tan^{-1}\frac{4}{3} = $ **53.13°**
Hence $(3 + j4) = 5\angle 53.13°$
Similarly, $(-3 + j4)$ is shown in Figure 63.4,

where $r = \sqrt{3^2 + 4^2} = 5$,
$\quad \theta' = \tan^{-1}\frac{4}{3} = 53.13°$ and
$\quad \theta = 180° - 53.13° = 126.87°$

Hence $(-3 + j4) = 5\angle 126.87°$
To change $5\angle -132°$ into Cartesian form:

$$5\angle -132° = 5[\cos(-132°) + j\sin(-132°)]$$

$$= 5\cos(-132°) + j5\sin(-132°) = -3.346 - j3.716$$

Multiplication and Division using Complex Numbers in Polar Form

(a) Multiplication

$$[r_1\angle\theta_1][r_2\angle\theta_2] = r_1 r_2\angle(\theta_1 + \theta_2)$$

For example, $\quad 3\angle 25° \times 2\angle 32° = 6\angle 57°,$

$\qquad\qquad\qquad 4\angle 11° \times 5\angle -18° = 20\angle -7°,$

and $\qquad\qquad 2\angle(\pi/3) \times 7\angle(\pi/6) = 14\angle(\pi/2)$

(b) Division

$$\frac{r_1\angle\theta_1}{r_2\angle\theta_2} = \frac{r_1}{r_2}\angle(\theta_1 - \theta_2)$$

For example, $\quad \dfrac{8\angle 58°}{2\angle 11°} = 4\angle 47°, \dfrac{9\angle 136°}{3\angle -60°} = 3\angle(136° - -60°)$

$$= 3\angle 196° \text{ or } 3\angle -164°$$

and $\qquad\qquad \dfrac{10\angle(\pi/2)}{5\angle(-\pi/4)} = 2\angle(3\pi/4)$

Conversion from Cartesian or rectangular form to polar form, and *vice versa*, may be achieved by using the $R \leftarrow P$ and $P \leftarrow R$ conversion facility that is available on most calculators with scientific notation. This allows, of course, a great saving of time.

De Moivres Theorem — Powers and Roots of Complex Numbers

De Moivre's theorem states: $\boxed{[r\angle\theta]^n = r^n\angle n\theta}$

This result is true for all positive, negative or fractional values of n. De Moivre's theorem is thus useful in determining powers and roots of complex numbers.

For example, $[2\angle 15°]^6 = 2^6\angle(6 \times 15°) = \mathbf{64\angle 90° = 0 + j\,64}$

A square root of a complex number is determined as follows:

$$\sqrt{[r\angle\theta]} = [r\angle\theta]^{\frac{1}{2}} = r^{\frac{1}{2}}\angle\tfrac{1}{2}\theta$$

However, it is important to realise that a real number has two square roots, equal in size but opposite in sign. On an Argand diagram the roots are $180°$ apart.

For example, the two square roots of the complex number $(12 + j5)$ is determined as follows:

In polar form $(12 + j5) = \sqrt{12^2 + 5^2}\angle\tan^{-1}\left(\dfrac{5}{12}\right)$, since $(12 + j5)$ is in the first quadrant of the Argand diagram, i.e. $(12 + j5) = 13\angle 22.62°$

Since we are finding the square roots of $13\angle 22.62°$ there will be two solutions. To obtain the second solution it is helpful to express $13\angle 22.62°$ also as: $13\angle(360° + 22.62°)$, i.e. $13\angle 382.62°$ (we have merely rotated one

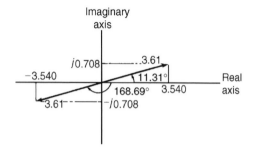

Figure 63.5

revolution to obtain this result). The reason for doing this is that when we divide the angle by 2 we still obtain an angle less than 360°, as shown below.

Hence $\sqrt{(12 + j5)} = \sqrt{13 \angle 22.62°}$ or $\sqrt{13 \angle 382.62°}$

$$= [13 \angle 22.62°]^{\frac{1}{2}} \text{ or } [13 \angle 382.62°]^{\frac{1}{2}}$$

$$= 13^{\frac{1}{2}} \angle (\tfrac{1}{2} \times 22.62°) \text{ or } 13^{\frac{1}{2}} \angle (\tfrac{1}{2} \times 382.62°)$$

from De Moivre's theorem,

$$= \sqrt{13} \angle 11.31° \text{ or } \sqrt{13} \angle 191.31°$$

$$= \mathbf{3.61 \angle 11.31° \text{ or } 3.61 \angle -168.69°}$$

These two solutions of $\sqrt{(12 + j5)}$ are shown in the Argand diagram of Figure 63.5.

$$3.61 \angle 11.31° = 3.61(\cos 11.31° + j \sin 11.31°)$$

$$= 3.540 + j0.708$$

$$3.61 \angle -168.69° = 3.61[\cos(-168.69°) + j \sin(-168.69°)]$$

$$= -3.540 - j0.708$$

Thus in Cartesian form the two roots are $\pm(\mathbf{3.540 + j0.708})$

From the Argand diagram the roots are seen to be 180° apart, i.e. they lie on a straight line. This is always true when finding square roots of complex numbers.

64 Application of Complex Numbers to Series a.c. Circuits

Introduction

Simple a.c. circuits may be analysed by using phasor diagrams. However, when circuits become more complicated analysis is considerably simplified by using complex numbers. It is essential that the basic operations used with complex numbers, as outlined in chapter 63, are thoroughly understood before proceeding with a.c. circuit analysis. The theory introduced in chapter 55 is relevant; in this chapter similar circuits are analysed using j notation and Argand diagrams.

Purely Resistive a.c. Circuit

In an a.c. circuit containing resistance R only (see Figure 64.1(a)), the current I_R is **in phase** with the applied voltage V_R as shown in the phasor diagram of Figure 64.1(b). The phasor diagram may be superimposed on the Argand diagram as shown in Figure 64.1(c). The impedance Z of the circuit is given by:

$$Z = \frac{V_R \angle 0°}{I_R \angle 0°} = R$$

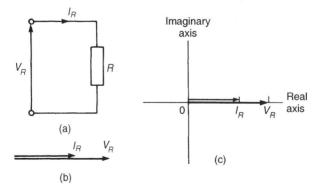

Figure 64.1

Purely Inductive a.c. Circuit

In an a.c. circuit containing pure inductance L only (see Figure 64.2(a)), the current I_L **lags** the applied voltage V_L by 90° as shown in the phasor diagram of Figure 64.2(b). The phasor diagram may be superimposed on the Argand diagram as shown in Figure 64.2(c). The impedance Z of the circuit is given by:

$$Z = \frac{V_L \angle 90°}{I_L \angle 0°} = \frac{V_L}{I_L} \angle 90° = X_L \angle 90° \quad \text{or} \quad jX_L$$

where X_L is the **inductive reactance** given by

$$\boxed{X_L = \omega L = 2\pi f L \text{ ohms}}$$

where f is the frequency in hertz and L is the inductance in henrys.

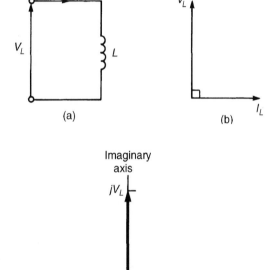

Figure 64.2

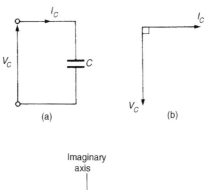

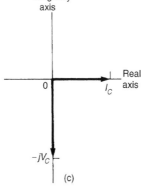

Figure 64.3

Purely Capacitive a.c. Circuit

In an a.c. circuit containing pure capacitance only (see Figure 64.3(a)), the current I_C **leads** the applied voltage V_C by 90° as shown in the phasor diagram of Figure 64.3(b). The phasor diagram may be superimposed on the Argand diagram as shown in Figure 64.3(c). The impedance Z of the circuit is given by:

$$Z = \frac{V_C \angle -90°}{I_C \angle 0°} = \frac{V_C}{I_C} \angle -90° = X_C \angle \mathbf{-90°} \quad \text{or} \quad -jX_C$$

where X_C is the **capacitive reactance** given by:

$$\boxed{X_C = \frac{1}{\omega C} = \frac{1}{2\pi f C} \text{ ohms}}$$

where C is the capacitance in farads.

R-L Series Circuit

In an a.c. circuit containing resistance R and inductance L in series (see Figure 64.4(a)), the applied voltage V is the phasor sum of V_R and V_L as shown in the phasor diagram of Figure 64.4(b). The current I lags the applied voltage V by an angle lying between $0°$ and $90°$ — the actual value depending on the values of V_R and V_L, which depend on the values of R and L. The circuit phase angle, i.e. the angle between the current and the applied voltage, is shown as angle ϕ in the phasor diagram. In any series circuit the current is common to all components and is thus taken as the reference phasor in Figure 64.4(b). The phasor diagram may be superimposed on the Argand diagram as shown in Figure 64.4(c), where it may be seen that in complex form the supply voltage V is given by: $V = V_R + jV_L$

Figure 64.5(a) shows the voltage triangle that is derived from the phasor diagram of Figure 64.4(b) (i.e. triangle Oab). If each side of the voltage triangle is divided by current I then the impedance triangle of Figure 64.5(b) is derived. The impedance triangle may be superimposed on the Argand diagram,

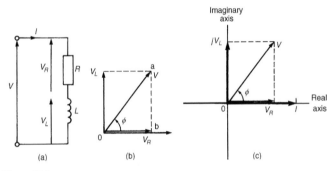

Figure 64.4

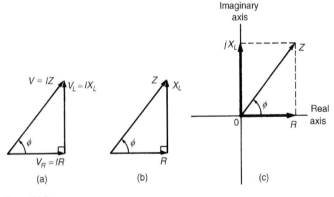

Figure 64.5

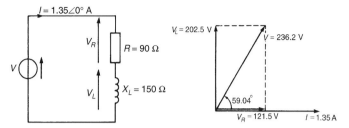

Figure 64.6   **Figure 64.7**

as shown in Figure 64.5(c), where it may be seen that in complex form the impedance Z is given by: $Z = R + jX_L$

For example, impedance expressed as $(3 + j4)$ Ω means that the resistance is 3 Ω and the inductive reactance is 4 Ω.

In polar form, $Z = |Z| \angle\phi$ where, from the impedance triangle, the modulus of impedance $|Z| = \sqrt{R^2 + X_L^2}$ and the circuit phase angle $\phi = \tan^{-1}(X_L/R)$ lagging.

For example, a circuit comprises a resistance of 90 Ω in series with an inductor of inductive reactance 150 Ω. The supply current is $1.35\angle0°$ A and the circuit diagram is shown in Figure 64.6.

Circuit impedance $Z = R + jX_L$

$$= (90 + j150) \text{ Ω} \quad \text{or} \quad 174.93\angle59.04° \text{ Ω}$$

Supply voltage, $V = IZ = (1.35\angle0°)(174.93\angle59.04°)$

$$= \mathbf{236.2\angle59.04°} \text{ V or } \mathbf{(121.5 + j\,202.5)} \text{ V}$$

Voltage across 90 Ω resistor, $V_R = 121.5$ V (since $V = V_R + jV_L$)
Voltage across inductance, $V_L = 202.5$ V leading V_R by 90°
Circuit phase angle is the angle between the supply current and voltage, i.e. **59.04°** lagging (i.e. current lags voltage). The phasor diagram is shown in Figure 64.7.

R-C Series Circuit

In an a.c. circuit containing resistance R and capacitance C in series (see Figure 64.8(a)), the applied voltage V is the phasor sum of V_R and V_C as shown in the phasor diagram of Figure 64.8(b). The current I leads the applied voltage V by an angle lying between 0° and 90° — the actual value depending on the values of V_R and V_C, which depend on the values of R and C. The circuit phase angle is shown as angle ϕ in the phasor diagram. The phasor diagram may be superimposed on the Argand diagram as shown in Figure 64.8(c), where it may be seen that in complex form the supply voltage V is given by:
$V = V_R - jV_C$

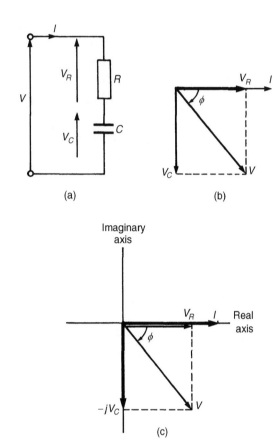

Figure 64.8

Figure 64.9(a) shows the voltage triangle that is derived from the phasor diagram of Figure 64.8(b). If each side of the voltage triangle is divided by current I, the impedance triangle is derived as shown in Figure 64.9(b). The impedance triangle may be superimposed on the Argand diagram as shown in Figure 64.9(c), where it may be seen that in complex form the impedance Z is given by:

$$Z = R - jX_C$$

For example, impedance expressed as $(9 - j14)$ Ω means that the resistance is 9 Ω and the capacitive reactance X_C is 14 Ω.
In polar form, $Z = |Z| \angle \phi$ where, from the impedance triangle,
$|Z| = \sqrt{R^2 + X_C^2}$ and $\phi = \arctan(X_C/R)$ leading.

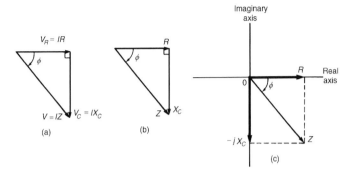

Figure 64.9

For **example**, if the impedance of an electrical circuit is $(30 - j50)$ ohms, and is connected to a 240 V, 50 Hz supply, then since impedance $Z = (30 - j50)$ Ω, **the resistance is 30 ohms**, and the capacitive reactance is 50 Ω

Since $\qquad X_C = \dfrac{1}{2\pi f C}$,

capacitance, $\qquad C = \dfrac{1}{2\pi f X_C} = \dfrac{1}{2\pi(50)(50)} = \textbf{63.66 } \boldsymbol{\mu}\textbf{F}$

The modulus of impedance,

$$|Z| = \sqrt{R^2 + X_C^2} = \sqrt{30^2 + 50^2}$$
$$= \textbf{58.31 } \boldsymbol{\Omega}$$

Impedance $\qquad Z = (30 - j50) \ \Omega = 58.31\angle \tan^{-1} \dfrac{X_C}{R}$

$$= 58.31\angle -59.04° \ \Omega$$

Hence current $\qquad I = \dfrac{V}{Z} = \dfrac{240\angle 0°}{58.31\angle -59.04°} = \textbf{4.12}\boldsymbol{\angle}\textbf{59.04° A}$

R-L-C Series Circuit

In an a.c. circuit containing resistance R, inductance L and capacitance C in series (see Figure 64.10(a)), the applied voltage V is the phasor sum of V_R, V_L and V_C as shown in the phasor diagram of Figure 64.10(b) (where the condition $V_L > V_C$ is shown). The phasor diagram may be superimposed on the Argand diagram as shown in Figure 64.10(c), where it may be seen that in complex form the supply voltage V is given by: $V = V_R + j(V_L - V_C)$

478

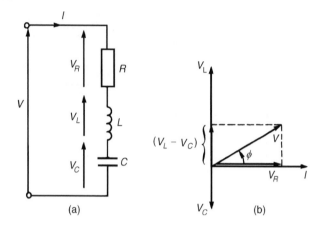

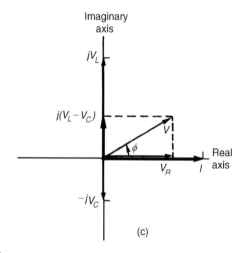

Figure 64.10

From the voltage triangle the impedance triangle is derived and superimposing this on the Argand diagram gives, in complex form,

impedance $Z = R + j(X_L - X_C)$ or $Z = |Z| \angle \phi$

where $|Z| = \sqrt{R^2 + (X_L - X_C)^2}$

and $\phi = \tan^{-1}\left(\dfrac{X_L - X_C}{R}\right)$

When $V_L = V_C$, $X_L = X_C$ and the applied voltage V and the current I are in phase. This effect is called **series resonance** and is discussed separately in chapter 68.

General Series Circuit

In an a.c. circuit containing several impedances connected in series, say, Z_1, Z_2, Z_3, ..., Z_n, then the total equivalent impedance Z_T is given by

$$Z_T = Z_1 + Z_2 + Z_3 + \ldots + Z_n$$

For example, a 240 V, 50 Hz voltage is applied across a series circuit comprising a coil of resistance 12 Ω and inductance 0.10 H, and 120 μF capacitor.
The circuit diagram is shown in Figure 64.11.

Inductive reactance, $X_L = 2\pi f L = 2\pi(50)(0.10) = 31.4$ Ω

Capacitive reactance, $X_C = \dfrac{1}{2\pi f C} = \dfrac{1}{2\pi(50)(120 \times 10^{-6})}$

$= 26.5$ Ω

Impedance $Z = R + j(X_L - X_C)$

i.e. $Z = 12 + j(31.4 - 26.5) = (12 + j4.9)\Omega$

$= 13.0\angle 22.20°$ Ω

Current flowing, $I = \dfrac{V}{Z} = \dfrac{240\angle 0°}{13.0\angle 22.2°} = \textbf{18.5}\angle \textbf{-22.20° A}$,

i.e. the current flowing is 18.5 A, lagging the voltage by 22.20°
The phasor diagram is shown on the Argand diagram in Figure 64.12.

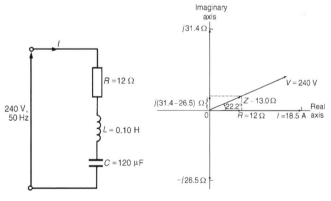

Figure 64.11   **Figure 64.12**

65 Application of Complex Numbers to Parallel a.c. Networks

Introduction

As with series circuits, parallel networks may be analysed by using phasor diagrams. However, with parallel networks containing more than two branches this can become very complicated. It is with parallel a.c. network analysis in particular that the full benefit of using complex numbers may be appreciated.

The theory for parallel a.c. networks introduced in chapter 56 is relevant; more advanced networks may be analysed using j notation.

Admittance, Conductance and Susceptance

Admittance is defined as the current I flowing in an a.c. circuit divided by the supply voltage V (i.e. it is the reciprocal of impedance Z). The symbol for admittance is Y. Thus

$$Y = \frac{I}{V} = \frac{1}{Z}$$

The unit of admittance is the **siemen, S**.

An impedance may be resolved into a real part R and an imaginary part X, giving $Z = R \pm jX$. Similarly, an admittance may be resolved into two parts — the real part being called the **conductance G**, and the imaginary part being called the **susceptance B** — and expressed in complex form.

Thus admittance

$$Y = G \pm jB$$

For example, if $Z = (3 - j2)\ \Omega$, then

$$\text{admittance } Y = \frac{1}{Z} = \frac{1}{(3 - j2)} = \frac{3 + j2}{3^2 + 2^2} = \frac{3}{13} + j\frac{2}{13}$$

$$= (0.231 + j0.154)\ \text{S}$$

Thus **conductance, $G = 0.231$ S** and **capacitive susceptance, $B_C = 0.154$ S**

Parallel a.c. Networks

Figure 65.1 shows a circuit diagram containing three impedances, Z_1, Z_2 and Z_3 connected in parallel. The potential difference across each impedance is the same, i.e. the supply voltage V. Current $I_1 = \dfrac{V}{Z_1}$, $I_2 = \dfrac{V}{Z_2}$ and $I_3 = \dfrac{V}{Z_3}$.

If Z_T is the total equivalent impedance of the circuit then $I = \dfrac{V}{Z_T}$.

481

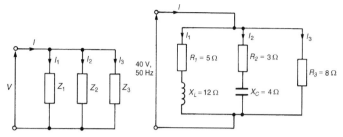

Figure 65.1　　　　　　**Figure 65.2**

The supply current, $I = I_1 + I_2 + I_3$ (phasorially).

Thus $\dfrac{V}{Z_T} = \dfrac{V}{Z_1} + \dfrac{V}{Z_2} + \dfrac{V}{Z_3}$ and $\boxed{\dfrac{1}{Z_T} = \dfrac{1}{Z_1} + \dfrac{1}{Z_2} + \dfrac{1}{Z_3}}$

or total admittance, $Y_T = Y_1 + Y_2 + Y_3$
In general, for n impedances connected in parallel,

$$\boxed{Y_T = Y_1 + Y_2 + Y_3 + \cdots + Y_n} \quad \text{(phasorially)}$$

It is in parallel circuit analysis that the use of admittance has its greatest advantage.

For example, for the parallel network shown in Figure 65.2, impedance $Z_1 = (5 + j12)\ \Omega$, $Z_2 = (3 - j4)\ \Omega$ and $Z_3 = 8\ \Omega$

Supply current $I = \dfrac{V}{Z_T} = VY_T$ where Z_T = total circuit impedance, and total circuit admittance

$$Y_T = Y_1 + Y_2 + Y_3 = \frac{1}{Z_1} + \frac{1}{Z_2} + \frac{1}{Z_3}$$

$$= \frac{1}{(5 + j12)} + \frac{1}{(3 - j4)} + \frac{1}{8}$$

$$= \frac{5 - j12}{5^2 + 12^2} + \frac{3 + j4}{3^2 + 4^2} + \frac{1}{8}$$

$$= (0.0296 - j0.0710) + (0.1200 + j0.1600) + (0.1250)$$

i.e. $Y_T = (0.2746 + j0.0890)$ S or $0.2887\angle17.96°$ S

Current $I = VY_T = (40\angle0°)(0.2887\angle17.96°) = 11.55\angle17.96°$ A

Hence the current I is 11.55 A and is leading the 40 V supply by 17.96°

Alternatively, current $I = I_1 + I_2 + I_3$

$$\text{Current } I_1 = \frac{40\angle0°}{5 + j12} = \frac{40\angle0°}{13\angle67.38°}$$

$$= 3.077\angle-67.38°\text{ A or }(1.183 - j2.840)\text{ A}$$

$$\text{Current } I_2 = \frac{40\angle 0°}{3 - j4} = \frac{40\angle 0°}{5\angle - 53.13°}$$

$$= 8\angle 53.13° \text{ A or } (4.80 + j6.40) \text{ A}$$

$$\text{Current } I_3 = \frac{40\angle 0°}{8\angle 0°} = 5\angle 0° \text{ A or } (5 + j0) \text{ A}$$

$$\text{Thus current } I = I_1 + I_2 + I_3$$

$$= (1.183 - j2.840) + (4.80 + j6.40) + (5 + j0)$$

$$= 10.983 + j3.560$$

$$= \mathbf{11.55\angle 17.96°} \text{ A, as previously obtained.}$$

Current Division in a.c. Circuits

For the special case of two impedances, Z_1 and Z_2, connected in parallel, as shown in Figure 65.3,

$$\boxed{I_1 = I\left(\frac{Z_2}{Z_1 + Z_2}\right)} \text{ and } \boxed{I_2 = I\left(\frac{Z_1}{Z_1 + Z_2}\right)}$$

For example, in the network of Figure 65.4, the total circuit impedance,

$$Z_T = 5 + \frac{(8)(j6)}{8 + j6} = 5 + \frac{(j48)(8 - j6)}{8^2 + 6^2} = 5 + \frac{j384 + 288}{100}$$

$$= (7.88 + j3.84) \ \Omega \text{ or } 8.77\angle 25.98° \ \Omega$$

$$\text{Current } I = \frac{V}{Z_T} = \frac{50\angle 0°}{8.77\angle 25.98°} = \mathbf{5.70\angle - 25.98°} \text{ A}$$

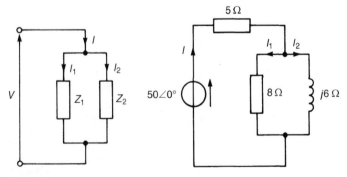

Figure 65.3   **Figure 65.4**

$$\text{Current } I_1 = I \left(\frac{j6}{8 + j6} \right) = (5.70\angle - 25.98°) \left(\frac{6\angle 90°}{10\angle 36.87°} \right)$$

$$= \mathbf{3.42\angle 27.15° \ A}$$

$$\text{Current } I_2 = I \left(\frac{8}{8 + j6} \right) = (5.70\angle - 25.98°) \left(\frac{8\angle 0°}{10\angle 36.87°} \right)$$

$$= \mathbf{4.56\angle - 62.85° \ A}$$

$$[\text{Note: } I = I_1 + I_2 = 3.42\angle 27.15° + 4.56\angle - 62.85°$$

$$= (3.043 + j1.561) + (2.081 - j4.058)$$

$$= (5.124 - j2.497) \text{ A} = \mathbf{5.70\angle - 25.98° \ A}]$$

In another example, for the series-parallel arrangement shown in Figure 65.5, the impedance, Z, of the two branches connected in parallel is given by:

$$Z = \frac{(5 + j7)(4 - j15)}{(5 + j7) + (4 - j15)} = \frac{20 - j75 + j28 - j^2 105}{9 - j8}$$

$$= \frac{125 - j47}{9 - j8} = \frac{133.54\angle - 20.61°}{12.04\angle - 41.63°}$$

$$= 11.09\angle 21.02° \ \Omega \text{ or } (10.35 + j3.98) \ \Omega$$

Equivalent series circuit impedance,

$$Z_T = (1.65 + j1.02) + (10.35 + j3.98)$$

$$= \mathbf{(12 + j5) \ \Omega \text{ or } 13\angle 22.62° \ \Omega}$$

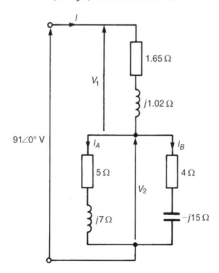

Figure 65.5

Supply current, $I = \dfrac{V}{Z} = \dfrac{91\angle 0°}{13\angle 22.62°}$

$\qquad\qquad\qquad = 7\angle - 22.62°$ A

Voltage $V_1 = IZ_1,$

where $Z_1 = (1.65 + j1.02)\ \Omega$ or $1.94\angle 31.72°\ \Omega.$

Hence $\pmb{V_1} = (7\angle - 22.62°)(1.94\angle 31.72°)$

$\qquad\qquad\quad = \pmb{13.58\angle 9.10°}$ V

Voltage $V_2 = IZ,$

where Z is the equivalent impedance of the two branches connected in parallel.

Hence, $\pmb{V_2} = (7\angle - 22.62°)(11.09\angle 21.02°)$

$\qquad\qquad\quad = \pmb{77.63\angle - 1.60°}$ V

Current $I_A = \dfrac{V_2}{Z_A},$

where $Z_A = (5 + j7)\ \Omega$ or $8.60\angle 54.46°\ \Omega$

Thus $\pmb{I_A} = \dfrac{77.63\angle - 1.60°}{8.60\angle 54.46°}$

$\qquad\qquad = \pmb{9.03\angle - 56.06°}$ A

Current $I_B = \dfrac{V_2}{Z_B},$

where $Z_B = (4 - j15)\ \Omega$ or $15.524\angle - 75.07°\ \Omega$

Thus $\pmb{I_B} = \dfrac{77.63\angle - 1.60°}{15.524\angle - 75.07°}$

$\qquad\qquad = \pmb{5.00\angle 73.47°}$ A

Alternatively, by current division,

$$\pmb{I_A} = I\left(\frac{Z_B}{Z_A + Z_B}\right) = (7\angle - 22.62°)\left(\frac{15.524\angle - 75.07°}{(5 + j7) + (4 - j15)}\right)$$

$$= (7\angle 22.62°)\left(\frac{15.524\angle - 75.07°}{9 - j8}\right)$$

$$= (7\angle - 22.62°)\left(\frac{15.524\angle - 75.07°}{12.04\angle - 41.63°}\right)$$

$$= \pmb{9.03\angle - 56.06°}\ \mathbf{A}$$

$$\pmb{I_B} = I\left(\frac{Z_A}{Z_A + Z_B}\right) = (7\angle - 22.62°)\left(\frac{8.60\angle 54.46°}{12.04\angle - 41.63°}\right)$$

$$= \pmb{5.00\angle 73.47°}\ \mathbf{A}$$

66 Power in a.c. Circuits and Power Factor Improvement

Power in a.c. Circuits

As shown in Chapter 55, in a purely resistive a.c. circuit, the average power P is given by:

$$P = VI = I^2R = \frac{V^2}{R} \text{ watts}$$

(where V and I are r.m.s. values)
In a purely inductive a.c. circuit, the average power is zero.
In a purely capacitive a.c. circuit, the average power is zero.
The average power P in a circuit containing resistance and inductance and/or capacitance, whether in series or in parallel, is given by:

$$\boxed{P = VI \cos\phi} \text{ or } \boxed{P = I_R^2 R} \text{ } (V, I \text{ and } I_R \text{ being r.m.s. values})$$

For example, for the circuit shown in Figure 66.1,

$$\text{circuit impedance, } Z = 5 + \frac{(3 + j4)(-j10)}{(3 + j4 - j10)}$$

$$= 5 + \frac{(40 - j30)}{(3 - j6)}$$

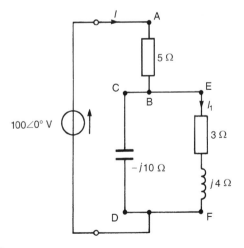

100∠0° V

I

A

5 Ω

C

E

B

I_1

3 Ω

$-j10$ Ω

$j4$ Ω

D

F

Figure 66.1

$$= 5 + \frac{50\angle - 36.87°}{6.71\angle - 63.43°}$$

$$= 5 + 7.45\angle 26.56°$$

$$= 5 + 6.66 + j3.33$$

$$= (11.66 + j3.33) \ \Omega \ \text{or} \ 12.13\angle 15.94° \ \Omega$$

$$\text{Current} \ I = \frac{V}{Z} = \frac{100\angle 0°}{12.13\angle 15.94°}$$

$$= 8.24\angle - 15.94° \ A$$

Active power developed between points A and B $= I^2R = (8.24)^2(5)$
$$= \textbf{339.5 W}$$

Active power developed between points C and D **is zero**, since no power is developed in a pure capacitor.

$$\text{Current,} \ I_1 = I\left(\frac{Z_{CD}}{Z_{CD} + Z_{EF}}\right) = 8.24\angle - 15.94°\left(\frac{-j10}{3 - j6}\right)$$

$$= 8.24\angle - 15.94°\left(\frac{10\angle - 90°}{6.71\angle - 63.43°}\right)$$

$$= 12.28\angle - 42.51° \ A$$

Hence the active power developed between points E and F

$$= I_1^2R = (12.28)^2(3) = \textbf{452.4 W}$$

Hence, from above, total active power developed $P = 339.5 + 452.4 = 791.9$ W or 792 W, correct to three significant figures.

Alternatively, total active power, $P = I^2R_T = (8.24)^2(11.66) = 792$ W (since 11.66 Ω is the total circuit equivalent resistance) or $P = VI \cos\phi = (100)(8.24) \cos 15.94° = 792$ W.

Power Triangle and Power Factor

A phasor diagram in which the current I lags the applied voltage V by angle ϕ (i.e. an inductive circuit) is shown in Figure 66.2(a). The horizontal component of V is $V \cos\phi$, and the vertical component of V is $V \sin\phi$. If each of the voltage phasors of triangle $0ab$ is multiplied by I, Figure 66.2(b) is produced and is known as the **'power triangle'.** Each side of the triangle represents a particular type of power:

> **True or active power** $P = VI \cos\phi$ **watts (W)**
>
> **Apparent power** $S = VI$ **voltamperes (VA)**
>
> **Reactive power** $Q = VI \sin\phi$ **vars (var)**

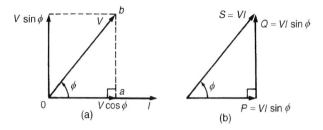

Figure 66.2

The power triangle is **not** a phasor diagram since quantities P, Q and S are mean values and not r.m.s. values of sinusoidally varying quantities.

Superimposing the power triangle on an Argand diagram produces a relationship between P, S and Q in complex form, i.e. $\boxed{S = P + jQ}$

Apparent power, S, is an important quantity since a.c. apparatus, such as generators, transformers and cables, is usually rated in voltamperes rather than in watts. The allowable output of such apparatus is usually limited not by mechanical stress but by temperature rise, and hence by the losses in the device. The losses are determined by the voltage and current and are almost independent of the power factor. Thus the amount of electrical equipment installed to supply a certain load is essentially determined by the voltamperes of the load rather than by the power alone.

The **rating** of a machine is defined as the maximum apparent power that it is designed to carry continuously without overheating.

The **reactive power,** Q, contributes nothing to the net energy transfer and yet it causes just as much loading of the equipment as if it did so. Reactive power is a term much used in power generation, distribution and utilization of electrical energy.

Inductive reactive power, by convention, is defined as positive reactive power; capacitive reactive power, by convention, is defined as negative reactive power.

The above relationships derived from the phasor diagram of an inductive circuit may be shown to be true for a capacitive circuit, the power triangle being as shown in Figure 66.3.

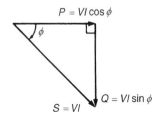

Figure 66.3

Power factor is defined as:

$$\text{power factor} = \frac{\text{active power } P}{\text{apparent power } S}$$

For sinusoidal voltages and currents,

$$\text{power factor} = \frac{P}{S} = \frac{VI \cos \phi}{VI}$$

$$= \cos \phi = \frac{R}{Z} \text{ (from the impedance triangle)}$$

A circuit in which current lags voltage (i.e. an inductive circuit) is said to have a lagging power factor, and indicates a lagging reactive power Q.

A circuit in which current leads voltage (i.e. a capacitive circuit) is said to have a leading power factor, and indicates a leading reactive power Q.

Use of Complex Numbers for Determination of Power

In an a.c. circuit, if voltage $V = (a + jb)$ and current $I = (c + jd)$, then:

power, $P = ac + bd$	(1)
reactive power, $Q = bc - ad$	(2)
apparent power, $S = P + jQ = VI^*$	(3)

where I^* is the conjugate of I, i.e. $(c - jd)$.

The above expressions provide an alternative method of determining true power P, reactive power Q and apparent power S when the voltage and current are complex quantities.

For example, a coil of resistance 5 Ω and inductive reactance 12 Ω is connected across a supply voltage of $52\angle 30°$ volts, as shown in Figure 66.4.

$$\text{Impedance } Z = (5 + j12) \ \Omega \text{ or } 13\angle 67.38° \ \Omega$$

$$\text{Voltage } V = 52\angle 30° \text{ V or } (45.03 + j26.0) \text{ V}$$

$$\text{Current } I = \frac{V}{Z} = \frac{52\angle 30°}{13\angle 67.38°}$$

$$= 4\angle -37.38° \text{ A or } (3.18 - j2.43) \text{ A}$$

There are three methods of calculating power.

Method 1. Active power, $P = VI \cos \phi$, where ϕ is the angle between voltage V and current I. Hence

$$P = (52)(4) \cos[30° - (-37.38°)]$$

$$= (52)(4) \cos 67.38° = \textbf{80 W}$$

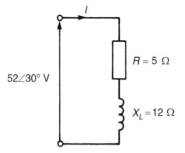

Figure 66.4

Method 2. Active power, $P = I_R^2 R = (4)^2(5) = $ **80 W**

Method 3. Since $V = (45.03 + j26.0)$ V and $I = (3.18 - j2.43)$ A, then active power,

$$P = (45.03)(3.18) + (26.0)(-2.43) \text{ from equation (1)},$$

i.e. $P = 143.2 - 63.2 = $ **80 W**

Also, apparent power, $S = VI = (52)(4) = $ **208 VA** and reactive power, $Q = VI \sin \phi = (52)(4) \sin 67.38° = $ **192 var**
  Alternatively,

$$S = VI^* = (45.03 + j26.0)(3.18 + j2.43)$$

$$= (\mathbf{80 + j\,192}) = P + jQ \text{ and}$$

$$|S| = \sqrt{(80^2 + 192^2)}$$

$$= \mathbf{208\ VA}$$

Power Factor Improvement

For a particular active power supplied, a high power factor reduces the current flowing in a supply system and therefore reduces the cost of cables, transformers, switchgear and generators. Supply authorities use tariffs that encourage consumers to operate at a reasonably high power factor. One method of improving the power factor of an inductive load is to connect a bank of capacitors in parallel with the load. Capacitors are rated in reactive voltamperes and the effect of the capacitors is to reduce the reactive power of the system without changing the active power. Most residential and industrial loads on a power system are inductive, i.e. they operate at a lagging power factor.

  A simplified circuit diagram is shown in Figure 66.5(a) where a capacitor C is connected across an inductive load. Before the capacitor is connected the circuit current is I_{LR} and is shown lagging voltage V by angle ϕ_1 in the phasor diagram of Figure 66.5(b). When the capacitor C is connected it takes

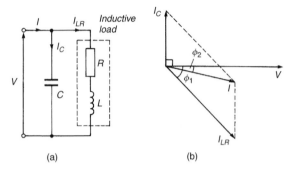

(a)   (b)

Figure 66.5

a current I_C which is shown in the phasor diagram leading voltage V by 90°. The supply current I in Figure 66.5(a) is now the phasor sum of currents I_{LR} and I_C as shown in Figure 66.5(b). The circuit phase angle, i.e. the angle between V and I, has been reduced from ϕ_1 to ϕ_2 and the power factor has been improved from $\cos\phi_1$ to $\cos\phi_2$.

Figure 66.6(a) shows the power triangle for an inductive circuit with a lagging power factor of $\cos\phi_1$. In Figure 66.6(b), the angle ϕ_1 has been reduced to ϕ_2, i.e. the power factor has been improved from $\cos\phi_1$ to $\cos\phi_2$ by introducing leading reactive voltamperes (shown as length ab) which is achieved by connecting capacitance in parallel with the inductive load. The power factor has been improved by reducing the reactive voltamperes; the active power P has remained unaffected.

Power factor correction results in the apparent power S decreasing (from $0a$ to $0b$ in Figure 66.6(b)) and thus the current decreasing, so that the power distribution system is used more efficiently.

For example, a circuit has an impedance $Z = (3 + j4)$ Ω and a source p.d. of $50\angle 30°$ V at a frequency of 1.5 kHz.

$$\text{Supply current, } I = \frac{V}{Z} = \frac{50\angle 30°}{(3 + j4)} = \frac{50\angle 30°}{5\angle 53.13°}$$

$$= 10\angle -23.13° \text{ A}$$

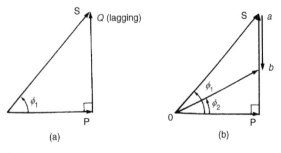

(a)   (b)

Figure 66.6

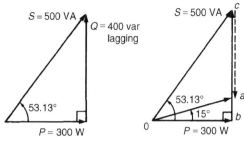

Figure 66.7   **Figure 66.8**

Apparent power, $S = VI^* = (50\angle30°)(10\angle23.13°)$

$$= 500\angle53.13° \text{ VA} = (300 + j400) \text{ VA}$$

$$= P + jQ$$

Hence **active power, $P = 300$ W,**

apparent power, $S = 500$ VA and

reactive power, $Q = 400$ var lagging

The power triangle is shown in Figure 66.7.

The capacitance needed to improve the power factor to, say, 0.966 lagging is determined as follows:

Since power factor $= 0.966$, then $\cos\phi = 0.966$ and

$$\text{angle } \phi = \cos^{-1} 0.966 = 15°.$$

To improve the power factor from $\cos 53.13°$, i.e. 0.60, to 0.966, the power triangle will need to change from $0cb$ in Figure 66.8, to $0ab$, the length ca representing the rating of a capacitor connected in parallel with the circuit. From Figure 66.8, $\tan 15° = ab/300$, from which, $ab = 300\tan 15° = 80.38$ var. Hence, rating of the capacitor, $ca = cb - ab = 400 - 80.38 = 319.6$ var leading

$$\text{Current in capacitor, } I_C = \frac{Q}{V} = \frac{319.6}{50} = 6.39 \text{ A}$$

$$\text{Capacitive reactance, } X_C = \frac{V}{I_C} = \frac{50}{6.39} = 7.82 = \frac{1}{2\pi f C}$$

from which,

required capacitance $C = \dfrac{1}{2\pi(1500)(7.82)}$ F $\equiv$ 13.57 μF

Another method of power factor improvement, besides the use of static capacitors, is by using synchronous motors; such machines can be made to operate at leading power factors.

67 A.c. Bridges

Introduction to a.c. Bridges

A.C. bridges are electrical networks, based upon an extension of the Wheatstone bridge principle, used for the determination of unknown impedance by comparison with known impedances and for the determination of frequency. In general, they contain four impedance arms, an a.c. power supply and a balance detector that is sensitive to alternating currents. It is more difficult to achieve balance in an a.c. bridge than in a d.c. bridge because both the magnitude and the phase angle of impedances are related to the balance condition. Balance equations are derived by using complex numbers. A.C. bridges provide precise methods of measurement of inductance and capacitance, as well as resistance.

Balance Conditions for an a.c. Bridge

The majority of well-known a.c. bridges are classified as four-arm bridges and consist of an arrangement of four impedances (in complex form, $Z = R \pm jX$) as shown in Figure 67.1. As with the d.c. Wheatstone bridge circuit, an a.c. bridge is said to be 'balanced' when the current through the detector is zero (i.e. when no current flows between B and D of Figure 67.1). If the current through the detector is zero, then the current I_1 flowing in impedance Z_1 must also flow in impedance Z_2. Also, at balance, the current I_4 flowing in impedance Z_4, must also flow through Z_3. At balance:

(i) the volt drop between A and B is equal to the volt drop between A and D, i.e.$V_{AB} = V_{AD}$

  i.e. $I_1 Z_1 = I_4 Z_4$ (both in magnitude and in phase)   (1)

(ii) the volt drop between B and C is equal to the volt drop between D and C, i.e. $V_{BC} = V_{DC}$

  i.e. $I_1 Z_2 = I_4 Z_3$ (both in magnitude and in phase)   (2)

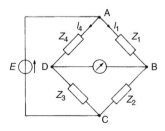

Figure 67.1

Dividing equation (1) by equation (2) gives:

$$\frac{I_1 Z_1}{I_1 Z_2} = \frac{I_4 Z_4}{I_4 Z_3}$$

from which, $\quad \dfrac{Z_1}{Z_2} = \dfrac{Z_4}{Z_3}$

or $\qquad \boxed{Z_1 Z_3 = Z_2 Z_4} \qquad\qquad\qquad (3)$

Equation (3) shows that at balance the products of the impedances of opposite arms of the bridge are equal.

If in polar form, $Z_1 = |Z_1| \angle \alpha_1$, $Z_2 = |Z_2| \angle \alpha_2$, $Z_3 = |Z_3| \angle \alpha_3$ and $Z_4 = |Z_4| \angle \alpha_4$, then, from equation (3): $(|Z_1| \angle \alpha_1)(|Z_2| \angle \alpha_2) = (|Z_3| \angle \alpha_3)(|Z_4| \angle \alpha_4)$ which shows that there are **two** conditions to be satisfied simultaneously for balance in a.c. bridge, i.e.

$$|Z_1|\,|Z_3| = |Z_2|\,|Z_4| \quad \text{and} \quad \alpha_1 + \alpha_3 = \alpha_2 + \alpha_4$$

When deriving balance equations of a.c. bridges, where at least two of the impedances are in complex form, it is important to appreciate that for a complex equation $(a + jb) = (c + jd)$ the real parts are equal, i.e. $a = c$, and the imaginary parts are equal, i.e. $b = d$, as stated in Chapter 63.

Usually one arm of an a.c. bridge circuit contains the unknown impedance while the other arms contain known fixed or variable components. Normally only two components of the bridge are variable. When balancing a bridge circuit, the current in the detector is gradually reduced to zero by successive adjustments of the two variable components. At balance, the unknown impedance can be expressed in terms of the fixed and variable components.

Procedure for determining the balance equations of any a.c. bridge circuit

(i) Determine for the bridge circuit the impedance in each arm in complex form and write down the balance equation as in equation (3). Equations are usually easier to manipulate if L and C are initially expressed as X_L and X_C, rather than ωL or $\dfrac{1}{\omega C}$

(ii) Isolate the unknown terms on the left-hand side of the equation in the form $(a + jb)$

(iii) Manipulate the terms on the right-hand side of the equation into the form $(c + jd)$

(iv) Equate the real parts of the equation, i.e. $a = c$, and equate the imaginary parts of the equation, i.e. $b = d$

(v) Substitute ωL for X_L and $\dfrac{1}{\omega C}$ for X_C where appropriate and express the final equations in their simplest form.

Types of Detector

The types of detector used with a.c. bridges vary with the type of bridge and with the frequency at which it is operated. Common detectors used include:

(i) a C.R.O., which is suitable for use with a very wide range of frequencies,

(ii) earphones (or telephone headsets), which are suitable for frequencies up to about 10 kHz and are used often at about 1 kHz, in which region the human ear is very sensitive,

(iii) various electronic detectors, which use tuned circuits to detect current at the correct frequency, and

(v) vibration galvanometers, which are usually used for mains-operated bridges; this type of detector consists basically of a narrow moving coil which is suspended on a fine phosphor bronze wire between the poles of a magnet. When a current of the correct frequency flows through the coil, it is set into vibration. This is because the mechanical resonant frequency of the suspension is purposely made equal to the electrical frequency of the coil current. A mirror attached to the coil reflects a spot of light on to a scale, and when the coil is vibrating the spot appears as an extended beam of light. When the band reduces to a spot the bridge is balanced. Vibration galvanometers are available in the frequency range 10 Hz to 300 Hz.

Types of a.c. Bridge Circuit

A large number of bridge circuits have been developed, each of which has some particular advantage under certain conditions. Some of the most important a.c. bridges include the Maxwell, Hay, Owen and Maxwell-Wien bridges for measuring inductance, the De Sauty and Schering bridges for measuring capacitance, and the Wien bridge for measuring frequency. Obviously a large number of combinations of components in bridges is possible.

In many bridges it is found that two of the balancing impedances will be of the same nature, and often consist of standard non-inductive resistors. For a bridge to balance quickly the requirement is either:

(i) the adjacent arms are both pure components (i.e. either both resistors, or both pure capacitors, or one of each) — this type of bridge being called a **ratio-arm bridge** (see, for example, paras (a), (c), (e) and (g) below); or

(ii) a pair of opposite arms are pure components, this type of bridge being called a **product-arm bridge** (see, for example, paras (b), (d) and (f) below).

A ratio-arm bridge can only be used to measure reactive quantities of the same type. When using a product-arm bridge, the reactive component of the balancing impedance must be of opposite sign to the unknown reactive component.

A commercial or universal bridge is available and can be used to measure resistance, inductance or capacitance.

(a) The simple Maxwell bridge

This bridge is used to measure the resistance and inductance of a coil having a high Q-factor (where Q-factor $= \dfrac{\omega L}{R}$, see chapters 55 and 68).

A coil having unknown resistance R_x and inductance L_x is shown in the circuit diagram of a simple Maxwell bridge in Figure 67.2. R_4 and L_4 represent a standard coil having known variable values. At balance, expressions for R_x and L_x may be derived in terms of known components R_2, R_3, R_4 and L_4.

The procedure for determining the balance equations is as follows:

(i) From Figure 67.2, $Z_x = R_x + jX_{L_x}$, $Z_2 = R_2$, $Z_3 = R_3$ and $Z_4 = R_4 + jX_{L_4}$

At balance, $\qquad (Z_x)(Z_3) = (Z_2)(Z_4)$ from equation (3)

i.e. $\qquad (R_x + jX_{L_x})(R_3) = (R_2)(R_4 + jX_{L_4})$

(ii) Isolating the unknown impedance on the left-hand side of the equation gives:

$$(R_x + jX_{L_x}) = \frac{R_2}{R_3}(R_4 + jX_{L_4})$$

(iii) Manipulating the right-hand side of the equation into $(a + jb)$ form gives

$$(R_x + jX_{L_x}) = \frac{R_2 R_4}{R_3} + j\frac{R_2 X_{L_4}}{R_3}$$

(iv) Equating the real parts gives: $R_x = \dfrac{R_2 R_4}{R_3}$

Equating the imaginary parts gives: $jX_{L_x} = \dfrac{R_2 X_{L_4}}{R_3}$

(v) Since $X_L = \omega L$, then $\omega L_x = \dfrac{R_2(\omega L_4)}{R_3}$ from which, $L_x = \dfrac{R_2 L_4}{R_3}$

Thus at balance the unknown components in the simple Maxwell bridge are given by:

$$\boxed{R_x = \frac{R_2 R_4}{R_3} \quad \text{and} \quad L_x = \frac{R_2 L_4}{R_3}}$$

These are known as the 'balance equations' for the bridge.

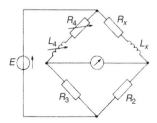

Figure 67.2

496

(b) The Hay bridge

This bridge is used to measure the resistance and inductance of a coil having a very high Q-factor. A coil having unknown resistance R_x and inductance L_x is shown in the circuit diagram of a Hay bridge in Figure 67.3. It may be shown that, at balance, the unknown components in the Hay bridge are given by:

$$R_x = \frac{\omega^2 C_3^2 R_2 R_3 R_4}{(1 + \omega^2 C_3^2 R_3^2)} \quad \text{and} \quad L_x = \frac{C_3 R_2 R_4}{(1 + \omega^2 C_3^2 R_3^2)}$$

Since $\omega\,(=2\pi f)$ appears in the balance equations, the bridge is **frequency-dependent**.

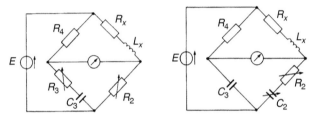

Figure 67.3   **Figure 67.4**

(c) The Owen bridge

This bridge, shown in Figure 67.4, is used to measure the resistance and inductance of coils possessing a large value of inductance. It may be shown that at balance the unknown components in the Owen bridge are given by:

$$R_x = \frac{R_4 C_3}{C_2} \quad \text{and} \quad L_x = R_2 R_4 C_3$$

(d) The Maxwell-Wien bridge

This bridge, shown in Figure 67.5, is used to measure the resistance and inductance of a coil having a low Q-factor. It may be shown that at balance the unknown components in the Maxwell-Wien bridge are given by:

$$R_x = \frac{R_2 R_4}{R_3} \quad \text{and} \quad L_x = C_3 R_2 R_4$$

497

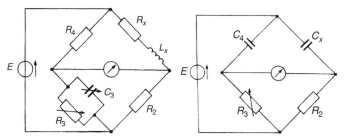

Figure 67.5   **Figure 67.6**

(e) The de Sauty bridge

This bridge, shown in Figure 67.6, provides a very simple method of measuring a capacitance by comparison with another known capacitance. It may be shown that, at balance:

$$C_x = \frac{R_3 C_4}{R_2}$$

This simple bridge is usually inadequate in most practical cases. The power factor of the capacitor under test is significant because of internal dielectric losses – these losses being the dissipation within a dielectric material when an alternating voltage is applied to a capacitor.

(f) The Schering bridge

This bridge, shown in Figure 67.7, is used to measure the capacitance and equivalent series resistance of a capacitor. From the measured values the power factor of insulating materials and dielectric losses may be determined. It may be shown that, at balance, the unknown components in the Schering bridge are given by:

$$R_x = \frac{C_3 R_4}{C_2} \quad \text{and} \quad C_x = \frac{C_2 R_3}{R_4}$$

The loss in a dielectric may be represented by either (a) a resistance in parallel with a capacitor, or (b) a loss-less capacitor in series with a resistor.

If the dielectric is represented by an R-C series circuit, as shown by R_x and C_x in Figure 67.7, the phasor diagram for the unknown arm is as shown in Figure 67.8.

$$\text{Angle } \phi = \tan^{-1} \frac{V_{C_x}}{V_{R_x}} = \tan^{-1} \frac{I_x X_{C_x}}{I_x R_x} = \tan^{-1} \left(\frac{1}{\omega C_x R_x} \right)$$

The power factor of the unknown arm is given by $\cos \phi$.

The angle $\delta \, (= 90° - \phi)$ is called the **loss angle** and is given by:

$$\delta = \tan^{-1} \frac{V_{R_x}}{V_{C_x}} = \mathbf{tan^{-1}} \, \omega C_x R_x$$

498

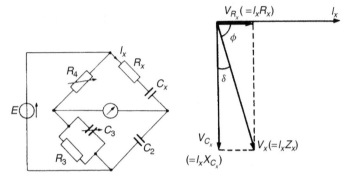

Figure 67.7   **Figure 67.8**

$$= \tan^{-1} \left[\omega \left(\frac{C_2 R_3}{R_4} \right) \left(\frac{C_3 R_4}{C_2} \right) \right]$$

$$= \tan^{-1} \omega R_3 C_3$$

(see also Chapter 78, page 587)

(g) The Wien bridge

This bridge, shown in Figure 67.9, is used to measure frequency in terms of known components (or, alternatively, to measure capacitance if the frequency is known). It may also be used as a frequency-stabilizing network. It may be shown that, at balance:

$$\boxed{\frac{R_3}{R_2} + \frac{C_2}{C_3} = \frac{R_4}{R_1} \quad \text{and} \quad \textbf{frequency}, \; f = \frac{1}{2\pi\sqrt{C_2 C_3 R_2 R_3}}}$$

Note that if $C_2 = C_3 = C$ and $R_2 = R_3 = R$, then frequency,

$$f = \frac{1}{2\pi\sqrt{C^2 R^2}} = \frac{1}{2\pi C R}$$

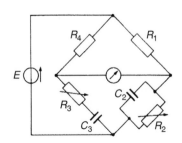

Figure 67.9

For example, for the Wien bridge shown in Figure 67.9, $R_2 = R_3 = 30$ kΩ, $R_4 = 1$ kΩ and $C_2 = C_3 = 1$ nF. When the bridge is balanced the value of resistance R_1 is determined from: $\dfrac{R_3}{R_2} + \dfrac{C_2}{C_3} = \dfrac{R_4}{R_1}$ i.e. $1 + 1 = \dfrac{1000}{R_1}$ since $R_2 = R_3$ and $C_2 = C_3$ from which,

$$\textbf{resistance } R_1 = \frac{1000}{2} = \textbf{500 } \boldsymbol{\Omega}$$

Also, at balance,

$$\textbf{frequency, } f = \frac{1}{2\pi\sqrt{C_2 C_3 R_2 R_3}}$$

$$= \frac{1}{2\pi\sqrt{(10^{-9})^2(30 \times 10^3)^2}} = \textbf{5.305 kHz}$$

68 Series Resonance and Q-factor

Introduction

When the voltage V applied to an electrical network containing resistance, inductance and capacitance is in phase with the resulting current I, the circuit is said to be **resonant**. The phenomenon of **resonance** is of great value in all branches of radio, television and communications engineering, since it enables small portions of the communications frequency spectrum to be selected for amplification independently of the remainder.

At resonance, the equivalent network impedance Z is purely resistive since the supply voltage and current are in phase. The power factor of a resonant network is unity (i.e. power factor $= \cos \phi = \cos 0 = 1$).

In electrical work there are two types of resonance — one associated with series circuits, which was introduced in chapter 55, when the input impedance is a minimum, (which is discussed further in this chapter), and the other associated with simple parallel networks, when the input impedance is a maximum (which was introduced in chapter 56).

Series Resonance

Figure 68.1 shows a circuit comprising a coil of inductance L and resistance R connected in series with a capacitor C. The R-L-C series circuit has a total impedance Z given by $Z = R + j(X_L - X_C)$ ohms, or $Z = R + j(\omega L - 1/\omega C)$ ohms where $\omega = 2\pi f$. The circuit is at resonance when $(X_L - X_C) = 0$, i.e. when $X_L = X_C$ or $\omega L = \dfrac{1}{\omega C}$. The phasor diagram for this condition is shown in Figure 68.2, where $|V_L| = |V_C|$. Since at resonance:

$$\omega_r L = \frac{1}{\omega_r C}, \quad \omega_r^2 = \frac{1}{LC} \text{ and } \omega_r = \frac{1}{\sqrt{LC}}$$

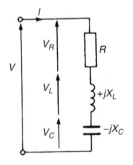

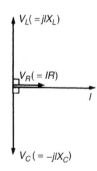

Figure 68.1   **Figure 68.2**

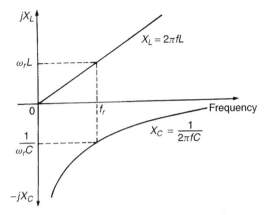

Figure 68.3

Thus resonant frequency, $\boxed{f_r = \dfrac{1}{2\pi\sqrt{LC}} \textbf{ hertz}}$ since $\omega_r = 2\pi f_r$

Figure 68.3 shows how inductive reactance X_L and capacitive reactance X_C vary with the frequency. At the resonant frequency f_r, $|X_L| = |X_C|$.

Since impedance, $Z = R + j(X_L - X_C)$ and, at resonance, $(X_L - X_C) = 0$, then **impedance $Z = R$ at resonance**. This is the **minimum** value possible for the impedance as shown in the graph of the modulus of impedance, $|Z|$, against frequency in Figure 68.4.

At frequencies less than f_r, $X_L < X_C$ and the circuit is capacitive; at frequencies greater than f_r, $X_L > X_C$ and the circuit is inductive.

Current $I = \dfrac{V}{Z}$; since impedance Z is a minimum value at resonance, the **current I has a maximum value**. At resonance, current $I = \dfrac{V}{R}$. A graph of current against frequency is also shown in Figure 68.4.

For example, a coil having a resistance of 10 Ω and an inductance of 75 mH is connected in series with a 40 μF capacitor across a 200 V a.c. supply. Then the **resonant frequency**,

$$f_r = \frac{1}{2\pi\sqrt{LC}} = \frac{1}{2\pi\sqrt{(75 \times 10^{-3})(40 \times 10^{-6})}} = \textbf{91.9 Hz}$$

and the current at resonance, $I = \dfrac{V}{R} = \dfrac{200}{10} = \textbf{20 A}$

Q-factor

Q-factor is a figure of merit for a resonant device such as an *L-C-R* circuit. Such a circuit resonates by cyclic interchange of stored energy, accompanied

502

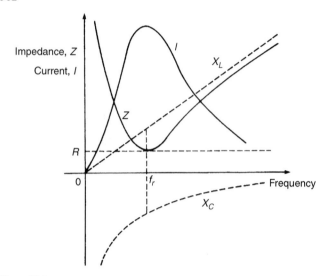

Figure 68.4

by energy dissipation due to the resistance. By definition, at resonance,

$$Q = 2\pi \left(\frac{\text{maximum energy stored}}{\text{energy loss per cycle}} \right)$$

$$= 2\pi \left(\frac{\text{maximum energy stored}}{\text{average power dissipated} \times \text{periodic time}} \right)$$

$$= 2\pi \left(\frac{\text{maximum energy stored}}{\text{average power dissipated} \times \dfrac{1}{f_r}} \right)$$

$$= 2\pi f_r \left(\frac{\text{maximum energy stored}}{\text{average power dissipated}} \right)$$

i.e. $\quad Q = \omega_r \left(\dfrac{\text{maximum energy stored}}{\text{average power dissipated}} \right)$

where ω_r is the angular frequency at resonance.

In an L-C-R circuit both of the reactive elements store energy during a quarter cycle of the alternating supply input and return it to the circuit source during the following quarter cycle. An inductor stores energy in its magnetic field, then transfers it to the electric field of the capacitor and then back to the magnetic field, and so on. Thus the inductive and capacitive elements transfer energy from one to the other successively with the source of supply ideally providing no additional energy at all. Practical reactors both store and dissipate energy. Q-factor is an abbreviation for **quality factor** and refers to the 'goodness' of a reactive component.

For an **inductor,** $\quad Q = \omega \left(\dfrac{\text{maximum energy stored}}{\text{average power dissipated}} \right)$

$$= \omega_r \left(\frac{\frac{1}{2} L I_m^2}{I^2 R} \right) = \left(\frac{\omega_r \left(\frac{1}{2} L I_m^2 \right)}{\left(\frac{I_m}{\sqrt{2}} \right)^2 R} \right) = \frac{\omega_r L}{R} \qquad (1)$$

For a **capacitor,** $\quad Q = \left(\dfrac{\omega_r \left(\frac{1}{2} C V_m^2 \right)}{\left(\dfrac{I_m}{\sqrt{2}} \right)^2 R} \right) = \left(\dfrac{\omega_r \left(\frac{1}{2} C \right) (I_m X_C)^2}{\left(\dfrac{I_m}{\sqrt{2}} \right)^2 R} \right)$

$$= \left(\frac{\omega_r \left(\frac{1}{2} C I_m^2 \right) \left(\dfrac{1}{\omega_r C} \right)^2}{\left(\dfrac{I_m}{\sqrt{2}} \right)^2 R} \right)$$

i.e. $\qquad Q = \dfrac{1}{\omega_r C R} \qquad (2)$

From expressions (1) and (2) it can be deduced that

$$Q = \frac{X_L}{R} = \frac{X_C}{R} = \frac{\text{reactance}}{\text{resistance}}$$

In fact, Q-factor can also be defined as

$$Q\text{-factor} = \frac{\text{reactive power}}{\text{active power}} = \frac{Q}{P}$$

where Q is the reactive power which is also the peak rate of energy storage, and P is the average energy dissipation rate. Hence

$$Q\text{-factor} = \frac{Q}{P} = \frac{I^2 X_L (\text{or } I^2 X_C)}{I^2 R} = \frac{X_L}{R} \left(\text{or } \frac{X_C}{R} \right)$$

i.e. $\qquad \boxed{Q = \dfrac{\textbf{reactance}}{\textbf{resistance}}}$

In an R-L-C series circuit the amount of energy stored at resonance is constant. When the capacitor voltage is a maximum, the inductor current is zero, and vice versa, i.e. $\frac{1}{2} L I_m^2 = \frac{1}{2} C V_m^2$

Thus the Q-factor at resonance, Q_r is given by

$$\boxed{Q_r = \frac{\omega_r L}{R} = \frac{1}{\omega_r C R}} \qquad (3)$$

However, at resonance $\omega_r = \dfrac{1}{\sqrt{LC}}$

Hence $\quad Q_r = \dfrac{\omega_r L}{R} = \dfrac{1}{\sqrt{LC}}\left(\dfrac{L}{R}\right)$ i.e. $\boxed{Q_r = \dfrac{1}{R}\sqrt{\dfrac{L}{C}}}$

It should be noted that when Q-factor is referred to, it is nearly always assumed to mean 'the Q-factor at resonance'

With reference to Figures 68.1 and 68.2, at resonance, $V_L = V_C$

$$V_L = IX_L = I\omega_r L = \dfrac{V}{R}\omega_r L = \left(\dfrac{\omega_r L}{R}\right)V = Q_r V$$

and $\quad V_C = IX_C = \dfrac{I}{\omega_r C} = \dfrac{V/R}{\omega_r C} = \left(\dfrac{1}{\omega_r CR}\right)V = Q_r V$

Hence, at resonance, $V_L = V_C = Q_r V$ or $\boxed{Q_r = \dfrac{V_L(\text{or } V_C)}{V}}$

The voltages V_L and V_C at resonance may be much greater than that of the supply voltage V. For this reason Q is often called the **circuit magnification factor**. It represents a measure of the number of times V_L or V_C is greater than the supply voltage.

The Q-factor at resonance can have a value of several hundreds. Resonance is usually of interest only in circuits of Q-factor greater than about 10; circuits considerably below this value are effectively merely operating at unity power factor.

For example, a series circuit comprises a 10 Ω resistance, a 5 μF capacitor and a variable inductance L. The supply voltage is $20\angle 0°$ volts at a frequency of 318.3 Hz. The inductance is adjusted until the p.d. across the 10 Ω resistance is a maximum.

The maximum voltage across the resistance occurs at resonance when the current is a maximum. At resonance, $\omega_r L = \dfrac{1}{\omega_r C}$, from which

inductance $\qquad\qquad L = \dfrac{1}{\omega_r^2 C} = \dfrac{1}{(2\pi 318.3)^2(5 \times 10^{-6})}$

$$= \textbf{0.050 H} \quad \text{or} \quad \textbf{50 mH}$$

Current at resonance $\qquad I_r = \dfrac{V}{R} = \dfrac{20\angle 0°}{10\angle 0°} = 2.0\angle 0° \text{ A}$

P.d. across resistance, $\qquad V_R = I_r R = (2.0\angle 0°)(10) = \textbf{20}\angle\textbf{0}° \text{ V}$

P.d. across inductance, $\qquad V_L = IX_L$

$$X_L = 2\pi(318.3)(0.050) = 100 \ \Omega$$

Hence $\qquad\qquad\qquad V_L = (2.0\angle 0°)(100\angle 90°) = \textbf{200}\angle\textbf{90}° \text{ V}$

P.d. across capacitor, $\qquad V_C = IX_C = (2.0\angle 0°)(100\angle -90°)$

$$= \textbf{200}\angle\textbf{-90}° \text{ V}$$

Q-factor at resonance, $\quad Q_r = \dfrac{V_L(\text{or } V_C)}{V} = \dfrac{200}{20} = \mathbf{10}$

[Alternatively, $\quad\quad Q_r = \dfrac{\omega_r L}{R} = \dfrac{100}{10} = \mathbf{10}$

or $\quad Q_r = \dfrac{1}{\omega_r CR} = \dfrac{1}{2\pi(318.3)(5 \times 10^{-6})(10)} = \mathbf{10}$

or $\quad Q_r = \dfrac{1}{R}\sqrt{\dfrac{L}{C}} = \dfrac{1}{10}\sqrt{\dfrac{0.050}{5 \times 10^{-6}}} = \mathbf{10}$]

Voltage Magnification

For a circuit with a high value of Q (say, exceeding 100), the maximum volt-drop across the coil, V_{COIL}, and the maximum volt-drop across the capacitor, V_C, coincide with the maximum circuit current at the resonant frequency f_r, as shown in Figure 68.5(a). However, if a circuit of low Q (say, less than 10) is used, it may be shown experimentally that the maximum value of V_C occurs at a frequency less than f_r while the maximum value of V_{COIL} occurs at a frequency higher than f_r, as shown in Figure 68.5(b). The maximum current, however, still occurs at the resonant frequency with low Q. It may be shown that:

$$f = f_r\sqrt{1 - \dfrac{1}{2Q^2}} \tag{4}$$

Hence the maximum p.d. across the capacitor does not occur at the resonant frequency, but at a frequency slightly less than f_r as shown in Figure 68.5(b). If Q is large, then $f \approx f_r$ as shown in Figure 68.5(a).

Also $\quad V_{C_m} = V_{L_m} = \dfrac{QV}{\sqrt{\left[1 - \left(\dfrac{1}{2Q}\right)^2\right]}}$ $\tag{5}$

From equation (5), when Q is large, $V_{C_m} = V_{L_m} \approx QV$.

For example, a series L-R-C circuit has a sinusoidal input voltage of maximum value 12 V. If inductance, $L = 20$ mH, resistance, $R = 80\ \Omega$, and capacitance, $C = 400$ nF, then the resonant frequency,

$$f_r = \dfrac{1}{2\pi\sqrt{LC}} = \dfrac{1}{2\pi\sqrt{(20 \times 10^{-3})(400 \times 10^{-9})}} = \mathbf{1779.4\ Hz}$$

$$V_C = QV \text{ and } Q = \dfrac{\omega_r L}{R} \left(\text{or } \dfrac{1}{\omega_r CR} \text{ or } \dfrac{1}{R}\sqrt{\dfrac{L}{C}}\right)$$

Hence $\quad Q = \dfrac{(2\pi)(1779.4)(20 \times 10^{-3})}{80} = 2.80$

Thus $\quad V_C = QV = (2.80)(12) = \mathbf{33.60\ V}$

506

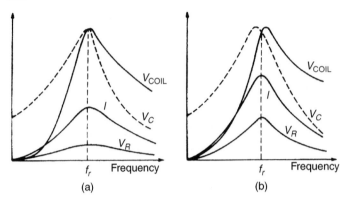

Figure 68.5

From equation (4), the frequency f at which V_C is a maximum value,

$$f = f_r\sqrt{\left(1 - \frac{1}{2Q^2}\right)} = (1779.4)\sqrt{\left(1 - \frac{1}{2(2.80)^2}\right)} = \mathbf{1721.7\ Hz}$$

From equation (5), the maximum value of the p.d. across the capacitor is given by:

$$V_{C_m} = \frac{QV}{\sqrt{\left[1 - \left(\frac{1}{2Q}\right)^2\right]}} = \frac{(2.80)(12)}{\sqrt{\left[1 - \left(\frac{1}{2(2.80)}\right)^2\right]}} = \mathbf{34.15\ V}$$

Overall Q-factor for Series Components

If the losses of a capacitor are not considered as negligible, the overall Q-factor of the circuit will depend on the Q-factor of the individual components. If the Q-factor of the inductor be Q_L and that of the capacitor be Q_C then the overall Q-factor,

$$\boxed{Q_T = \frac{Q_L Q_C}{Q_L + Q_C}}$$

For example, if an inductor of Q-factor 60 is connected in series with a capacitor having a Q-factor of 390, then the overall Q-factor,

$$Q_T = \frac{Q_L Q_C}{Q_L + Q_C} = \frac{(60)(390)}{60 + 390} = \frac{232\,400}{450} = \mathbf{52}$$

Bandwidth

Figure 68.6 shows how current I varies with frequency f in an R-L-C series circuit. At the resonant frequency f_r, current is a maximum value, shown as I_r. Also shown are the points A and B where the current is 0.707 of the maximum value at frequencies f_1 and f_2. The power delivered to the circuit is $I^2 R$. At $I = 0.707\ I_r$, the power is $(0.707\ I_r)^2 R = 0.5\ I_r^2 R$, i.e. half the power that occurs at frequency f_r. The points corresponding to f_1 and f_2 are called the **half-power points**. The distance between these points, i.e. $(f_2 - f_1)$, is called the **bandwidth**.

When the ratio of two powers P_1 and P_2 is expressed in decibel units, the number of decibels N is given by:

$$N = 10\ \lg \left(\frac{P_2}{P_1} \right)\ \text{dB}$$

Let the power at the half-power points be $(0.707\ I_r)^2 R = \dfrac{I_r^2 R}{2}$ and let the peak power be $I_r^2 R$, then the ratio of the power in decibels is given by:

$$10\ \lg \left[\frac{\dfrac{I_r^2 R}{2}}{I_r^2 R} \right] = 10\ \lg \frac{1}{2} = -3\ \textbf{dB}$$

It is for this reason that the half-power points are often referred to as **'the −3 dB points'**

At the half-power frequencies, $I = 0.707\ I_r$, thus

$$\text{impedance } Z = \frac{V}{I} = \frac{V}{0.707 I_r} = 1.414 \left(\frac{V}{I_r} \right) = \sqrt{2}\ Z_r = \sqrt{2}\ R$$

$$\text{(since at resonance } Z_r = R)$$

It may be shown that for a series R-L-C circuit:

$$\boxed{Q_r = \frac{f_r}{f_2 - f_1}\ \text{ and } f_r = \sqrt{f_1 f_2}} \tag{6}$$

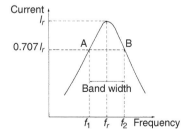

Figure 68.6

Selectivity

Selectivity is the ability of a circuit to respond more readily to signals of a particular frequency to which it is tuned than to signals of other frequencies. The response becomes progressively weaker as the frequency departs from the resonant frequency. Discrimination against other signals becomes more pronounced as circuit losses are reduced, i.e. as the Q-factor is increased. Thus $Q_r = f_r/(f_2 - f_1)$ is a measure of the circuit selectivity in terms of the points on each side of resonance where the circuit current has fallen to 0.707 of its maximum value reached at resonance. The higher the Q-factor, the narrower the bandwidth and the more selective is the circuit. Circuits having high Q-factors (say, in the order of 300) are therefore useful in communications engineering. A high Q-factor in a series power circuit has disadvantages in that it can lead to dangerously high voltages across the insulation and may result in electrical breakdown.

For example, suppose that the working voltage of a capacitor is stated as 1 kV and is used in a circuit having a supply voltage of 240 V. The maximum value of the supply will be $\sqrt{2}$ (240), i.e. 340 V. The working voltage of the capacitor would appear to be ample. However, if the Q-factor is, say, 10, the voltage across the capacitor will reach 2.4 kV. Since the capacitor is rated only at 1 kV, dielectric breakdown is more than likely to occur.

Low Q-factors, say, in the order of 5 to 25, may be found in power transformers using laminated iron cores.

A capacitor-start induction motor, as used in domestic appliances such as washing machines and vacuum cleaners, having a Q-factor as low as 1.5 at starting would result in a voltage across the capacitor 1.5 times that of the supply voltage; hence the cable joining the capacitor to the motor would require extra insulation.

For example, an R-L-C series circuit has a resonant frequency of 1.2 kHz and a Q-factor at resonance of 30. If the impedance of the circuit at resonance is 50 Ω, then

at resonance the circuit impedance, $\quad Z = R$, i.e. $R = 50\ \Omega$

and the Q-factor at resonance, $\qquad Q_r = \dfrac{\omega_r L}{R}$

Hence, **inductance,** $\qquad L = \dfrac{Q_r R}{\omega_r} = \dfrac{(30)\,(50)}{(2\pi 1200)}$

$$= \textbf{0.199 H or 199 mH}$$

At resonance $\qquad \omega_r L = \dfrac{1}{\omega_r C}$

Hence, **capacitance,** $\qquad C = \dfrac{1}{\omega_r^2 L} = \dfrac{1}{(2\pi 1200)^2\,(0.199)}$

$$= \textbf{0.088 μF or 88 nF}$$

Q-factor at resonance is also given by: $Q_r = f_r/(f_2 - f_1)$, from which,

$$\textbf{bandwidth, } (f_2 - f_1) = \dfrac{f_r}{Q_r} = \dfrac{1200}{30} = \textbf{40 Hz}$$

From equation (6), resonant frequency, $f_r = \sqrt{f_1 f_2}$

i.e. $\qquad\qquad\qquad 1200 = \sqrt{f_1 f_2}$

from which, $\qquad\quad f_1 f_2 = (1200)^2 = 1.44 \times 10^6$ $\qquad$ (7)

From above, $\qquad\quad f_2 - f_1 = 40$ $\qquad\qquad\qquad\qquad$ (8)

From equation (7), $\qquad\quad f_1 = \dfrac{1.44 \times 10^6}{f_2}$

Substituting in equation (8) gives: $f_2 - \dfrac{1.44 \times 10^6}{f_2} = 40$

Multiplying throughout by f_2 gives: $f_2^2 - 1.44 \times 10^6 = 40 f_2$

i.e. $\qquad f_2^2 - 40 f_2 - 1.44 \times 10^6 = 0$

Then $\quad f_2 = \dfrac{40 \pm \sqrt{[(40)^2 - 4(-1.44 \times 10^6)]}}{2} = \dfrac{40 \pm 2400}{2}$

$\qquad\quad = \dfrac{40 + 2400}{2}$ (since f_2 cannot be negative)

Hence **the upper half-power frequency, $f_2 = 1220$ Hz**
From equation (8), **the lower half-power frequency,**

$$f_1 = f_2 - 40 = 1220 - 40 = \mathbf{1180 \ Hz}$$

Note that the upper and lower half-power frequency values are symmetrically placed about the resonance frequency. This is usually the case when the Q-factor has a high value (say, > 10).

Small Deviations from the Resonant Frequency

Let ω_1 be a frequency below the resonant frequency ω_r in an L-R-C series circuit, and ω_2 be a frequency above ω_r by the same amount as ω_1 is below, i.e. $\omega_r - \omega_1 = \omega_2 - \omega_r$

Let the fractional deviation from the resonant frequency be δ where

$$\delta = \frac{\omega_r - \omega_1}{\omega_r} = \frac{\omega_2 - \omega_r}{\omega_r}$$

It may be shown that **at frequency ω_2,**

$$\boxed{\frac{I}{I_r} = \frac{1}{1 + j\delta Q}} \qquad (9)$$

and

$$\boxed{\frac{Z}{Z_r} = 1 + j 2\delta Q} \qquad (10)$$

It may also be shown that **at frequency ω_1,**

$$\boxed{\frac{I}{I_r} = \frac{1}{1 + j2\delta Q}}$$

and

$$\boxed{\frac{Z}{Z_r} = 1 - j2\delta Q}$$

For example, in an L-R-C series network, the inductance, $L = 8$ mH, the capacitance, $C = 0.3\ \mu F$, and the resistance, $R = 15\ \Omega$. To determine the current flowing in the circuit when the input voltage is $7.5\angle 0°$ V and the frequency is (a) the resonant frequency, (b) a frequency 3% above the resonant frequency:

(a) At resonance, $Z_r = R = 15\ \Omega$

Current at resonance, $I_r = \dfrac{V}{Z_r} = \dfrac{7.5\angle 0°}{15\angle 0°} = \mathbf{0.5\angle 0°}$ **A**

(b) If the frequency is 3% above the resonant frequency, then $\delta = 0.03$

From equation (9),

$$\frac{I}{I_r} = \frac{1}{1 + j2\delta Q}$$

$$Q = \frac{1}{R}\sqrt{\frac{L}{C}} = \frac{1}{15}\sqrt{\frac{8 \times 10^{-3}}{0.3 \times 10^{-6}}} = 10.89$$

Hence $\dfrac{I}{0.5\angle 0°} = \dfrac{1}{1 + j2\,(0.03)\,(10.89)}$

$$= \frac{1}{1 + j0.6534} = \frac{1}{1.1945\angle 33.16°}$$

and $I = \dfrac{0.5\angle 0°}{1.1945\angle 33.16°} = \mathbf{0.4186\angle -33.16°}$ **A**

The impedance of the circuit when the frequency is 3% above the resonant frequency is determined as follows:

From equation (10), $\dfrac{Z}{Z_r} = 1 + j2\delta Q$ hence,

$$Z = Z_r(1 + j2\delta Q) = R(1 + j2\delta Q) = 15(1 + j2(0.03)(10.89))$$

$$= 15(1 + j0.6534) = 15(1.1945\angle 33.16°) = \mathbf{17.92\angle 33.16°\ \Omega}$$

Alternatively,

$$Z = \frac{V}{I} = \frac{7.5\angle 0°}{0.4186\angle -33.16°} = \mathbf{17.92\angle 33.16°\ \Omega}$$

69 Parallel Resonance and Q-factor

Introduction to Parallel Resonance

A parallel network containing resistance R, pure inductance L and pure capacitance C connected in parallel is shown in Figure 69.1. Since the inductance and capacitance are considered as pure components, this circuit is something of an 'ideal' circuit. However, it may be used to highlight some important points regarding resonance, which are applicable to any parallel circuit. From Figure 69.1,

$$\text{the admittance of the resistive branch, } G = \frac{1}{R}$$

$$\left.\begin{array}{c}\text{the admittance of the}\\ \text{inductive branch,}\end{array}\right\} B_L = \frac{1}{jX_L} = \frac{-j}{\omega L}$$

$$\left.\begin{array}{c}\text{the admittance of the}\\ \text{capacitive branch,}\end{array}\right\} B_C = \frac{1}{-jX_C}$$

$$= \frac{j}{1/\omega C} = j\omega C$$

$$\text{Total circuit admittance, } Y = G + j(B_C - B_L)$$

$$\text{i.e. } Y = \frac{1}{R} + j\left(\omega C - \frac{1}{\omega L}\right)$$

The circuit is at resonance when the imaginary part is zero, i.e. when $\left(\omega C - \dfrac{1}{\omega L}\right) = 0$. Hence at resonance $\omega_r C = \dfrac{1}{\omega_r L}$ and $\omega_r^2 = \dfrac{1}{LC}$ from which $\omega_r = \dfrac{1}{\sqrt{LC}}$ and the resonant frequency

$$\boxed{f_r = \frac{1}{2\pi\sqrt{LC}} \text{ hertz}}$$

the same expression as for a series R-L-C circuit.

Figure 64.2 shows typical graphs of B_C, B_L, G and Y against frequency f for the circuit shown in Figure 69.1. At resonance, $B_C = B_L$ and admittance

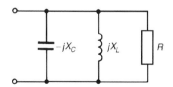

Figure 69.1

$Y = G = \dfrac{1}{R}$. This represents the condition of **minimum admittance** for the circuit and thus **maximum impedance**.

Since current $I = V/Z = VY$, the **current** is also at a **minimum** value at resonance in a parallel network.

From the ideal circuit of Figure 69.2 we have therefore established the following facts which apply to any parallel circuit. At resonance:

(i) admittance Y is a minimum
(ii) impedance Z is a maximum
(iii) current I is a minimum
(iv) an expression for the resonant frequency f_r may be obtained by making the 'imaginary' part of the complex expression for admittance equal to zero.

The *LR-C* Parallel Network

A more practical network, containing a coil of inductance L and resistance R in parallel with a pure capacitance C, is shown in Figure 69.3.

Admittance of coil, $\qquad Y_{\text{COIL}} = \dfrac{1}{R + jX_L} = \dfrac{R - jX_L}{R^2 + X_L^2}$

$$= \dfrac{R}{R^2 + \omega^2 L^2} - \dfrac{j\omega L}{R^2 + \omega^2 L^2}$$

Admittance of capacitor, $\quad Y_C = \dfrac{1}{-jX_C} = \dfrac{j}{X_C} = j\omega C$

Total circuit admittance, $\qquad Y = Y_{\text{COIL}} + Y_C$

$$= \dfrac{R}{R^2 + \omega^2 L^2} - \dfrac{j\omega L}{R^2 + \omega^2 L^2} + j\omega C \qquad (1)$$

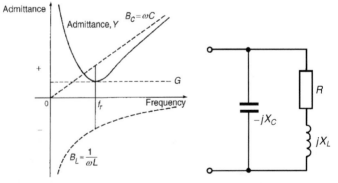

Figure 69.2   **Figure 69.3**

At resonance, the total circuit admittance Y is real $\left(Y = \dfrac{R}{R^2 + \omega^2 L^2}\right)$, i.e. the imaginary part is zero.

Hence, at resonance: $\qquad \dfrac{-\omega_r L}{R^2 + \omega_r^2 L^2} + \omega_r C = 0$

Therefore $\qquad \dfrac{\omega_r L}{R^2 + \omega_r^2 L^2} = \omega_r C$ and $\dfrac{L}{C} = R^2 + \omega_r^2 L^2$

Thus $\qquad \omega_r^2 L^2 = \dfrac{L}{C} - R^2$

and $\qquad \omega_r^2 = \dfrac{L}{CL^2} - \dfrac{R^2}{L^2} = \dfrac{1}{LC} - \dfrac{R^2}{L^2}$ (2)

Hence $\qquad \omega_r = \sqrt{\dfrac{1}{LC} - \dfrac{R^2}{L^2}}$

and

$$\boxed{\text{resonant frequency, } f_r = \frac{1}{2\pi}\sqrt{\frac{1}{LC} - \frac{R^2}{L^2}} \text{ Hz}}$$ (3)

Note that when $\dfrac{R^2}{L^2} \ll \dfrac{1}{LC}$ then $f_r = \dfrac{1}{2\pi\sqrt{LC}}$, as for the series R-L-C circuit. Equation (3) is the same as obtained in chapter 56, page 368; however, the above method may be applied to any parallel network.

Dynamic Resistance

Since the current at resonance is in phase with the voltage, the impedance of the network acts as a resistance. This resistance is known as the **dynamic resistance, R_D**. Impedance at resonance, $R_D = \dfrac{V}{I_r}$, where I_r is the current at resonance.

$I_r = VY_r = V\left(\dfrac{R}{R^2 + \omega_r^2 L^2}\right)$ from equation (1) with the j terms equal to zero.

Hence $\quad R_D = \dfrac{V}{I_r} = \dfrac{V}{\dfrac{VR}{R^2 + \omega_r^2 L^2}} = \dfrac{R^2 + \omega_r^2 L^2}{R}$

$$= \dfrac{R^2 + L^2\left(\dfrac{1}{LC} - \dfrac{R^2}{L^2}\right)}{R} \text{ from equation (2)}$$

$$= \frac{R^2 + \dfrac{L}{C} - R^2}{R} = \frac{\dfrac{L}{C}}{R} = \frac{L}{CR}$$

Hence $\boxed{\textbf{dynamic resistance, } R_D = \dfrac{L}{CR}}$ (4)

Q-factor in a Parallel Network

The Q-factor in the series R-L-C circuit is a measure of the voltage magnification. In a parallel circuit, currents higher than the supply current can circulate within the parallel branches of a parallel resonant network, the current leaving the capacitor and establishing the magnetic field of the inductance, this then collapsing and recharging the capacitor, and so on. The Q-factor of a parallel resonant circuit is the ratio of the current circulating in the parallel branches of the circuit to the supply current, i.e. in a parallel circuit, Q-factor is a measure of the **current magnification**.

Circulating currents may be several hundreds of times greater than the supply current at resonance. For the parallel network of Figure 69.4, the Q-factor at resonance is given by:

$$\boxed{Q_r = \frac{\text{circulating current}}{\text{current at resonance}} = \frac{\text{capacitor current}}{\text{current at resonance}} = \frac{I_C}{I_r}}$$

Current in capacitor, $\quad I_C = V/X_C = V\omega_r C$

Current at resonance, $\quad I_r = \dfrac{V}{R_D} = \dfrac{V}{L/CR} = \dfrac{VCR}{L}$

Hence $\qquad Q_r = \dfrac{I_C}{I_r} = \dfrac{V\omega_r C}{VCR/L}$ i.e. $\boxed{Q_r = \dfrac{\omega_r L}{R}}$

the same expression as for series resonance.

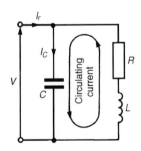

Figure 69.4

For example, a coil of inductance 5 mH and resistance 10 Ω is connected in parallel with a 250 nF capacitor across a 50 V variable-frequency supply. Then the resonance frequency

$$f_r = \frac{1}{2\pi} \sqrt{\frac{1}{LC} - \frac{R^2}{L^2}} \text{ from equation (3)}$$

$$= \frac{1}{2\pi} \sqrt{\frac{1}{5 \times 10^{-3} \times 250 \times 10^{-9}} - \frac{10^2}{(5 \times 10^{-3})^2}}$$

$$= \frac{1}{2\pi} \sqrt{(800 \times 10^6 - 4 \times 10^6)}$$

$$= \frac{1}{2\pi} \sqrt{(796 \times 10^6)} = \mathbf{4490 \ Hz}$$

From equation (4), dynamic resistance,

$$R_D = \frac{L}{CR} = \frac{5 \times 10^{-3}}{(250 \times 10^{-9})(10)} = \mathbf{2000 \ \Omega}$$

Current at resonance, $I_r = \dfrac{V}{R_D} = \dfrac{50}{2000} = \mathbf{25 \ mA}$

Q-factor at resonance, $Q_r = \dfrac{\omega_r L}{R} = \dfrac{(2\pi 4490)(5 \times 10^{-3})}{10} = \mathbf{14.1}$

Natural and Forced Resonant Frequency

The difference between the resonant frequency of a series circuit and that of a parallel circuit can be quite small. The resonant frequency of a coil in parallel with a capacitor is shown in equation (3); however, around the closed loop comprising the coil and capacitor the energy would naturally resonate at a frequency given by that for a series R-L-C circuit, as shown in chapter 68. This latter frequency is termed the **natural frequency**, f_n, and the frequency of resonance seen at the terminals of Figure 69.4 is often called the **forced resonant frequency**, f_r. (For a series circuit, the forced and natural frequencies coincide).

From the coil-capacitor loop of Figure 69.4 it may be shown that

$$\boxed{f_r = f_n \sqrt{\left(1 - \frac{1}{Q^2}\right)}}$$

Thus it is seen that even with small values of Q the difference between f_r and f_n tends to be very small. A high value of Q makes the parallel resonant frequency tend to the same value as that of the series resonant frequency. The expressions already obtained in chapter 68 for bandwidth and resonant frequency, also apply to parallel circuits,

i.e.
$$Q_r = \frac{f_r}{(f_2 - f_1)} \tag{5}$$

and
$$f_r = \sqrt{f_1 f_2} \tag{6}$$

The *LR-CR* Parallel Network

A more general network comprising a coil of inductance L and resistance R_L in parallel with a capacitance C and resistance R_C in series is shown in Figure 69.5.

Admittance of inductive branch,

$$Y_L = \frac{1}{R_L + jX_L} = \frac{R_L - jX_L}{R_L^2 + X_L^2} = \frac{R_L}{R_L^2 + X_L^2} - \frac{jX_L}{R_L^2 + X_L^2}$$

Admittance of capacitive branch,

$$Y_C = \frac{1}{R_C - jX_C} = \frac{R_C + jX_C}{R_C^2 + X_C^2} = \frac{R_C}{R_C^2 + X_C^2} + \frac{jX_C}{R_C^2 + X_C^2}$$

Total network admittance,

$$Y = Y_L + Y_C = \frac{R_L}{R_L^2 + X_L^2} - \frac{jX_L}{R_L^2 + X_L^2} + \frac{R_C}{R_C^2 + X_C^2} + \frac{jX_C}{R_C^2 + X_C^2}$$

At resonance the admittance is a minimum, i.e. when the imaginary part of Y is zero.

Hence, at resonance,
$$\frac{-X_L}{R_L^2 + X_L^2} + \frac{X_C}{R_C^2 + X_C^2} = 0$$

i.e.
$$\frac{\omega_r L}{R_L^2 + \omega^2 L^2} = \frac{1/(\omega_r C)}{R_C^2 + (1/\omega_r^2 C^2)} \tag{7}$$

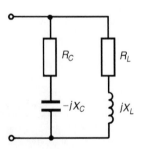

Figure 69.5

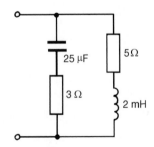

Figure 69.6

Rearranging gives: $\omega_r L \left(R_C^2 + \dfrac{1}{\omega_r^2 C^2} \right) = \dfrac{1}{\omega_r C}(R_L^2 + \omega_r^2 L^2)$

$$\omega_r L R_C^2 + \dfrac{L}{\omega_r C^2} = \dfrac{R_L^2}{\omega_r C} + \dfrac{\omega_r L^2}{C}$$

Multiplying throughout by $\omega_r C^2$ gives:

$$\omega_r^2 C^2 L R_C^2 + L = R_L^2 C + \omega_r^2 L^2 C$$

$$\omega_r^2 (C^2 L R_C^2 - L^2 C) = R_L^2 C - L$$

$$\omega_r^2 C L (C R_C^2 - L) = R_L^2 C - L$$

Hence $\qquad \omega_r^2 = \dfrac{(C R_L^2 - L)}{LC(C R_C^2 - L)}$

i.e. $\qquad \omega_r = \dfrac{1}{\sqrt{(LC)}} \sqrt{\dfrac{R_L^2 - (L/C)}{R_C^2 - (L/C)}}$

Hence $\qquad \boxed{\textbf{resonant frequency, } f_r = \dfrac{1}{2\pi\sqrt{(LC)}} \sqrt{\dfrac{R_L^2 - (L/C)}{R_C^2 - (L/C)}}}$ $\qquad$ (8)

It is clear from equation (7) that parallel resonance may be achieved in such a circuit in several ways — by varying either the frequency f, the inductance L, the capacitance C, the resistance R_L or the resistance R_C.

For example, a two-branch parallel network is shown in Figure 69.6. From equation (8), the resonant frequency,

$$f_r = \dfrac{1}{2\pi\sqrt{[(2 \times 10^{-3})(25 \times 10^{-6})]}} \sqrt{\dfrac{5^2 - ((2 \times 10^{-3})/(25 \times 10^{-6}))}{3^2 - ((2 \times 10^{-3})/(25 \times 10^{-6}))}}$$

$$= \dfrac{1}{2\pi\sqrt{5 \times 10^{-8}}} \sqrt{\dfrac{25 - 80}{9 - 80}} = \dfrac{10^4}{2\pi\sqrt{5}} \sqrt{\dfrac{-55}{-71}} = \textbf{626.5 Hz}$$

Q-factor of Parallel Components

The overall Q-factor Q_T of two parallel components having different Q-factors is given by:

$$\boxed{Q_T = \dfrac{Q_L Q_C}{Q_L + Q_C}} \qquad (9)$$

as for the series circuit.

For example, if a capacitor having a Q-factor of 300 is connected in parallel with a coil having a Q-factor of 60,

the overall Q-factor $Q_T = \dfrac{Q_L Q_C}{Q_L + Q_C} = \dfrac{(60)(300)}{60 + 300} = \dfrac{18\,000}{360} = \mathbf{50}$

Small Deviations from the Resonant Frequency

It may be shown that for a parallel network:

$$\boxed{\dfrac{Y}{Y_r} = \dfrac{R_D}{Z} = 1 + j2\delta Q} \tag{10}$$

where Y is the circuit admittance, Y_r is the admittance at resonance, Z is the network impedance, R_D is the dynamic resistance (i.e. the impedance at resonance) and δ is the fractional deviation from the resonant frequency.

For example, in an LR-C network, the capacitance is 10.61 nF, the bandwidth is 500 Hz and the resonant frequency is 150 kHz. Then, from equation (5),

$$Q = \dfrac{f_r}{f_2 - f_1} = \dfrac{150 \times 10^3}{500} = \mathbf{300}$$

and from equation (4), dynamic resistance, $R_D = \dfrac{L}{CR}$

Also, in an LR-C network, $Q = \dfrac{\omega_r L}{R}$ from which, $R = \dfrac{\omega_r L}{Q}$

Hence, $\quad R_D = \dfrac{L}{CR} = \dfrac{L}{C\left(\dfrac{\omega_r L}{Q}\right)} = \dfrac{LQ}{C\omega_r L} = \dfrac{Q}{\omega_r C}$

$$= \dfrac{300}{(2\pi 150 \times 10^3)(10.61 \times 10^{-9})} = \mathbf{30 \ k\Omega}$$

The magnitude of the impedance when the supply frequency is, say, 0.4% greater than the tuned frequency is determined as follows:

From equation (10), $\dfrac{R_D}{Z} = 1 + j2\delta Q$

from which, $\quad Z = \dfrac{R_D}{1 + j2\delta Q} \qquad \delta = 0.4\% = 0.004$

hence $\quad Z = \dfrac{30 \times 10^3}{1 + j2(0.004)(300)} = \dfrac{30 \times 10^3}{1 + j2.4}$

$$= \dfrac{30 \times 10^3}{2.6\angle 67.38°} = 11.54\angle -67.38° \ k\Omega$$

Hence **the magnitude of the impedance** when the frequency is 0.4% greater than the tuned frequency is **11.54 kΩ.**

70 Introduction to Network Analysis

Introduction

Voltage sources in series-parallel networks cause currents to flow in each branch of the circuit and corresponding volt-drops occur across the circuit components. A.c. circuit (or network) analysis involves the determination of the currents in the branches and/or the voltages across components.

The laws which determine the currents and voltage drops in a.c. networks are:

(a) **current, $I = \dfrac{V}{Z}$**, where Z is the complex impedance and V the voltage across the impedance,

(b) **the laws for impedances in series and parallel**, i.e. total impedance, $Z_T = Z_1 + Z_2 + Z_3 + \ldots + Z_n$ for n impedances connected in series, and $\dfrac{1}{Z_T} = \dfrac{1}{Z_1} + \dfrac{1}{Z_2} + \dfrac{1}{Z_3} + \ldots + \dfrac{1}{Z_n}$ for n impedances connected in parallel, and

(c) **Kirchhoff's laws,** which may be stated as:

(i) 'At any point in an electrical circuit the phasor sum of the currents flowing towards that junction is equal to the phasor sum of the currents flowing away from the junction'.

(ii) 'In any closed loop in a network, the phasor sum of the voltage drops (i.e. the products of current and impedance) taken around the loop is equal to the phasor sum of the e.m.f.'s acting in that loop'.

In any circuit the currents and voltages at any point may be determined by applying Kirchhoff's laws (as demonstrated in this chapter), or by extensions of Kirchhoff's laws, called mesh-current analysis and nodal analysis (see chapter 71).

However, for more complicated circuits, a number of circuit theorems have been developed as alternatives to the use of Kirchhoff's laws to solve problems involving both d.c. and a.c. electrical networks. These include:

(a) the superposition theorem (see chapter 72)
(b) Thévenin's theorem (see chapter 73)
(c) Norton's theorem (see chapter 73),
(d) the maximum power transfer theorems (see chapter 75).

In addition to these theorems, and often used as a preliminary to using circuit theorems, star-delta (or $T - \pi$) and delta-star (or $\pi - T$) transformations provide a method for simplifying certain circuits (see chapter 74).

In a.c. circuit analysis involving Kirchhoff's laws or circuit theorems, the use of complex numbers is essential.

Solution of Simultaneous Equations using Determinants

When Kirchhoff's laws are applied to electrical circuits, simultaneous equations result which require solution. If two loops are involved, two simultaneous equations containing two unknowns need to be solved; if three loops are involved, three simultaneous equations containing three unknowns need to be solved and so on. The elimination and substitution methods of solving simultaneous equations may be used to solve such equations. However a more convenient method is to use **determinants.**

Two unknowns

When solving linear simultaneous equations in two unknowns using determinants:

(i) the equations are initially written in the form:

$$a_1x + b_1y + c_1 = 0$$

$$a_2x + b_2y + c_2 = 0$$

(ii) the solution is given by: $\dfrac{x}{D_x} = \dfrac{-y}{D_y} = \dfrac{1}{D}$

where
$$D_x = \begin{vmatrix} b_1 & c_1 \\ b_2 & c_2 \end{vmatrix}$$

i.e. the determinant of the coefficients left when the x-column is 'covered up',

$$D_y = \begin{vmatrix} a_1 & c_1 \\ a_2 & c_2 \end{vmatrix}$$

i.e. the determinant of the coefficients left when the y-column is 'covered up', and

$$D = \begin{vmatrix} a_1 & b_1 \\ a_2 & b_2 \end{vmatrix}$$

i.e. the determinant of the coefficients left when the constants-column is 'covered up'

A '2 by 2' determinant $\begin{vmatrix} a & b \\ c & d \end{vmatrix}$ is evaluated as $ad - bc$.

Three unknowns

When solving linear simultaneous equations in three unknowns using determinants:

(i) the equations are initially written in the form:

$$a_1x + b_1y + c_1z + d_1 = 0$$

$$a_2x + b_2y + c_2z + d_2 = 0$$

$$a_3x + b_3y + c_3z + d_3 = 0$$

(ii) the solution is given by: $\dfrac{x}{D_x} = \dfrac{-y}{D_y} = \dfrac{z}{D_z} = \dfrac{-1}{D}$

where $\quad D_x = \begin{vmatrix} b_1 & c_1 & d_1 \\ b_2 & c_2 & d_2 \\ b_3 & c_3 & d_3 \end{vmatrix} \qquad D_y = \begin{vmatrix} a_1 & c_1 & d_1 \\ a_2 & c_2 & d_2 \\ a_3 & c_3 & d_3 \end{vmatrix}$

$D_z = \begin{vmatrix} a_1 & b_1 & d_1 \\ a_2 & b_2 & d_2 \\ a_3 & b_3 & d_3 \end{vmatrix}$ and $D = \begin{vmatrix} a_1 & b_1 & c_1 \\ a_2 & b_2 & c_2 \\ a_3 & b_3 & c_3 \end{vmatrix}$

To evaluate a 3 × 3 determinant :

(a) The **minor** of an element of a 3 by 3 matrix is the value of the 2 by 2 determinant obtained by covering up the row and column containing that element.

Thus for the matrix $\begin{pmatrix} 1 & 2 & 3 \\ 4 & 5 & 6 \\ 7 & 8 & 9 \end{pmatrix}$ the minor of element 4 is the determinant

$\begin{vmatrix} 2 & 3 \\ 8 & 9 \end{vmatrix}$ i.e. $(2 \times 9) - (3 \times 8) = 18 - 24 = -6$. Similarly, the minor of

element 3 is $\begin{vmatrix} 4 & 5 \\ 7 & 8 \end{vmatrix}$ i.e. $(4 \times 8) - (5 \times 7) = 32 - 35 = -3$.

(b) The sign of the minor depends on its position within the matrix, the sign

pattern being $\begin{pmatrix} + & - & + \\ - & + & - \\ + & - & + \end{pmatrix}$. Thus the signed minor of element 4 in the

above matrix is $-\begin{vmatrix} 2 & 3 \\ 8 & 9 \end{vmatrix} = -(-6) = +6$.

The signed-minor of an element is called the **cofactor** of the element.

Thus the cofactor of element 2 is $-\begin{vmatrix} 4 & 6 \\ 7 & 9 \end{vmatrix} = -(36 - 42) = +6$.

(c) **The value of a 3 by 3 determinant is the sum of the products of the elements and their cofactors of any row or any column of the corresponding 3 by 3 matrix.**

Thus a 3 by 3 determinant $\begin{vmatrix} a & b & c \\ d & e & f \\ g & h & j \end{vmatrix}$

is evaluated as:

$$a \begin{vmatrix} e & f \\ h & j \end{vmatrix} - b \begin{vmatrix} d & f \\ g & j \end{vmatrix} + c \begin{vmatrix} d & e \\ g & h \end{vmatrix} \quad \text{using the top row,}$$

$$\text{or } -b \begin{vmatrix} d & f \\ g & j \end{vmatrix} + e \begin{vmatrix} a & c \\ g & j \end{vmatrix} - h \begin{vmatrix} a & c \\ d & f \end{vmatrix} \quad \text{using the second column.}$$

There are thus six ways of evaluating a 3 by 3 determinant.

522

Network Analysis using Kirchhoff's Laws

Kirchhoff's laws may be applied to both d.c. and a.c. circuits. The laws are introduced in chapter 53 for d.c. circuits. To demonstrate the method of analysis, consider the d.c. network shown in Figure 70.1. If the current flowing in each branch is required, the following three-step procedure may be used:

(i) Label branch currents and their directions on the circuit diagram. The directions chosen are arbitrary but, as a starting-point, a useful guide is to assume that current flows from the positive terminals of the voltage sources. This is shown in Figure 70.2 where the three branch currents are expressed in terms of I_1 and I_2 only, since the current through resistance R, by Kirchhoff's current law, is $(I_1 + I_2)$.

(ii) Divide the circuit into loops — two in this ease (see Figure 70.2) and then apply Kirchhoff's voltage law to each loop in turn. From loop ABEF, and moving in a clockwise direction (the choice of loop direction is arbitrary), $E_1 = I_1 r + (I_1 + I_2)R$ (note that the two voltage drops are positive since the loop direction is the same as the current directions involved in the volt drops). Hence

$$8 = I_1 + 5(I_1 + I_2)$$

or $\quad 6I_1 + 5I_2 = 8$ $\qquad\qquad$ (1)

From loop BCDE in Figure 70.2, and moving in an anticlockwise direction, (note that the direction does not have to be the same as that used for the first loop),

$$E_2 = I_2 r_2 + (I_1 + I_2)R$$

i.e. $\quad 3 = 2I_2 + 5(I_1 + I_2)$

or $\quad 5I_1 + 7I_2 = 3$ $\qquad\qquad$ (2)

(iii) Solve simultaneous equations (1) and (2) for I_1 and I_2

Multiplying equation (1) by 7 gives: $\quad 42I_1 + 35I_2 = 56$ $\qquad$ (3)

Multiplying equation (2) by 5 gives: $\quad 25I_1 + 35I_2 = 15$ $\qquad$ (4)

Equation (3) − equation (4) gives: $\quad 17I_1 = 41$

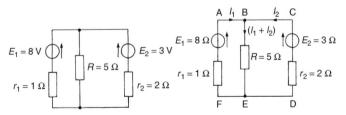

Figure 70.1 $\qquad\qquad\qquad\qquad$ **Figure 70.2**

from which, current

$$I_1 = \frac{41}{17} = 2.412 \text{ A} = \textbf{2.41 A}, \text{ correct to two decimal places.}$$

From equation (1): $6(2.412) + 5I_2 = 8$ from which, current,

$$I_2 = \frac{8 - 6(2.412)}{5} = -1.294 \text{ A}$$

$$= \textbf{-1.29 A}, \text{ correct to two decimal places.}$$

The minus sign indicates that current I_2 flows in the opposite direction to that shown in Figure 70.2.
The current flowing through resistance R is $(I_1 + I_2) =$
$2.412 + (-1.294) = 1.118 \text{ A} = \textbf{1.12 A}$, correct to two decimal places.
[A third loop may be selected in Figure 70.2, (just as a check), moving clockwise around the outside of the network. Then

$$E_1 - E_2 = I_1 r_1 - I_2 r_2 \quad \text{i.e.} \quad 8 - 3 = I_1 - 2I_2$$

Thus $\qquad\qquad\qquad\qquad\qquad\qquad 5 = 2.412 - 2(-1.294) = 5]$

An alternative method of solving equations (1) and (2) is shown below using determinants. Since

$$6I_1 + 5I_2 - 8 = 0 \qquad\qquad (1)$$

$$5I_1 + 7I_2 - 3 = 0 \qquad\qquad (2)$$

then

$$\frac{I_1}{\begin{vmatrix} 5 & -8 \\ 7 & -3 \end{vmatrix}} = \frac{-I_2}{\begin{vmatrix} 6 & -8 \\ 5 & -3 \end{vmatrix}} = \frac{1}{\begin{vmatrix} 6 & 5 \\ 5 & 7 \end{vmatrix}}$$

i.e. $\qquad \dfrac{I_1}{-15 + 56} = \dfrac{-I_2}{-18 + 40} = \dfrac{1}{42 - 25}$

$$\frac{I_1}{41} = \frac{-I_2}{22} = \frac{1}{17}$$

from which, $I_1 = \dfrac{41}{17} = \textbf{2.41 A}$ and $I_2 = -\dfrac{22}{17} = \textbf{-1.29 A}$, as obtained previously.

The above procedure is shown for a simple d.c. circuit having two unknown values of current. The procedure however applies equally well to a.c. networks and/or to circuits where three unknown currents are involved.
For example, in the network shown in Figure 70.3, the magnitude of the current in the $(4 + j3)$ Ω impedance using Kirchhoff's laws is determined as follows:

(i) Currents I_1, I_2 and I_3 with their directions are shown in Figure 70.4. The current in the $(4 + j3)$ Ω impedance is specified by one symbol only (i.e. I_3), which means that the three equations formed need to be solved for only one unknown current.

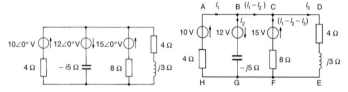

Figure 70.3   **Figure 70.4**

(ii) Three loops are chosen. From loop ABGH, and moving clockwise,

$$4I_1 - j5I_2 = 10 + 12 \qquad (1)$$

From loop BCFG, and moving anticlockwise,

$$-j5I_2 - 8(I_1 - I_2 - I_3) = 15 + 12 \qquad (2)$$

From loop CDEF, and moving clockwise,

$$-8(I_1 - I_2 - I_3) + (4 + j3)(I_3) = 15 \qquad (3)$$

Hence
$$4I_1 - j5I_2 + 0I_3 - 22 = 0$$

$$-8I_1 + (8 - j5)I_2 + 8I_3 - 27 = 0$$

$$-8I_1 + 8I_2 + (12 + j3)I_3 - 15 = 0$$

Solving for I_3 using determinants gives:

$$\frac{I_3}{\begin{vmatrix} 4 & -j5 & -22 \\ -8 & (8-j5) & -27 \\ -8 & 8 & -15 \end{vmatrix}} = \frac{-1}{\begin{vmatrix} 4 & -j5 & 0 \\ -8 & (8-j5) & 8 \\ -8 & 8 & (12+j3) \end{vmatrix}}$$

Thus

$$\frac{I_3}{4\begin{vmatrix} (8-j5) & -27 \\ 8 & -15 \end{vmatrix} + j5\begin{vmatrix} -8 & -27 \\ -8 & -15 \end{vmatrix} - 22\begin{vmatrix} -8 & (8-j5) \\ -8 & 8 \end{vmatrix}}$$

$$= \frac{-1}{4\begin{vmatrix} (8-j5) & 8 \\ 8 & (12+j3) \end{vmatrix} + j5\begin{vmatrix} -8 & 8 \\ -8 & (12+j3) \end{vmatrix}}$$

Hence $\dfrac{I_3}{384 + j700} = \dfrac{-1}{308 - j304}$ from which,

$$I_3 = \frac{-(384 + j700)}{(308 - j304)}$$

$$= \frac{798.41\angle -118.75}{432.76\angle -44.63°}$$

$$= 1.85\angle -74.12° \text{ A}$$

Hence the magnitude of the current flowing in the $(4 + j3)$ Ω impedance is 1.85 A

71 Mesh-current and Nodal Analysis

Mesh-current Analysis

Mesh-current analysis is merely an extension of the use of Kirchhoff's laws, explained in chapter 70. Figure 71.1 shows a network whose circulating currents I_1, I_2 and I_3 have been assigned to closed loops in the circuit rather than to branches. Currents I_1, I_2 and I_3 are called **mesh-currents** or **loop-currents**.

In mesh-current analysis the loop-currents are all arranged to circulate in the **same** direction (in Figure 71.1, shown as clockwise direction). Kirchhoff's second law is applied to each of the loops in turn, which in the circuit of Figure 71.1 produces three equations in three unknowns that may be solved for I_1, I_2 and I_3. The three equations produced from Figure 71.1 are:

$$I_1(Z_1 + Z_2) - I_2 Z_2 = E_1$$

$$I_2(Z_2 + Z_3 + Z_4) - I_1 Z_2 - I_3 Z_4 = 0$$

$$I_3(Z_4 + Z_5) - I_2 Z_4 = -E_2$$

The branch currents are determined by taking the phasor sum of the mesh currents common to that branch. For example, the current flowing in impedance Z_2 of Figure 71.1 is given by $(I_1 - I_2)$ phasorially. The method of mesh-current analysis is often called **Maxwell's theorem**.

For example, for the a.c. network shown in Figure 71.2, using mesh-current analysis:

For the first loop: $(5 - j4)I_1 - (-j4I_2) = 100\angle 0°$ (1)

For the second loop: $(4 + j3 - j4)I_2 - (-j4I_1) = 0$ (2)

Rewriting equations (1) and (2) gives:

$$(5 - j4)I_1 + j4I_2 - 100 = 0 \qquad (1')$$

$$j4I_1 + (4 - j)I_2 + 0 = 0 \qquad (2')$$

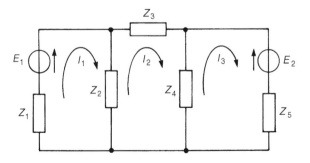

Figure 71.1

526

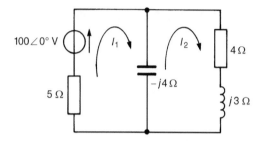

Figure 71.2

Using determinants:

$$\frac{I_1}{\begin{vmatrix} j4 & -100 \\ (4-j) & 0 \end{vmatrix}} = \frac{-I_2}{\begin{vmatrix} (5-j4) & -100 \\ j4 & 0 \end{vmatrix}}$$

$$= \frac{1}{\begin{vmatrix} (5-j4) & j4 \\ j4 & (4-j) \end{vmatrix}}$$

$$\frac{I_1}{(400-j100)} = \frac{-I_2}{j400} = \frac{1}{(32-j21)}$$

Hence $\quad I_1 = \dfrac{(400-j100)}{(32-j21)} = \dfrac{412.31\angle -14.04°}{38.28\angle -33.27°}$

$$= 10.77\angle 19.23° \text{ A}$$

$$= \mathbf{10.8\angle -19.2°A}, \text{ correct to one decimal place.}$$

$$I_2 = \frac{400\angle -90°}{38.28\angle -33.27°}$$

$$= 10.45\angle -56.73° \text{ A}$$

$$= \mathbf{10.5\angle -56.7° \ A}, \text{ correct to one decimal place.}$$

Current flowing in the capacitor $= I_1 - I_2$

$$= 10.77\angle 19.23° - 10.45\angle -56.73°$$

$$= 4.44 + j12.28 = 13.1\angle 70.13° \text{ A}$$

i.e. **the current in the capacitor is 13.1 A.**

Source power $\quad P = VI\cos\phi = (100)(10.77)\cos 19.23°$

$$= \mathbf{1016.9 \ W}$$

$$= 1020 \text{ W, correct to three significant figures.}$$

[Check: power in 5 Ω resistor $= I_1^2(5) = (10.77)^2(5) = 579.97$ W

and power in 4 Ω resistor $= I_2^2(4) = (10.45)^2(4) = 436.81$ W

Thus total power dissipated $= 579.97 + 436.81 = 1016.8$ W

$$= 1020 \text{ W},$$

correct to three significant figures]

Nodal Analysis

A **node** of a network is defined as a point where two or more branches are joined. If three or more branches join at a node, then that node is called a **principal node** or **junction**. In Figure 71.3, points 1, 2, 3, 4 and 5 are nodes, and points 1, 2 and 3 are principal nodes.

A node voltage is the voltage of a particular node with respect to a node called the reference node. If in Figure 71.3, for example, node 3 is chosen as the reference node then V_{13} is assumed to mean the voltage at node 1 with respect to node 3 (as distinct from V_{31}). Similarly, V_{23} would be assumed to mean the voltage at node 2 with respect to node 3, and so on. However, since the node voltage is always determined with respect to a particular chosen reference node, the notation V_1 for V_{13} and V_2 for V_{23} would always be used in this instance.

The object of nodal analysis is to determine the values of voltages at all the principal nodes with respect to the reference node, for example, to find voltages V_1 and V_2 in Figure 71.3. When such voltages are determined, the currents flowing in each branch can be found.

Kirchhoff's current law is applied to nodes 1 and 2 in turn in Figure 71.3 and two equations in unknowns V_1 and V_2 are obtained which may be simultaneously solved using determinants.

The branches leading to node 1 are shown separately in Figure 71.4. Let us assume that all branch currents are leaving the node as shown. Since the sum of currents at a junction is zero,

$$\frac{V_1 - V_x}{Z_A} + \frac{V_1}{Z_D} + \frac{V_1 - V_2}{Z_B} = 0 \tag{1}$$

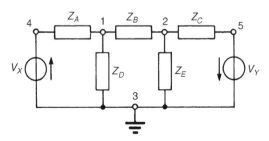

Figure 71.3

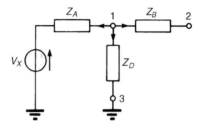

Figure 71.4

Similarly, for node 2, assuming all branch currents are leaving the node as shown in Figure 71.5,

$$\frac{V_2 - V_1}{Z_B} + \frac{V_2}{Z_E} + \frac{V_2 + V_Y}{Z_C} = 0 \qquad (2)$$

In equations (1) and (2), the currents are all assumed to be leaving the node. In fact, any selection in the direction of the branch currents may be made –the resulting equations will be identical. (For example, if for node 1 the current flowing in Z_B is considered as flowing towards node 1 instead of away, then the equation for node 1 becomes

$$\frac{V_1 - V_x}{Z_A} + \frac{V_1}{Z_D} = \frac{V_2 - V_1}{Z_B}$$

which if rearranged is seen to be exactly the same as equation (1)).
Rearranging equations (1) and (2) gives:

$$\left(\frac{1}{Z_A} + \frac{1}{Z_B} + \frac{1}{Z_D}\right) V_1 - \left(\frac{1}{Z_B}\right) V_2 - \left(\frac{1}{Z_A}\right) V_x = 0 \qquad (3)$$

$$-\left(\frac{1}{Z_B}\right) V_1 + \left(\frac{1}{Z_B} + \frac{1}{Z_C} + \frac{1}{Z_E}\right) V_2 + \left(\frac{1}{Z_C}\right) V_Y = 0 \qquad (4)$$

Equations (3) and (4) may be solved for V_1 and V_2 by using determinants. Current equations, and hence voltage equations, may be written at each principal node of a network with the exception of a reference node. The number of equations necessary to produce a solution for a circuit is, in fact, always one less than the number of principal nodes.

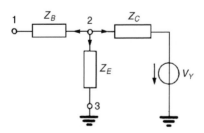

Figure 71.5

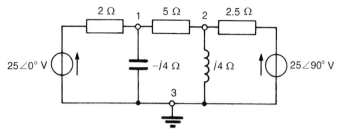

Figure 71.6

For example, in the network of Figure 71.6 using nodal analysis:

At node 1,
$$\frac{V_1 - 25\angle 0°}{2} + \frac{V_1}{-j4} + \frac{V_1 - V_2}{5} = 0$$

Rearranging gives:
$$\left(\frac{1}{2} + \frac{1}{-j4} + \frac{1}{5}\right)V_1 - \left(\frac{1}{5}\right)V_2 - \frac{25\angle 0°}{2} = 0$$

i.e.
$$(0.7 + j0.25)V_1 - 0.2V_2 - 12.5 = 0 \qquad (1)$$

At node 2,
$$\frac{V_2 - 25\angle 90°}{2.5} + \frac{V_2}{j4} + \frac{V_2 - V_1}{5} = 0$$

Rearranging gives:
$$-\left(\frac{1}{5}\right)V_1 + \left(\frac{1}{2.5} + \frac{1}{j4} + \frac{1}{5}\right)V_2 - \frac{25\angle 90°}{2.5} = 0$$

i.e.
$$-0.2V_1 + (0.6 - j0.25)V_2 - j10 = 0 \qquad (2)$$

Thus two simultaneous equations have been formed with two unknowns, V_1 and V_2. Using determinants, if

$$(0.7 + j0.25)V_1 - 0.2V_2 - 12.5 = 0 \qquad (1)$$

and
$$-0.2V_1 + (0.6 - j0.25)V_2 - j10 = 0 \qquad (2)$$

then

$$\frac{V_1}{\begin{vmatrix} -0.2 & -12.5 \\ (0.6 - j0.25) & -j10 \end{vmatrix}} = \frac{-V_2}{\begin{vmatrix} (0.7 + j0.25) & -12.5 \\ -0.2 & -j10 \end{vmatrix}}$$

$$= \frac{1}{\begin{vmatrix} (0.7 + j0.25) & -0.2 \\ -0.2 & (0.6 - j0.25) \end{vmatrix}}$$

i.e.
$$\frac{V_1}{(j2 + 7.5 - j3.125)} = \frac{-V_2}{(-j7 + 2.5 - 2.5)}$$

$$= \frac{1}{(0.42 - j0.175 + j0.15 + 0.0625 - 0.04)}$$

and $$\frac{V_1}{7.584\angle -8.53°} = \frac{-V_2}{-7\angle 90°} = \frac{1}{0.443\angle -3.23°}$$

Thus **voltage, V_1** $= \dfrac{7.584\angle -8.53°}{0.443\angle -3.23°} = 17.12\angle -5.30°\,V$

$$= \mathbf{17.1\angle -5.3°V},$$

correct to one decimal place,

and **voltage, V_2** $= \dfrac{7\angle 90°}{0.443\angle -3.23°} = 15.80\angle 93.23°\ V$

$$= \mathbf{15.8\angle 93.2°\ V},$$

correct to one decimal place.

The current in the $j4\ \Omega$ inductance is given by:

$$\frac{V_2}{j4} = \frac{15.80\angle 93.23°}{4\angle 90°} = \mathbf{3.95\angle 3.23°\ A}\ \text{flowing toward node 3}$$

The current in the $5\ \Omega$ resistance is given by:

$$I_5 = \frac{V_1 - V_2}{5} = \frac{17.12\angle -5.30° - 15.80\angle 93.23°}{5}$$

i.e. $I_5 = \dfrac{(17.05 - j1.58) - (-0.89 + j15.77)}{5} = \dfrac{17.94 - j17.35}{5}$

$$= \frac{24.96\angle -44.04°}{5} = \mathbf{4.99\angle -44.04°\ A}$$

flowing from node 1 to node 2

The active power dissipated in the $2.5\ \Omega$ resistor is given by:

$$P_{2.5} = (I_{2.5})^2(2.5) = \left(\frac{V_2 - 25\angle 90°}{2.5}\right)^2 (2.5)$$

$$= \frac{(-0.89 + j15.77 - j25)^2}{2.5}$$

$$= \frac{(9.273\angle -95.51°)^2}{2.5} = \frac{85.99\angle -191.02°}{2.5}$$

by de Moivre's theorem

$$= 34.4\angle 169°\ W$$

Thus the magnitude of the active power dissipated in the $2.5\ \Omega$ resistance is 34.4 W.

Whether mesh-current analysis or nodal analysis is used to determine currents in circuits depends on the number of loops and nodes the circuit contains. Basically, the method that requires the least number of equations is used.

72 The Superposition Theorem

Introduction

The superposition theorem states:

> 'In any network made up of linear impedances and containing more than one source of e.m.f. the resultant current flowing in any branch is the phasor sum of the currents that would flow in that branch if each source were considered separately, all other sources being replaced at that time by their respective internal impedances'

Using the Superposition Theorem

The superposition theorem, which was introduced in chapter 53 for d.c. circuits, may be applied to both d.c. and a.c. networks. A d.c. network is shown in Figure 72.1 and will serve to demonstrate the principle of application of the superposition theorem.

To find the current flowing in each branch of the circuit, the following six-step procedure can be adopted:

(i) Redraw the original network with one of the sources, say E_2, removed and replaced by r_2 only, as shown in Figure 72.2.

(ii) Label the current in each branch and its direction as shown in Figure 72.2, and then determine its value. The choice of current direction for I_1 depends on the source polarity which, by convention, is taken as flowing from the positive terminal as shown.

R in parallel with r_2 gives an equivalent resistance of $\dfrac{5 \times 2}{5 + 2} = 1.429 \ \Omega$

as shown in the equivalent network of Figure 72.3.

From Figure 72.3,

$$\text{current } I_1 = \frac{E_1}{(r_1 + 1.429)} = \frac{8}{2.429} = 3.294 \text{ A}$$

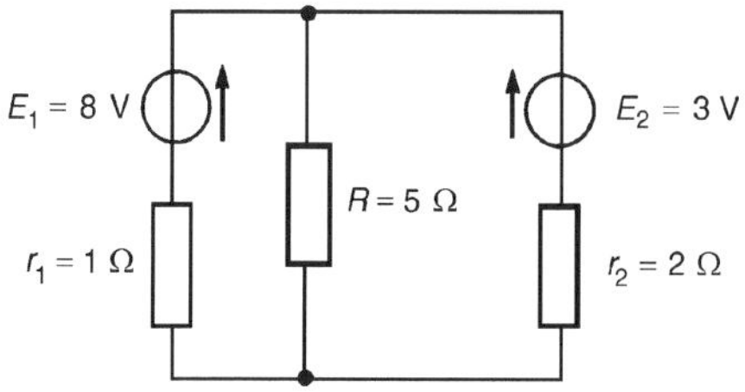

Figure 72.1

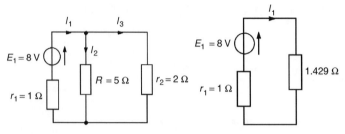

Figure 72.2   **Figure 72.3**

From Figure 72.2,

$$\text{current } I_2 = \left(\frac{r_2}{R + r_2}\right)(I_1) = \left(\frac{2}{5+2}\right)(3.294) = 0.941 \text{ A}$$

$$\text{and current } I_3 = \left(\frac{5}{5+2}\right)(3.294) = 2.353 \text{ A}$$

(iii) Redraw the original network with source E_1 removed and replaced by r_1 only, as shown in Figure 72.4.

(iv) Label the currents in each branch and their directions as shown in Figure 72.4, and determine their values.

R and r_1 in parallel gives an equivalent resistance of $\dfrac{5 \times 1}{5+1} = 0.833 \ \Omega$ as shown in the equivalent network of Figure 72.5.

From Figure 72.5,

$$\text{current } I_4 = \frac{E_2}{r_2 + 0.833} = \frac{3}{2.833} = 1.059 \text{ A}$$

From Figure 72.4,

$$\text{current } I_5 = \left(\frac{1}{1+5}\right)(1.059) = 0.177 \text{ A}$$

$$\text{and current } I_6 = \left(\frac{5}{1+5}\right)(1.059) = 0.8825 \text{ A}$$

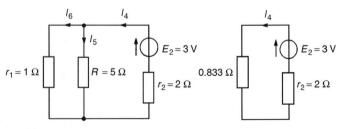

Figure 72.4   **Figure 72.5**

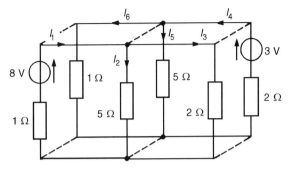

Figure 72.6

(v) Superimpose Figure 72.2 on Figure 72.4, as shown in Figure 72.6.
(vi) Determine the algebraic sum of the currents flowing in each branch. (Note
    that in an a.c. circuit it is the phasor sum of the currents that is required).
    From Figure 72.6, the resultant current flowing through the 8 V source is
    given by $I_1 - I_6 = 3.294 - 0.8825 = \mathbf{2.41\ A}$ (discharging, i.e. flowing
    from the positive terminal of the source).
    The resultant current flowing in the 3 V source is given by $I_3 - I_4 =$
    $2.353 - 1.059 = \mathbf{1.29\ A}$ (charging, i.e. flowing into the positive terminal
    of the source).
    The resultant current flowing in the 5 Ω resistance is given by $I_2 + I_5 =$
    $0.941 + 0.177 = \mathbf{1.12\ A}$

Using the superposition theorem in Figure 72.7 will demonstrate its use with an
a.c. network.

 (i) The network is redrawn with the $30\angle90°$ V source removed, as shown in
     Figure 72.8.
(ii) Currents I_1 to I_5 are shown labelled in Figure 72.8. From Figure 72.8,
     two 8 Ω resistors in parallel give an equivalent resistance of 4 Ω. Hence

$$I_1 = \frac{50\angle0°}{20 + \dfrac{5(4 - j3)}{5 + 4 - j3}} = \frac{50\angle0°}{20 + \dfrac{5(5\angle - 36.87°)}{9.487\angle - 18.43°}}$$

$$= \frac{50\angle0°}{20 + 2.635\angle - 18.44°} = \frac{50\angle0°}{22.50 - j0.833}$$

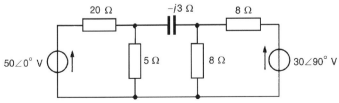

Figure 72.7

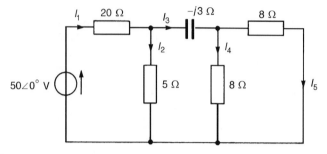

Figure 72.8

$$= \frac{50\angle 0°}{22.52\angle - 2.12°} = 2.220\angle 2.12° \text{ A}$$

$$I_2 = \frac{(4 - j3)}{(5 + 4 - j3)}I_1 = \left(\frac{5\angle - 36.87°}{9.487\angle - 18.43°}\right)(2.220\angle 2.12°)$$

$$= 1.170\angle - 16.32° \text{ A}$$

$$I_3 = \left(\frac{5}{5 + 4 - j3}\right)I_1 = \left(\frac{5}{9.487\angle - 18.43°}\right)(2.220\angle 2.12°)$$

$$= 1.170\angle 20.55° \text{ A}$$

$$I_4 = \left(\frac{8}{8 + 8}\right)I_3 = \frac{1}{2}(1.170\angle 20.55°)$$

$$= 0.585\angle 20.55° \text{ A} = I_5$$

(iii) The original network is redrawn with the $50\angle 0°$ V source removed, as shown in Figure 72.9.

(iv) Currents I_6 to I_{10} are shown labelled in Figure 72.9. From Figure 72.9, 20 Ω in parallel with 5 Ω gives an equivalent resistance of: $\frac{20 \times 5}{20 + 5} = 4$ Ω.

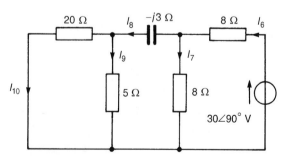

Figure 72.9

Hence

$$I_6 = \frac{30\angle 90°}{8 + \dfrac{8(4-j3)}{8+4-j3}} = \frac{30\angle 90°}{8 + \dfrac{8(5\angle -36.87°)}{12.369\angle -14.04°}}$$

$$= \frac{30\angle 90°}{8 + 3.234\angle -22.83°} = \frac{30\angle 90°}{10.981 - j1.255}$$

$$= \frac{30\angle 90°}{11.052\angle -6.52°} = 2.714\angle 96.52° \text{ A}$$

$$I_7 = \frac{(4-j3)}{(8+4-j3)}I_6$$

$$= \left(\frac{5\angle -36.87°}{12.369\angle -14.04°}\right)(2.714\angle 96.52°) = 1.097\angle 73.69° \text{ A}$$

$$I_8 = \left(\frac{8}{8+4-j3}\right)I_6$$

$$= \left(\frac{8}{12.369\angle -14.04°}\right)(2.714\angle 96.52°) = 1.755\angle 110.56° \text{ A}$$

$$I_9 = \left(\frac{20}{20+5}\right)I_8 = 1.404\angle 110.56° \text{ A}$$

$$\text{and } I_{10} = \left(\frac{5}{20+5}\right)I_8 = 0.351\angle 110.56° \text{ A}$$

The current flowing in the capacitor is given by:

$$(I_3 - I_8) = 1.170\angle 20.55° - 1.755\angle 110.56°$$

$$= (1.712 - j1.233)\text{A or } 2.11\angle -35.76° \text{ A}$$

i.e. **the magnitude of the current in the capacitor is 2.11 A.**
The p.d. across the 5 Ω resistance is given by $(I_2 + I_9)(5)$.

$$(I_2 + I_9) = 1.170\angle -16.32° + 1.404\angle 110.56°$$

$$= (0.630 + j0.986) \text{ A or } 1.17\angle 57.42° \text{ A}$$

Hence **the magnitude of the p.d. across the 5 Ω resistance** is $(1.17)(5) =$
5.85 V.
Active power dissipated in the 20 Ω resistance is given by:

$$(I_1 - I_{10})^2(20).$$

$$(I_1 - I_{10}) = 2.220\angle 2.12° - 0.351\angle 110.56°$$

$$= (2.342 - j0.247) \text{ A or } 2.355\angle -6.02° \text{ A.}$$

Hence **the active power dissipated in the 20 Ω resistance** is $(2.355)^2(20) =$
111 W
Active power developed by the $50\angle 0°$ V source

$$P_1 = V(I_1 - I_{10})\cos\phi_1 = (50)(2.355)\cos(6.02° - 0°) = 117.1 \text{ W}$$

Active power developed by $30\angle 90°$ V source, $P_2 = 30(I_6 - I_5)\cos\phi_2$

$$(I_6 - I_5) = 2.714\angle 96.52° - 0.585\angle 20.55°$$

$$= (-0.856 + j2.491) \text{ A or } 2.634\angle 108.96° \text{ A}$$

Hence $P_2 = (30)(2.634)\cos(108.96° - 90°) = 74.7$ W

Total power developed, $P = P_1 + P_2 = 117.1 + 74.7 = \mathbf{191.8}$ **W**

(This value may be checked by summing the I^2R powers dissipated in the four resistors).

The superposition theorem is straightforward to apply, but is lengthy. Thévenin's and Norton's theorems (described in chapter 73) produce the same results more quickly.

73 Thévenin's and Norton's Theorems

Introduction

Many of the networks analysed in Chapters 70 to 72 using Kirchhoff's laws, mesh-current and nodal analysis and the superposition theorem can be analysed more quickly and easily by using Thévenin's or Norton's theorems. Each of these theorems involves replacing what may be a complicated network of sources and linear impedances with a simple equivalent circuit. A set procedure may be followed when using each theorem, the procedures themselves requiring a knowledge of basic circuit theory. (It may be worth checking some general d.c. circuit theory in chapter 53, page 331, before proceeding).

Thévenin's Theorem

Thévenin's theorem states:

> 'The current which flows in any branch of a network is the same as that which would flow in the branch if it were connected across a source of electrical energy, the e.m.f. of which is equal to the potential difference which would appear across the branch if it were open-circuited, and the internal impedance of which is equal to the impedance which appears across the open-circuited branch terminals when all sources are replaced by their internal impedances'

The theorem applies to any linear active network ('linear' meaning that the measured values of circuit components are independent of the direction and magnitude of the current flowing in them, and 'active' meaning that it contains a source, or sources, of e.m.f.).

The above statement of Thévenin's theorem simply means that a complicated network with output terminals AB, as shown in Figure 73.1(a), can be replaced by a single voltage source E in series with an impedance z, as shown in Figure 73.1(b). E is the open-circuit voltage measured at terminals AB and z is the equivalent impedance of the network at the terminals AB when all internal sources of e.m.f. are made zero. The polarity of voltage E is chosen so that the current flowing through an impedance connected between A and B will have the same direction as would result if the impedance had been connected between A and B of the original network. Figure 73.1(b) is known as the **Thévenin equivalent circuit,** and was initially introduced in chapter 53, page 335, for d.c. networks.

The following four-step **procedure** can be adopted when determining, by means of Thévenin's theorem, the current flowing in a branch containing impedance Z_L of an active network:

  (i) remove the impedance Z_L from that branch,
  (ii) determine the open-circuit voltage E across the break,

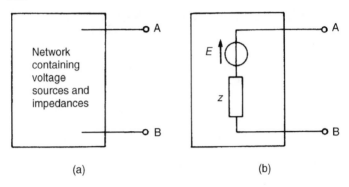

(a)   (b)

Figure 73.1

(iii) remove each source of e.m.f. and replace each by its internal impedance
(if it has zero internal impedance then replace it by a short-circuit), and
then determine the internal impedance, z, 'looking in' at the break,
(iv) determine the current from the Thévenin equivalent circuit shown in
Figure 73.2, i.e. **current, $i_L = \dfrac{E}{Z_L + z}$**

A simple d.c. network (Figure 73.3) serves to demonstrate how the above
procedure is applied to determine the current flowing in the 5 Ω resistance by
using Thévenin's theorem. Using the above procedure:

(i) The 5 Ω resistor is removed, as shown in Figure 73.4(a).
(ii) The open-circuit voltage E across the break is now required; the network
of Figure 73.4(a) is redrawn for convenience as shown in Figure 73.4(b),
where current, $I_1 = \dfrac{E_1 - E_2}{r_1 + r_2} = \dfrac{8 - 3}{1 + 2} = \dfrac{5}{3}$ or $1\dfrac{2}{3}$ A

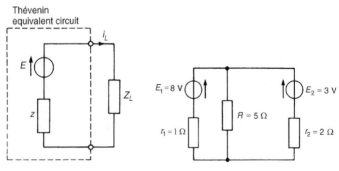

Figure 73.2   **Figure 73.3**

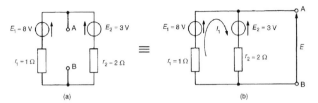

Figure 73.4

Hence the open-circuit voltage E is given by:

$$E = E_1 - I_1 r_1 \quad \text{i.e.} \quad E = 8 - (1\tfrac{2}{3})(1) = 6\tfrac{1}{3} \text{ V}$$

(Alternatively, $E = E_2 - (-I_1)r_2 = 3 + (1\tfrac{2}{3})(2) = 6\tfrac{1}{3}\text{V}$)

(iii) Removing each source of e.m.f. gives the network of Figure 73.5. The impedance, z, 'looking in' at the break AB is given by: $z = \dfrac{1 \times 2}{1 + 2} = \dfrac{2}{3}\ \Omega$

(iv) The Thévenin equivalent circuit is shown in Figure 73.6, where current i_L is given by:

$$i_L = \frac{E}{Z_L + z} = \frac{6\tfrac{1}{3}}{5 + \tfrac{2}{3}} = 1.1177 = \textbf{1.12 A},$$

correct to two decimal places.

To determine the currents flowing in the other two branches of the circuit of Figure 73.3, basic circuit theory is used. Thus, from Figure 73.7, voltage $V = (1.1177)(5) = 5.5885$ V.

Then $V = E_1 - I_A r_1$ i.e. $5.5885 = 8 - I_A$ (1) from which,

$$\text{current } I_A = 8 - 5.5885 = \textbf{2.41 A}$$

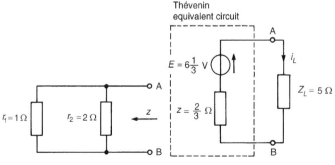

Figure 73.5　　　　**Figure 73.6**

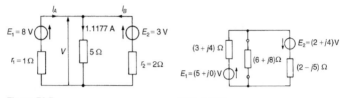

Figure 73.7   **Figure 73.8**

Similarly, $V = E_2 - I_B r_2$ i.e. $5.5885 = 3 - I_B$ (2) from which,

$$\text{current } I_B = \frac{3 - 5.5885}{2} = -1.29 \text{ A}$$

(i.e. flowing in the direction opposite to that shown in Figure 73.7).

The Thévenin theorem procedure used above may be applied to a.c. as well as d.c. networks, as shown below.

An a.c. network is shown in Figure 73.8 where it is required to find the current flowing in the $(6 + j8)$ Ω impedance by using Thévenin's theorem.

Using the above procedure:

(i) The $(6 + j8)$ Ω impedance is removed, as shown in Figure 73.9(a).
(ii) The open-circuit voltage across the break is now required. The network is redrawn for convenience as shown in Figure 73.9(b), where current,

$$I_1 = \frac{(5 + j0) + (2 + j4)}{(3 + j4) + (2 - j5)} = \frac{7 + j4}{5 - j} = 1.581\angle 41.05° \text{ A}$$

Hence open-circuit voltage across AB, $E = E_1 - I_1(3 + j4)$ i.e.
$E = (5 + j0) - (1.581\angle 41.05°)(5\angle 53.13°)$ from which,
$E = 9.657\angle - 54.73°$ V

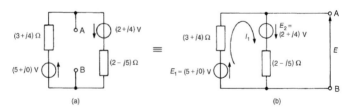

Figure 73.9

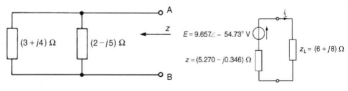

Figure 73.10   **Figure 73.11**

(iii) From Figure 73.10, the impedance z 'looking in' at terminals AB is given by: $z = \dfrac{(3 + j4)(2 - j5)}{(3 + j4) + (2 - j5)} = 5.281\angle - 3.76° \; \Omega$ or

$$(5.270 - j0.346) \; \Omega$$

(iv) The Thévenin equivalent circuit is shown in Figure 73.11, from which the current in the $(6 + j8) \; \Omega$ impedance,

$$i_L = \frac{E}{Z_L + z} = \frac{9.657\angle - 54.73°}{(6 + j8) + (5.270 - j0.346)}$$

$$= \frac{9.657\angle - 54.73°}{13.623\angle 34.18°} = \mathbf{0.71\angle - 88.91° \; A}$$

Norton's Theorem

A source of electrical energy can be represented by a source of e.m.f. in series with an impedance. In the previous section, the Thévenin constant-voltage source consisted of a constant e.m.f. E, which may be alternating or direct, in series with an internal impedance z. However, this is not the only form of representation. A source of electrical energy can also be represented by a constant-current source, which may be alternating or direct, in parallel with an impedance. The two forms are in fact equivalent.

Norton's theorem states:

'The current that flows in any branch of a network is the same as that which would flow in the branch if it were connected across a source of electrical energy, the short-circuit current of which is equal to the current that would flow in a short-circuit across the branch, and the internal impedance of which is equal to the impedance which appears across the open-circuited branch terminals'

The above statement simply means that any linear active network with output terminals AB, as shown in Figure 73.12(a), can be replaced by a current source in parallel with an impedance z as shown in Figure 73.12(b). The equivalent current source I_{SC} (note the symbol in Figure 73.12(b) as per BS 3939: 1985) is the current through a short-circuit applied to the terminals of the network. The impedance z is the equivalent impedance of the network at the terminals AB when all internal sources of e.m.f. are made zero. Figure 73.12(b) is known as the **Norton equivalent circuit** and was initially introduced in chapter 53, page 336, for d.c. networks.

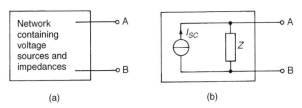

(a)   (b)

Figure 73.12

The following four-step procedure may be adopted when determining the current flowing in an impedance Z_L of a branch AB of an active network, using Norton's theorem:

(i) short-circuit branch AB,
(ii) determine the short-circuit current, I_{SC},
(iii) remove each source of e.m.f. and replace it by its internal impedance (or, if a current source exists, replace with an open circuit), then determine the impedance, z, 'looking in' at a break made between A and B,
(iv) determine the value of the current i_L flowing in impedance Z_L from the Norton equivalent network shown in Figure 73.13, i.e.

$$i_L = \left(\frac{z}{Z_L + z}\right) I_{SC}$$

A simple d.c. network (Figure 73.14) serves to demonstrate how the above procedure is applied to determine the current flowing in the 5 Ω resistance by using Norton's theorem:

(i) The 5 Ω branch is short-circuited, as shown in Figure 73.15.
(ii) From Figure 73.15, $I_{SC} = I_1 + I_2 = \frac{8}{1} + \frac{3}{2} = 9.5$ A.
(iii) If each source of e.m.f. is removed the impedance 'looking in' at a break made between A and B is given by:

$$z = \frac{1 \times 2}{1 + 2} = \frac{2}{3} \ \Omega$$

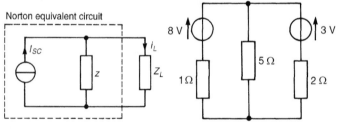

Norton equivalent circuit

Figure 73.13

Figure 73.14

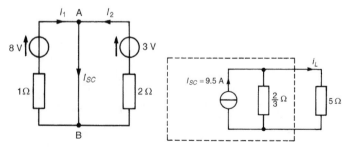

Figure 73.15

Figure 73.16

(iv) From the Norton equivalent network shown in Figure 73.16, the current in the 5 Ω resistance is given by:

$$I_L = \left(\frac{\frac{2}{3}}{5 + \frac{2}{3}} \right) (9.5) = \mathbf{1.12 \ A}$$

As with Thévenin's theorem, Norton's theorem may be used with a.c. as well as d.c. networks, as shown below.

An a.c. network is shown in Figure 73.17 where it is required to find the current flowing in the $(6 + j8)$ Ω impedance by using Norton's theorem.

Using the above procedure:

(i) Initially the $(6 + j8)$ Ω impedance is short-circuited, as shown in Figure 73.18

(ii) From Figure 73.18,

$$I_{SC} = I_1 + I_2 = \frac{(5 + j0)}{(3 + j4)} + \frac{-(2 + j4)}{(2 - j5)}$$

$$= 1\angle - 53.13° - \frac{4.472\angle 63.43°}{5.385\angle - 68.20°}$$

$$= 1\angle - 53.13° - 0.8305\angle 131.63°$$

$$= (1.152 - j1.421) \ \text{A or } \mathbf{1.829\angle - 50.97° \ A}$$

(iii) If each source of e.m.f. is removed, the impedance, z, 'looking in' at a break made between A and B is given by:

$$z = \frac{(3 + j4)(2 - j5)}{(3 + j4) + (2 - j5)}$$

$$= 5.281\angle - 3.76° \ \Omega \text{ or } \mathbf{(5.270 - j0.346) \ \Omega}$$

(iv) From the Norton equivalent network shown in Figure 73.19, the current i_L is given by:

$$i_L = \left(\frac{z}{Z_L + z} \right) I_{SC}$$

$$= \left(\frac{5.281\angle - 3.76°}{(6 + j8) + (5.270 - j0.346)} \right) (1.829\angle - 50.97°)$$

i.e. **current in the $(6 + j8)$ Ω impedance, $i_L = 0.71\angle - 88.91° \ A$**

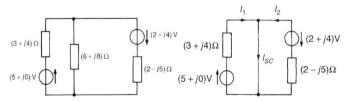

Figure 73.17   **Figure 73.18**

544

Thévenin and Norton Equivalent Networks

It is seen earlier that when Thévenin's and Norton's theorems are applied to the same circuit, identical results are obtained. Thus the Thévenin and Norton networks shown in Figure 73.20 are equivalent to each other. The impedance 'looking in' at terminals AB is the same in each of the networks, i.e. z.

If terminals AB in Figure 73.20(a) are short-circuited, the short-circuit current is given by $\dfrac{E}{z}$. If terminals AB in Figure 73.20(b) are short-circuited, the short-circuit current is I_{SC}. Thus $I_{SC} = \dfrac{E}{z}$.

For example, the circuit of Figure 73.21(a) is equivalent to that of Figure 73.21(b).

Similarly, the circuit of Figure 73.22(a) is equivalent to that of Figure 73.22(b).

In another example, a Thévenin equivalent circuit for terminals AB of Figure 73.23 is determined as follows:

For the branch containing the 5 V source, converting to a Norton equivalent network gives: $I_{SC} = \dfrac{5}{1000} = 5$ mA and $z = 1$ kΩ.

For the branch containing the 10 V source, converting to a Norton equivalent network gives: $I_{SC} = \dfrac{10}{4000} = 2.5$ mA and $z = 4$ kΩ.

Thus the circuit of Figure 73.23 converts to that of Figure 73.24.

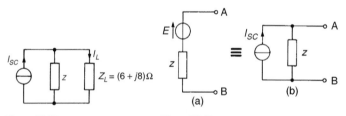

Figure 73.19   **Figure 73.20**

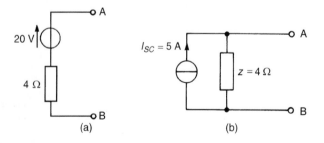

Figure 73.21

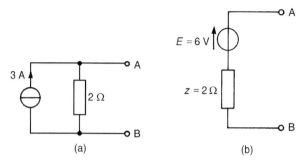

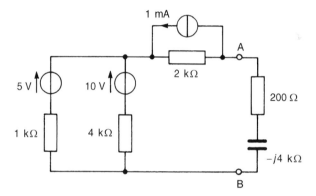

Figure 73.22

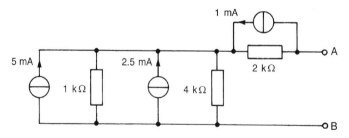

Figure 73.23

Figure 73.24

The above two Norton equivalent networks shown in Figure 73.24 may be combined, since the total short-circuit current is $(5 + 2.5) = 7.5$ mA and the total impedance z is given by $\dfrac{1 \times 4}{1 + 4} = 0.8$ kΩ. This results in the network of Figure 73.25.

546

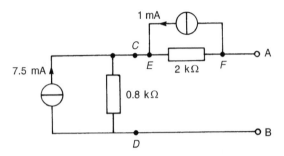

Figure 73.25

Both of the Norton equivalent networks shown in Figure 73.25 may be converted to Thévenin equivalent circuits. Open-circuit voltage across CD is: $(7.5 \times 10^{-3})(0.8 \times 10^3) = 6$ V and the impedance 'looking in' at CD is 0.8 kΩ. Open-circuit voltage across EF is $(1 \times 10^{-3})(2 \times 10^3) = 2$ V and the impedance 'looking in' at EF is 2 kΩ. Thus Figure 73.25 converts to Figure 73.26.

Combining the two Thévenin circuits gives e.m.f. $E = 6 - 2 = \mathbf{4}$ **V**, and impedance $z = (0.8 + 2) = 2.8$ kΩ. Thus the Thévenin equivalent circuit for terminals AB of Figure 73.23 is as shown in Figure 73.27.

If an impedance of, say, $(200 - j4000)$ Ω is connected across terminals AB, then the current I flowing is given by:

$$I = \frac{4}{2800 + (200 - j4000)} = \frac{4}{5000\angle - 53.13°}$$

$$= 0.80\angle 53.13° \text{ mA}$$

i.e. **the current in the capacitive branch in 0.80 mA.**

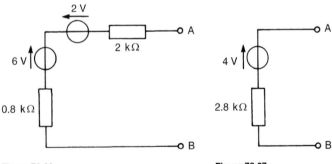

Figure 73.26   **Figure 73.27**

74 Delta-star and Star-delta Transformations

Introduction

By using Kirchhoff's laws, mesh-current analysis, nodal analysis or the super-position theorem, currents and voltages in many network can be determined as shown in chapters 70 to 72. Thevenin's and Norton's theorems, introduced in chapter 73, provide an alternative method of solving networks and often with considerably reduced numerical calculations. Also, these latter theorems are especially useful when only the current in a particular branch of a complicated network is required. Delta-star and star-delta transformations may be applied in certain types of circuit to simplify them before application of circuit theorems.

Delta and Star Connections

The network shown in Figure 74.1(a) consisting of three impedances Z_A, Z_B and Z_C is said to be **π-connected**. This network can be redrawn as shown in Figure 74.1(b), where the arrangement is referred to as **delta-connected** or **mesh-connected**.

The network shown in Figure 74.2(a), consisting of three impedances, Z_1, Z_2 and Z_3, is said to be **T-connected**. This network can be redrawn as shown in Figure 74.2(b), where the arrangement is referred to as **star-connected**.

Delta-star Transformation

It is possible to replace the delta connection shown in Figure 74.3(a) by an equivalent star connection as shown in Figure 74.3(b) such that the impedance

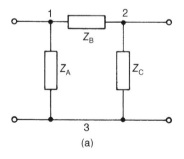
(a)

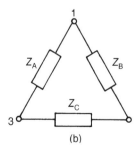
(b)

Figure 74.1

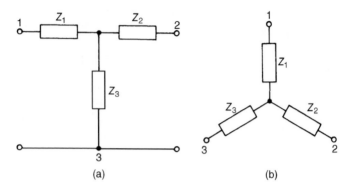

Figure 74.2

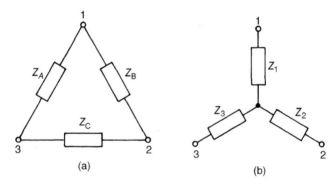

Figure 74.3

measured between any pair of terminals (1–2, 2–3 or 3–1) is the same in star as in delta. The equivalent star network will consume the same power and operate at the same power factor as the original delta network. A delta-star transformation may alternatively be termed 'π to T transformation'.

The star section shown in Figure 74.3(b) is equivalent to the delta section shown in Figure 74.3(a) when

$$Z_1 = \frac{Z_A Z_B}{Z_A + Z_B + Z_C} \tag{1}$$

$$Z_2 = \frac{Z_B Z_C}{Z_A + Z_B + Z_C} \tag{2}$$

and

$$Z_3 = \frac{Z_A Z_C}{Z_A + Z_B + Z_C} \tag{3}$$

549

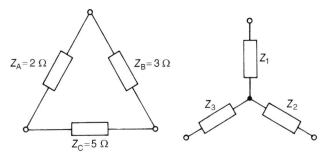

Figure 74.4

For example, the star equivalent of the resistive delta network shown in Figure 74.4 is given by:

$$Z_1 = \frac{(2)(3)}{2+3+5} = \textbf{0.6 } \boldsymbol{\Omega},$$

$$Z_2 = \frac{(3)(5)}{2+3+5} = \textbf{1.5 } \boldsymbol{\Omega}$$

and $\quad Z_3 = \frac{(2)(5)}{2+3+5} = \textbf{1.0 } \boldsymbol{\Omega}$

In another example, the equivalent circuit impedance across terminals AB in the network of Figure 74.5 is determined as follows:

The network of Figure 74.5 is redrawn, as in Figure 74.6, showing more clearly the part of the network 1, 2, 3 forming a delta connection. This may he transformed into a star connection as shown in Figure 74.7

From equation (1), $\quad Z_1 = \dfrac{Z_A Z_B}{Z_A + Z_B + Z_C} = \dfrac{(j10)(j15)}{j10 + j15 + j25}$

$$= \frac{(j10)(j15)}{(j50)} = j\mathbf{3} \; \boldsymbol{\Omega}$$

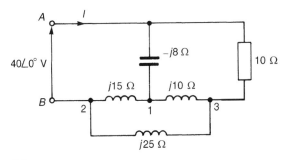

Figure 74.5

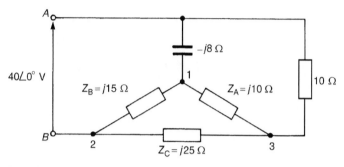

Figure 74.6

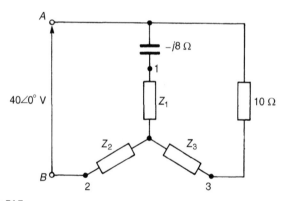

Figure 74.7

From equation (2), $\quad Z_2 = \dfrac{Z_B Z_C}{Z_A + Z_B + Z_C} = \dfrac{(j15)(j25)}{(j50)} = j7.5 \ \Omega$

From equation (3), $\quad Z_3 = \dfrac{Z_A Z_C}{Z_A + Z_B + Z_C} = \dfrac{(j10)(j25)}{(j50)} = j5 \ \Omega$

The equivalent network is shown in Figure 74.8 and is further simplified in Figure 74.9. $(10 + j5) \ \Omega$ in parallel with $-j5 \ \Omega$ gives an equivalent impedance of

$$\frac{(10 + j5)(-j5)}{(10 + j5 - j5)} = (2.5 - j5) \ \Omega$$

Hence the total circuit equivalent impedance across terminals AB is given by:

$$Z_{AB} = (2.5 - j5) + j7.5 = \mathbf{(2.5 + j2.5)} \ \mathbf{\Omega} \text{ or } \mathbf{3.54\angle 45°} \ \mathbf{\Omega}$$

The supply current $I = \dfrac{V}{Z_{AB}} = \dfrac{40\angle 0°}{3.54\angle 45°} = \mathbf{11.3\angle -45°} \ \mathbf{A}$

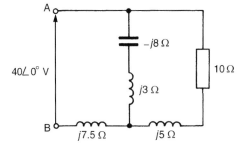

Figure 74.8

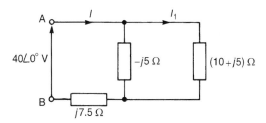

Figure 74.9

The power P dissipated in, say, the 10 Ω resistance of Figure 74.5 is given by: $(I_1)^2(10)$, where I_1 (see Figure 74.9) is given by:

$$I_1 = \left(\frac{-j5}{10 + j5 - j5} \right)(11.3\angle - 45°) = 5.65\angle - 135° \text{ A}$$

Hence power $P = (5.65)^2(10) = \mathbf{319 \text{ W}}$

Star-delta Transformation

It is possible to replace the star section shown in Figure 74.10(a) by an equivalent delta section as shown in Figure 74.10(b). Such a transformation is also known as 'T to π transformation'.

The delta section shown in Figure 74.10(b) is equivalent to the star section shown in Figure 74.10(a) when

$$Z_A = \frac{Z_1 Z_2 + Z_2 Z_3 + Z_3 Z_1}{Z_2} \qquad (4)$$

$$Z_B = \frac{Z_1 Z_2 + Z_2 Z_3 + Z_3 Z_1}{Z_3} \qquad (5)$$

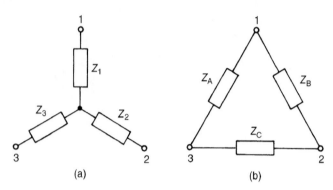

Figure 74.10

and
$$Z_C = \frac{Z_1 Z_2 + Z_2 Z_3 + Z_3 Z_1}{Z_1} \qquad (6)$$

For example, the delta equivalent of the resistive star circuit shown in Figure 74.11 is given by:

$$Z_A = \frac{(0.6)(1.5) + (1.5)(1.0) + (1.0)(0.6)}{1.5} = \frac{3.0}{1.5} = 2 \ \Omega,$$

$$Z_B = \frac{3.0}{1.0} = 3 \ \Omega,$$

$$Z_C = \frac{3.0}{0.6} = 5 \ \Omega$$

In another example, the delta-connected equivalent network for the star-connected impedances shown in Figure 74.12 is determined as follows:
Figure 74.13(a) shows the network of Figure 74.12 redrawn and Figure 74.13(b) shows the equivalent delta connection containing impedances Z_A, Z_B

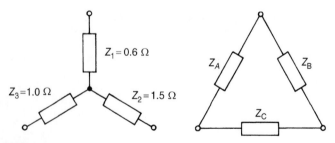

Figure 74.11

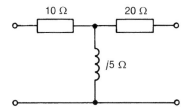

Figure 74.12

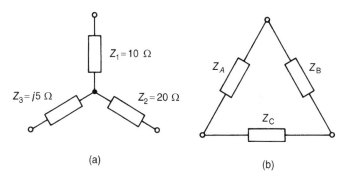

(a)

(b)

Figure 74.13

and Z_C. From equation (4),

$$Z_A = \frac{Z_1 Z_2 + Z_2 Z_3 + Z_3 Z_1}{Z_2}$$

$$= \frac{(10)(20) + (20)(j5) + (j5)(10)}{20}$$

$$= \frac{200 + j150}{20}$$

$$= (10 + j7.5)\ \Omega$$

From equation (5),

$$Z_B = \frac{(200 + j150)}{Z_3} = \frac{(200 + j150)}{j5}$$

$$= \frac{-j5(200 + j150)}{25} = (30 - j40)\ \Omega$$

From equation (6),

$$Z_C = \frac{(200 + j150)}{Z_1} = \frac{(200 + j150)}{10} = (20 + j15)\ \Omega$$

75 Maximum Power Transfer Theorems and Impedance Matching

Maximum Power Transfer Theorems

A network that contains linear impedances and one or more voltage or current sources can be reduced to a Thévenin equivalent circuit as shown in chapter 73. When a load is connected to the terminals of this equivalent circuit, power is transferred from the source to the load.

A Thevenin equivalent circuit is shown in Figure 75.1 with source internal impedance, $z = (r + jx)$ Ω and complex load $Z = (R + jX)$ Ω.

The maximum power transferred from the source to the load depends on the following four conditions:

1. When the load is purely resistive (i.e. $X = 0$) and adjustable, maximum power transfer is achieved when

$$\boxed{R = |z| = \sqrt{r^2 + R^2}}$$

2. When both the load and the source impedance are purely resistive (i.e. $X = x = 0$), maximum power transfer is achieved when $\boxed{R = r}$ (This is, in fact, the d.c. condition explained in Chapter 53, page 340).

3. When the load resistance R and reactance X are both independently adjustable, maximum power transfer is achieved when

$$\boxed{X = -x \text{ and } R = r}$$

4. When the load resistance R is adjustable with reactance X fixed, maximum power transfer is achieved when

$$\boxed{R = \sqrt{[r^2 + (x + X)^2]}}$$

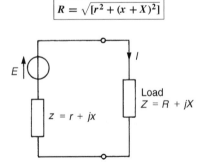

Figure 75.1

The maximum power transfer theorems are primarily important where a small source of power is involved — such as, for example, the output from a telephone system.

For example, for the circuit shown in Figure 75.2 the load impedance Z is a pure resistance. The value of R for maximum power to be transferred from the source to the load is determined as follows:

From condition 1, maximum power transfer occurs when $R = |z|$, i.e. when

$$R = |15 + j20| = \sqrt{15^2 + 20^2} = \mathbf{25 \ \Omega}$$

Current I flowing in the load is given by $I = \dfrac{E}{Z_T}$, where the total circuit impedance $Z_T = z + R = 15 + j20 + 25 = (40 + j20)$ Ω or $44.72\angle 26.57°$ Ω

Hence $\quad I = \dfrac{120\angle 0°}{44.72\angle 26.57°} = 2.683\angle - 26.57°$ A

Thus **maximum power delivered**, $P = I^2 R = (2.683)^2(25) = \mathbf{180 \ W}$

In another example, if the load impedance Z in Figure 75.2 consists of variable resistance R and variable reactance X, the value of Z that results in maximum power transfer is determined as follows:

From condition 3, maximum power transfer occurs when $X = -x$ and $R = r$. Thus if $z = r + jx = (15 + j20)$ Ω then $\mathbf{Z = (15 - j\,20) \ \Omega}$ or $\mathbf{25\angle - 53.13° \ \Omega}$

Total circuit impedance at maximum power transfer condition,

$$Z_T = z + Z, \text{ i.e. } Z_T = (15 + j20) + (15 - j20) = 30 \ \Omega$$

Hence current in load,

$$I = \frac{E}{Z_T} = \frac{120\angle 0°}{30} = 4\angle 0° \text{ A}$$

and **maximum power** delivered to the load, $P = I^2 R = (4)^2(15) = \mathbf{240 \ W}$

In a further example, in the network shown in Figure 75.3 the load consists of a fixed capacitive reactance of 7 Ω and a variable resistance R. The value of R for which the power transferred to the load is a maximum is determined as follows:

From condition (4), maximum power transfer is achieved when

$$\mathbf{R} = \sqrt{r^2 + (x + X)^2} = \sqrt{4^2 + (10 - 7)^2}$$

$$= \sqrt{4^2 + 3^2} = \mathbf{5 \ \Omega}$$

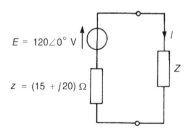

Figure 75.2

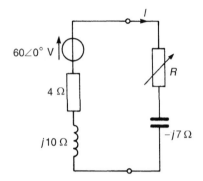

Figure 75.3

Current $\quad I = \dfrac{60\angle 0°}{(4+j10)+(5-j7)} = \dfrac{60\angle 0°}{(9+j3)} = \dfrac{60\angle 0°}{9.487\angle 18.43°}$

$\qquad\qquad = 6.324\angle - 18.43° \text{ A}$

Thus the **maximum power** transferred, $P = I^2 R = (6.324)^2(5) = \textbf{200 W}$

Impedance Matching

It is seen from the previous section that when it is necessary to obtain the maximum possible amount of power from a source, it is advantageous if the circuit components can be adjusted to give equality of impedances. This adjustment is called '**impedance matching**' and is an important consideration in electronic and communications devices, which normally involve small amounts of power. Examples where matching is important include coupling an aerial to a transmitter or receiver, or coupling a loudspeaker to an amplifier.

The mains power supply is considered as infinitely large compared with the demand upon it, and under such conditions it is unnecessary to consider the conditions for maximum power transfer. With transmission lines (see chapter 83), the lines are 'matched', ideally, i.e. terminated in their characteristic impedance.

With d.c. generators, motors or secondary cells, the internal impedance is usually very small and in such cases, if an attempt is made to make the load impedance as small as the source internal impedance, overloading of the source results.

A method of achieving maximum power transfer between a source and a load is to adjust the value of the load impedance to match the source impedance, which can be done using a '**matching-transformer**'.

A transformer is represented in Figure 75.4 supplying a load impedance Z_L. Small transformers used in low power networks are usually regarded as ideal (i.e. losses are negligible), such that

$$\frac{V_1}{V_2} = \frac{N_1}{N_2} = \frac{I_2}{I_1}$$

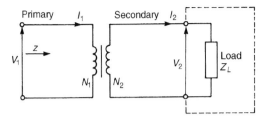

Figure 75.4

From Figure 75.4, the primary input impedance $|z|$ is given by

$$|z| = \frac{V_1}{I_1} = \frac{\left(\dfrac{N_1}{N_2}\right) V_2}{\left(\dfrac{N_2}{N_1}\right) I_2} = \left(\frac{N_1}{N_2}\right)^2 \frac{V_2}{I_2}$$

Since the load impedance, $|Z_L| = \dfrac{V_2}{I_2}$,

$$\boxed{|z| = \left(\frac{N_1}{N_2}\right)^2 |Z_L|} \tag{1}$$

If the input impedance of Figure 75.4 is purely resistive (say, r) and the load impedance is purely resistive (say, R_L) then equation (1) becomes

$$\boxed{r = \left(\frac{N_1}{N_2}\right)^2 R_L} \tag{2}$$

(This is the case introduced in chapter 60, page 419)

Thus by varying the value of the transformer turns ratio, the equivalent input impedance of the transformer can be 'matched' to the impedance of a source to achieve maximum power transfer.

For example, a generator has an output impedance of $(450 + j60)$ Ω. The turns ratio of an ideal transformer necessary to match the generator to a load of $(40 + j19)$ Ω for maximum transfer of power is determined as follows:

Let the output impedance of the generator be z, where $z = (450 + j60)$ Ω or $453.98\angle 7.59°$ Ω and the load impedance be Z_L, where $Z_L = (40 + j19)$ Ω or $44.28\angle 25.41°$ Ω. From Figure 70.4 and equation (1), $|z| = \left(\dfrac{N_1}{N_2}\right)^2 |Z_L|$.

Hence transformer turns ratio $\left(\dfrac{N_1}{N_2}\right) = \sqrt{\dfrac{|z|}{|Z_L|}} = \sqrt{\dfrac{453.98}{44.28}} = \sqrt{(10.25)}$
$$= \mathbf{3.20}$$

In another example, let an ac. source of $30\angle 0°$ V and internal resistance 20 kΩ be matched to a load by a $20 : 1$ ideal transformer. The network diagram is shown in Figure 75.5. The value of the load resistance for maximum power transfer is determined as follows:

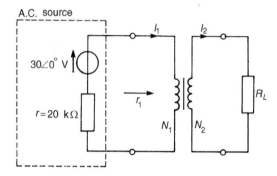

Figure 75.5

For maximum power transfer, r_1 must he equal to 20 kΩ. From equation (2),

$$r_1 = \left(\frac{N_1}{N_2}\right)^2 R_L \text{ from which,}$$

load resistance $R_L = r_1 \left(\frac{N_2}{N_1}\right)^2 = (20\,000)\left(\frac{1}{20}\right)^2 = \mathbf{50\ \Omega}$

The power dissipated in the load is determined as follows:

The total input resistance when the source is connected to the matching transformer is $(r + r_1)$, i.e. 20 kΩ + 20 kΩ = 40 kΩ

Primary current, $I_1 = \dfrac{V}{40\,000} = \dfrac{30}{40\,000} = 0.75$ mA

$\dfrac{N_1}{N_2} = \dfrac{I_2}{I_1}$ from which, $I_2 = I_1 \left(\dfrac{N_1}{N_2}\right) = (0.75 \times 10^{-3})\left(\dfrac{20}{1}\right)$

$$= 15 \text{ mA}$$

Power dissipated in load resistance R_L is given by:

$$P = I_2^2 R_L = (15 \times 10^{-3})^2 (50) = \mathbf{0.01125\ W} \text{ or } \mathbf{11.25\ mW}$$

76 Complex Waveforms

Introduction

In preceding chapters a.c. supplies have been assumed to be sinusoidal, this being a form of alternating quantity commonly encountered in electrical engineering. However, many supply waveforms are **not** sinusoidal. For example, saw-tooth generators produce ramp waveforms, and rectangular waveforms may be produced by multivibrators. A waveform that is not sinusoidal is called a **complex wave**. Such a waveform may be shown to be composed of the sum of a series of sinusoidal waves having various interrelated periodic times.

A function $f(t)$ is said to be **periodic** if $f(t + T) = f(t)$ for all values of t, where T is the interval between two successive repetitions and is called the **period** of the function. A sine wave having a period of $\dfrac{2\pi}{\omega}$ is a familiar example of a periodic function.

A typical complex periodic-voltage waveform, shown in Figure 76.1, has period T seconds and frequency f hertz. A complex wave such as this can be resolved into the sum of a number of sinusoidal waveforms, and each of the sine waves can have a different frequency, amplitude and phase.

The initial, major sine wave component has a frequency f equal to the frequency of the complex wave and this frequency is called the **fundamental frequency**. The other sine wave components are known as **harmonics**, these having frequencies which are integer multiples of frequency f. Hence the second harmonic has a frequency of $2f$, the third harmonic has a frequency of $3f$, and so on. Thus if the fundamental (i.e. supply) frequency of

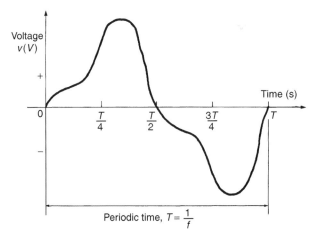

Figure 76.1

a complex wave is 50 Hz, then the third harmonic frequency is 150 Hz, the fourth harmonic frequency is 200 Hz, and so on.

The General Equation for a Complex Waveform

The instantaneous value of a complex voltage wave acting in a linear circuit may be represented by the general equation:

$$v = V_{1m} \sin(\omega t + \psi_1) + V_{2m} \sin(2\omega t + \psi_2)$$
$$+ \ldots + V_{nm} \sin(n\omega t + \psi_n) \text{ volts} \qquad (1)$$

Here $V_{1m} \sin(\omega t + \psi_1)$ represents the fundamental component of which V_{1m} is the maximum or peak value, frequency $f = \dfrac{\omega}{2\pi}$, and ψ_1 is the phase angle with respect to time, $t = 0$.

Similarly, $V_{2m} \sin(2\omega t + \psi_2)$ represents the second harmonic component, and $V_{nm} \sin(n\omega t + \psi_n)$ represents the n th harmonic component, of which V_{nm} is the peak value, frequency $= \dfrac{n\omega}{2\pi} (= nf)$ and ψ_n is the phase angle.

In the same way, the instantaneous value of a complex current i may be represented by the general equation:

$$i = I_{1m} \sin(\omega t + \theta_1) + I_{2m} \sin(2\omega t + \theta_2)$$
$$+ \ldots + I_{nm} \sin(n\omega t + \theta_n) \text{ amperes} \qquad (2)$$

Where equations (1) and (2) refer to the voltage across and the current flowing through a given linear circuit, the phase angle between the fundamental voltage and current is $\phi_1 = (\psi_1 - \theta_1)$, the phase angle between the second harmonic voltage and current is $\phi_2 = (\psi_2 - \theta_2)$, and so on.

It often occurs that not all harmonic components are present in a complex waveform. Sometimes only the fundamental and odd harmonics are present, and in others only the fundamental and even harmonics are present.

Harmonic Synthesis

Harmonic analysis is the process of resolving a complex periodic waveform into a series of sinusoidal components of ascending order of frequency. Many of the waveforms met in practice can be represented by mathematical expressions similar to those of equations (1) and (2), and the magnitude of their harmonic components together with their phase may be calculated using **Fourier series** (see *Mathematics Pocket Book*). **Numerical methods** are used to analyse waveforms for which simple mathematical expressions cannot be obtained. A numerical method of harmonic analysis is explained in chapter 77. In a laboratory, waveform analysis may be performed using a **waveform analyser** which produces a direct readout of the component waves present in a complex wave.

By adding the instantaneous values of the fundamental and progressive harmonics of a complex wave for given instants in time, the shape of a complex waveform can be gradually built up. This graphical procedure is known as **harmonic synthesis** (synthesis meaning 'the putting together of parts or elements so as to make up a complex whole').

(i) A complex waveform comprising the sum of the fundamental and a third harmonic of about half the amplitude of the fundamental is shown in Figure 76.2(a), both waveforms being initially in phase with each other. If further odd harmonic waveforms of the appropriate amplitudes are added, a good approximation to a square wave results. In Figure 76.2(b), the third harmonic is shown having an initial phase displacement from the fundamental. The positive and negative half cycles of each of the

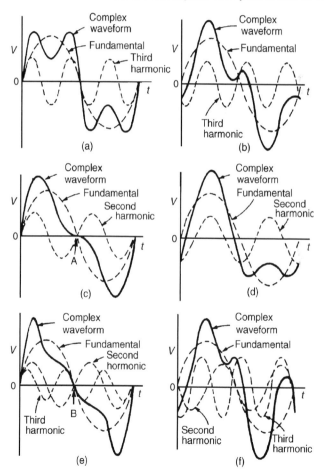

Figure 76.2

562

complex waveforms shown in Figures 76.2(a) and (b) are identical in shape, and this is a feature of waveforms containing the fundamental and only odd harmonics.

(ii) A complex waveform comprising the sum of the fundamental and a second harmonic of about half the amplitude of the fundamental is shown in Figure 76.2(c), each waveform being initially in phase with each other. If further even harmonics of appropriate amplitudes are added, a good approximation to a triangular wave results. In Figure 76.2(c), the negative cycle, if reversed, appears as a mirror image of the positive cycle about point A. In Figure 76.2(d) the second harmonic is shown with an initial phase displacement from the fundamental and the positive and negative half cycles are dissimilar.

(iii) A complex waveform comprising the sum of the fundamental, a second harmonic and a third harmonic is shown in Figure 76.2(e), each waveform being initially 'in-phase'. The negative half cycle, if reversed, appears as a mirror image of the positive cycle about point B. In Figure 76.2(f), a complex waveform comprising the sum of the fundamental, a second harmonic and a third harmonic are shown with initial phase displacement. The positive and negative half cycles are seen to be dissimilar.

The features mentioned relative to Figures 76.2(a) to (f) make it possible to recognize the harmonics present in a complex waveform displayed on a CRO.

R.m.s. Value, Mean Value and the Form Factor of a Complex Wave

R.m.s. value

For a complex current waveform represented by:

$$i = I_0 + I_{1m}\sin(\omega t + \theta_1) + I_{2m}\sin(2\omega t + \theta_2)$$
$$+ \ldots + I_{nm}\sin(n\omega t + \theta_n) \text{ amperes}$$

the r.m.s. current I is given by:

$$I = \sqrt{\left(I_0^2 + \frac{I_{1m}^2 + I_{2m}^2 + \ldots + I_{nm}^2}{2}\right)}$$

Similarly, for a complex voltage waveform represented by:

$$v = V_0 + V_{1m}\sin(\omega t + \psi_1) + V_{2m}\sin(2\omega t + \psi_2)$$
$$+ \ldots + V_{nm}\sin(n\omega t + \psi_n) \text{ volts}$$

the r.m.s. value of voltage, V, is given by:

$$V = \sqrt{\left(V_0^2 + \frac{V_{1m}^2 + V_{2m}^2 + \ldots + V_{nm}^2}{2}\right)}$$

For example, if a current waveform is represented by:

$$i = 25 + 80 \sin \omega t + 30 \sin \left(3\omega t + \frac{\pi}{6} \right) + 10 \sin \left(5\omega t + \frac{\pi}{3} \right) \text{ mA}$$

then the **r.m.s. current**,

$$I = \sqrt{ \left(25^2 + \frac{80^2 + 30^2 + 10^2}{2} \right) } = \textbf{65.8 mA}$$

Mean value

The mean or average value of a complex quantity whose negative half-cycle is similar to its positive half-cycle is given, for current, by:

$$I_{av} = \frac{1}{\pi} \int_0^\pi i \, \mathrm{d}(\omega t)$$

and for voltage by:

$$V_{av} = \frac{1}{\pi} \int_0^\pi v \, \mathrm{d}(\omega t)$$

each waveform being taken over half a cycle.

Unlike r.m.s. values, mean values **are** affected by the relative phase angles of the harmonic components.

For example, for the above current,

$$I_{av} = \frac{1}{\pi} \int_0^\pi \left(25 + 80 \sin \omega t + 30 \sin \left(3\omega t + \frac{\pi}{6} \right) \right.$$

$$\left. + 10 \sin \left(5\omega t + \frac{\pi}{3} \right) \right) \mathrm{d}(\omega t)$$

$$= \frac{1}{\pi} \left[25\omega t - 80 \cos \omega t - \frac{30}{3} \cos \left(3\omega t + \frac{\pi}{6} \right) \right.$$

$$\left. - \frac{10}{5} \cos \left(5\omega t + \frac{\pi}{3} \right) \right]_0^\pi$$

$$= \frac{1}{\pi} \left\{ \left[25\pi - 80 \cos \pi - 10 \cos \left(3\pi + \frac{\pi}{6} \right) - 2 \cos \left(5\pi + \frac{\pi}{3} \right) \right] \right.$$

$$\left. - \left[0 - 80 \cos 0 - 10 \cos \frac{\pi}{6} - 2 \cos \frac{\pi}{3} \right] \right\}$$

$$= \frac{1}{\pi} [(78.54 + 80 + 8.66 + 1.00) - (0 - 80 - 8.66 - 1.00)]$$

$$= \textbf{82.1 mA}$$

Form factor

The form factor of a complex waveform whose negative half-cycle is similar in shape to its positive half-cycle is defined as:

$$\text{form factor} = \frac{\text{r.m.s. value of the waveform}}{\text{mean value}}$$

where the mean value is taken over half a cycle.

Changes in the phase displacement of the harmonics may appreciably alter the form factor of a complex waveform.

Thus, for the above current, **formfactor** $= \dfrac{65.8}{82.1} = \mathbf{0.80}$

Power Associated with Complex Waves

The average power P supplied for one cycle of the fundamental is given by:

$$P = V_0 I_0 + V_1 I_1 \cos\phi_1 + V_2 I_2 \cos\phi_2 + \ldots + V_n I_n \cos\phi_n$$

where V_1 and I_1 are the r.m.s. values of the fundamental voltage and current respectively, ϕ_1 is the phase difference between the fundamental voltage and current, and so on.

Alternatively, if R is the equivalent series resistance of a circuit then the total power is given by:

$$P = I^2 R$$

where I is the r.m.s. value of current.

For example, if a complex voltage given by:

$$v = 25 + 100 \sin\omega t + 40 \sin\left(3\omega t + \frac{\pi}{6}\right) + 20 \sin\left(5\omega t + \frac{\pi}{12}\right) \text{ volts}$$

is applied to a circuit, and the resulting current is given by:

$$i = 5 + 14.14 \sin(\omega t - 0.785) + 2.53 \sin(3\omega t - 0.726)$$
$$+ \, 0.784 \sin(5\omega t - 1.112) \text{ amperes,}$$

the **average power dissipated** in the circuit is given by:

$$P = (25)(5) + \left(\frac{100}{\sqrt{2}}\right)\left(\frac{14.14}{\sqrt{2}}\right) \cos(0 - 0.785)$$

$$+ \left(\frac{40}{\sqrt{2}}\right)\left(\frac{2.53}{\sqrt{2}}\right) \cos\left(\frac{\pi}{6} - -0.726\right)$$

$$+ \left(\frac{20}{\sqrt{2}}\right)\left(\frac{0.784}{\sqrt{2}}\right) \cos\left(\frac{\pi}{12} - -1.112\right)$$

$$= 125 + 500.12 + 15.97 + 1.53 = \mathbf{642.6 \text{ W}}$$

Power factor

When harmonics are present in a waveform the overall circuit power factor is defined as:

$$\text{Overall power factor} = \frac{\text{total power supplied}}{\text{total r.m.s. voltage} \times \text{total r.m.s. current}}$$

$$= \frac{\text{total power}}{\text{volt amperes}}$$

i.e. $$\boxed{\text{p.f.} = \frac{V_1 I_1 \cos\phi_1 + V_2 I_2 \cos\phi_2 + \dots}{VI}}$$

With a sinusoidal waveform, power factor

$$= \frac{\text{power}}{\text{volt-amperes}} = \frac{VI \cos\phi}{VI} = \cos\phi$$

Thus power factor depends upon the value of phase angle ϕ, and is lagging for an inductive circuit and leading for a capacitive circuit. However, with a complex waveform, power factor is not given by $\cos\phi$. In the expression for power, there are n phase-angle terms, $\phi_1, \phi_2 \dots, \phi_n$, all of which may be different. It is for this reason that it is not possible to state whether the overall power factor is lagging or leading when harmonics are present. For example, for the above voltage and current,

$$I = \sqrt{\left(5^2 + \frac{14.14^2 + 2.53^2 + 0.784^2}{2}\right)} = 11.33 \text{ A}$$

and $$V = \sqrt{\left(25^2 + \frac{100^2 + 40^2 + 20^2}{2}\right)} = 81.39 \text{ V}$$

hence, the **overall power factor** $= \dfrac{642.6}{(81.39)(11.33)} = \mathbf{0.697}$

Harmonics in Single-phase Circuits

When a complex alternating voltage wave, i.e. one containing harmonics, is applied to a single-phase circuit containing resistance, inductance and/or capacitance (i.e. linear circuit elements), then the resulting current will also be complex and contain harmonics.

Let a complex voltage be represented by:

$$v = V_{1m} \sin\omega t + V_{2m} \sin 2\omega t + V_{3m} \sin 3\omega t + \dots$$

(a) Pure resistance

The impedance of a pure resistance R is independent of frequency and the current and voltage are in phase for each harmonic. Thus the general expression for current is given by:

$$i = \frac{v}{R} = \frac{V_{1m}}{R} \sin \omega t + \frac{V_{2m}}{R} \sin 2\omega t + \frac{V_{3m}}{R} \sin 3\omega t + \ldots \quad (1)$$

The percentage harmonic content in the current wave is the same as that in the voltage wave. For example, the percentage second harmonic content from equation (1) is:

$$\frac{\frac{V_{2m}}{R}}{\frac{V_{1m}}{R}} \times 100\% \text{ i.e. } \frac{V_{2m}}{V_{1m}} \times 100\%$$

the same as for the voltage wave. The current and voltage waveforms will therefore be identical in shape.

(b) Pure inductance

The impedance of a pure inductance L, i.e. inductive reactance $X_L(= 2\pi f L)$, varies with the harmonic frequency when voltage is applied to it. Also, for every harmonic term, the current will lag the voltage by 90° or $\frac{\pi}{2}$ rad. The current i is given by:

$$i = \frac{v}{X_L} = \frac{V_{1m}}{\omega L} \sin \left(\omega t - \frac{\pi}{2} \right) + \frac{V_{2m}}{2\omega L} \sin \left(2\omega t - \frac{\pi}{2} \right)$$
$$+ \frac{V_{3m}}{3\omega L} \sin \left(3\omega t - \frac{\pi}{2} \right) + \ldots \quad (2)$$

since for the n'th harmonic the reactance is $n\omega L$

Equation (2) shows that for, say, the n'th harmonic, the percentage harmonic content in the current waveform is only $\frac{1}{n}$ of the corresponding harmonic content in the voltage waveform.

If a complex current contains a d.c. component then the direct voltage drop across a pure inductance is zero.

(c) Pure capacitance

The impedance of a pure capacitance C, i.e. capacitive reactance $X_C \left(= \frac{1}{2\pi f C} \right)$, varies with the harmonic frequency when voltage is applied to it. Also, for each harmonic term the current will lead the voltage by 90° or

$\dfrac{\pi}{2}$ rad. The current i is given by:

$$i = \frac{v}{X_C} = \frac{V_{1m}}{\dfrac{1}{\omega C}} \sin\left(\omega t + \frac{\pi}{2}\right) + \frac{V_{2m}}{\dfrac{1}{2\omega C}} \sin\left(2\omega t + \frac{\pi}{2}\right)$$

$$+ \frac{V_{3m}}{\dfrac{1}{3\omega C}} \sin\left(3\omega t + \frac{\pi}{2}\right) + \ldots,$$

since for the n'th harmonic the reactance is $\dfrac{1}{n\omega C}$. Hence current,

$$\boxed{\begin{aligned} i = V_{1m}(\omega C)\sin\left(\omega t + \frac{\pi}{2}\right) + V_{2m}(2\omega C)\sin\left(2\omega t + \frac{\pi}{2}\right) \\ + V_{3m}(3\omega C)\sin\left(3\omega t + \frac{\pi}{2}\right) + \ldots \end{aligned}} \quad (3)$$

Equation (3) shows that the percentage harmonic content of the current waveform is n times larger for the n'th harmonic than that of the corresponding harmonic voltage.

For example, a complex voltage waveform represented by:

$$v = 100\sin\omega t + 30\sin\left(3\omega t + \frac{\pi}{3}\right) + 10\sin\left(5\omega t - \frac{\pi}{6}\right) \text{ volts}$$

is applied across (a) a pure 40 Ω resistance, (b) a pure 7.96 mH inductance, and (c) a pure 25 μF capacitor. The current flowing if the fundamental frequency is 1 kHz is determined for each case as follows:
(a) From equation (1),

$$\text{current} \quad i = \frac{v}{R} = \frac{100}{40}\sin\omega t + \frac{30}{40}\sin\left(3\omega t + \frac{\pi}{3}\right)$$

$$+ \frac{10}{40}\sin\left(5\omega t - \frac{\pi}{6}\right)$$

i.e. $\quad i = 2.5\sin\omega t + 0.75\sin\left(3\omega t + \frac{\pi}{3}\right)$

$$+ 0.25\sin\left(5\omega t - \frac{\pi}{6}\right) \text{ amperes}$$

(b) At the fundamental frequency, $\omega L = 2\pi(1000)(7.96 \times 10^{-3}) = 50$ Ω.
From equation (2),

$$\text{current, } i = \frac{100}{50}\sin\left(\omega t - \frac{\pi}{2}\right) + \frac{30}{3 \times 50}\sin\left(3\omega t + \frac{\pi}{3} - \frac{\pi}{2}\right)$$

$$+ \frac{10}{5 \times 50}\sin\left(5\omega t - \frac{\pi}{6} - \frac{\pi}{2}\right)$$

i.e. **current,** $i = 2\sin\left(\omega t - \dfrac{\pi}{2}\right) + 0.20\sin\left(3\omega t - \dfrac{\pi}{6}\right)$

$$+ 0.04\sin\left(5\omega t - \dfrac{2\pi}{3}\right) \text{ A}$$

(c) At the fundamental frequency, $\omega C = 2\pi(1000)(25 \times 10^{-6}) = 0.157$
From equation (3),

current, $\quad i = 100(0.157)\sin\left(\omega t + \dfrac{\pi}{2}\right)$

$$+ 30(3 \times 0.157)\sin\left(3\omega t + \dfrac{\pi}{3} + \dfrac{\pi}{2}\right)$$

$$+ 10(5 \times 0.157)\sin\left(5\omega t - \dfrac{\pi}{6} + \dfrac{\pi}{2}\right)$$

i.e. $\quad\quad i = 15.70\sin\left(\omega t + \dfrac{\pi}{2}\right) + 14.13\sin\left(3\omega t + \dfrac{5\pi}{6}\right)$

$$+ 7.85\sin\left(5\omega t + \dfrac{\pi}{3}\right) \text{ A}$$

In another example, an e.m.f. is represented by:

$$e = 50 + 200\sin\omega t + 40\sin\left(2\omega t - \dfrac{\pi}{2}\right) + 5\sin\left(4\omega t + \dfrac{\pi}{4}\right) \text{ volts,}$$

the fundamental frequency being 50 Hz. The e.m.f. is applied across a circuit comprising a 100 μF capacitor connected in series with a 50 Ω resistor. The expression for the current flowing is determined as follows:

d.c. component

In a d.c. circuit no current will flow through a capacitor. The current waveform will not possess a d.c. component even though the e.m.f. waveform has a 50 V d.c. component. Hence $i_0 = 0$

Fundamental

Capacitive reactance,

$$X_{C1} = \frac{1}{2\pi f C} = \frac{1}{2\pi(50)(100 \times 10^{-6})} = 31.83 \ \Omega$$

Impedance, $Z_1 = (50 - j31.83) \ \Omega = 59.27\angle-32.48° \ \Omega$

$$I_{1m} = \frac{V_{1m}}{Z_1} = \frac{200\angle0°}{59.27\angle-32.48°} = 3.374\angle32.48° \text{ A} = 3.374\angle0.567 \text{ A}$$

Hence the fundamental current, $i_1 = 3.374\sin(\omega t + 0.567)$ A

Second harmonic

Capacitive reactance, $X_{C2} = \dfrac{1}{2(2\pi f C)} = \dfrac{31.83}{2} = 15.92 \ \Omega$

Impedance, $Z_2 = (50 - j15.92) \ \Omega = 52.47\angle -17.66° \ \Omega$

$$I_{2m} = \frac{V_{2m}}{Z_2} = \frac{40\angle -\dfrac{\pi}{2}}{52.47\angle -17.66°} = 0.762\angle \left(-\frac{\pi}{2} - (-17.66°)\right)$$

$$= 0.762\angle -72.34° \ A$$

Hence the second harmonic current,

$$i_2 = 0.762\sin(2\omega t - 72.34°) = 0.762\sin(2\omega t - 1.263) \ A$$

Fourth harmonic

Capacitive reactance, $X_{C4} = \dfrac{1}{4}X_{C1} = \dfrac{31.83}{4} = 7.958 \ \Omega$

Impedance, $Z_4 = (50 - j7.958) \ \Omega = 50.63\angle -9.04° \ \Omega$

$$I_{4m} = \frac{V_{4m}}{Z_4} = \frac{5\angle \dfrac{\pi}{4}}{50.63\angle -9.04°} = 0.099\angle \left(\frac{\pi}{4} - (-9.04°)\right)$$

$$= 0.099\angle 54.04° \ A$$

Hence the fourth harmonic current,

$$i_4 = 0.099\sin(4\omega t + 54.04°) \ A = 0.099\sin(4\omega t + 0.943) \ A$$

An expression for current flowing is therefore given by: $i = i_0 + i_1 + i_2 + i_4$

i.e. $\quad i = 3.374\sin(\omega t + 0.567) + 0.762\sin(2\omega t - 1.263)$

$$+ 0.099\sin(4\omega t + 0.943) \ A$$

From earlier, r.m.s. current, $\quad I = \sqrt{\left(\dfrac{3.374^2 + 0.762^2 + 0.099^2}{2}\right)}$

$$= 2.45 \ A$$

Resonance due to Harmonics

In industrial circuits at power frequencies the typical values of L and C involved make resonance at the fundamental frequency very unlikely. (An exception to this is with the capacitor-start induction motor where the start winding can achieve unity power factor during run-up).

However, if the voltage waveform is not a pure sine wave it is quite possible for the resonant frequency to be near the frequency of one of the harmonics. In this case the magnitude of the particular harmonic in the current waveform is greatly increased and may even exceed that of the fundamental. The effect of this is a great distortion of the resultant current waveform so that dangerous volt drops may occur across the inductance and capacitance in the circuit.

When a circuit resonates at one of the harmonic frequencies of the supply voltage, the effect is called **selective or harmonic resonance**.

For resonance with the fundamental, the condition is: $\omega L = \dfrac{1}{\omega C}$; for resonance at, say, the third harmonic, the condition is: $3\omega L = \dfrac{1}{3\omega C}$; for resonance at the n'th harmonic, the condition is:

$$n\omega L = \frac{1}{n\omega C}$$

For example, a voltage waveform having a fundamental of maximum value 400 V and a third harmonic of maximum value 10 V is applied to the circuit shown in Figure 76.3. The fundamental frequency for resonance with the third harmonic is determined as follows:

Resonance with the third harmonic means that $3\omega L = \dfrac{1}{3\omega C}$

from which, $\omega = \sqrt{\left(\dfrac{1}{9LC}\right)} = \dfrac{1}{3\sqrt{(0.5)(0.2 \times 10^{-6})}} = 1054$ rad/s

and **fundamental frequency**, $f = \dfrac{\omega}{2\pi} = \dfrac{1054}{2\pi} = 167.7$ **Hz**

At the fundamental frequency,

$$\text{impedance, } Z_1 = R + j\left(\omega L - \frac{1}{\omega C}\right)$$

$$= 2 + j\left[(1054)(0.5) - \frac{1}{(1054)(0.2 \times 10^{-6})}\right]$$

$$= (2 - j4217)\ \Omega = 4217\angle{-89.97°}\ \Omega$$

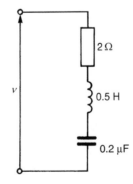

Figure 76.3

Maximum value of current at the fundamental frequency,

$$I_{1m} = \frac{V_{1m}}{Z_1} = \frac{400}{4217} = \textbf{0.095 A}$$

At the third harmonic frequency,

$$Z_3 = R + j\left(3\omega L - \frac{1}{3\omega C}\right) = R$$

since resonance occurs at the third harmonic, i.e. $Z_3 = 2 \; \Omega$
Maximum value of current at the third harmonic frequency,

$$I_{3m} = \frac{V_{3m}}{Z_3} = \frac{10}{2} = \textbf{5 A}$$

Note that the magnitude of I_{3m} compared with I_{1m} is

$$\frac{5}{0.095} = \textbf{52.6 greater}.$$

Sources of Harmonics

(i) Harmonics may be produced in the output waveform of an a.c. generator. This may be due either to 'tooth-ripple', caused by the effect of the slots that accommodate the windings, or to the non-sinusoidal air-gap flux distribution.

Great care is taken to ensure a sinusoidal output from generators in large supply systems; however, non-linear loads will cause harmonics to appear in the load current waveform. Thus harmonics are produced in devices that have a non-linear response to their inputs. Non-linear circuit elements (i.e. those in which the current flowing through them is not proportional to the applied voltage) include rectifiers and any large-signal electronic amplifier in which diodes, transistors, valves or iron-cored inductors are used.

(ii) A **rectifier** is a device for converting an alternating or an oscillating current into a unidirectional or approximate direct current. A rectifier has a low impedance to current flow in one direction and a nearly infinite impedance to current flow in the opposite direction. Thus when an alternating current is applied to a rectifier, current will flow through it during the positive half-cycles only; the current is zero during the negative half-cycles. A typical current waveform is shown in Figure 76.4. Using a single diode produces this 'half-wave rectification', as shown in Chapter 51. The waveform shown in Figure 76.4 is typical of one containing a fairly large second harmonic.

(iii) **Transistors** and **valves** are non-linear devices in that sinusoidal input results in different positive and negative half-cycle amplifications. This means that the output half-cycles have different amplitudes. Since they have a different shape, even harmonic distortion is suggested (see earlier on harmonic synthesis).

Figure 76.4

(iv) **Ferromagnetic-cored coils** are a source of harmonic generation in a.c. circuits because of the non-linearity of the curve and the hysteresis loop, especially if saturation occurs. Let a sinusoidal voltage be applied to a ferromagnetic-cored coil (having low resistance relative to inductive reactance) of cross-section area A square metres and possessing N turns. If ϕ is the flux produced in the core then the magnitude of the instantaneous voltage is given by: $v = N\dfrac{\mathrm{d}\phi}{\mathrm{d}t}$ (from Chapter 48). If B is the flux density of the core, then since $\Phi = BA$, $v = N\dfrac{\mathrm{d}}{\mathrm{d}t}(BA) = NA\dfrac{\mathrm{d}B}{\mathrm{d}t}$ since area A is a constant for a particular core.

Separating the variables gives: $\displaystyle\int \mathrm{d}B = \frac{1}{NA}\int v\,\mathrm{d}t$

i.e.
$$B = \frac{1}{NA}\int V_m \sin \omega t\,\mathrm{d}t$$

$$= -\frac{V_m}{\omega NA}\cos \omega t$$

Since $-\cos \omega t = \sin(\omega t - 90°)$,

$$B = \frac{V_m}{\omega NA}\sin(\omega t - 90°) \tag{4}$$

Equation (4) shows that if the applied voltage is sinusoidal, the flux density in the iron core must also be sinusoidal but lagging by 90°. The condition of low resistance relative to inductive reactance, giving a sinusoidal flux from a sinusoidal supply voltage, is called **free magnetisation**. Consider the application of a sinusoidal voltage to a coil wound on a core with a hysteresis loop as shown in Figure 76.5(a). The horizontal axis of a hysteresis loop is magnetic field strength H, but since $H = \dfrac{Ni}{l}$ and N and l (the length of the flux path) are constant, the axis may be directly scaled as current i $\left(\text{i.e. } i = \dfrac{Hl}{N}\right)$. Figure 76.5(b) shows sinusoidal voltage v and flux density B waveforms, B lagging v by 90°.

The current waveform is shown in Figure 76.5(c) and is derived as follows. At time t_1, point a on the voltage curve corresponds to point b on the flux density curve and the point c on the hysteresis loop. The current at time t_1 is given by the distance dc. Plotting this current on a vertical

573

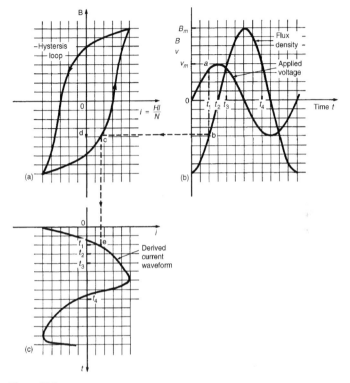

Figure 76.5

time-scale gives the derived point e on the current curve. A similar procedure is adopted for times t_2, t_3 and so on over one cycle of the voltage.
(Note that it is important to move around the hysteresis loop in the correct direction). It is seen from the current curve that it is non-sinusoidal
and that the positive and negative half-cycles are identical. This indicates
that the waveform contains only odd harmonics.

(v) If, in a circuit containing a ferromagnetic-cored coil, the resistance is high
compared with the inductive reactance, then the current flowing from a
sinusoidal supply will tend to be sinusoidal. This means that the flux
density B of the core cannot be sinusoidal since it is related to the current
by the hysteresis loop. This means, in turn, that the induced voltage due
to the alternating flux $\left(\text{i.e. } v = NA\dfrac{\mathrm{d}B}{\mathrm{d}t}\right)$ will not be sinusoidal. This
condition is called **forced magnetisation**.

The shape of the induced voltage waveform under forced magnetisation is
obtained as follows. The current waveform is shown on a vertical axis in
Figure 76.6(a). The hysteresis loop corresponding to the maximum value
of circuit current is drawn as shown in Figure 76.7(b). The flux density

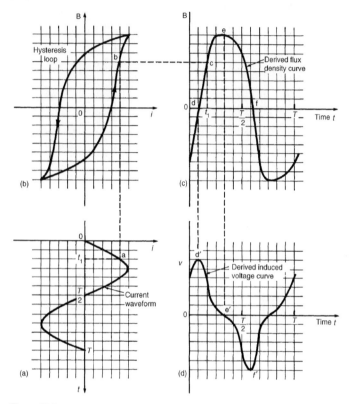

Figure 76.6

curve which is derived from the sinusoidal current waveform is shown in Figure 76.6(c). Point a on the current wave at time t_1 corresponds to point b on the hysteresis loop and to point c on the flux density curve. By taking other points throughout the current cycle the flux density curve is derived as shown.

The relationship between the induced voltage v and the flux density B is given by: $v = NA\dfrac{dB}{dt}$. Here $\dfrac{dB}{dt}$ represents the rate of change of flux density with respect to time, i.e. the gradient of the B/t curve. At point d the gradient of the B/t curve is a maximum in the positive direction. Thus v will be maximum positive as shown by point d' in Figure 76.6(d). At point e the gradient $\left(\text{i.e. } \dfrac{dB}{dt}\right)$ is zero, thus v is zero, as shown by point e'. At point f the gradient is maximum in a negative direction, thus v is maximum negative, as shown by point f'. If all such points are taken around the B/t curve, the curve representing induced voltage, shown

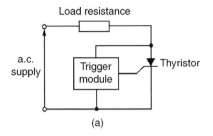

(a)

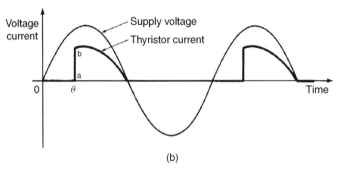

(b)

Figure 76.7

in Figure 76.6(d), is produced. The resulting voltage waveform is non-sinusoidal. The positive and negative half-cycles are identical in shape, indicating that the waveform contains a fundamental and a prominent third harmonic.

(vi) The amount of power delivered to a load can be controlled using a **thyristor**, which is a semi-conductor device. Examples of applications of controlled rectification include lamp and heater controls and the control of motor speeds. A basic circuit used for single-phase power control is shown in Figure 76.7(a). The trigger module contains circuitry to produce the necessary gate current to turn the thyristor on. If the pulse is applied at time $\dfrac{\theta}{\omega}$ where θ is the firing or triggering angle, then the current flowing in the load resistor has a waveform as shown in Figure 76.7(b). The sharp rise-time (shown as ab in Figure 76.7(b)), however, gives rise to harmonics.

(vii) In **microelectronic systems** rectangular waveforms are common. Again, fast rise-times give rise to harmonics, especially at high frequency. These harmonics can be fed back to the mains if not filtered.

There are thus a large number of sources of harmonics.

77 A Numerical Method of Harmonic Analysis

Introduction

Many practical waveforms can be represented by simple mathematical expressions, and, by using Fourier series, the magnitude of their harmonic components determined. For waveforms not in this category, analysis may be achieved by numerical methods. **Harmonic analysis** is the process of resolving a periodic, non-sinusoidal quantity into a series of sinusoidal components of ascending order of frequency.

Harmonic Analysis on data given in Tabular or Graphical Form

A Fourier series is merely a trigonometric series of the form:

$$f(x) = a_0 + a_1 \cos x + a_2 \cos 2x + \ldots + b_1 \sin x + b_2 \sin 2x + \ldots$$

i.e. $f(x) = a_0 + \sum_{n=1}^{\infty} (a_n \cos nx + b_n \sin nx)$

The Fourier coefficients a_0, a_n and b_n all require functions to be integrated.

$$a_0 = \frac{1}{2\pi} \int_{-\pi}^{\pi} f(x) \, dx \, dx = \frac{1}{2\pi} \int_{0}^{2\pi} f(x) \, dx$$

$\quad = $ mean value of $f(x)$ in the range $-\pi$ to π or 0 to 2π

$$a_n = \frac{1}{\pi} \int_{-\pi}^{\pi} f(x) \cos nx \, dx = \frac{1}{\pi} \int_{0}^{2\pi} f(x) \cos nx \, dx$$

$\quad = $ twice the mean value of $f(x) \cos nx$ in the range 0 to 2π

$$b_n = \frac{1}{\pi} \int_{-\pi}^{\pi} f(x) \sin nx \, dx = \frac{1}{\pi} \int_{0}^{2\pi} f(x) \sin nx \, dx$$

$\quad = $ twice the mean value of $f(x) \sin nx$ in the range 0 to 2π

However, irregular waveforms are not usually defined by mathematical expressions and thus the Fourier coefficients cannot be determined by using calculus. In these cases, approximate methods, such as the **trapezoidal rule,** can be used to evaluate the Fourier coefficients.

Most practical waveforms to be analysed are periodic. Let the period of a waveform be 2π and be divided into p equal parts as shown in

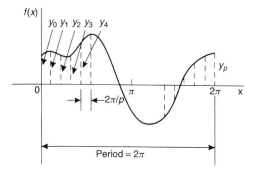

Figure 77.1

Figure 77.1. The width of each interval is thus $\dfrac{2\pi}{p}$. Let the ordinates be labelled $y_0, y_1, y_2, \ldots y_p$ (note that $y_0 = y_p$). The trapezoidal rule states:

$$\text{Area} = (\text{width of interval})[\tfrac{1}{2}(\text{first} + \text{last ordinate})$$
$$+\text{sum of remaining ordinates}]$$

$$\approx \frac{2\pi}{p}[\tfrac{1}{2}(y_0 + y_p) + y_1 + y_2 + y_3 + \ldots]$$

Since $y_0 = y_p$, then $\tfrac{1}{2}(y_0 + y_p) = y_0 = y_p$.

Hence area $\approx \dfrac{2\pi}{p} \sum_{k=1}^{p} y_k$

Mean value $= \dfrac{\text{area}}{\text{length of base}} \approx \dfrac{1}{2\pi}\left(\dfrac{2\pi}{p}\right) \sum_{k=1}^{p} y_k \approx \dfrac{1}{p} \sum_{k=1}^{p} y_k$

However, $a_0 = $ mean value of $f(x)$ in the range 0 to 2π

Thus $\boxed{a_0 \approx \dfrac{1}{p} \sum_{k=1}^{p} y_k}$ (1)

Similarly, $a_n = $ twice the mean value of $f(x)\cos nx$ in the range 0 to 2π,

thus $\boxed{a_n \approx \dfrac{2}{p} \sum_{k=1}^{p} y_k \cos nx_k}$ (2)

and $b_n = $ twice the mean value of $f(x)\sin nx$ in the range 0 to 2π,

thus $\boxed{b_n \approx \dfrac{2}{p} \sum_{k=1}^{p} y_k \sin nx_k}$ (3)

For example, the values of the voltage v volts at different moments in a cycle are given by:

$\theta°$ degrees	30	60	90	120	150	180	210	240	270	300	330	360
V (volts)	62	35	−38	−64	−63	−52	−28	24	80	96	90	70

Analysing the voltage into its first three constituent harmonics is achieved as follows:

The graph of voltage V against angle θ is shown in Figure 77.2. The range 0 to 2π is divided into 12 equal intervals giving an interval width of $\dfrac{2\pi}{12}$, i.e. $\dfrac{\pi}{6}$ rad or 30°. The values of the ordinates y_1, y_2, y_3, ... are 62, 35, −38, ... from the given table of values. If a larger number of intervals are used, results having a greater accuracy are achieved. The data is tabulated in the proforma shown in Table 77.1.

From equation (1), $\quad a_0 \approx \dfrac{1}{p}\sum\limits_{k=1}^{p} y_k$

$$= \frac{1}{12}(212) = 17.67 \text{ (since } p = 12)$$

From equation (2), $\quad a_n \approx \dfrac{2}{p}\sum\limits_{k=1}^{p} y_k \cos nx_k$

hence $\qquad a_1 \approx \dfrac{2}{12}(417.94) = 69.66,$

$$a_2 \approx \frac{2}{12}(-39) = -6.50 \text{ and}$$

$$a_3 \approx \frac{2}{12}(-49) = -8.17$$

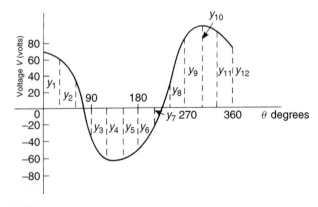

Figure 77.2

Table 77.1

Ordinates	θ	V	$\cos\theta$	$V\cos\theta$	$\sin\theta$	$V\sin\theta$	$\cos 2\theta$	$V\cos 2\theta$	$\sin 2\theta$	$V\sin 2\theta$	$\cos 3\theta$	$V\cos 3\theta$	$\sin 3\theta$	$V\sin 3\theta$
y_1	30	62	0.866	53.69	0.5	31	0.5	31	0.866	53.69	0	0	1	62
y_2	60	35	0.5	17.5	0.866	30.31	-0.5	-17.5	0.866	30.31	-1	-35	0	0
y_3	90	-38	0	0	1	-38	-1	38	0	0	0	0	-1	38
y_4	120	-64	-0.5	32	0.866	-55.42	-0.5	32	-0.866	55.42	1	-64	0	0
y_5	150	-63	-0.866	54.56	0.5	-31.5	0.5	-31.5	-0.866	54.56	0	0	1	-63
y_6	180	-52	-1	52	0	0	1	-52	0	0	-1	52	0	0
y_7	210	-28	-0.866	24.25	-0.5	14	0.5	-14	0.866	-24.25	0	0	-1	28
y_8	240	24	-0.5	-12	-0.866	-20.78	-0.5	-12	0.866	20.78	1	24	0	0
y_9	270	80	0	0	-1	-80	-1	-80	0	0	0	0	1	80
y_{10}	300	96	0.5	48	-0.866	-83.14	-0.5	-48	-0.866	-83.14	-1	-96	0	0
y_{11}	330	90	0.866	77.94	-0.5	-45	0.5	45	-0.866	-77.94	0	0	-1	-90
y_{12}	360	70	1	70	0	0	1	70	0	0	1	70	0	0
$\displaystyle\sum_{k=1}^{12} y_k = 212$			$\displaystyle\sum_{k=1}^{12} y_k\cos\theta_k$ $= 417.94$		$\displaystyle\sum_{k=1}^{12} y_k\sin\theta_k$ $= -278.53$		$\displaystyle\sum_{k=1}^{12} y_k\cos 2\theta_k$ $= -39$		$\displaystyle\sum_{k=1}^{12} y_k\sin 2\theta_k$ $= 29.43$		$\displaystyle\sum_{k=1}^{12} y_k\cos 3\theta_k$ $= -49$		$\displaystyle\sum_{k=1}^{12} y_k\sin 3\theta_k$ $= 55$	

From equation (3), $b_n \approx \dfrac{2}{p} \displaystyle\sum_{k=1}^{p} y_k \sin nx_k$

hence $b_1 \approx \dfrac{2}{12}(-278.53) = -46.42,$

$b_2 \approx \dfrac{2}{12}(29.43) = 4.91$ and

$b_3 \approx \dfrac{2}{12}(55) = 9.17$

Substituting these values into the Fourier series:

$$f(x) = a_0 + \sum_{n=1}^{\infty}(a_n \cos nx + b_n \sin nx)$$

gives: $v = 17.67 + 69.66 \cos \theta - 6.50 \cos 2\theta - 8.17 \cos 3\theta + \ldots$

$- 46.42 \sin \theta + 4.91 \sin 2\theta + 9.17 \sin 3\theta + \ldots$ (4)

Note that in equation (4), $-46.42 \sin \theta + 69.66 \cos \theta$ comprises the fundamental, $(4.91 \sin 2\theta - 6.50 \cos 2\theta)$ comprises the second harmonic and $(9.17 \sin 3\theta - 8.17 \cos 3\theta)$ comprises the third harmonic.

It is shown in *Engineering Mathematics Pocket Book* that:

$$a \sin \omega t + b \cos \omega t = R \sin(\omega t + \alpha)$$

where $a = R \cos \alpha$, $b = R \sin \alpha$, $R = \sqrt{a^2 + b^2}$ and $\alpha = \tan^{-1}\dfrac{b}{a}$

For the fundamental, $R = \sqrt{(-46.42)^2 + (69.66)^2} = 83.71$

If $a = R \cos \alpha$, then $\cos \alpha = \dfrac{a}{R} = \dfrac{-46.42}{83.71}$ which is negative,

and if $b = R \sin \alpha$, then $\sin \alpha = \dfrac{b}{R} = \dfrac{69.66}{83.71}$ which is positive.

The only quadrant where $\cos \alpha$ is negative <u>and</u> $\sin \alpha$ is positive is the second quadrant.

Hence $\alpha = \tan^{-1}\dfrac{b}{a} = \tan^{-1}\dfrac{69.66}{-46.42} = 123.68°$ or 2.16 rad.

Thus $(-46.42 \sin \theta + 69.66 \cos \theta) = 83.71 \sin(\theta + 2.16)$

By a similar method it may be shown that the second harmonic $(4.91 \sin 2\theta - 6.50 \cos 2\theta) = 8.15 \sin(2\theta - 0.92)$, and the third harmonic $(9.17 \sin 3\theta - 8.17 \cos 3\theta) = 12.28 \sin(3\theta - 0.73)$

Hence equation (4) may be re-written as:

$v = 17.67 + 83.71 \sin(\theta + 2.16) + 8.15 \sin(2\theta - 0.92)$

$+ 12.28 \sin(3\theta - 0.73)$ volts

which is the form used in chapter 76 with complex waveforms.

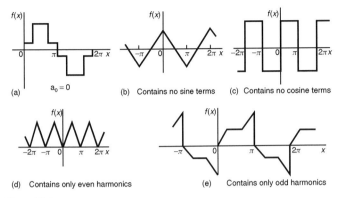

(a) $a_0 = 0$

(b) Contains no sine terms

(c) Contains no cosine terms

(d) Contains only even harmonics

(e) Contains only odd harmonics

Figure 77.3

Complex Waveform Considerations

It is sometimes possible to predict the harmonic content of a waveform on inspection of particular waveform characteristics; the result of this is reduced calculation.

(i) If a periodic waveform is such that the area above the horizontal axis is equal to the area below then the mean value is zero. Hence $a_0 = 0$ (see Figure 77.3(a)).

(ii) An **even function** is symmetrical about the vertical axis and contains **no sine terms** (see Figure 77.3(b)).

(iii) An **odd function** is symmetrical about the origin and contains **no cosine terms** (see Figure 77.3(c)).

(iv) $f(x) = f(x + \pi)$ represents a waveform which repeats after half a cycle and **only even harmonics** are present (see Figure 77.3(d)).

(v) $f(x) = -f(x + \pi)$ represents a waveform for which the positive and negative cycles are identical in shape and **only odd harmonics** are present (see Figure 77.3(e)).

78 Dielectrics and Dielectric Loss

Electric Fields, Capacitance and Permittivity

Any region in which an electric charge experiences a force is called an electrostatic field. Electric fields, Coulombs law, capacitance and permittivity are discussed in chapter 44 — refer back to page 233. Summarising the main formulae:

$$\text{Electric field strength, } E = \frac{V}{d} \text{ volts/metre,}$$

$$\text{Capacitance } C = \frac{Q}{V} \text{ farads}$$

$$\text{Electric flux density, } D = \frac{Q}{A} \text{ coulombs/metre}^2,$$

$$\frac{D}{E} = \varepsilon_0 \varepsilon_r = \varepsilon$$

$$\text{Relative permittivity } \varepsilon_r = \frac{\text{flux density in material}}{\text{flux density in vacuum}}$$

The insulating medium separating charged surfaces is called a **dielectric**. Compared with conductors, dielectric materials have very high resistivities $\left(\text{and hence low conductance, since } \rho = \frac{1}{\sigma} \right)$. They are therefore used to separate conductors at different potentials, such as capacitor plates or electric power lines.

For a parallel-plate capacitor, **capacitance** $C = \dfrac{\varepsilon_0 \varepsilon_r A (n - 1)}{d}$

Polarisation

When a dielectric is placed between charged plates, the capacitance of the system increases. The mechanism by which a dielectric increases capacitance is called **polarisation**. In an electric field the electrons and atomic nuclei of the dielectric material experience forces in opposite directions. Since the electrons in an insulator cannot flow, each atom becomes a tiny dipole (i.e. an arrangement of two electric charges of opposite polarity) with positive and negative charges slightly separated, i.e. the material becomes polarised.

Within the material this produces no discernible effects. However, on the surfaces of the dielectric, layers of charge appear. Electrons are drawn towards the positive potential, producing a negative charge layer, and away from the negative potential, leaving positive surface charge behind. Therefore the dielectric becomes a volume of neutral insulator with surface charges of opposite polarity on opposite surfaces. The result of this is that the electric field inside the dielectric is less than the electric field causing the polarisation, because these two charge layers give rise to a field, which opposes the electric field causing it. Since electric field strength, $E = \dfrac{V}{d}$, the p.d. between the plates, $V = Ed$. Thus, if E decreases when the dielectric is inserted, then V falls too and this drop in p.d. occurs without change of charge on the plates. Thus, since capacitance $C = \dfrac{Q}{V}$, capacitance increases, this increase being by a factor equal to ε_r above that obtained with a vacuum dielectric.

There are two main ways in which polarisation takes place:

(i) The electric field, as explained above, pulls the electrons and nucleii in opposite directions because they have opposite charges, which makes each atom into an electric dipole. The movement is only small and takes place very fast since the electrons are very light. Thus, if the applied electric field is varied periodically, the polarisation, and hence the permittivity due to these induced dipoles, is independent of the frequency of the applied field.
(ii) Some atoms have a permanent electric dipole as a result of their structure and, when an electric field is applied, they turn and tend to align along the field. The response of the permanent dipoles is slower than the response of the induced dipoles and that part of the relative permittivity which arises from this type of polarisation decreases with increase of frequency.

Most materials contain both induced and permanent dipoles, so the relative permittivity usually tends to decrease with increase of frequency.

Dielectric Strength

The maximum amount of field strength that a dielectric can withstand is called the dielectric strength of the material. When an electric field is established across the faces of a material, molecular alignment and distortion of the electron orbits around the atoms of the dielectric occur. This produces a mechanical stress, which in turn generates heat. The production of heat represents a dissipation of power, such a loss being present in all practical dielectrics, especially when used in high-frequency systems where the field polarity is continually and rapidly changing.

A dielectric whose conductivity is not zero between the plates of a capacitor provides a conducting path along which charges can flow and thus discharge the capacitor. The resistance R of the dielectric is given by: $R = \dfrac{\rho l}{a}$, l being the thickness of the dielectric film (which may be as small as 0.001 mm) and a being the area of the capacitor plates. The resistance R of the dielectric may be represented as a leakage resistance across an ideal capacitor (see dielectric loss later). The required lower limit for acceptable resistance between the plates varies with the use to which the capacitor is put. High-quality capacitors have high shunt-resistance values. A measure of dielectric quality is the time taken for a capacitor to discharge a given amount through the resistance of the dielectric. This is related to the product CR.

Capacitance, $C \propto \dfrac{\text{area}}{\text{thickness}}$ and $\dfrac{1}{R} \propto \dfrac{\text{area}}{\text{thickness}}$ thus CR is a characteristic of a given dielectric. In practice, circuit design is considerably simplified if the shunt conductance of a capacitor can be ignored (i.e. $R \to \infty$) and the capacitor therefore regarded as an open circuit for direct current.

Since capacitance C of a single parallel plate capacitor is given by: $C = \dfrac{\varepsilon_0 \varepsilon_r A}{d}$, reducing the thickness d of a dielectric film increases the capacitance, but decreases the resistance. It also reduces the voltage the capacitor can withstand without breakdown $\left(\text{since } V = \dfrac{Q}{V}\right)$. Any material will eventually break down, usually destructively, when subjected to a sufficiently large electric field. A spark may occur at breakdown, which produces a hole through the film. The metal film forming the metal plates may be welded together at the point of breakdown.

Breakdown depends on electric field strength E $\left(\text{where } E = \dfrac{V}{d}\right)$, so thinner films will break down with smaller voltages across them. This is the main reason for limiting the voltage that may be applied to a capacitor. All practical capacitors have a safe working voltage stated on them, generally at a particular maximum temperature. Figure 78.1 shows the typical shapes of graphs expected for electric field strength E plotted against thickness and for breakdown voltage plotted against thickness. The shape of the curves depend on a number of factors, and these include:

(i) the type of dielectric material,
(ii) the shape and size of the conductors associated with it,
(iii) the atmospheric pressure,
(iv) the humidity/moisture content of the material,
(v) the operating temperature.

Dielectric strength is an important factor in the design of capacitors as well as transformers and high voltage insulators, and in motors and generators.

Dielectrics vary in their ability to withstand large fields. Some typical values of dielectric strength, together with resistivity and relative permittivity

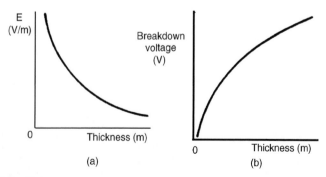

(a)

(b)

Figure 78.1

Table 78.1 Dielectric properties of some common materials

Material	Resistivity, ρ (Ωm)	Relative permittivity, ε_r	Dielectric strength (V/m)
Air		1.0	3×10^6
Paper	10^{10}	3.7	1.6×10^7
Mica	5×10^{11}	5.4	10^8–10^9
Titaniumdioxide	10^{12}	100	6×10^6
Polythene	$>10^{11}$	2.3	4×10^7
Polystyrene	$>10^{13}$	2.5	2.5×10^7
Ceramic (type 1)	4×10^{11}	6–500	4.5×10^7
Ceramic (type 2)	10^6–10^{13}	500–1000	2×10^6–10^7

are shown in Table 78.1. The ceramics have very high relative permittivities and they tend to be 'ferroelectric', i.e. they do not lose their polarities when the electric field is removed. When ferroelectric effects are present, the charge on a capacitor is given by: $Q = (CV) +$ (remanent polarisation). These dielectrics often possess an appreciable negative temperature coefficient of resistance. Despite this, a high permittivity is often very desirable and ceramic dielectrics are widely used.

Thermal Effects

As the temperature of most dielectrics is increased, the insulation resistance falls rapidly. This causes the leakage current to increase, which generates further heat. Eventually a condition known as **thermal avalanche** or **thermal runaway** may develop, when the heat is generated faster than it can be dissipated to the surrounding environment. The dielectric will burn and thus fail. Thermal effects may often seriously influence the choice and application of insulating materials. Some important factors to be considered include:

   (i) the melting-point (for example, for waxes used in paper capacitors),
   (ii) aging due to heat,
  (iii) the maximum temperature that a material will withstand without serious deterioration of essential properties,
  (iv) flash-point or ignitability,
   (v) resistance to electric arcs,
  (vi) the specific heat capacity of the material,
 (vii) thermal resistivity,
(viii) the coefficient of expansion,
  (ix) the freezing-point of the material.

Mechanical Properties

Mechanical properties determine, to varying degrees, the suitability of a solid material for use as an insulator: tensile strength, transverse strength, shearing strength and compressive strength are often specified. Most solid insulations have a degree of inelasticity and many are quite brittle, thus it is often necessary to consider features such as compressibility, deformation under bending stresses, impact strength and extensibility, tearing strength, machinability and the ability to fold without damage.

Types of Practical Capacitor

Practical types of capacitor are characterised by the material used for their dielectric. The main types include: variable air, mica, paper, ceramic, plastic, titanium oxide and electrolytic. Refer back to chapter 44, page 241, for a description of each type.

Liquid Dielectrics and Gas Insulation

Liquid dielectrics used for insulation purposes include refined mineral oils, silicone fluids and synthetic oils such as chlorinated diphenyl. The principal uses of liquid dielectrics are as a filling and cooling medium for transformers, capacitors and rheostats, as an insulating and arc-quenching medium in switchgear such as circuit breakers, and as an impregnant of absorbent insulations — for example, wood, slate, paper and pressboard, used mainly in transformers, switchgear, capacitors and cables.

Two **gases** used as insulation are nitrogen and sulphur hexafluoride. Nitrogen is used as an insulation medium in some sealed transformers and in power cables, and sulphur hexafluoride is finding increasing use in switchgear both as an insulant and as an arc-extinguishing medium.

Dielectric Loss and Loss Angle

In capacitors with solid dielectrics, losses can be attributed to two causes:

(i) **dielectric hysteresis**, a phenomenon by which energy is expended and heat produced as the result of the reversal of electrostatic stress in a dielectric subjected to alternating electric stress — this loss is analogous to hysteresis loss in magnetic materials;
(ii) **leakage currents** that may flow through the dielectric and along surface paths between the terminals.

The total dielectric loss may be represented as the loss in an additional resistance connected between the plates. This may be represented as either a small resistance in series with an ideal capacitor or as a large resistance in parallel with an ideal capacitor.

Series representation

The circuit and phasor diagrams for the series representation are shown in Figure 78.2. The circuit phase angle is shown as angle ϕ. If resistance R_S is

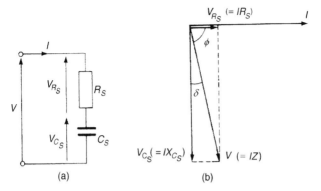

Figure 78.2

zero then current I would lead voltage V by 90°, this being the case of a perfect capacitor. The difference between 90° and the circuit phase angle ϕ is the angle shown as δ. This is known as the **loss angle** of the capacitor, i.e.

$$\boxed{\text{loss angle, } \delta = (90° - \phi)}$$

For the equivalent series circuit,

$$\tan \delta = \frac{V_{R_S}}{V_{C_S}} = \frac{IR_S}{IX_{C_S}} = \frac{R_S}{\dfrac{1}{\omega C_S}} = R_S \omega C_S$$

Since from chapter 68, $Q = \dfrac{1}{\omega CR}$ then

$$\boxed{\tan \delta = R_S \,\omega C_S = \frac{1}{Q}} \qquad (1)$$

Power factor of capacitor,

$$\cos \phi = \frac{V_{R_S}}{V} = \frac{IR_S}{IZ_S} = \frac{R_S}{Z_S} \approx \frac{R_S}{X_{C_S}} \text{ since } X_{C_S} \approx Z_S \text{ when } \delta \text{ is small}$$

Hence **power factor** $= \cos \phi \approx R_S \omega C_S$

i.e. $$\boxed{\cos \phi \approx \tan \delta} \qquad (2)$$

Dissipation factor, D is defined as the reciprocal of Q-factor and is an indication of the quality of the dielectric,

i.e. $$\boxed{D = \frac{1}{Q} = \tan \delta} \qquad (3)$$

For example, the equivalent series circuit for a particular capacitor consists of a 1.5 Ω resistance in series with a 400 pF capacitor at a frequency of 8 MHz.

From equation (1), for a series equivalent circuit,

$$\tan \delta = R_S \omega C_S = (1.5)(2\pi \times 8 \times 10^6)(400 \times 10^{-12}) = 0.030159$$

Hence **loss angle,** $\delta = \tan^{-1}(0.030159) = \textbf{1.727°}$ **or 0.030 rad.**
From equation (2),

$$\textbf{power factor} = \cos \phi \approx \tan \delta = \textbf{0.030}$$

From equation (1),

$$\tan \delta = \frac{1}{Q} \text{ hence } Q = \frac{1}{\tan \delta} = \frac{1}{0.030159} = 33.16$$

From equation (3), **dissipation factor,**

$$D = \frac{1}{Q} = 0.030159 \text{ or } \textbf{0.030,} \text{ correct to 3 decimal places.}$$

Parallel representation

The circuit and phasor diagrams for the parallel representation are shown in Figure 78.3. From the phasor diagram,

$$\tan \delta = \frac{I_{R_P}}{I_{C_P}} = \frac{V/R_P}{V/X_{C_P}} = \frac{X_{C_P}}{R_P}$$

i.e.

$$\boxed{\tan \delta = \frac{1}{R_P \omega C_P}} \qquad (4)$$

Power factor of capacitor,

$$\cos \phi = \frac{I_{R_P}}{I} = \frac{V/R_P}{V/Z_P}$$

$$= \frac{Z_P}{R_P} \approx \frac{X_{C_P}}{R_P} \text{ since } X_{C_P} \approx Z_P, \text{ when } \delta \text{ is small}$$

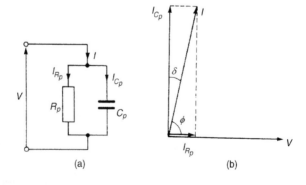

(a)   (b)

Figure 78.3

Hence, power factor $= \cos\phi \approx \dfrac{1}{R_P\omega C_P}$

i.e. $\boxed{\cos\phi \approx \tan\delta}$

(For equivalence between the series and the parallel circuit representations,

$$C_S \approx C_P = C \text{ and } R_S\omega C_S \approx \frac{1}{R_P\omega C_P} \text{ from which, } R_S \approx \frac{1}{R_P\omega^2 C^2})$$

Power loss

Power loss in the dielectric $= VI\cos\phi$
From the phasor diagram of Figure 78.3,

$$\cos\delta = \frac{I_{C_P}}{I} = \frac{V/X_{C_P}}{I} = \frac{V\omega C}{I} \text{ or } I = \frac{V\omega C}{\cos\delta}$$

Hence, power loss $= VI\cos\phi = V\left(\dfrac{V\omega C}{\cos\delta}\right)\cos\phi$

However, $\cos\phi = \sin\delta$ (complementary angles), thus

$$\text{power loss} = V\left(\frac{V\omega C}{\cos\delta}\right)\sin\delta = V^2\omega C\tan\delta \left(\text{since } \frac{\sin\delta}{\cos\delta} = \tan\delta\right)$$

Hence, $\boxed{\text{dielectric power loss} = V^2\omega C\tan\delta}$ (5)

For example, a capacitor has a loss angle of 0.025 rad, and when it is connected across a 5 kV, 50 Hz supply, the power loss is 20 W. The component values of the equivalent parallel circuit is determined as follows:

From equation (5), power loss $= V^2\omega C\tan\delta$

i.e. $20 = (5000)^2(2\pi 50)(C)\tan(0.025)$

from which, **capacitance, $C = \dfrac{20}{(5000)^2(2\pi 50)\tan(0.025)}$**

$= \mathbf{0.102\ \mu F}$

(Note $\tan(0.025)$ means 'the tangent of 0.025 rad').
From equation (4), for a parallel equivalent circuit, $\tan\delta = \dfrac{1}{R_P\omega C_P}$

from which, $R_P = \dfrac{1}{\omega C_P\tan\delta} = \dfrac{1}{(2\pi 50)(0.102\times 10^{-6})\tan 0.025}$

i.e. **parallel resistance, $R_P = 1.248\ M\Omega$**

79 Field Theory

Introduction

Electric fields, magnetic fields and conduction fields (i.e. a region in which an electric current flows) are analogous, i.e. they all exhibit similar characteristics. Thus they may all be analysed by similar processes. In the following the electric field is analysed.

Figure 79.1 shows two parallel plates A and B. Let the potential on plate A be $+V$ volts and that on plate B be $-V$ volts. The force acting on a point charge of 1 coulomb placed between the plates is the electric field strength E. It is measured in the direction of the field and its magnitude depends on the p.d. between the plates and the distance between the plates. In Figure 79.1, moving along a line of force from plate B to plate A means moving from $-V$ to $+V$ volts. The p.d. between the plates is therefore $2V$ volts and this potential changes linearly when moving from one plate to the other. Hence a potential gradient is followed which changes by equal amounts for each unit of distance moved.

Lines may be drawn connecting together all points within the field having equal potentials. These lines are called **equipotential lines** and these have been drawn in Figure 79.1 for potentials of $\frac{2}{3}$V, $\frac{1}{3}$V, 0, $-\frac{1}{3}$V, and $-\frac{2}{3}$V. The zero equipotential line represents earth potential and the potentials on plates A and B are respectively above and below earth potential. Equipotential lines form part of an equipotential surface. Such surfaces are parallel to the plates shown in Figure 79.1 and the plates themselves are equipotential surfaces. There can be no current flow between any given points on such a surface since all points on an equipotential surface have the same potential. Thus a line of force (or flux) must intersect an equipotential surface at right angles. A line of force in an electrostatic field is often termed a **streamline**.

An electric field distribution for a concentric cylinder capacitor is shown in Figure 79.2. An electric field is set up in the insulating medium between two good conductors. Any volt drop within the conductors can usually be neglected compared with the p.d.'s across the insulation since the conductors have a high conductivity. All points in the conductors are thus at the same potential so that the conductors form the boundary equipotentials for the electrostatic field.

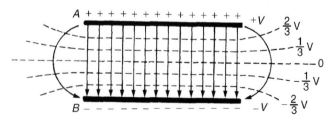

Figure 79.1

591

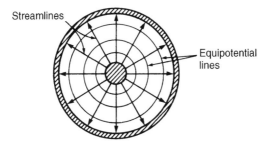

Figure 79.2

Streamlines (or lines of force) which must cut all equipotentials at right angles leave one boundary at right angles, pass across the field, and enter the other boundary at right angles.

In a magnetic field, a streamline is a line so drawn that its direction is everywhere parallel to the direction of the magnetic flux. An equipotential surface in a magnetic field is the surface over which a magnetic pole may be moved without the expenditure of work or energy.

In a conduction field, a streamline is a line drawn with a direction that is everywhere parallel to the direction of the current flow.

Capacitance between Concentric Cylinders

A **concentric cable** is one that contains two or more separate conductors, arranged concentrically (i.e. having a common centre), with insulation between them. In a **coaxial cable**, the central conductor, which may be either solid or hollow, is surrounded by an outer tubular conductor, the space in between being occupied by a dielectric. If air is the dielectric then concentric insulating discs are used to prevent the conductors touching each other. The two kinds of cable serve different purposes. The main feature they have in common is a complete absence of external flux and therefore a complete absence of interference with and from other circuits.

The electric field between two concentric cylinders (i.e. a coaxial cable) is shown in the cross-section of Figure 79.3. The conductors form the boundary equipotentials for the field, the boundary equipotentials in Figure 79.3 being concentric cylinders of radii a and b. The streamlines, or lines of force, are radial lines cutting the equipotentials at right angles.

The capacitance C between concentric cylinders (or coaxial cable) is given by:

$$C = \frac{2\pi\varepsilon_0\varepsilon_r}{\ln\dfrac{b}{a}} \text{ farads/metre}$$ (1)

where a = inner conductor radius, and b = outer conductor radius.

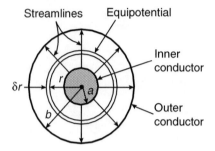

Figure 79.3

For example, a coaxial cable has an inner core radius of 0.5 mm, an outer conductor of internal radius 6.0 mm, and a relative permittivity of 2.7. Hence

$$\text{capacitance, } C = \frac{2\pi\varepsilon_0\varepsilon_r}{\ln\dfrac{b}{a}} = \frac{2\pi(8.85 \times 10^{-12})(2.7)}{\ln\left(\dfrac{6.0}{0.5}\right)} = \mathbf{60.4\ pF}$$

Dielectric Stress

Dielectric stress E is given by:

$$E = \frac{V}{r\ln\dfrac{b}{a}} \text{ volts/metre} \tag{2}$$

where V = core potential, r = conductor radius, a = inner conductor radius and b = outer conductor radius.

$$\text{Also} \quad E_{\text{max}} = \frac{V}{a\ln\dfrac{b}{a}} \tag{3}$$

$$\text{and} \quad E_{\text{min}} = \frac{V}{b\ln\dfrac{b}{a}} \tag{4}$$

For example, a concentric cable has a core diameter of 32 mm and an inner sheath diameter of 80 mm. The core potential is 40 kV and the relative permittivity of the dielectric is 3.5.

From equation (1), capacitance per metre length,

$$C = \frac{2\pi\varepsilon_0\varepsilon_r}{\ln\dfrac{b}{a}} = \frac{2\pi(8.85 \times 10^{-12})(3.5)}{\ln\left(\dfrac{40}{16}\right)}$$

$$= 212.4 \times 10^{-12}$$

F/m = 212 nF/km or 0.212 μF/km

From equation (2), dielectric stress at a radius of, say, 30 mm is:

$$E = \frac{V}{r\ln\dfrac{b}{a}} = \frac{40 \times 10^3}{(30 \times 10^{-3})\ln\left(\dfrac{40}{16}\right)}$$

$$= 1.46 \times 10^6 \text{ V/m or } 1.46 \text{ MV/m}$$

From equation (3), maximum dielectric stress,

$$E_{max} = \frac{V}{a\ln\dfrac{b}{a}} = \frac{40 \times 10^3}{(16 \times 10^{-3})\ln\left(\dfrac{40}{16}\right)} = 2.73 \text{ MV/m}$$

From equation (4), minimum dielectric stress,

$$E_{min} = \frac{V}{b\ln\dfrac{b}{a}} = \frac{40 \times 10^3}{(40 \times 10^{-3})\ln\left(\dfrac{40}{16}\right)} = 1.09 \text{ MV/m}$$

Dimensions of most Economical Cable

It is important to obtain the most economical dimensions when designing a cable. For the most economical cable,

inner conductor radius, $\boxed{a = \dfrac{V}{E_{max}}}$ (5)

and   outer conductor radius, $\boxed{b = a\,e}$ (6)

where $e = 2.718$ correct to 4 significant figures.

For example, a single-core concentric cable is to be manufactured for a 60 kV, 50 Hz transmission system. The dielectric used is paper which has a maximum permissible safe dielectric stress of 10 MV/m r.m.s. and a relative permittivity of 3.5. The core and inner sheath radii for the most economical cable is given by: from equation (5), core radius,

$$a = \frac{V}{E_{max}} = \frac{60 \times 10^3 \text{ V}}{10 \times 10^6 \text{ V/m}} = 6 \times 10^{-3} \text{ m} = 6.0 \text{ mm}$$

and from equation (6), internal sheath radius,

$$b = ae = 6.0e = 16.3 \text{ mm}$$

Capacitance of an Isolated Twin Line

The capacitance C of an isolated twin line is given by:

$$C = \frac{\pi \varepsilon_0 \varepsilon_r}{\ln \dfrac{D}{a}} \ \text{ farads/metre}$$
(7)

where D = distance between the centres of the two conductors, and a = radius of each conductor.

For example, two parallel wires, each of diameter 5 mm, are uniformly spaced in air at a distance of 50 mm between centres. The capacitance of the line if the total length is 200 m is determined as follows:

From equation (7), capacitance per metre length,

$$C = \frac{\pi \varepsilon_0 \varepsilon_r}{\ln \dfrac{D}{a}} = \frac{\pi(8.85 \times 10^{-12})(1)}{\ln \left(\dfrac{50}{5/2} \right)} \quad \text{since } \varepsilon_r = 1 \text{ for air,}$$

$$= \frac{\pi(8.85 \times 10^{-12})}{\ln 20} = 9.28 \times 10^{-12} \ \text{F}$$

Hence the capacitance of, say, a 500 m length

$$= (9.28 \times 10^{-12} \times 500) \ \text{F}$$

$$= \textbf{4640 pF} \text{ or } \textbf{4.64 nF}$$

Energy Stored in an Electric Field

The energy stored in the electric field of a capacitor is given by (from Chapter 44):

$$W_f = \tfrac{1}{2}CV^2 \ \text{ joules}$$
(8)

For example, the energy stored in a 10 nF capacitor when charged to 1 kV is given by:

$$W_f = \tfrac{1}{2}CV^2 = \tfrac{1}{2}(10 \times 10^{-9})(10^3)^2 = \textbf{5 mJ}$$

The **energy stored per unit volume of dielectric** is given by:

$$\omega_f = \frac{1}{2}DE = \frac{1}{2}\varepsilon_0 \varepsilon_r E^2 = \frac{D^2}{2\varepsilon_0 \varepsilon_r} \ \text{ joules/cubic metre}$$
(9)

For example, a 400 pF capacitor is charged to a p.d. of 100 V. The dielectric has a cross-sectional area of 200 cm² and a relative permittivity of 2.3. The energy stored per cubic metre of the dielectric is determined as follows:

From equation (9), energy stored per unit volume of dielectric,

$$\omega_f = \frac{D^2}{2\varepsilon_0\varepsilon_r}$$

Electric flux density

$$D = \frac{Q}{A} = \frac{CV}{A} = \frac{(400 \times 10^{-12})(100)}{200 \times 10^{-12}} = 2 \times 10^{-6} \text{ C/m}^2$$

Hence, energy stored,

$$\omega_f = \frac{D^2}{2\varepsilon_0\varepsilon_r} = \frac{(2 \times 10^{-6})^2}{2(8.85 \times 10^{-12})(2.3)}$$

$$= \mathbf{0.0983 \ J/m^3} \text{ or } \mathbf{98.3 \ mJ/m^3}$$

Skin Effect

When a direct current flows in a uniform conductor the current will tend to distribute itself uniformly over the cross-section of the conductor. However, with alternating current, particularly if the frequency is high, the current carried by the conductor is not uniformly distributed over the available cross-section, but tends to be concentrated at the conductor surface. This is called **skin effect**. When current is flowing through a conductor, the magnetic flux that results is in the form of concentric circles. Some of this flux exists within the conductor and links with the current more strongly near the centre. The result is that the inductance of the central part of the conductor is greater than the inductance of the conductor near the surface. This is because of the greater number of flux linkages existing in the central region. At high frequencies the reactance ($X_L = 2\pi f L$) of the extra inductance is sufficiently large to seriously affect the flow of current, most of which flows along the surface of the conductor where the impedance is low rather than near the centre where the impedance is high.

Inductance of a Concentric Cylinder (or Coaxial Cable)

The inductance L of a pair of concentric cylinders (or coaxial cable) is given by:

$$\boxed{L = \frac{\mu}{2\pi} \left(\frac{1}{4} + \ln \frac{b}{a} \right) \text{ henry/metre}} \tag{10}$$

where a = inner conductor radius and b = outer conductor radius

For example, a coaxial cable has an inner core of radius 1.0 mm and an outer sheath of internal radius 4.0 mm. The inductance of the cable per metre length, assuming that the relative permeability is unity, is given by:

$$\text{Inductance, } L = \frac{\mu}{2\pi} \left(\frac{1}{4} + \ln \frac{b}{a} \right)$$

$$= \frac{\mu_0 \mu_r}{2\pi} \left(\frac{1}{4} + \ln \frac{4.0}{1.0} \right) = \frac{(4\pi \times 10^{-7})(1)}{2\pi} (0.25 + \ln 4)$$

$$= 3.27 \times 10^{-7} \text{ H/m or } 0.327 \text{ } \mu\text{H/m}$$

Inductance of an Isolated Twin Line

The inductance of an isolated twin line (i.e. the loop inductance) is given by:

$$\boxed{L = \frac{\mu_0 \mu_r}{\pi} \left(\frac{1}{4} + \ln \frac{D}{a} \right) \text{ henry/metre}} \qquad (11)$$

where D = distance between the centers of the two conductors, and a = radius of each conductor.
In most practical lines the relative permeability, $\mu_r = 1$
For example, the loop inductance of a 1 km length of single-phase twin line having conductors of diameter 10 mm and spaced 800 mm apart in air is determined as follows:
From equation (11), total inductance per loop metre

$$= \frac{\mu_0 \mu_r}{\pi} \left(\frac{1}{4} + \ln \frac{D}{a} \right) = \frac{(4\pi \times 10^{-7})(1)}{\pi} \left(\frac{1}{4} + \ln \frac{800}{10/2} \right)$$

$$= 4 \times 10^{-7} (0.25 + \ln 160) = 21.3 \times 10^{-7} \text{ H/m}$$

Hence, loop inductance of, say, 1 km length of line

$$= 21.3 \times 10^{-7} \text{ H/m} \times 10^3 \text{ m}$$

$$= 21.3 \times 10^{-4} \text{ H or } 2.13 \text{ mH}$$

Energy Stored in an Electromagnetic Field

The **magnetic energy in a nonmagnetic medium** is given by:

$$\boxed{\omega_f = \frac{1}{2} HB = \frac{1}{2} \mu_0 H^2 = \frac{B^2}{2\mu_0} \text{ joules/metre}^3} \qquad (12)$$

For example, the air gap of a moving coil instrument is 2.0 mm long and has a cross-sectional area of 500 mm². If the flux density is 50 mT, the total energy stored in the magnetic field of the air gap is determined as follows:

From equation (26), energy stored,

$$W_f = \frac{B^2}{2\mu_0} = \frac{(50 \times 10^{-3})^2}{2(4\pi \times 10^{-7})} = 9.95 \times 10^2 \text{ J/m}^3$$

Volume of air gap $= Al = (500 \times 2.0) \text{ mm}^3 = 500 \times 2.0 \times 10^{-9} \text{ m}^3$
Hence the energy stored in the air gap,

$$W_f = 9.95 \times 10^2 \text{ J/m}^3 \times 500 \times 2.0 \times 10^{-9} \text{ m}^3$$

$$= 9.95 \times 10^{-4} \text{ J} = 0.995 \text{ mJ} = 995 \text{ } \mu\text{J}$$

The **magnetic energy stored in an inductor** is given by (from Chapter 48):

$$\boxed{W_f = \tfrac{1}{2}LI^2 \text{ joules}} \tag{13}$$

For example, the energy stored when a current of 50 mA is flowing in a coil of inductance 200 mH is given by:

$$W_f = \tfrac{1}{2}LI^2 = \tfrac{1}{2}(200 \times 10^{-3})(50 \times 10^{-3})^2$$

$$= 2.5 \times 10^{-4} \text{ J} \text{ or } 0.25 \text{ mJ} \text{ or } 250 \text{ } \mu\text{J}$$

80 Attenuators

Introduction

An attenuator is a device for introducing a specified loss between a signal source and a matched load without upsetting the impedance relationship necessary for matching. The loss introduced is constant irrespective of frequency; since reactive elements (L or C) vary with frequency, it follows that ideal attenuators are networks containing pure resistances. A fixed attenuator section is usually known as a 'pad'.

Attenuation is a reduction in the magnitude of a voltage or current due to its transmission over a line or through an attenuator. Any degree of attenuation may be achieved with an attenuator by suitable choice of resistance values but the input and output impedances of the pad must be such that the impedance conditions existing in the circuit into which it is connected are not disturbed. Thus an attenuator must provide the correct input and output impedances as well as providing the required attenuation.

Two-port Networks

Networks in which electrical energy is fed in at one pair of terminals and taken out at a second pair of terminals are called **two-port networks**. Thus an attenuator is a two-port network, as are transmission lines, transformers and electronic amplifiers. If a network contains only passive circuit elements, such as in an attenuator, the network is said to be **passive**; if a network contains a source of e.m.f., such as in an electronic amplifier, the network is said to be **active**.

Figure 80.1(a) shows a T-network, which is termed symmetrical if $Z_A = Z_B$ and Figure 80.1(b) shows a π-network which is symmetrical if $Z_E = Z_F$. If $Z_A \neq Z_B$ in Figure 80.1(a), and $Z_E \neq Z_F$ in Figure 80.1(b), the sections are termed **asymmetrical**. Both networks shown have one common terminal, which may be earthed, and are therefore said to be **unbalanced**. The **balanced**

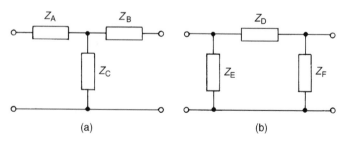

(a)   (b)

Figure 80.1

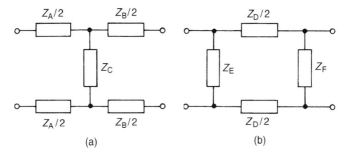

Figure 80.2

form of the T-network is shown in Figure 80.2(a) and the balanced form of the π-network is shown in Figure 80.2(b).

Characteristic Impedance

The input impedance of a network is the ratio of voltage to current (in complex form) at the input terminals. With a two-port network the input impedance often varies according to the load impedance across the output terminals. For any passive two-port network it is found that a particular value of load impedance can always be found which will produce an input impedance having the same value as the load impedance. This is called the **iterative impedance** for an asymmetrical network and its value depends on which pair of terminals is taken to be the input and which the output (there are thus two values of iterative impedance, one for each direction). For a symmetrical network there is only one value for the iterative impedance and this is called the **characteristic impedance** of the symmetrical two-port network.

Logarithmic Ratios

The ratio of two powers P_1 and P_2 may be expressed in logarithmic form as shown in chapter 50. Let P_1 be the input power to a system and P_2 the output power.

If logarithms to base 10 are used then

$$\text{power ratio in decibels} = 10 \lg \frac{P_2}{P_1} = 20 \lg \frac{V_2}{V_1} = 20 \lg \frac{I_2}{I_1} \qquad (1)$$

For example, if, say, 5% of the power supplied to a cable appears at the output terminals then the attenuation in decibels is determined as follows:

From equation (1),

$$\text{power ratio in decibels} = 10 \; \lg \frac{P_2}{P_1} = 10 \; \lg \frac{5}{100}$$

$$= 10 \; \lg 0.05 = -13 \text{ dB}.$$

Hence the attenuation (i.e. power loss) is 13 dB.

In another example, if an amplifier has a gain of 15 dB and the input power is 12 mW, the output power is determined as follows:

From equation (1),

$$15 = 10 \; \lg \left(\frac{P_2}{12} \right) \quad \begin{array}{l} \text{where } P_2 \text{ is the output} \\ \text{power in milliwatts} \end{array}$$

$$1.5 = \lg \left(\frac{P_2}{12} \right)$$

Hence $\left(\dfrac{P_2}{12} \right) = 10^{1.5} \quad \begin{array}{l} \text{from the definition} \\ \text{of a logarithm} \end{array}$

Thus **the output power,** $P_2 = 12(10)^{1.5} = \textbf{379.5 mW}$

If **logarithms to base** e (i.e. natural or Napierian logarithms) are used then

$$\boxed{\text{power ratio in nepers} = \frac{1}{2} \ln \frac{P_2}{P_1} = \ln \frac{V_2}{V_1} = \ln \frac{I_2}{I_1}} \qquad (2)$$

Thus when the power ratio $\dfrac{P_2}{P_1} = 5$ the power ratio in nepers $= \frac{1}{2} \ln 5 = 0.805$ Np.

The attenuation of filter sections and along a transmission line are of an exponential form and it is in such applications that the unit of the neper is used (see chapters 81 and 83).

Symmetrical T-attenuator

For the symmetrical T-pad attenuator is shown in Figure 80.3, the characteristic impedance R_0 is given by:

$$\boxed{R_0 = \sqrt{R_1^2 + 2R_1 R_2}} \qquad (3)$$

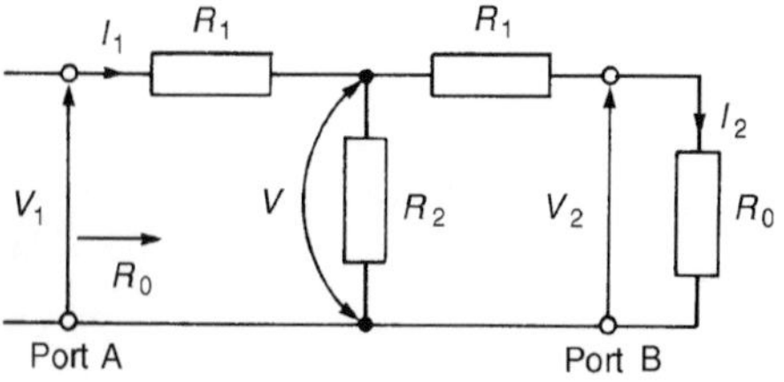

Figure 80.3

or $\boxed{R_0 = \sqrt{R_{OC} R_{SC}}}$ (4)

For example, if in Figure 80.3, $R_1 = 20\ \Omega$ and $R_2 = 15\ \Omega$, then the characteristic impedance R_0 is given by either:

(i) $R_0 = \sqrt{R_1^2 + 2R_1 R_2} = \sqrt{20^2 + 2(20)(15)} = \textbf{31.62 } \boldsymbol{\Omega}$, or

(ii) with the output terminal open circuited, resistance at input

$$R_{OC} = 20 + 15 = 35\ \Omega$$

and with the output terminal short circuited the resistance at input

$$R_{SC} = 20 + \frac{(20)(15)}{20 + 15} = 28.57\ \Omega$$

from which, $R_0 = \sqrt{R_{OC} R_{SC}} = \sqrt{(35)(28.57)} = \textbf{31.62 } \boldsymbol{\Omega}$

Designing a symmetrical T-attenuator

If the characteristic impedance R_0 and the attenuation $N\left(= \dfrac{V_1}{V_2}\right)$ are known for the symmetrical T-network of Figure 80.3 then:

$$\boxed{R_1 = R_0 \left(\frac{N - 1}{N + 1}\right) \text{ and } R_2 = R_0 \left(\frac{2N}{N^2 - 1}\right)}$$ (5)

For example, a T-section symmetrical attenuator pad to provide a voltage attenuation of 20 dB and having a characteristic impedance of 600 Ω is designed as follows:

Voltage attenuation in decibels $= 20\ \lg \dfrac{V_1}{V_2}$.

Attenuation, $N = \dfrac{V_1}{V_2}$, hence $20 = 20\ \lg N$, from which, $N = 10$.

Characteristic impedance, $R_0 = 600\ \Omega$.

From equation (5), resistance

$$R_1 = R_0 \left(\frac{N - 1}{N + 1}\right) = (600)\left(\frac{10 - 1}{10 + 1}\right) = \textbf{491 } \boldsymbol{\Omega}$$

and resistance

$$R_2 = R_0 \left(\frac{2N}{N^2 - 1}\right) = (600)\left(\frac{2(10)}{10^2 - 1}\right) = \textbf{121 } \boldsymbol{\Omega}$$

[Check: From equation (3)),

$$R_0 = \sqrt{R_1^2 + 2R_1 R_2} = \sqrt{491^2 + 2(491)(121)} = 600\ \Omega]$$

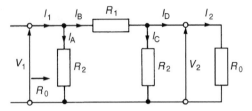

Figure 80.4

Symmetrical π-attenuator

For the symmetrical π-attenuator shown in Figure 80.4 the characteristic impedance R_0 is given by:

$$R_0 = \sqrt{\left(\frac{R_1 R_2^2}{R_1 + 2R_2}\right)} \tag{6}$$

or $\boxed{R_0 = \sqrt{R_{OC} R_{SC}}}$ (7)

For example, if in Figure 80.4, $R_1 = 20\ \Omega$ and $R_2 = 15\ \Omega$, then the characteristic impedance R_0 is given by either:

(i) $R_0 = \sqrt{\dfrac{R_1 R_2^2}{R_1 + 2R_2}} = \sqrt{\dfrac{(20)(15)^2}{10 + 2(15)}} = \mathbf{9.49\ \Omega}$, or

(ii) with the output terminal open circuited, resistance at input

$$R_{OC} = \frac{(15)(20 + 15)}{15 + (20 + 15)} = 10.5\ \Omega$$

and with the output terminal short circuited the resistance at input

$$R_{SC} = \frac{(15)(20)}{15 + 20} = 8.57\ \Omega$$

from which, $R_0 = \sqrt{R_{OC} R_{SC}} = \sqrt{(10.5)(8.57)} = \mathbf{9.49\ \Omega}$

Designing a symmetrical π-attenuator

If the characteristic impedance R_0 and the attenuation $N\ \left(=\dfrac{V_1}{V_2}\right)$ are known for the symmetrical T-network of Figure 80.4 then:

$$\boxed{R_1 = R_0 \left(\frac{N^2 - 1}{2N}\right) \text{ and } R_2 = R_0 \left(\frac{N + 1}{N - 1}\right)} \tag{8}$$

For **example**, if a π-section symmetrical attenuator is required to provide a voltage attenuation of 25 dB and have a characteristic impedance of 600 Ω, then:

$$25 = 20 \ \lg \frac{V_1}{V_2} = 20 \ \lg N \quad \text{i.e.} \quad \lg N = \frac{25}{20} = 1.25$$

from which, attenuation $N = 10^{1.25} = 17.78$

hence, series resistance $\quad R_1 = R_0 \left(\frac{N^2 - 1}{2N}\right) = (600)\left(\frac{17.78^2 - 1}{2(17.78)}\right)$

$$= \textbf{5.317 k}\boldsymbol{\Omega}$$

and shunt resistance $\quad R_2 = R_0 \left(\frac{N + 1}{N - 1}\right) = (600)\left(\frac{17.78 + 1}{17.78 - 1}\right)$

$$= \textbf{671.5 }\boldsymbol{\Omega}$$

[Check: From equation (6),

$$R_0 = \sqrt{\left(\frac{R_1 R_2^2}{R_1 + 2R_2}\right)} = \sqrt{\left(\frac{(5317)(671.5)^2}{5317 + 2(671.5)}\right)} = 600 \ \Omega]$$

Insertion Loss

Figure 80.5(a) shows a generator E connected directly to a load Z_L. Let the current flowing be I_L and the p.d. across the load V_L. z is the internal impedance of the source.

Figure 80.5(b) shows a two-port network connected between the generator E and load Z_L. The current through the load, shown as I_2, and the p.d. across the load, shown as V_2 will generally be less than current I_L and voltage V_L of Figure 80.5(a), as a result of the insertion of the two-port network between generator and load.

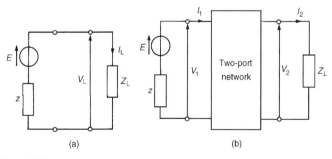

(a)   (b)

Figure 80.5

The **insertion loss ratio, A_L**, is defined as:

$$A_L = \frac{\text{voltage across load when connected directly to the generator}}{\text{voltage across load when the two-port network is connected}}$$

i.e.

$$\boxed{A_L = \frac{V_L}{V_2} = \frac{I_L}{I_2}} \tag{9}$$

since $V_L = I_L Z_L$ and $V_2 = I_2 Z_L$. Since both V_L and V_2 refer to p.d.'s across the same impedance Z_L, the insertion loss ratio may also be expressed as:

$$\boxed{\text{insertion loss ratio} = 20 \lg \left(\frac{V_L}{V_2} \right) \text{dB or } 20 \lg \left(\frac{I_L}{I_2} \right) \text{dB}} \tag{10}$$

When the two-port network is terminated in its characteristic impedance Z_0 the network is said to be **matched**. In such circumstances the input impedance is also Z_0, thus the insertion loss is simply the ratio of input to output voltage $\left(\text{i.e.} \dfrac{V_1}{V_2} \right)$. Thus, **for a network terminated in its characteristic impedance,**

$$\boxed{\text{insertion loss} = 20 \lg \left(\frac{V_1}{V_2} \right) \text{dB or } 20 \lg \left(\frac{I_1}{I_2} \right) \text{dB}}$$

For example, a 0–3 kΩ rheostat is connected across the output of a signal generator of internal resistance 500 Ω. If a load of 2 kΩ is connected across the rheostat, the insertion loss at a tapping of, say, 2 kΩ is determined as follows:

The circuit diagram is shown in Figure 80.6. Without the rheostat in the circuit the voltage across the 2 kΩ load, V_L (see Figure 80.7), is given by:

$$V_L = \left(\frac{2000}{2000 + 500} \right) E = 0.8\,E$$

With the 2 kΩ tapping, the network of Figure 80.7 may be redrawn as shown in Figure 80.8, which in turn is simplified as shown in Figure 80.9. From Figure 80.9,

$$\text{voltage } V_2 = \left(\frac{1000}{1000 + 1000 + 500} \right) E = 0.4\,E$$

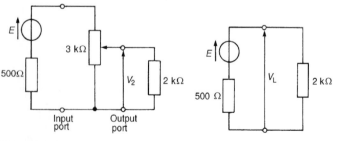

Figure 80.6   **Figure 80.7**

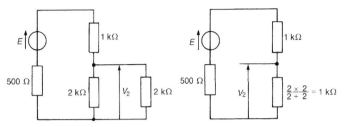

Figure 80.8   Figure 80.9

Hence, from equation (9), **insertion loss ratio**, $A_L = \dfrac{V_L}{V_2} = \dfrac{0.8\,E}{0.4\,E} = \mathbf{2}$ or, from

equation (10), **insertion loss ratio** $= 20\ \lg\left(\dfrac{V_L}{V_2}\right) = 20\ \lg 2 = \mathbf{6.02\ dB}$

Asymmetrical T- and π-sections

Figure 80.10(a) shows an asymmetrical T-pad section where resistance $R_1 \neq R_3$. Figure 80.10(b) shows an asymmetrical π-section where $R_2 \neq R_3$.

When viewed from port A, in each of the sections, the output impedance is R_{OB}; when viewed from port B, the input impedance is R_{OA}. Since the sections are asymmetrical R_{OA} does not have the same value as R_{OB}.

Iterative impedance is the term used for the impedance measured at one port of a two-port network when the other port is terminated with an impedance of the same value. For example, the impedance looking into port 1 of Figure 80.11(a) is, say, 500 Ω when port 2 is terminated in 500 Ω and the impedance looking into port 2 of Figure 80.11(b) is, say, 600 Ω when port 1 is terminated in 600 Ω. (In symmetric T- and π-sections the two iterative impedances are equal, this value being the characteristic impedance of the section).

An **image impedance** is defined as the impedance which, when connected to the terminals of a network, equals the impedance presented to it at the opposite terminals. For example, the impedance looking into port 1 of Figure 80.12(a) is, say, 400 Ω when port 2 is terminated in, say, 750 Ω, and the impedance seen looking into port 2 (Figure 80.12(b)) is 750 Ω when port

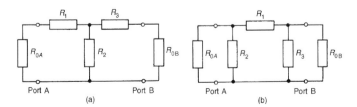

Figure 80.10

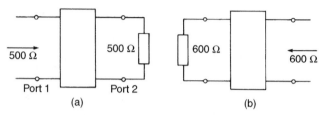

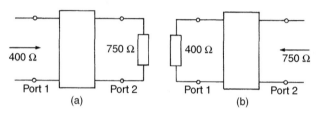

Figure 80.11

Figure 80.12

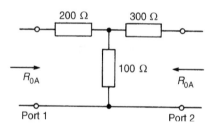

Figure 80.13

1 is terminated in 400 Ω. An asymmetrical network is correctly terminated when it is terminated in its image impedance. (If the image impedances are equal, the value is the characteristic impedance).

For example, an asymmetrical T-section attenuator is shown in Figure 80.13. The image and iterative impedances are determined as follows:

The image impedance R_{OA} seen at port 1 in Figure 80.13 is given by equation (4): $R_{OA} = \sqrt{R_{OC}R_{SC}}$ where R_{OC} and R_{SC} refer to port 2 being respectively open-circuited and short circuited.

$$R_{OC} = 200 + 100 = 300 \ \Omega \quad \text{and} \quad R_{SC} = 200 + \frac{(100)(300)}{100 + 300}$$

$$= 275 \ \Omega$$

Hence $\quad R_{OA} = \sqrt{(300)(275)} = \mathbf{287.2 \ \Omega}$

Similarly, $R_{OB} = \sqrt{R_{OC}R_{SC}}$ where R_{OC} and R_{SC} refer to port 1 being respectively open-circuited and short-circuited.

$$R_{OC} = 300 + 100 = 400 \ \Omega \text{ and } R_{SC} = 300 + \frac{(200)(100)}{200 + 100}$$

$$= 366.7 \ \Omega$$

Hence, $\quad R_{OB} = \sqrt{(400)(366.7)} = \mathbf{383 \ \Omega}$

Thus the image impedances are 287.2 Ω and 383 Ω

[Checking: $R_{OA} = 200 + \dfrac{(100)(300 + 383)}{100 + 300 + 383} = 287.2 \ \Omega$

and $\quad R_{OB} = 300 + \dfrac{(100)(200 + 287.2)}{100 + 200 + 287.2} = 383 \ \Omega$]

The iterative impedance at port 1 in Figure 80.14, is shown as R_1. Hence

$$R_1 = 200 + \frac{(100)(300 + R_1)}{100 + 300 + R_1} = 200 + \frac{30\,000 + 100R_1}{400 + R_1}$$

$$= \frac{80\,000 + 200R_1 + 30\,000 + 100R_1}{400 + R_1}$$

from which, $\quad 400R_1 + R_1^2 = 80\,000 + 200R_1 + 30\,000 + 100R_1$

and $R_1^2 + 100R_1 - 110\,000 = 0$

Solving by the quadratic formula gives:

$$R_1 = \frac{-100 \pm \sqrt{100^2 - (4)(1)(-110\,000)}}{2}$$

$$= \frac{-100 \pm 670.8}{2} = \mathbf{285.4 \ \Omega} \text{ (neglecting the negative value).}$$

The iterative impedance at port 2 in Figure 80.15 is shown as R_2.

Hence $\qquad R_2 = 300 + \dfrac{(100)(200 + R_2)}{100 + 200 + R_2}$

$$= 300 + \frac{20\,000 + 100R_2}{300 + R_2}$$

from which, $\quad 300R_2 + R_2^2 = 90\,000 + 300R_2 + 20\,000 + 100R_2$

and $\qquad R_2^2 - 100R_2 - 110\,000 = 0$

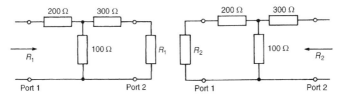

Figure 80.14   **Figure 80.15**

Thus,
$$R_2 = \frac{100 \pm \sqrt{(-100)^2 - (4)(1)(-110\,000)}}{2}$$

$$= \frac{100 \pm 670.8}{2} = \mathbf{385.4\ \Omega}$$

Thus the iterative impedances of the section shown in Figure 80.13 are 285.4 Ω and 385.4 Ω.

The *L*-section Attenuator

For the *L*-section attenuator shown in Figure 80.16,

$$R_1 = \sqrt{R_{OA}(R_{OA} - R_{OB})} \quad \text{and} \quad R_2 = \sqrt{\left(\frac{R_{OA}R_{OB}^2}{R_{OA} - R_{OB}}\right)} \qquad (11)$$

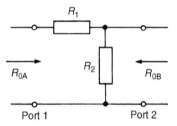

Figure 80.16

For example, if a generator having an internal resistance of 400 Ω is connected to a 100 Ω load via an *L*-section attenuator, as shown in Figure 80.16, then from equation (11),

$$R_1 = \sqrt{R_{OA}(R_{OA} - R_{OB})} = \sqrt{400(400 - 100)} = \mathbf{346.4\ \Omega}$$

$$\text{and} \quad R_2 = \sqrt{\left(\frac{R_{OA}R_{OB}^2}{R_{OA} - R_{OB}}\right)} = \sqrt{\left(\frac{400(100)^2}{400 - 100}\right)} = \mathbf{115.5\ \Omega}$$

Two-port Networks in Cascade

Often two-port networks are connected in cascade, i.e. the output from the first network becomes the input to the second network, and so on, as shown in Figure 80.17. Thus an attenuator may consist of several cascaded sections so as to achieve a particular desired overall performance.

If the cascade is arranged so that the impedance measured at one port and the impedance with which the other port is terminated have the same

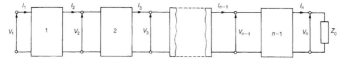

Figure 80.17

value, then each section (assuming they are symmetrical) will have the same characteristic impedance Z_0 and the last network will be terminated in Z_0. Thus each network will have a matched termination and hence the attenuation in decibels of section 1 in Figure 80.17 is given by $a_1 = 20 \lg \dfrac{V_1}{V_2}$. Similarly, the attenuation of section 2 is given by $a_2 = 20 \lg \dfrac{V_2}{V_3}$, and so on.

The overall attenuation is given by:

$$a = 20 \lg \frac{V_1}{V_n} = 20 \lg \left(\frac{V_1}{V_2} \times \frac{V_2}{V_3} \times \frac{V_3}{V_4} \times \ldots \times \frac{V_{n-1}}{V_n} \right)$$
$$= 20 \lg \frac{V_1}{V_2} + 20 \lg \frac{V_2}{V_3} + \ldots + 20 \lg \frac{V_{n-1}}{V_n}$$

by the laws of logarithms,

i.e. $\boxed{\textbf{overall attenuation, } a = a_1 + a_2 + \ldots + a_{n-1}}$ (12)

Thus the overall attenuation is the sum of the attenuations (in decibels) of the matched sections.

For example, five identical attenuator sections are connected in cascade. The overall attenuation is 70 dB and the voltage input to the first section is 20 mV.

From equation (12), the overall attenuation is equal to the sum of the attenuations of the individual sections and, since in this case each section is identical, **the attenuation of each section $= \dfrac{70}{5} = 14$ dB.**

If $V_1 =$ the input voltage to the first stage and $V_0 =$ the output of the final stage, then the overall attenuation $= 20 \lg \left(\dfrac{V_1}{V_0} \right)$

i.e. $70 = 20 \lg \left(\dfrac{20}{V_0} \right)$ where V_0 is in millivolts

$3.5 = \lg \left(\dfrac{20}{V_0} \right)$ and $10^{3.5} = \dfrac{20}{V_0}$

from which, **output voltage of final stage, $V_0 = \dfrac{20}{10^{3.5}} = 6.32 \times 10^{-3}$ mV**

$= \textbf{6.32 } \boldsymbol{\mu}\textbf{V}$

81 Filter Networks

Introduction

A **filter** is a network designed to pass signals having frequencies within certain bands (called **passbands**) with little attenuation, but greatly attenuates signals within other bands (called **attenuation bands** or **stopbands**).

A filter is frequency sensitive and is thus composed of reactive elements. Since certain frequencies are to be passed with minimal loss, ideally the inductors and capacitors need to be pure components since the presence of resistance results in some attenuation at all frequencies.

Between the pass band of a filter, where ideally the attenuation is zero, and the attenuation band, where ideally the attenuation is infinite, is the **cut-off frequency**, this being the frequency at which the attenuation changes from zero to some finite value.

A filter network containing no source of power is termed **passive**, and one containing one or more power sources is known as an **active** filter network.

Filters are used for a variety of purposes in nearly every type of electronic communications and control equipment. The bandwidths of filters used in communications systems vary from a fraction of a hertz to many megahertz, depending on the application.

There are four basic types of filter sections:

(a) low-pass
(b) high-pass
(c) band-pass
(d) band-stop

Low-pass Filters

Figure 81.1 shows simple unbalanced T- and π-section filters using series inductors and shunt capacitors. If either section is connected into a network and a continuously increasing frequency is applied, each would have a frequency-attenuation characteristic as shown in Figure 81.2. This is an ideal characteristic and assumes pure reactive elements. All frequencies are seen

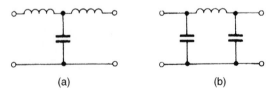

(a)   (b)

Figure 81.1

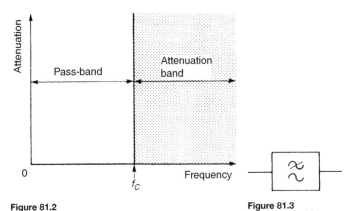

Figure 81.2   **Figure 81.3**

to be passed from zero up to a certain value without attenuation, this value being shown as f_C, the cut-off frequency; all values of frequency above f_C are attenuated. It is for this reason that the networks shown in Figures 81.1(a) and (b) are known as **low-pass filters**.

The electrical circuit diagram symbol for a low-pass filter is shown in Figure 81.3.

Summarising, **a low-pass filter is one designed to pass signals at frequencies below a specified cut-off frequency.**

In practise, the characteristic curve of a low-pass prototype filter section looks more like that shown in Figure 81.4. The characteristic may be improved somewhat closer to the ideal by connecting two or more identical sections in cascade. This produces a much sharper cut-off characteristic, although the attenuation in the pass band is increased a little.

When rectifiers are used to produce the d.c. supplies of electronic systems, a large ripple introduces undesirable noise and may even mask the effect of the

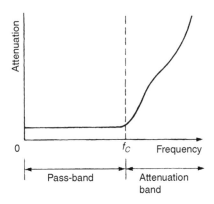

Figure 81.4

signal voltage. Low-pass filters are added to smooth the output voltage wave-form, this being one of the most common applications of filters in electrical circuits.

Filters are employed to isolate various sections of a complete system and thus to prevent undesired interactions. For example, the insertion of low-pass decoupling filters between each of several amplifier stages and a common power supply reduces interaction due to the common power supply impedance.

High-pass Filters

Figure 81.5 shows simple unbalanced T- and π-section filters using series capacitors and shunt inductors. If either section is connected into a network and a continuously increasing frequency is applied, each would have a frequency-attenuation characteristic as shown in Figure 81.6.

Once again this is an ideal characteristic assuming pure reactive elements. All frequencies below the cut-off frequency f_c are seen to be attenuated and all frequencies above f_c are passed without loss. It is for this reason that the networks shown in Figures 81.5(a) and (b) are known as **high-pass filters**.

The electrical circuit diagram symbol for a high-pass filter is shown in Figure 81.7.

Summarising, **a high-pass filter is one designed to pass signals at frequencies above a specified cut-off frequency.**

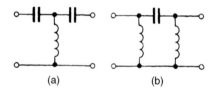

(a)　　　　　　(b)

Figure 81.5

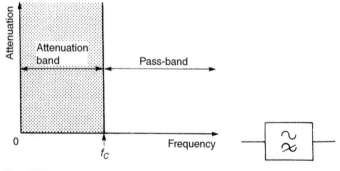

Figure 81.6　　　　　　　　　　　　　**Figure 81.7**

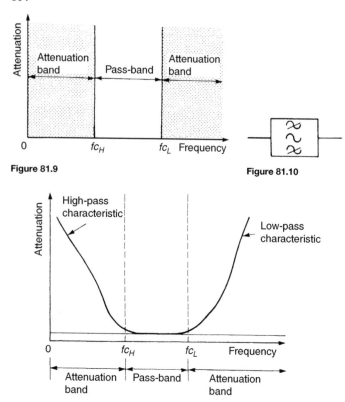

Figure 81.9

Figure 81.10

Figure 81.11

A typical practical characteristic for a band-pass filter is shown in Figure 81.11.

Crystal and ceramic devices are used extensively as band-pass filters. They are common in the intermediate-frequency amplifiers of v.h.f. radios where a precisely defined bandwidth must be maintained for good performance.

Band-stop Filters

A **band-stop filter is one designed to pass signals with all frequencies except those between two specified cut-off frequencies**. The characteristic of an ideal band-stop filter is shown in Figure 81.12.

Such a filter may be formed by connecting a high-pass and a low-pass filter in parallel. As can be seen, for a band-stop filter $f_{C_H} > f_{C_L}$, the stop-band being given by the difference between these values.

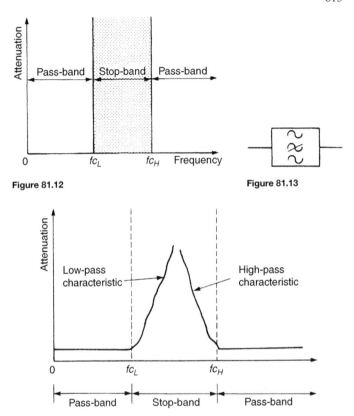

Figure 81.12

Figure 81.13

Figure 81.14

The electrical circuit diagram symbol for a band-stop filter is shown in Figure 81.13.

A typical practical characteristic for a band-stop filter is shown in Figure 81.14.

Sometimes, as in the case of interference from 50 Hz power lines in an audio system, the exact frequency of a spurious noise signal is known. Usually such interference is from an odd harmonic of 50 Hz, for example, 250 Hz. A sharply tuned band-stop filter, designed to attenuate the 250 Hz noise signal, is used to minimise the effect of the output. A high-pass filter with cut-off frequency greater than 250 Hz would also remove the interference, but some of the lower frequency components of the audio signal would be lost as well.

82 Modulation

Introduction to Modulation

The transmission of information such as speech, music and data over long distances requires the use of a carrier channel. It is common practise to 'carry' different communications, called **signals**, at different frequencies to stop one signal from interfering with another. A signal can be shifted bodily from its original band to another, this being achieved by 'modulating' one waveform with another.

The mean frequency level to which a signal is moved is called the **carrier frequency** and the process of superimposing the information signal on the carrier is called **modulation**. The resultant signal is called the **modulated signal**. Many signals, such as telephone conversations, can be transmitted simultaneously along a single pair of lines by using modulation techniques. Modulation of a band of low frequencies on to a higher frequency carrier is fundamental to radio communications, and using different carrier frequencies leads to numerous programmes being transmitted simultaneously. The carrier frequency is the frequency to which the receiver has to be tuned, for example, 97.6 to 99.8 MHz for BBC Radio 1, the signal which is heard being obtained from the modulated carrier by a process called **demodulation**.

Amplitude Modulation

The carrier frequency must have one or more of its characteristics (i.e. amplitude, frequency and/or phase) varied by the information signal. When the amplitude of the carrier is changed by the information signal, the process is called **amplitude modulation**. To illustrate amplitude modulation, consider the signal to be a sine wave of frequency f_m, as shown in Figure 82.1(a), and the carrier to be a sine wave of frequency f_c, as shown in Figure 82.1(b). The result of amplitude modulation is shown in Figure 82.1(c), the signal information being duplicated on both sides of the carrier, as shown by the broken lines, which are construction lines outlining the pattern of change of amplitude of the modulated waveform. This results in a band of frequencies over a range $(f_c - f_m)$ to $(f_c + f_m)$, i.e. the carrier frequency $\pm$ the signal frequency band. The frequency range between the highest and lowest of these frequencies is called the **bandwidth**.

Frequency Modulation

Instead of varying the amplitude of the carrier waveform, the modulating signal may be used to vary the frequency of the carrier. An increase in signal

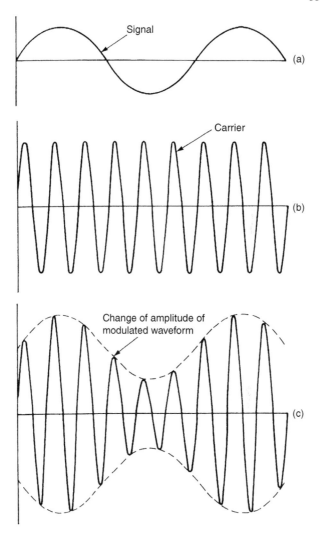

Signal

(a)

Carrier

(b)

Change of amplitude of
modulated waveform

(c)

Figure 82.1

amplitude then causes a change in the modulated signal frequency, which is proportional to the amplitude of the modulating signal. This is called **frequency modulation** and is shown for a cosine wave signal in Figure 82.2.

When the signal amplitude is positive, the frequency of the carrier is modulated to be less than it was originally, shown as (a). The original carrier is shown for reference. The modulated wave is in the same position as the

618

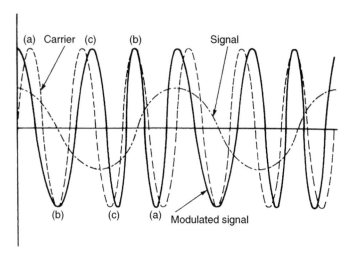

Figure 82.2

original carrier when the signal amplitude is zero, as shown at (b). When the signal amplitude is negative, the frequency of the carrier is modulated to be greater than that of the original carrier, as shown at (c).

Frequency Deviation, Frequency Swing and Modulating Index

Frequency deviation is a term used in frequency modulation and is defined as the peak difference between the instantaneous frequency of the modulated wave and the carrier frequency during one cycle of modulation.

Frequency swing is the difference between the maximum and minimum values of the instantaneous frequency of a frequency-modulated wave.

The **modulating index** for a sinusoidal modulating waveform is the ratio of frequency deviation to the frequency of the modulating wave. Thus the modulating index is the ratio of the frequency deviation caused by a particular signal to the frequency of that signal.

Phase Modulation

The modulating signal can be used to advance or retard the phase of the carrier in proportion to the amplitude of the modulating signal. This technique is called **phase modulation** and this also involves a variation of frequency. In this case it depends on the rate of change of phase and thus on both the

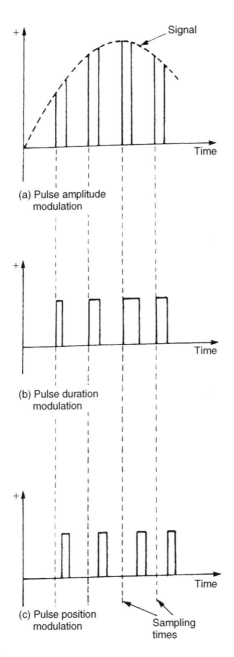

(a) Pulse amplitude
    modulation

(b) Pulse duration
    modulation

(c) Pulse position
    modulation

Signal

Time

Time

Time

Sampling
times

Figure 82.3

620

amplitude and frequency of the modulating signal. The waveform is similar to that shown in Figure 82.2.

Pulse Modulation

In pulse modulation, the signal is sampled at a frequency that is at least twice that of the highest frequency present in the signal. Thus for speech, which has frequencies ranging from about 300 Hz to 3.4 kHz, a typical sampling frequency is 8 kHz. Various forms of pulse modulation are used and include pulse amplitude modulation, pulse position modulation and pulse duration modulation.

The principle of pulse modulation is shown in Figure 82.3(a), in which the amplitude of the pulse is proportional to the amplitude of the signal. The amplitude of the pulse may change during the 'on' period or alternatively it may be kept constant, resulting in the stepped waveform as shown in Figure 82.3(a).

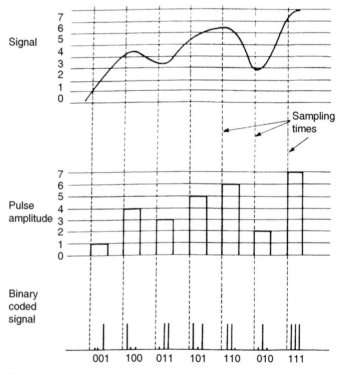

Figure 82.4

Figure 82.3(b) shows the principle of pulse duration modulation, the duration of the pulse being proportional to the amplitude of the signal.

The position of the pulse relative to some datum (such as the sampling time), is made proportional to the amplitude of the signal in pulse position modulation, as shown in Figure 82.3(c).

Pulse Code Modulation

In **pulse code modulation,** the signal amplitude is divided into a number of equal increments, each increment being designated by a number. For example, an amplitude divided into eight increments can have the instantaneous value of the amplitude transmitted by using the natural binary numbering system with three bits, the level being transmitted as: 000, 001, 011,..., 111. This concept is shown in Figure 82.4.

Thus in pulse code modulation, an analogue signal is converted into a digital signal. Since the analogue signal can have any value between certain limits, but the resulting digital signal has only discrete values, some distortion of the signal results. The greater the number of increments, the more closely the digital signal resembles the analogue signal.

83 Transmission Lines

Introduction

A transmission line is a system of conductors connecting one point to another and along which electromagnetic energy can be sent. Thus telephone lines and power distribution lines are typical examples of transmission lines; in electronics, however, the term usually implies a line used for the transmission of radio-frequency (r.f.) energy such as that from a radio transmitter to the antenna.

An important feature of a transmission line is that it should guide energy from a source at the sending end to a load at the receiving end without loss by radiation. One form of construction often used consists of two similar conductors mounted close together at a constant separation. The two conductors form the two sides of a balanced circuit and any radiation from one of them is neutralised by that from the other. Such twin-wire lines are used for carrying high r.f. power, for example, at transmitters. The coaxial form of construction is commonly employed for low power use, one conductor being in the form of a cylinder that surrounds the other at its centre, and thus acts as a screen. Such cables are often used to couple f.m. and television receivers to their antennas.

At frequencies greater than 1000 MHz, transmission lines are usually in the form of a waveguide, which may be regarded as coaxial lines without the centre conductor, the energy being launched into the guide or abstracted from it by probes or loops projecting into the guide.

Transmission Line Primary Constants

Let an a.c. generator be connected to the input terminals of a pair of parallel conductors of infinite length. A sinusoidal wave will move along the line and a finite current will flow into the line. The variation of voltage with distance along the line will resemble the variation of applied voltage with time. The moving wave, sinusoidal in this case, is called a voltage **travelling wave.** As the wave moves along the line the capacitance of the line is charged up and the moving charges cause magnetic energy to be stored. Thus the propagation of such an **electromagnetic wave** constitutes a flow of energy.

After sufficient time the magnitude of the wave may be measured at any point along the line. The line does not therefore appear to the generator as an open circuit but presents a definite load Z_0. If the sending-end voltage is V_S and the sending end current is I_S then $Z_0 = \dfrac{V_S}{I_S}$. Thus, the line absorbs all of the energy and the line behaves in a similar manner to the generator, as would a single 'lumped' impedance of value Z_0 connected directly across the generator terminals.

There are **four parameters** associated with transmission lines, these being resistance, inductance, capacitance and conductance.

(i) **Resistance R** is given by $R = \rho l/A$, where ρ is the resistivity of the conductor material, A is the cross-sectional area of each conductor and l is the length of the conductor (for a two-wire system, l represents twice the length of the line). Resistance is stated in ohms per metre length of a line and represents the imperfection of the conductor. A resistance stated in ohms per loop metre is a little more specific since it takes into consideration the fact that there are two conductors in a particular length of line.

(ii) **Inductance L** is due to the magnetic field surrounding the conductors of a transmission line when a current flows through them. The inductance of an isolated twin line is considered in chapter 79. From equation (11), page 596, the inductance L is given by: $L = \dfrac{\mu_0\mu_r}{\pi}\left(\dfrac{1}{4} + \ln\dfrac{D}{a}\right)$ henry/metre, where D is the distance between centres of the conductor and a is the radius of each conductor. In most practical lines $\mu_r = 1$. An inductance stated in henrys per loop metre takes into consideration the fact that there are two conductors in a particular length of line.

(iii) **Capacitance C** exists as a result of the electric field between conductors of a transmission line. The capacitance of an isolated twin line is considered in chapter 79. From equation (7), page 594, the capacitance between the two conductors is given by: $C = \dfrac{\pi\varepsilon_0\varepsilon_r}{\ln\dfrac{D}{a}}$ farads/metre. In most practical lines $\varepsilon_r = 1$.

(iv) **Conductance G** is due to the insulation of the line allowing some current to leak from one conductor to the other. Conductance is measured in siemens per metre length of line and represents the imperfection of the insulation. Another name for conductance is leakance.

Each of the four transmission line constants, R, L, C and G, known as the **primary constants**, are uniformly distributed along the line.

From chapter 80, when a symmetrical T-network is terminated in its characteristic impedance Z_0, the input impedance of the network is also equal to Z_0. Similarly, if a number of identical T-sections are connected in cascade, the input impedance of the network will also be equal to Z_0.

A transmission line can be considered to consist of a network of a very large number of cascaded T-sections each a very short length (δl) of transmission line, as shown in Figure 83.1. This is an approximation of the uniformly distributed line; the larger the number of lumped parameter sections, the nearer

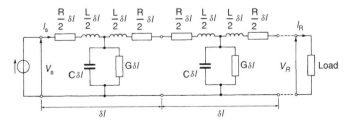

Figure 83.1

it approaches the true distributed nature of the line. When the generator V_S is connected, a current I_S flows which divides between that flowing through the leakage conductance G, which is lost, and that which progressively charges each capacitor C and which sets up the voltage travelling wave moving along the transmission line. The loss or attenuation in the line is caused by both the conductance G and the series resistance R.

Phase Delay, Wavelength and Velocity of Propagation

Each section of that shown in Figure 83.1 is simply a low-pass filter possessing losses R and G. If losses are neglected, and R and G are removed, the circuit simplifies and the infinite line reduces to a repetitive T-section low-pass filter network as shown in Figure 83.2. Let a generator be connected to the line as shown and let the voltage be rising to a maximum positive value just at the instant when the line is connected to it. A current I_S flows through inductance L_1 into capacitor C_1. The capacitor charges and a voltage develops across it. The voltage sends a current through inductance L_1' and L_2 into capacitor C_2. The capacitor charges and the voltage developed across it sends a current through L_2' and L_3 into C_3, and so on. Thus all capacitors will in turn charge up to the maximum input voltage. When the generator voltage falls, each capacitor is charged in turn in opposite polarity, and as before the input charge is progressively passed along to the next capacitor. In this manner voltage and current waves travel along the line together and depend on each other.

The process outlined above takes time; for example, by the time capacitor C_3 has reached its maximum voltage, the generator input may be at zero or moving towards its minimum value. There will therefore be a time, and thus a phase difference between the generator input voltage and the voltage at any point on the line.

Phase delay

Since the line shown in Figure 83.2 is a ladder network of low-pass T-section filters, it may be shown that the phase delay, β, is given by:

$$\boxed{\beta = \omega\sqrt{LC} \text{ radians/metre}} \tag{1}$$

where L and C are the inductance and capacitance per metre of the line.

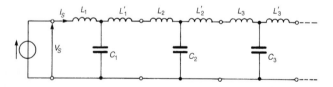

Figure 83.2

Wavelength

The wavelength λ on a line is the distance between a given point and the next point along the line at which the voltage is the same phase, the initial point leading the latter point by 2π radian. Since in one wavelength a phase change of 2π radians occurs, the phase change per metre is $\dfrac{2\pi}{\lambda}$. Hence, phase change per metre,

$$\beta = \frac{2\pi}{\lambda} \quad \text{or} \quad \boxed{\textbf{wavelength, } \lambda = \frac{2\pi}{\beta} \textbf{ metres}} \tag{2}$$

Velocity of propagation

The velocity of propagation, u, is given by $u = f\lambda$, where f is the frequency and λ the wavelength. Hence

$$\boxed{u = f\lambda = f\left(\frac{2\pi}{\beta}\right) = \frac{2\pi f}{\beta} = \frac{\omega}{\beta}} \tag{3}$$

The velocity of propagation of free space is the same as that of light, i.e. approximately 300×10^6 m/s. The velocity of electrical energy along a line is always less than the velocity in free space. The wavelength λ of radiation in free space is given by $\lambda = \dfrac{c}{f}$ where c is the velocity of light. Since the velocity along a line is always less than c, the wavelength corresponding to any particular frequency is always shorter on the line than it would be in free space.

For example, a transmission line has an inductance of 4 mH/loop km and a capacitance of 0.004 µF/km. For a frequency of operation of 1 kHz, from equation (1), phase delay, $\beta = \omega\sqrt{LC} = (2\pi 1000)\sqrt{(4 \times 10^{-3})(0.004 \times 10^{-6})}$ $= \textbf{0.025 rad/km}$, from equation (2), wavelength, $\lambda = \dfrac{2\pi}{\beta} = \dfrac{2\pi}{0.025} = \textbf{251 km}$ and from equation (3), velocity of propagation,

$$u = f\lambda = (1000)(251) \text{ km/s} = \textbf{251} \times \textbf{10}^6 \textbf{ m/s}$$

Current and Voltage Relationships

Figure 83.3 shows a voltage source V_S applied to the input terminals of an infinite line, or a line terminated in its characteristic impedance, such that a current I_S flows into the line. At a point, say, 1 km down the line let the current be I_1. The current I_1 will not have the same magnitude as I_S because of line attenuation; also I_1 will lag I_S by some angle β. The ratio $\dfrac{I_S}{I_1}$ is therefore a phasor quantity. Let the current a further 1km down the line be I_2, and so on, as shown in Figure 83.3. Each unit length of line can be treated as a section of a repetitive network, and the attenuation is in the form of a

626

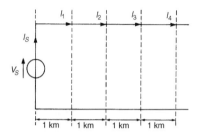

Figure 83.3

logarithmic decay:

$$\frac{I_S}{I_1} = \frac{I_1}{I_2} = \frac{I_2}{I_3} = e^{\gamma}$$

where γ is the **propagation constant**. γ has no unit.

The propagation constant is a complex quantity given by $\gamma = \alpha + j\beta$, where α is the **attenuation constant**, whose unit is the neper, and β is the **phase shift coefficient**, whose unit is the radian. For n such 1 km sections, $\frac{I_S}{I_R} = e^{n\gamma}$ where I_R is the current at the receiving end.

Hence $\dfrac{I_S}{I_R} = e^{n(\alpha + j\beta)} = e^{(n\alpha + jn\beta)} = e^{n\alpha} \angle n\beta$

from which $\boxed{I_R = I_S e^{-n\gamma} = I_S e^{-n\alpha} \angle - n\beta}$ (4)

In equation (4), the attenuation on the line is given by $n\alpha$ nepers and the phase shift is $n\beta$ radians.

At all points along an infinite line, the ratio of voltage to current is Z_0, the characteristic impedance. Thus from equation (4) it follows that: **receiving end voltage,**

$$\boxed{V_R = V_S e^{-n\gamma} = V_S e^{-n\alpha} \angle - n\beta}$$ (5)

Z_0, γ, α and β are referred to as the **secondary line constants** or **coefficients**.

For example, a transmission line 5 km long has a characteristic impedance of $800 \angle - 25°$ Ω. At a particular frequency, the attenuation coefficient of the line is 0.5 Np/km and the phase shift coefficient is 0.25 rad/km. If the sending end voltage is $2.0 \angle 0°$ V r.m.s., then the receiving end voltage (from equation (5)) is given by:

$$V_R = V_S e^{-n\gamma} = V_S e^{-n\alpha} \angle - n\beta = (2.0 \angle 0°) e^{-(5)(0.5)} \angle - (5)(0.25)$$

$$= 2.0 e^{-2.5} \angle - 1.25 = \mathbf{0.1642 \angle - 71.62° \ V}$$

Receiving end current,

$$I_R = \frac{V_R}{Z_0} = \frac{0.1642 \angle - 71.62°}{800 \angle - 25°} = 2.05 \times 10^{-4} \angle (-71.62° - (-25°)) \ A$$

$$= \mathbf{0.205 \angle - 46.62° \ mA}$$

Characteristic Impedance and Propagation Coefficient in Terms of the Primary Line Constants

At all points along an infinite line, the ratio of voltage to current is called the characteristic impedance Z_0. The value of Z_0 is independent of the length of the line; it merely describes a property of a line that is a function of the physical construction of the line. Since a short length of line may be considered as a ladder of identical low-pass filter sections, the characteristic impedance may be determined from chapter 80, i.e.

$$Z_0 = \sqrt{Z_{OC} Z_{SC}} \qquad (6)$$

since the open-circuit impedance Z_{OC} and the short-circuit impedance Z_{SC} may be easily measured.

The characteristic impedance of a transmission line may also be expressed in terms of the primary constants, R, L, G and C. Measurements of the primary constants may be obtained for a particular line and manufacturers usually state them for a standard length.

It may be shown that the characteristic impedance Z_0 is given by:

$$Z_0 = \sqrt{\frac{R + j\omega L}{G + j\omega C}} \text{ ohms} \qquad (7)$$

and if losses R and G are neglected, then

$$Z_0 = \sqrt{\frac{L}{C}} \text{ ohms} \qquad (8)$$

and the **propagation coefficient** γ is given by:

$$\gamma = \sqrt{(R + j\omega L)(G + j\omega C)} \qquad (9)$$

and for a loss less line, $R = G = 0$ and

$$\gamma = \sqrt{(j\omega L)(j\omega C)} = j\omega\sqrt{LC} \qquad (10)$$

For example, at a frequency of 1 kHz the primary constants of a transmission line are resistance $R = 25\ \Omega$/loop km, inductance $L = 5$ mH/loop km, capacitance $C = 0.04\ \mu$F/km and conductance $G = 80\ \mu$S/km. Then from equation (7), characteristic impedance $Z_0 = \sqrt{\dfrac{R + j\omega L}{G + j\omega C}}$ ohms

$$R + j\omega L = 25 + j(2\pi 1000)(5 \times 10^{-3}) = (25 + j31.42)$$

$$= 40.15\angle 51.49° \ \Omega$$

$$G + j\omega C = 80 \times 10^{-6} + j(2\pi 1000)(0.04 \times 10^{-6})$$

$$= (80 + j251.33)10^{-6}$$

$$= 263.76 \times 10^{-6}\angle 72.34° \text{ S}$$

Thus characteristic impedance, $Z_0 = \sqrt{\dfrac{40.15\angle 51.49°}{263.76 \times 10^{-6}\angle 72.34°}}$

$$= \mathbf{390.2\angle -10.43° \; \Omega}$$

From equation (10), propagation coefficient

$$\gamma = \sqrt{(R + j\omega L)(G + j\omega C)}$$

$$= \sqrt{(40.15\angle 51.49°)(263.76 \times 10^{-6}\angle 72.34°)}$$

$$= \sqrt{0.01059\angle 123.83°} = \mathbf{0.1029\angle 61.92°}$$

Distortion on Transmission Lines

If the waveform at the receiving end of a transmission line is not the same shape as the waveform at the sending end, **distortion** is said to have occurred.

In designing a transmission line, if $\boxed{LG = CR}$ no distortion is introduced. This means that the signal at the receiving end is the same as the sending-end signal except that it is reduced in amplitude and delayed by a fixed time. Also, with no distortion, the attenuation on the line is a minimum.

In practice, however, $\dfrac{R}{L} \gg \dfrac{G}{C}$. The inductance is usually low and the capacitance is large and not easily reduced. Thus if the condition $LG = CR$ is to be achieved in practice, either L or G must be increased since neither C or R can really be altered. It is undesirable to increase G since the attenuation and power losses increase. Thus the inductance L is the quantity that needs to be increased and such an artificial increase in the line inductance is called **loading**. This is achieved either by inserting inductance coils at intervals along the transmission line — this being called '**lumped loading**' — or by wrapping the conductors with a high-permeability metal tape — this being called '**continuous loading**'.

For example, an underground cable has the following primary constants: resistance $R = 10$ Ω/loop km, inductance $L = 1.5$ mH/loop km, conductance $G = 1.2$ μS/km and capacitance $C = 0.06$ μF/km.

The condition for minimum distortion is given by: $LG = CR$, from which,

inductance $L = \dfrac{CR}{G} = \dfrac{(0.06 \times 10^{-6})(10)}{(1.2 \times 10^{-6})} = 0.5$ H or 500 mH

Thus the inductance should be increased by $(500 - 1.5)$ mH, i.e. **498.5 mH** per loop km, for minimum distortion.

Wave Reflection and Reflection Coefficient

In earlier sections of this chapter it was assumed that the transmission line had been properly terminated in its characteristic impedance or regarded as an infinite line. In practice, of course, all lines have a definite length and often the terminating impedance does not have the same value as the characteristic impedance of the line. When this is the case, the transmission line is said to have a '**mismatched load**'.

The forward-travelling wave moving from the source to the load is called the **incident wave** or the sending-end wave. With a mismatched load the termination will absorb only a part of the energy of the incident wave, the remainder being forced to return back along the line toward the source. This latter wave is called the **reflected wave.**

A transmission line transmits electrical energy; when such energy arrives at a termination that has a value different from the characteristic impedance, it experiences a sudden change in the impedance of the medium. When this occurs, some reflection of incident energy occurs and the reflected energy is lost to the receiving load. Reflections commonly occur in nature when a change of transmission medium occurs; for example, sound waves are reflected at a wall, which can produce echoes (see Chapter 17), and mirrors reflect light rays (see Chapter 19).

If a transmission line is terminated in its characteristic impedance, no reflection occurs; if terminated in an open circuit or a short circuit, total reflection occurs, i.e. the whole of the incident wave reflects along the line. Between these extreme possibilities, all degrees of reflection are possible.

Energy associated with a travelling wave

A travelling wave on a transmission line may be thought of as being made up of electric and magnetic components. Energy is stored in the magnetic field due to the current (energy $= \frac{1}{2}LI^2$ — see page 597) and energy is stored in the electric field due to the voltage (energy $= \frac{1}{2}CV^2$ — see page 594). It is the continual interchange of energy between the magnetic and electric fields, and vice versa, that causes the transmission of the total electromagnetic energy along the transmission line.

When a wave reaches an open-circuited termination the magnetic field collapses since the current I is zero. Energy cannot be lost, but it can change form. In this case it is converted into electrical energy, adding to that already caused by the existing electric field. The voltage at the termination consequently doubles and this increased voltage starts the movement of a reflected wave back along the line. This movement will set up a magnetic field and the total energy of the reflected wave will again be shared between the magnetic and electric field components.

When a wave meets a short-circuited termination, the electric field collapses and its energy changes form to the magnetic energy. This results in a doubling of the current.

Reflection coefficient

The ratio of the reflected current to the incident current is called the **reflection coefficient** and is often given the symbol ρ, i.e.

$$\boxed{\frac{I_r}{I_i} = \rho = \frac{Z_0 - Z_R}{Z_0 + Z_R} = -\frac{V_r}{V_i}} \qquad (11)$$

where I_i = incident or sending-end current,
$\quad\ V_i$ = incident or sending-end voltage,

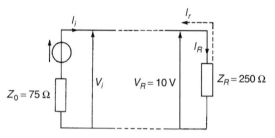

Figure 83.4

I_r = reflected current, V_r = reflected voltage,
Z_0 = characteristic impedance
and Z_R = impedance of termination.
When $Z_R = Z_0$, $\rho = 0$ and there is no reflection.

For example, a cable, which has a characteristic impedance of 75 Ω, is terminated in a 250 Ω resistive load. Assume that the cable has negligible losses and the voltage measured across the terminating load is 10 V.
From equation (11), **reflection coefficient,**

$$\rho = \frac{Z_0 - Z_R}{Z_0 + Z_R} = \frac{75 - 250}{75 + 250} = \frac{-175}{325} = \mathbf{-0.538}$$

The circuit is shown in Figure 83.4. Current flowing in the terminating load,

$$I_R = \frac{V_R}{Z_R} = \frac{10}{250} = 0.04 \text{ A}$$

However, current $I_R = I_i + I_r$. From equation (11), $I_r = \rho I_i$

Thus $I_R = I_i + \rho I_i = I_i(1 + \rho)$ from which

incident current, $I_i = \dfrac{I_R}{(1 + \rho)} = \dfrac{0.04}{1 + (-0.538)}$

$$= \mathbf{0.0866} \text{ A or } \mathbf{86.6 \text{ mA}}$$

The **incident voltage,** $V_i = I_i Z_0 = (0.0866)(75) = \mathbf{6.50 \text{ V}}$
Since $I_R = I_i + I_r$, **reflected current,**

$$I_r = I_R - I_i = 0.04 - 0.0866 = \mathbf{-0.0466} \text{ A or } \mathbf{-46.6 \text{ mA}}$$

The **reflected voltage,** $V_r = -I_r Z_0 = -(-0.0466)(75) = \mathbf{3.50 \text{ V}}$

Standing Waves and Standing Wave Ratio

Consider a loss free transmission line **open-circuited** at its termination. An incident current waveform is completely reflected at the termination, and, as stated in earlier, the reflected current is of the same magnitude as the incident

631

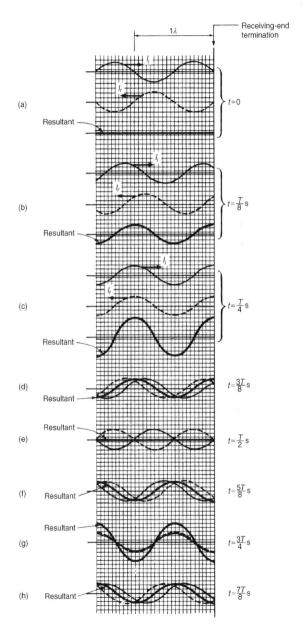

Figure 83.5

current but is 180° out of phase. Figure 83.5(a) shows the incident and reflected current waveforms drawn separately (shown as I_i moving to the right and I_r moving to the left respectively) at a time $t = 0$, with $I_i = 0$ and decreasing at the termination.

The resultant of the two waves is obtained by adding them at intervals. In this case the resultant is seen to be zero. Figures 83.5(b) and (c) show the incident and reflected waves drawn separately as times $t = T/8$ seconds and $t = T/4$, where T is the periodic time of the signal. Again, the resultant is obtained by adding the incident and reflected waveforms at intervals. Figures 83.5(d) to (h) show the incident and reflected current waveforms plotted on the same axis, together with their resultant waveform, at times $t = 3T/8$ to $t = 7T/8$ at intervals of $T/8$.

If the resultant waveforms shown in Figures 83.5(a) to (g) are superimposed one upon the other, Figure 83.6 results. (Note that the scale has been doubled for clarity.) The waveforms show clearly that waveform (a) moves to (b) after $T/8$, then to (c) after a further period of $T/8$, then to (d), (e), (f), (g) and (h) at intervals of $T/8$. It is noted that at any particular point the current varies sinusoidally with time, but the amplitude of oscillation is different at different points on the line.

Whenever two waves of the same frequency and amplitude travelling in opposite directions are superimposed on each other as above, interference takes place between the two waves and a **standing** or **stationary wave** is produced. The points at which the current is always zero are called **nodes** (labelled N in Figure 83.6). The standing wave does not progress to the left or right and the nodes do not oscillate. Those points on the wave that undergo maximum disturbance are called **antinodes** (labelled A in Figure 83.6). The distance between adjacent nodes or adjacent antinodes is $\lambda/2$, where λ is the wavelength. A standing wave is therefore seen to be a periodic variation in the vertical plane taking place on the transmission line without travel in either direction

The resultant of the incident and reflected voltage for the open-circuit termination may be deduced in a similar manner to that for current. However, as stated earlier, when the incident voltage wave reaches the termination it is reflected without phase change. Figure 83.7 shows the resultant waveforms of incident and reflected voltages at intervals of $t = T/8$. Figure 83.8 shows all the resultant waveforms of Figure 83.7(a) to (h) superimposed; again, standing waves are seen to result. Nodes (labelled N) and antinodes (labelled A) are

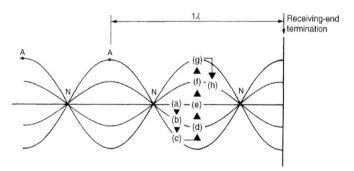

Figure 83.6

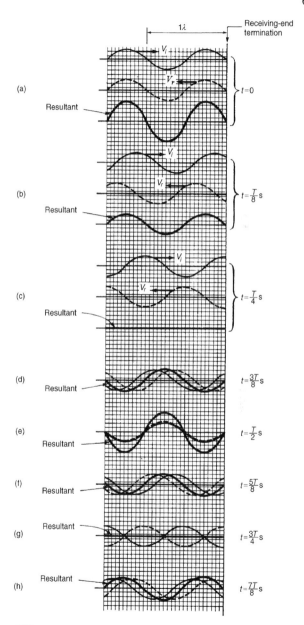

Figure 83.7

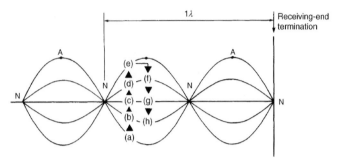

Figure 83.8

shown in Figure 83.8 and, in comparison with the current waves, are seen to occur 90° out of phase.

If the transmission line is short-circuited at the termination, it is the incident current that is reflected without phase change and the incident voltage that is reflected with a phase change of 180°. Thus the diagrams shown in Figures 83.5 and 83.6 representing current at an open-circuited termination may be used to represent voltage conditions at a short-circuited termination and the diagrams shown in Figures 83.7 and 83.8 representing voltage at an open-circuited termination may be used to represent current conditions at a short-circuited termination.

Figure 83.9 shows the r.m.s. current and voltage waveforms plotted on the same axis against distance for the case of total reflection, deduced from Figures 83.6 and 83.8. The r.m.s. values are equal to the amplitudes of the waveforms shown in Figures 83.6 and 83.8, except that they are each divided by $\sqrt{2}$ (since, for a sine wave, r.m.s. value $= \dfrac{1}{\sqrt{2}} \times$ maximum value). With total reflection, the standing-wave patterns of r.m.s. voltage and current consist of a succession of positive sine waves with the voltage node located at the current antinode and the current node located at the voltage antinode. The termination is a current nodal point. The r.m.s. values of current and voltage may be recorded on a suitable r.m.s. instrument moving along the line. Such measurements of the maximum and minimum voltage and current can provide a reasonably accurate indication of the wavelength, and also provide information regarding the amount of reflected energy relative to the incident energy that is absorbed at the termination, as shown below.

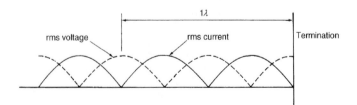

Figure 83.9

Standing-wave ratio

Let the incident current flowing from the source of a mismatched low-loss transmission line be I_i and the current reflected at the termination be I_r. If I_{MAX} is the sum of the incident and reflected current, and I_{MIN} is their difference, then the **standing-wave ratio** (symbol **s**) on the line is defined as:

$$s = \frac{I_{MAX}}{I_{MIN}} = \frac{I_i + I_r}{I_i - I_r} \qquad (12)$$

It may also be shown that:

$$\frac{I_r}{I_i} = \left(\frac{s-1}{s+1}\right) \qquad (13)$$

$$\frac{P_r}{P_t} = \left(\frac{s-1}{s+1}\right)^2 \qquad (14)$$

and

$$s = \frac{1 + |\rho|}{1 - |\rho|} \qquad (15)$$

For example, a transmission line has a characteristic impedance of $600\angle 0°$ Ω and negligible loss. If the terminating impedance of the line is $(400 + j250)$ Ω, then from equation (11),

reflection coefficient, $\quad \rho = \dfrac{Z_0 - Z_R}{Z_0 + Z_R} = \dfrac{600\angle 0° - (400 + j250)}{600\angle 0° + (400 + j250)}$

$$= \frac{200 - j250}{1000 + j250} = \frac{320.16\angle - 51.34°}{1030.78\angle 14.04°}$$

$$= \mathbf{0.3106\angle - 65.38°}$$

Hence, $|\rho| = 0.3106$. Thus from equation (15),

$$\text{standing-wave ratio, } \mathbf{s} = \frac{1 + |\rho|}{1 - |\rho|} = \frac{1 + 0.3106}{1 - 0.3106} = \mathbf{1.901}$$

If the incident power arriving at the termination is, say, 200 mW, then from equation (14):

$$\frac{P_r}{P_t} = \left(\frac{s-1}{s+1}\right)^2$$

$$\text{i.e. } \frac{P_r}{0.200} = \left(\frac{1.901 - 1}{1.901 + 1}\right)^2 = 0.09646$$

from which, reflected power, $P_r = \mathbf{19.29\ mW}$

INDEX

638

642

644

Printed and bound by CPI Group (UK) Ltd, Croydon, CR0 4YY

23/10/2024

01778238-0001